PHYSIOLOGICAL LIMNOLOGY

An Approach to the Physiology of Lake Ecosystems

DEVELOPMENTS IN WATER SCIENCE, 2

advisory editor
VEN TE CHOW
Professor of Hydraulic Engineering
Hydrosystems Laboratory
University of Illinois
Urbana, Ill., U.S.A.

FURTHER TITLES IN THIS SERIES

PHYSIOLOGICAL LIMNOLOGY

An Approach to the Physiology of Lake Ecosystems

H.L. GOLTERMAN

Limnological Institute
Nieuwersluis
The Netherlands

With the assistance of Dr. K.E. Clymo and Dr. R.S. Clymo

ELSEVIER SCIENTIFIC PUBLISHING COMPANY

AMSTERDAM — OXFORD — NEW YORK — 1975

ELSEVIER SCIENTIFIC PUBLISHING COMPANY
335 Jan van Galenstraat
P.O. Box 211, Amsterdam, The Netherlands

AMERICAN ELSEVIER PUBLISHING COMPANY, INC.
52 Vanderbilt Avenue
New York, New York 10017

Library of Congress Card Number: 74-21857

ISBN: 0-444-41270-0

With 131 illustrations and 62 tables

Copyright © 1975 by Elsevier Scientific Publishing Company, Amsterdam

Printed in The Netherlands

Lago Maggiore

To Nel

PREFACE

Every teacher of limnology faces the problem of which textbook to recommend to his students. There is a large gap between the classical books of Ruttner and Hutchinson and both were published some years ago. No textbook has yet dealt with questions such as: "Why is this organism living in this particular environment and what controls its numbers?" Nor with the questions which face the practical water manager. This book is a first approach to relating the chemical environment with the biochemistry and physiology of the organisms in a quantitative way.

The book is intended primarily for post-graduate students; no attempt is made to give complete coverage of all the fields that are discussed. The main aim is educational. Sometimes, therefore, a newer paper is cited, although an older one expressed the same ideas. Chronological development is not attempted. For the same reason many examples come from two research sites of the Netherlands Limnological Institute: the 11 m deep sandpit Lake Vechten, and the shallow polder reservoir, the Tjeukemeer. Many phenomena that are illustrated with data from these lakes were first reported from other lakes, but the coherence of many data from one lake has an educational force lacking in examples taken from scattered sources.

The book is written for both chemists and biologists; both groups will find some material well known to them, but some parts may be too difficult. For this reason the physiology of, for example, photosynthesis, is given in some detail, even though there are excellent books concerned with this topic alone. I hope I have selected what is important for understanding of the ecology of the aquatic organisms.

Because comparison between different lakes with different populations is a powerful tool in ecology, much attention is given to similarities of bacterial and blue-green algal physiology. Here is, however, an open field for much ecologically oriented research. I hope this book may stimulate such research. At present this field is neglected both by ecologists and by physiologists, who prefer to work with cultures rich in nutrients, although in nature nutrient concentrations are mostly much lower.

It is impossible to thank all those friends and colleagues who have contributed to this book. Many ideas thought to be the author's own were probably picked up in discussions with others. If somebody recognises his own ideas, he has probably been seeding in fertile soil (furthermore, it should not be forgotten that the art of printing was invented at several different places at about the same time). Those colleagues who have read through the different chapters and made their critical remarks can, however, be acknowledged properly. They are: Chapter 2, A. Powell; Chapters 3 and 12, H. de Haan;

Chapter 4, J. Talling; Chapter 9, H. Verdouw; Chapters 7 and 13, J. Moed; Chapter 14, J.W.G. Lund; Chapter 15, R. Gulati (section 15.5, M.M. Burgiss and J. Thorpe); Chapter 17, J. Vallentyne; Chapter 19, M. Owens.

I do not know how to express my thanks to Dr. K.E. Clymo and Dr. R.S. Clymo. They were asked to revise the so-called English, which they did. But while translating it into their concise and precise language (which English so often is), they expressed criticism and forced me to reconsider many problems, too easily taken for granted. Without their help and stimulating criticism, the book would never have been written. I still wonder who really is the author of the book, as the Clymo's have apparently adopted the words of Madame Curie: "En science, nous devons nous intéresser aux choses, non aux personnes."

It is a pleasure to acknowledge help from those who made the physical appearance of the book possible:

Mr. J. Landstra, who typed and retyped the manuscript with endless patience and great ability and who kept order in the chaos which the author made. His patience, optimism and good humour were stimulating.

Mrs. G.H. Würtz-Schulz, who, with her extraordinary knowledge as librarian, found all the lost literature, prepared the lists of references and in doing so made many useful critical remarks.

Mr. H. van Tol, who oils the machinery of the Netherlands Limnological Institute (without the existence of which this work would never have been possible) and who made several of the drawings and took care of all the figures.

Mr. E. Marien, who photographed the figures, and Drs. E. Mols, who provided the microphotographs.

Finally, Maarten, Roeland, Karen and Berend Han, who helped in typing and correcting the many drafts.

Answers to questions in ecology can be amazingly simple, for example:

Das Aesthetische Wiesel

Ein Wiesel
Sasz auf einen Kiesel
Inmitten Bachgeriesel.
  Wiszt ihr,
Weshalb?
Das Mondkalb
Verriet es mir
Im Stillen:
Das raffinier-
te Tier
Tats um des Reimes willen.   (Morgenstern, 1950)

May this book have a Mondkalb function.

CONTENTS

CHAPTER 1

GENERAL INTRODUCTION

Limnology can be defined in two ways, either as aquatic ecology or as the science of freshwater systems or "inland oceanology". Limnology — from the greek prefix "limn-" meaning "marsh" — is an interdisciplinary science combining aspects of hydrobiology, hydrochemistry, hydrophysics and geology. Odum (1971), envisaging the structure of biology as a cylindrical block, considered disciplines such as entomology, microbiology, botany to be vertically cut wedges and taxonomy, physiology and ecology to be horizontal layers, of which ecology is the basic one.

Aquatic ecology is thus the science of mutual relationships between organisms and of interactions between organisms and their environment. In this respect limnology is very much like ecology in general. There are differences in the nature of the ecosystem, however. The aquatic ecosystem is usually a closed one. Organisms cannot normally leave it and, with the exception of some insects, are confined within it throughout their entire life cycle. This closed structure results from the fact that water combines the functions of soil, air and water of the terrestrial ecosystem. In lakes and rivers all organisms use water for their transport, either actively as in the case of swimming fish or passively in the case of some algae. Algae take their nutrients, both minerals and CO_2, from the water, and release the products of their metabolism back into the water, either immediately or eventually during post-mortem mineralisation. Other organisms hunt for their food through the water and excrete their waste products into it. Heat and light are transmitted through the water which, due to its physicochemical properties, absorbs them more strongly than does the air. Great diversity — the occurrence of a large number of different organisms — is a feature common to both the terrestrial and the aquatic ecosystem. In the aquatic habitat high turnover rates of individual organisms and also of chemicals result in a rapid succession of types of organisms, this effect being called the periodicity of phyto- and zooplankton. These high turnover rates and periodicity effects are due to the fact that the great majority of the organisms in the system are of only microscopically visible size and also because of the variations in the physical and chemical environment which result from seasonal changes.

In a study of the aquatic ecosystem it is useful to subdivide it into several component features. Because this subdivision will be used throughout this book, the general outline is given here.

Abiotic features. These include physical phenomena such as radiant energy, temperature and other aspects of water physics such as density, viscosity, surface tension etc., together with properties of the water molecule itself which result in the peculiar characteristics of water such as its freezing point and certain of its chemical properties. Also included here are the chemical elements and compounds which are dissolved or suspended in the water and which enable most natural waters to support life. The suspended matter is termed allochthonous if it originates from sources outside the lake or autochthonous if it derives from sources within the lake system.

Abiotic factors are rarely uniform with depth. Irradiance decreases with depth, and temperature profiles often show layers of relatively rapid change. It is often useful to consider effects as occurring in a water column of e.g. 1 m^2. Loadings can be expressed in quantity per volume or per unit surface per year. The loading per surface is the loading per volume times depth. If biological processes must be expressed as units per surface, but have been measured per unit volume, then the unit per surface can be found by summing results in all layers, or mathematically by integrating the function describing the process over depth.

Photosynthetic yield can then be expressed as e.g. mg m^{-2}d^{-1} of C and can thus be related to incoming light, which is expressed in e.g. cal. cm^{-2}d^{-1}, or to fish yield, which is expressed in kg ha^{-1}yr^{-1}.

Primary producers (see Plates I—III). Primary producers are those organisms which use solar radiation as their only energy source and in this context include the macrophytes, the algae, and the photosynthetic bacteria. Except within the littoral zone of lakes, the algae are quantitatively the most important of these groups. Many algae of the open water are unicellular but multicellular colonies and filaments do occur. Morphological and functional differentiation into colonies is fairly common amongst the multicellular forms, but no highly differentiated organs such as leaves, stems, or roots are found. In some colonies, e.g. *Pediastrum*, the outer cells have a markedly different shape from the inner ones, and this feature imparts to the colonies an enlarged surface area which allows them to float better and increases the efficiency of gas exchange and nutrient uptake. In all colonies or filaments each single cell retains the capacity for reproduction (cell multiplication). The algal groups mentioned below will often be referred to throughout the different chapters of this book, amongst these the diatoms, the green algae and the blue-green algae are quantitatively the most important groups. A more detailed list of most genera mentioned is given as Appendix I of this book.

The diatoms (Bacillariophyceae) contain both chlorophyll and brownish-green pigments which often mask the green chlorophyll colour. When wetted with ethanol or methanol they turn green. The silicified cell wall is characteristic, and it may easily constitute more than 50% of the dry weight of the organism. (The structure of the cell wall is described in more detail in Chapter 7.) The skeletons are often beautifully ornamented with transverse lines, projections, or depressions, the detailed patterns of which provide a basis for species recognition. They remain identifiable in stratigraphic deposits of sediments of great age and enable details of species successions and inferred environmental changes to be discerned, often for a long time span in the history of a lake. The diatoms commonly occur during spring time in temperate waters, either as free-floating cells or chains (filaments) or attached to plants or stones. They are divided into two broad groups: Centrales, with valves usually circular, and Pennales, with elongated valves. The freshwater species of the genus *Melosira* are probably amongst the commonest diatoms occurring in Lake Baikal (Siberia) and in shallow ponds or ditches.

The green algae (Chlorophyceae) in which the chlorophyll is not masked by other pigments are bright green and have rigid cell walls made of cellulose-like carbohydrate. A common genus is the flagellate *Chlamydomonas*, a free-swimming single-celled organism with two flagella and an eyespot; it can develop rapidly in ditch water which has been left in the laboratory or classroom. *Volvox* is one of the genera which forms the largest colonies, sometimes with 20 000 cells. Beating of the flagella, which are possessed only by the outermost cells of the *Volvox* colony, enables this organism to move by slow rotations through the water. One of the smallest algae is *Chlorella*, a single cell. It is often used in physiological studies because it can be easily maintained in culture and because of its rapid growth rate. *Cladophora*, *Oedogonium* and *Spirogira* are familiar green filamentous algae growing on stones, poles, etc.

Asexual cell division is the commonest method of reproduction by green algae. Sexual reproduction, which occurs usually in adverse conditions, is often rather complicated.

The blue-green algae (Cyanophyceae or Myxophyceae) are the "simplest" of all algae and are perhaps more closely related to the bacteria with their procaryotic cells. They also resemble bacteria in some of their chemical properties. Their characteristic colour is due to the pigment phycocyanin but other pigments can also occur and impart a violet, brown, or yellow colour to any species possessing them. The pigments are evenly distributed throughout the protoplasm, although some infrastructure has been discerned by use of the electron microscope. The cells have no nucleus and reproduce by simple cell division. Certain schools classify them as bacteria and not as algae.

One of the more easily recognisable genera is *Microcystis*, which forms masses of free-floating cells readily visible to the naked eye. The cells multiply abundantly and may form a scum on the surface of smaller lakes or bays. *Oscillatoria* is a filamentous type and occurs in large quantities during the summer in many eutrophic lakes, thus causing many problems in water management. Other genera which commonly cause blooms of nuisance proportions are *Nostoc* and *Anabaena*. Unlike most autotrophic organisms, many of the blue-green algae are able to use N_2 in addition to, or instead of, the more usual NH_4^+ or NO_3^- ions, a property referred to as nitrogen fixation. (They are discussed in detail in Chapter 6.)

Other important groups are the Cryptophyceae, the Dinoflagellatae and the Chrysophyceae.

Consumers. These organisms derive their energy from other organisms. Included in this group are the zooplankton (Plate IV), the insects, fish, etc. The main groups are outlined in Chapter 15. Primary consumers are those consumers that feed on the producers, whereas the secondary consumers feed on the primary ones. Terminal consumers are those organisms that are not used by any other organism, except after their death. They consist mainly of the larger fishes, which are also called predators, although this word is often used for any organism that actively hunts its prey. (Feeding relations between aquatic organisms are outlined and discussed in Chapter 15.)

Decomposers. These are the microorganisms that mineralise all dead organic material. Included in this category are bacteria, fungi, and certain yeasts. Their main function in the ecosystem is to mineralise the organic matter back into simple inorganic compound which are then recycled.

Strictly speaking, decomposers are consumers, but since they mostly digest dead organisms without ingesting them, it is useful to distinguish them.

Each organism has a habitat, i.e. the type of place where it occurs. For example, a young trout has the cool mountain stream for its habitat. Each type of organism is also said to live in a niche, which could be described as its physiological habitat, i.e. the "place" where it can carry out its essential physiological processes. In this context the word "place" should not be taken only in its physical sense but also as place in a food chain, a place restricted by tolerance limits and historical and geographical factors. An association of plants and animals is called a community or biocoenose. A biotope is the geographic entity in which a biocoenose can be found.

In the present state of limnological knowledge, quantitative investigations are important since many aquatic processes must first be quantified in order to understand how they work. Although qualitative descriptions of organisms and processes are necessary as a first stage, the aim of many current studies is to achieve an understanding of the nature and rate of changes, the quantitative interrelationships in competition and the rate of predation. As a result of the study of factors affecting the regulation of quantities in biological yields and rates, two important principles have been formulated: the law of Liebig (1840) and the law of tolerance (Shelford, 1913).

Liebig's law of the minimum factor states that the yield of a plant crop is determined by that element of which relatively the smallest quantity is available for that plant. If e.g. a wheat crop could be as large as 1000 tonnes based on the amounts of C, N and other elements present, but only 500 tonnes, because no more iron is available, then 500 tonnes it will be. A minimum factor is nowadays often called a limiting factor, i.e. the factor which controls the rate of a process such as the growth of an algal culture. (The difference is extensively discussed in Chapter 10.) Much confusion is caused because the two concepts are muddled. Results of an experiment in which growth rate was measured are often used in terms of, or to predict, the total yield of the experiment. Thus in the study of eutrophication (Chapter 17), the results of a 4-h bottle test for measuring the photosynthetic rate, are often used to predict the density of the final algal crop. (Growth rate and yield are discussed in Chapter 10.)

The mere presence of a substance may not necessarily allow growth of an organism. The compound must be present in such a form that the organism can obtain and utilise it, that form being said to be "available".

The second principle in quantitative ecology was called by Shelford (1913) the law of tolerance. Factors are not necessarily present only in a minimum quantity; they may in fact be present "in excess". Many organisms can only tolerate certain quantities of some substances, and larger quantities than the tolerance level may be toxic and the growth rate may decrease. There is often an optimal quantity of any substance which permits the greatest growth of an organism and the presence of lesser or greater quantities than this both result in decreased growth rate or yield.

As an example, in Fig. 1.1 growth rates of *Chlorella* and *Nostoc* are shown as a function of temperature. It can be seen that growth takes place above a minimum "threshold" temperature, T_{min}, and below a maximum temperature, T_{max}. At one temperature within this range growth is optimal. Above the op-

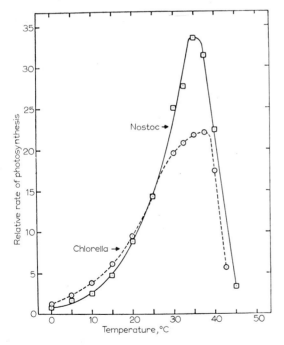

Fig. 1.1. Temperature relations of photosynthesis in *Nostoc muscorum* and *Chlorella pyrenoidosa*. (From Clendenning et al., 1956.)

6

timum "the excess is tolerated". It should be noted that growth optimum curves are seldom bilaterally symmetrical. The decrease in growth is frequently much steeper above the optimum than below it. Temperature optima are often quite different for different species of alga; they determine relative yields of different species in different circumstances and operate to produce certain features of competition and succession. For example the diatom *Fragilaria crotonensis* has an optimal temperature of 13—16°C. Wesenberg-Lund (1904) found that in the cold summer of 1902 this species was not replaced by blue-green algae as was normally the case, the blue-green algae having a higher optimum temperature than the diatom. Competition as one of the factors in algal succession is probably much more complicated than this example suggests (see Chapter 14). A similar effect may be seen in the response of organisms to different pH conditions (Fig. 1.2). E.g., the green alga *Scenedesmus* cannot grow at pH = 4 but can grow excellently even above pH = 11, while *Chlamydomonas* may thrive at pH = 4. Therefore, in a mixed community or culture the pH can have a selective effect, being optimal for one species at a given value, whereas it is minimal for another one. The pH has been shown to rise to high values during bloom conditions (e.g. Tjeukemeer, Loch Leven; subsections 14.2.2 and 15.5). It may therefore be a factor through which one algal species may inhibit other ones. Complications arise here also because the pH is not a property in itself, but is related to the CO_2—HCO_3^-—CO_3^{2-} system, so that, if the pH changes, the whole balance of the carbondioxide/bicarbonate system is automatically changed.

As a final example a harmless compound such as phosphate may be men-

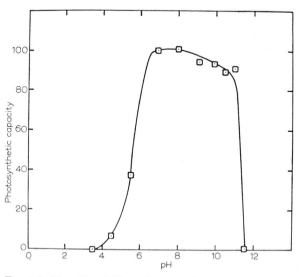

Fig. 1.2. The pH stability relations of photosynthesis in *Nostoc muscorum*. (From Clendenning et al., 1956.)

tioned. Rodhe (1948) noted that *Asterionella formosa* is never found in lakes with a PO_4-P concentration above 10 mg m^{-3}. Even in his cultures no growth took place at or above this phosphate concentration, but it is still difficult to prove that phosphate actually inhibits growth. In order to prove this, one must have a good growing culture and measure inhibition following the addition of further quantities of phosphate. If, however, the PO_4-P optimum lies near 2 or 5 mg m^{-3}, it is rather difficult to measure growth accurately because of the small quantity of algae which is present at that concentration. The situation may actually be more complicated in lake conditions and when *Asterionella* was found to disappear from Lake Windermere when phosphate increased above 10 mg m^{-3}, it seemed too good to be true. Possible explanations in this case could be that phosphate concentrations above 10 mg m^{-3} inactivated the iron by forming insoluble $FePO_4$. Recently Moed (1973) achieved a good growing culture of *Asterionella formosa* with PO_4-P = 1 g m^{-3}, but it could be argued that this may be an adaptation to unnatural conditions. Growth rate measurements of this culture as a function of the phosphate concentration may provide convincing proof of the regulating role of phosphate for this species of alga.

Limnology may be regarded therefore as a constant attempt to explain field observations by measuring the isolated process both in natural conditions and in culture. A culture is an oversimplified model, but nevertheless its use can indicate whether a process is possible or not, while sometimes — if one is lucky — information about the rate can also be obtained. One should be extremely cautious, however, in attempting to use rates obtained from experiments in culture (or in vitro as this is often referred to) as a basis for predicting or describing rates occurring in natural conditions.

The aims of ecological research are twofold. It is done firstly as an attempt to satisfy some of our curiosity about our environment, and secondly to attain understanding and perhaps control over part of that exciting but often unfriendly environment which relates directly to man's life and well-being. This may enable us to utilise our natural resources more effectively and may prevent us from harming them. The splendour of limnological research is that both of these aims can be pursued simultaneously, and perhaps there are few other scientific disciplines in which pure and applied research are so strongly interwoven.

REFERENCES

Clendenning, K.A., Brown, T.E. and Eyster, H.C., 1956. Comparative studies of photosynthesis in *Nostoc muscorum* and *Chlorella pyrenoidosa*. Can. J. Bot., 34: 943—966.
Moed, J.R., 1973. Effect of combined action of light and silicon depletion on *Asterionella formosa* Hass. Verh. Int. Ver. Theor. Angew. Limnol., 18 (prt. 3): 1367—1374.
Odum, E.P., 1971. *Fundamentals of Ecology*. Saunders, Philadelphia, Pa., 3rd ed., 574 pp.

Rodhe, W., 1948. Environmental requirements of fresh-water plankton algae. Experimental studies in the ecology of phytoplankton. *Symb. Bot. Ups.*, 10(1): 149 pp.

Shelford, V.E., 1913. *Animal Communities in Temperate America.* University of Chicago Press, Chicago, Ill.

Wesenberg-Lund, C., 1904. Studies over de Danske søers plankton. Specielle del. Copenhagen.

PLATE I (p. 9)

Blue-green algae (photographs by J.W.G. Lund)
1. *Anabaena flos aquae.*
2. *Anabaena flos aquae*, detail; *h*, heterocyst; *s*, spore (akinete).
3. a. *Microcystis aeruginosa.*
   b. *Gomphosphaeria naegeliana.*
   c. *Aphanizomenon flos aquae.*
4. a. *Anabaena flos aquae.*
   b. *Oscillatoria agardhii* var. *isothrix.*
5. a. *Aphanizomenon flos aquae.*
   b. *Gomphosphaeria naegeliana.*
6. a. *Oscillatoria agardhii* var. *isothrix.*
   b. *Oscillatoria bourrellyi.*

PLATE II (p. 10)

Diatoms (photographs by E. Mols)
1. *Diatoma elongatum*, chain, girdle view.
2. Periphytonic diatom community: *Gomphonema olivaceum, G. parvulum, G. constrictum, Synedra acus, Diatoma elongatum, Navicula gracilis, Achnanthes microcephala, Cymbella ventricosa.* Light microscope.
3. Same as 2, electron scanning microscope.
4. *Stephanodiscus astraea.*
5. *Gomphonema olivaceum; G. longiceps* var. *subclavata.*
6. *Gomphonema constrictum.*

PLATE III (p. 11)

Green algae (photographs by E. Mols)
1. a. *Scenedesmus quadricauda.* b. As a, abnormal 8-celled coenobia and large swollen mature cells, cultivated in Rodhe 8, continuous light, 30 000 lux.
2. Phytoplankton of the Tjeukemeer, i.a. *Pediastrum boryanum, Scenedesmus acuminatus, S. opoliensis, Oocystis crassa, Micractinium pusillum, Ankistrodesmus arcuatus.*
3. *Spirogyra* species, conjugating.
4. *Pediastrum boryanum*, old and just-formed young coenobium still laying in its vesicle.

PLATE IV (p. 12)

Primary consumers (photographs by E. Mols)
1. *Diaptomus gracilis* ♂, a common calanoid copepod in the plankton.
2. *Bosmina coregoni*, a common cladocer in large lakes.
3. *Keratella quadrata*, free-living rotifer with egg.
4. *Collotheca monoceros*, attached rotifer.

PLATE I

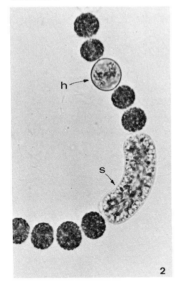

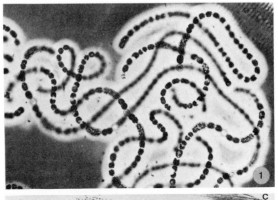

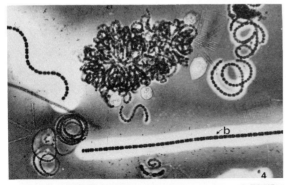

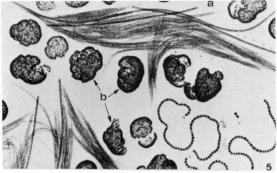

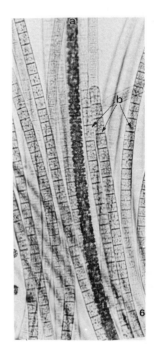

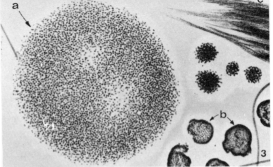

PLATE II

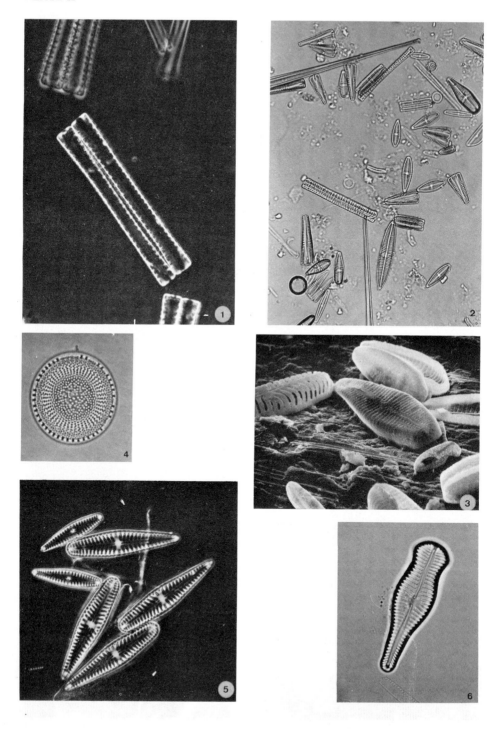

PLATE III

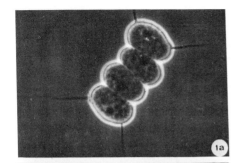

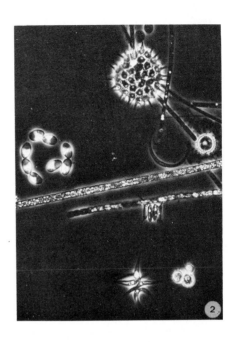

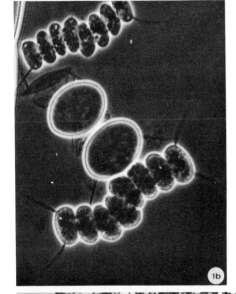

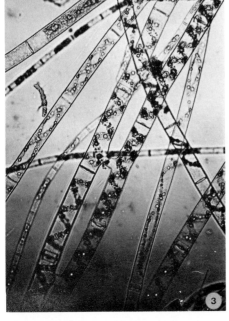

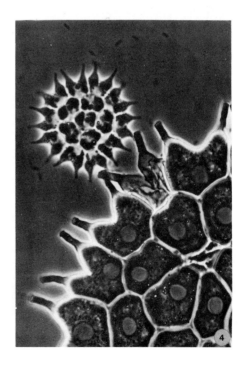

12

PLATE IV

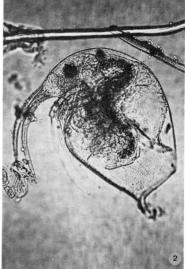

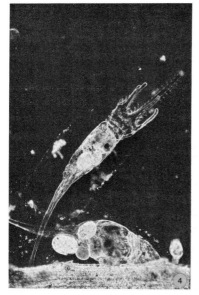

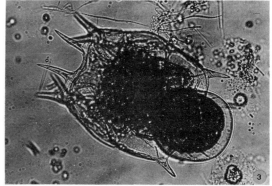

CHAPTER 2

LAKES, LAKE DISTRICTS AND THEIR ORIGIN

2.1. INTRODUCTION

Most lakes have a well-defined time and type of origin and are not always a constant feature of the landscape. Their origin is usually related to a catastrophic event such as glacial activity, volcanism, or an earthquake. A geological or geographical area containing several similar lakes may arise as the result of a single lake-forming episode. Lakes within one of these lake districts often have features in common such as morphology, geology of watershed, and chemical composition. Thus lakes in the English Lake District have calcium-poor waters, whereas the European alpine lakes — although of similar geological origin — have calcium-rich waters. The comparison between individual lakes and between lake districts is an interesting subject, although real knowledge is scarce because of a lack of data suitable for comparison.

The nature of the geological origin of a lake determines its morphometry and thus its water retention time and also, to some extent, its chemical composition. One would expect that the type of origin of a lake would be an important factor in determining its flora and fauna. This influence is modified however by the geographical position of the lake, since light regime and temperature will vary according to altitude and latitude.

Erosion, in its broadest sense, is another important modifying factor determining the chemical composition and biology of the aquatic habitat. Erosion from the watershed brings in dissolved salts and other solutes such as nitrogen, phosphorus and silica compounds, which are important for the growth of algae. It is strongly dependent on the geographical position of the lakes, on the geology of the watershed, and on man's use of the watershed. The importance of erosion on the supply of dissolved components and sediments will be discussed more fully in Chapter 18.

Lake origin, erosion, and geographical location are therefore the three main factors in determining the kinds and quantities of the various organisms which occur in a given lake. In the following paragraphs a few different types of lake origin will be considered.

The ages of the geological periods referred to are given in Appendix II.

A detailed study of the origin of lakes is given by Hutchinson (1957), who distinguished 11 different groups of lakes with 76 types.

2.2. TECTONIC ORIGIN (NOT INCLUDING VOLCANIC ACTIVITY)

The forces which give rise to deformations of the earth's crust such as mountain formation and other "uplifting" phenomena are called tectonic forces. Typical examples of lakes formed by these tremendous changes in the relief of the earth's crust are the Caspian Sea and the Aral Sea (U.S.S.R.). In Early Tertiary times both seas were part of the Tethys Sea which connected with the Atlantic Ocean and the Indo-Pacific Ocean. During this period large layers of marine limestone were deposited, where later — after the uplifting — the Alps were formed. Similar layers are found in arid areas such as in Tennessee (U.S.A.). The formation of the Alps during the Miocene Period severed the connection with the Atlantic Ocean. Subsequent glaciations and earth movements have had a modifying influence, causing drastic changes in the water level, though many details are still obscure. Archaeological evidence indicates that the water level of the Caspian Sea was at one time as much as 40 m above recent levels, but Berg (1950) has not found shells of the cockle *Cardium edule* more than 5 m above present-day levels. Further examples of tectonic lakes are Lake Baikal in Siberia and the lakes in the East African Rift Valley, e.g. Lake Tanganyika. Lake Tanganyika lies in a lake district which includes in the north Lake Kinneret (Israel), the Dead Sea and the Red Sea, and which extends as far south as Lake Malawi. All these lakes are situated in a fault in the earth's crust formed by the westward drift of the African continent. The geological history of the Dead Sea area has been summarised by Bentor (1961) as follows:

"Until the end of Mesozoic times the region of the present Jordan Arava graben was part of the mountain area of Palestine, but separated from the Transjordan block in the east by the very old geosuture, now forming the eastern border fault of the Rift. The Jordan Arava graben came into existence during the earlier Tertiary, probably in Oligocene times, through tensional movements which pushed the Transjordan block eastward and led to a downbreak of the western border of the graben along a complicated system of faults."

From the study of the sediments and fossils, Bentor concluded that the Rift area drained at least periodically toward the west, into the Tethys Sea. Lakes in the Jordan—Dead Sea graben were at some periods a series of freshwater lakes draining to the west, at some periods part of an extended arm of the Tethys Sea, and at other times a closed inland basin in which alternating periods of brackish, saline fluviatile, and aeolian deposition took place.

Both Lake Baikal and Lake Tanganyika show the typical morphology of tectonic lakes. Thus they have great depth, steep shores, and mountains rising steeply from the water. A few morphometric data illustrate this point (Table 2.1). Data for Lake Victoria have been added to demonstrate the difference from "shallow" lakes. Lake Victoria occupies a slightly depressed area in a plateau bounded by the Albert—Edward—Tanganyika system of rifts for which it is evident that there has been considerable tilting.

The enormous dimensions of lakes such as Baikal and Tanganyika greatly affect their hydrology, chemistry, and biology. Lake Tanganyika for example has an average water retention time of 1 500 years; if Lake Baikal had to be refilled by the river Rhine, it would take 400 years. Lake Tanganyika which is thermally stratified (see Chapters 3 and 9) has 90% of its water below the

TABLE 2.1

Morphometric data of some tectonic lakes

	Baikal	Tanganyika	Victoria	Dead Sea	Black Sea	Caspian Sea
Max. depth (m)	1 600	1 435	80	400	2 245	1 000
Surface area (km^2)	31 500	31 900	66 000	1 000	412 000	440 000
Max. width/max. depth	45	20	2 500	40	114	300

stable chemocline and this part, which does not mix with the upper layers, is permanently deoxygenated. The low productivity of Lake Tanganyika might be due to the loss of organic matter and nutrients to these lower layers (Beauchamp, 1964).

2.3. VOLCANIC ORIGIN

Lakes of volcanic origin include both those formed in extinct craters and those originating after an eruption has dammed a riverbed. The crater lakes, of which typical examples are found in France (Auvergne, Puy-de-Dôme), in Germany (Eifel, the so-called "Maare"), and several in Uganda, have characteristic bathymetric features. Thus they often cover a small area, are rather deep, and have contours usually forming concentric circles.

Crater lakes, especially in tropical areas such as Uganda, frequently contain high concentrations of salts. In Lake Mahega the following concentrations of anions were found: sulphate 1200—2400 meq. l^{-1}, chloride 1700—3300 meq. l^{-1}, alkalinity 200—400 meq. l^{-1}, sodium 2600—5600 meq. l^{-1}, and potassium 310—700 meq. l^{-1}, the higher concentrations occurring at greater depths (Melack and Kilham, 1972). Carbonate and bicarbonate concentrations are sometimes so high that during part of the year, or of the day, the lakes are covered with salt crusts. Nevertheless, the lakes occasionally contain dense populations, e.g. in Lake Mahega a population of an orange-coloured species of *Oscillatoria* (see Chapter 13) was found.

Examples of lakes formed by volcanic damming are Lac de Chambon and several other lakes in Auvergne, France, Lake Bunyoni in Uganda (Worthington, 1932; Denny, 1972), and several lakes in Japan (e.g. Lake Haruna-ko and Biua-ko). In the Asian part of the continent of Eurasia, the earth's crust is. more unstable than elsewhere in the world. Many active volcanoes extend from Kamchatka to Indonesia through Japan and the Philippines. There was relatively little glacial activity in this region.

Dams may be formed by lava or by a mud flow. In the latter case the origin is rather similar to that of lakes which are formed by landslides. There is often a serious risk of the dams bursting if the lake water level becomes high. Volcanic damming is common in Japan; silicate concentrations in these Japanese volcanic lakes are always high, and silicates are often the most abundant constituents (e.g. in Haruna-ko SiO_2 = 25 mg l^{-1}). Tectonic and volcanic activities often interact. Thus Lake Kivu (African W. Rift) drained originally to the north into Lake Edward, but since the volcanic dam formation it now drains to the south into Lake Tanganyika. Probably after this event Lake Tanganyika developed the present outlet, the Lukuya River, draining into the Congo. Before this Lake Tanganyika may have been the source of the Nile — a belief which was widely held until about 100 years ago.

2.4. GLACIAL ACTIVITY

Two main types of lakes may be formed by glaciers or glacial activity. The first type forms in a valley eroded by the mechanical violence of the glacier itself (e.g. the alpine lakes), whilst the second type is formed behind dams of morainic matter (soil, sand, stones) deposited by the glacier (terminal moraines). A moraine is often deposited at the mouth of a glacially eroded valley. Lakes were also formed at the end of the last ice age as steep-sided kettle holes were scoured out of the glacial drift by large lumps of ice debris which eroded and melted in situ. There are many of these in the midlands of England whilst Loch Leven in Scotland is one of the best known European ones (see section 15.5). Another well-studied example is Linsley Pond (Connecticut, U.S.A.) (see Chapter 18).

Lakes also occur on glaciers. These lakes are usually small and they may contain considerable amounts of suspended matter which impart a greyish colour to the water. Their temperature is, of course, rather low. Many lakes in the Vosges and Jura mountain ranges originated from the erosion of a valley by a glacier, sometimes combined with the closure of the valley by morainic debris or by a terminal moraine. The surroundings are less wild and rocky than those of the lakes in the Alps, which do not always have a morainic dam. Subsequent events may have drastically modified lakes of glacial origin. Subsidence of the eastern part of the Lake of Geneva for example redirected the flow of the river Rhône eastwards to its present course (Serruya, 1969), and large quantities of sediments have been deposited in the area of subsidence.

Much information on the origin of the alpine lakes comes from the study of their sediments, e.g. for the Lake of Constance by Wagner (1962) and by Müller and co-workers (see e.g. Müller, 1966, 1971; Förstner et al., 1968; Müller and Gees, 1970). The lake basin was formed in Tertiary times and now contains sediments consisting of three layers. The oldest (lowest) are comprised of a metamorphosed sandstone (molasse) of Miocene age. Subsequently the basin was scoured by glaciers during the successive glaciations of Pleistocene times. The deposits remaining are those left behind by the most recent (Würm) glaciation, which form the middle layer (Pleistocene sediments about 80 m), and the upper layer of sediments (Holocene sediments about 90 m) deposited in the last 15 000 years since the glaciers retreated. The present basin itself was formed by the Rhine glacier which had reached as far as the Danube during the Würm glaciation, and which left behind the middle layer of sediments. Lakes were formed between the terminal moraine and the retreating edge of the glacier approximately 18 000 years ago. Originally the lakes drained from the northern end through the old glacier into the Danube River, but as the glacier continued to retreat, the Rhine was formed and the meltwater was drained off to the west. After further retreat of the ice, the basin filled with water. The Lake of Constance was double its present size 15 000 years ago, and was connected with the Walensee and the Lake of Zürich. Later on, the three lakes became separated by the delta deposits of the Seez River. The Rhine now supplies two-thirds of the water and more than 90% of the allochthonous deposits of the Lake of Constance.

Most of the alpine lakes have calcium-rich water, since the valleys are formed in old marine limestone sediments. This results in biogenic sediments of calcium carbonate being formed, and these may have an effect on the algal productivity due to adsorbtion of phosphate by the calcium carbonate (see section 9.4.4).

Other examples of glacier lakes are the four Canadian "great" lakes, Great Bear, Great Slave, Athabasca, and Winnipeg, and many other smaller lakes in the same region (Larkin, 1964). They lie in the strata of the Precambrian Canadian Shield which was covered by ice during the last glaciation. The age of these lakes (10 000 years) is therefore similar to that of the alpine ones.

Drastic changes in drainage have occurred since their original formation, so that they are now relicts from much larger lakes. They receive enormous amounts of sediment: the Slave River discharges roughly 100 000 tonnes of sediment per day into the Great Slave Lake, a quantity which is roughly equal to 3 g of PO_4-P per m^2 per year.

The concentration of phytoplankton is roughly 2 g per m^2. An explanation of why this lake still remains oligotrophic is given in Chapter 5. However, Rawson (1956a, b), who studied the phytoplankton of this lake extensively, suggested that the combination of phytoplankton species found, *Melosira islandica, Asterionella formosa* and *Dinobryon divergens*, is different from which is to be expected using the European criteria for predicting oligotrophic species content.

Most of the flora and fauna in these glacier lakes belong to species which have become established during the last 10 000 years, but the presence of *Pantoporeia affinis* and *Mysis relicta* may reflect the preglacial populations. (For a more detailed treatment of this subject see Larkin, 1964.)

The origin of the St. Lawrence Great Lakes (North America, 48—50°N and 74—93°E) is thought to be related to glacier activity although there is some dispute about the complexity of these events.

The lakes were born when the retreating glaciers halted and built the Valparaiso—Fort Wayne moraine. Early lakes Chicago and Maumee drained southwards through the Mississippi, transporting considerable quantities of

18

clay to the Gulf of Mexico. After this first retreat, several re-advances and retreats occurred, the last re-advance taking place about 12 500 years ago.

During these movements the Great Lake basins often existed, but in a form and size very different from the present. When the Wisconsin glacier permanently retired from the Great Lakes region, the earth's crust began to rise slowly upwards, the so-called crustal rebound. Before this rebound the North Bay outlet was near sea level, whereas now it is about 200 m above this level, These events are summarised by Kelley and Farrand (1967).

The present basins may have been formed by preglacial waters draining across rock strata subject to erosion. During the Pleistocene these preglacial drainage paths were occupied several times by glacier lobes which formed

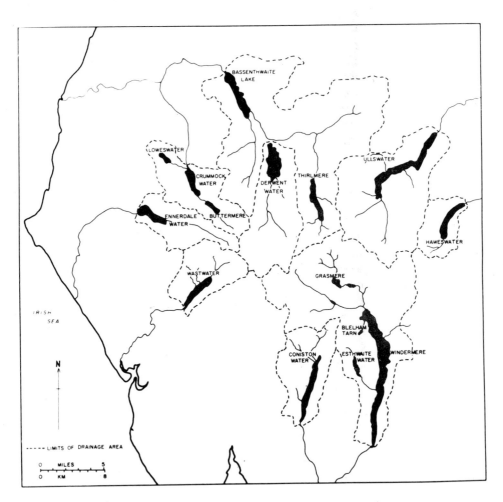

Fig. 2.1. The English Lake District.

deep channels. Most of Lake Superior and the northern part of Lake Huron lie in the Precambrian Canadian Shield, while the remaining bedrock is covered by glacial deposits (Chandler, 1964). The glaciated lake region of North America reaches as far south as the northeastern part of Indiana, where calcium-rich lakes prevail. In the Great Lakes Ca^{2+} (see Chapter 3) ranges from 0.5 meq. l^{-1} in Lake Superior to 1 meq. l^{-1} in Lake Ontario, with pH values ranging between 7.4 and 8.5. Total alkalinity is highest in Lake Michigan (2 meq. l^{-1}).

One of the best known lake districts is the English Lake District (northern England). Fourteen lakes with many similar features occur here in nine valleys within an area of 50 km radius. The lake areas radiate from a point which is the centre of an elevated dome of Ordovician and Silurian rocks (Fig. 2.1).

In the Carboniferous and Triassic periods large quantities of shallow marine sediments (mostly $CaCO_3$) were deposited upon the older rocks when these were below the sea. These younger calcareous rocks have been almost entirely eroded away from the dome after it was uplifted with the result that the lake waters contain less calcium than those of the alpine lakes. The present topography is the result of the different "hardness" (solubility) of the different rocks (see Chapter 18). Due to differences in catchment area, history of management, and erosion, lakes with differing trophic status occur, for example Esthwaite Water and Windermere. Esthwaite is much richer in phytoplankton than Windermere as a result of greater nutrient supply from the watershed. A controversy exists as to whether the lakes originated by glacier excavation or from damming by terminal moraines. Probably both processes were involved. Wastwater is an example of a glacier-eroded lake, whereas Windermere shows signs of damming by terminal moraines. (For a more complete geological survey see Hervey and Barnes (1970) and Macan (1970).)

Pearsall and Pennington (1947) described the interactions between archaeological and ecological events in the history of the English Lake District. Early forestry and sheep husbandry had a large influence on the erosion and thus sedimentation in the lake district. Pollen analysis of lake sediments has shown clear evidence of the following phases of activity: (1) clearing of the older valley swamps by neolithic people; (2) destruction of lowland woodlands containing oak and hazel; (3) great extension of grasslands as the result of this destruction and from sheep-walk exploitation by Norse settlers and later by monastic communities; and (4) extension of pine. These events have affected such features as carbon and nitrogen content of the different layers of lake sediments (see Chapter 18).

The lake beds are now eroded down into the older Ordovician—Silurian rocks, but initially the glaciers probably eroded the younger overlying limestone strata; they probably also eroded cave-like hollows into the older rocks underneath the limestone layers. When the limestone was eroded away much later, the original bottom part of the caves became the present lake basins.

2.5. SOLUTION OF ROCKS

Where soluble rocks are percolated by rain or underground water, cavities are often formed. In many limestone areas the rock is dissolved by water containing carbon dioxide, which results in the formation of caves, a feature well known to tourists. If the roof of such a cavity collapses, a surface lake may be formed. Such lakes are called karstic as they occur frequently in the Karst region of Yugoslavia near the Dalmatian coast. Many such lakes are found in the Alps, where considerable quantities of marine limestone sediments still occur over quite large areas. Examples of other soluble rocks are gypsum ($CaSO_4$) and NaCl, which will have a predictable effect on the water chemistry. Meromictic lakes, e.g. Lac de la Girotte (French Alps), often contain much H_2S, which is believed to derive from reduction of dissolved calcium sulphate by the bacterium *Desulphovibrio desulphuricans* (see Chapters 9 and 16).

2.6. RIVER ACTION AND LANDSLIDES

Lakes may be formed after a river has been obstructed by landslides caused either by the eroding activity of the river itself or as a result of earthquakes. In many cases in historic times a landslide dam has proved unable to support the pressure from the water accumulating behind it. A well-known example is a lake in Kashmir, which was formed by the Indus in 1840 or 1841. The lake, 64 km long and about 300 m deep, caused a flood in the Indus Valley when the dam collapsed half a year later. Many other floods occurring in India may have had a similar origin. More stable situations can occur if the water finds an alternative outlet, but usually the overflow above the dam together with the enormous pressure behind it makes the dam unstable and the lake only temporary.

Other types of lakes formed by "river activity" are those in delta regions, or after the river has widened out and the gradient has become less steep in the plain in its lower reaches. The rate of water flow decreases and sedimentation often occurs over a large triangular area to form a delta. The coastline moves progressively seawards, especially when the river outflow meets relatively calm coastal sea water (see Chapter 18) such as occurs in the river Rhine Delta area. At the mouth of the Amazon, however, ocean currents move large quantities of silt along the coast. In these deltaic sediments large shallow lakes form due to sinking and rising of both lake and sea level. The same phenomenon occurs to a smaller extent where rivers enter a lake as for example the Rhône flowing into the Lake of Geneva.

As a representative area the Rhine Delta will now be discussed in more detail; similarities to the Rhine system are found in the Nile, Danube, Mississippi, Amazon, Ganges and Rhône.

The present Rhine Delta is about 10 000 years old. The phase of silt deposition began in the Holocene (or Recent) period after the most recent glaciation. Although the Holocene spans only one-hundredth of the time of the

Quaternary Period, it is nevertheless very important for the formation of the present soil, especially as the sea level was 100 m lower during the Würm glaciation than at present. The melting of enormous quantities of ice (glaciers and ice sheets) caused a steady rise in sea level until about 5 000 years ago. This rise due to melting ice is called eustatic rise, and is a worldwide occurrence. Where the ice caps melted, the pressure on the underlying rocks was reduced and the land moved upwards (isostatic rebound). Due to this process Sweden is still rising nowadays, as can be seen from the present inland position of old villages originally on the shoreline. The Netherlands are however sinking, may be as a result of Sweden's rise. This occurs also in Great Britain: the northwest part of this island is still rising while the southeast is sinking. It has been also suggested that the weight of sedimentation in the North Sea is causing the southeast of England and The Netherlands to sink.

There have been other shorter-term changes in sea level during various periods due to processes not wholly understood.

The rise in sea level, resulting from the melting ice increased the amounts of sediments which were deposited, and the Rhine delta received not only sand and clay but also marine sand and silt. As a result of alternate rising and sinking of sea and land level more than twenty different layers of sand and clay have been formed (Golterman, 1973). Between some of these, layers of peat were formed. This is due to an incomplete decomposition of organic material when plants growing on the low-lying land during warmer periods were submerged. New layers of sand were then deposited upon the peat surface. It is interesting to note that the peat layers are often on top of clay layers, which may have provided the nutrients for the plant growth.

Lakes in delta regions are formed both as a result of changes in the course of a river and by waters draining into local — man-made or natural — hollows. Peat cutting and nowadays sand and gravel digging are the main processes giving rise to the present lakes in the Rhine delta (Gulati, 1972). Peat cutting has also been shown to have given rise to some of the Norfolk Broads in England. These Broads are situated in alluvial plains in East Anglia formed by the rivers Bure, Yare and Waveney. Before their confluence, these rivers drain through a region of subdued relief which declines from about 70 m above sea level, where the rivers rise, to about 15 m near the coast. In the west the rivers drain through chalk hills and in the east, near the shore of the North Sea, through shelly sand of the Pliocene crags. The bedrocks are extensively covered with Pleistocene boulder clays, loams and gravels. The lakes are shallow, ranging between 1 and 2 m in depth. Some of the lakes occupy the floors of subsidiary valleys and follow the line of these river courses, while others are connected with the river only by narrow openings. Water circulation between river and Broad is often nowadays restricted and controlled by gateways, dykes or sluices. In many cases the water level is maintained by pumps.

The history of this region is obscure and may go back to the twelfth century, about the same time that reclamation from the sea began in The Netherlands. Although there was clay subsoil, this often carried a layer of peat which was frequently drained by channel cutting, a situation closely resembling the Dutch one. The area was also frequently cut for litter and hay for animals and for sedge and reed for thatching. The area and its vegetation is described in detail by Jennings and Lambert (1951) and Lambert (1952). Outbreaks of *Prymnesium* occasionally occur in the Broads causing large fish kills. This poisonous alga (see Chapter 6) only flourishes in water with a high salt content. Salinity is high in some of the Broads which are tidal or where leakage from the nearby sea can occur. The salinity is a necessary but not a sufficient cause for *Prymnesium* growth; it may be that the special weather conditions of the region are particularly suited to it.

In the U.S.A., lakes occur as a result of surface coal exploitation. They are often deep, tend to be long, and are bounded on either side by a high bank of spoil and soil that was skimmed off. These lakes stratify because they are deep and sheltered. The waters are often acid and contain high concentrations of sulphate and sulphides.

In nearly all densely populated areas of the world, sand, gravel or even clay are dug for use as building material. With good management the resulting pits may be converted into excellent recreation sites. One of these pits is Lake Vechten (The Netherlands), which has been the subject of detailed continuous study since 1960. It has a surface area of about 5 000 m^2, a maximum depth of about 11.5 m and shows thermal stratification during the summer months. In England it is estimated that the gravel industry excavates more than 4 000 acres per year, about 80% of these being wet pits (Powell, 1975). They are often rich in calcium and silicate and are slightly alkaline. If the pits are properly managed, the phosphate and nitrate concentrations can be low so that blooms of algae are rare.

The character of the sediments controls the chemical composition of the water of shallow lakes. Thus, most of the Dutch lakes are humus-rich due to the presence of peat in the bottom and alkaline due to the presence of calcium carbonate in the clays. Phosphate, silicate, nitrogen, and iron originating from the decomposition of peat may be present in rather high concentrations. The biological implications of the sediment/water relationship are discussed in Chapter 18.

In delta regions large differences may be found between lakes situated near each other, owing to heterogeneous composition of the deposits. Thus running across the alluvial deposits in the Rhine delta are two chains of hills (Drenthe and Utrecht) formed from diluvial or glacial drift. Lakes occurring here often form in cracks (or faults) and have a low pH (see Table 3.4) because their only source of water is rain. This situation contrasts with the adjacent alkaline lakes which lie on alluvial deposits.

REFERENCES

Beauchamp, R.S.A., 1964. The Rift Valley Lakes of Africa. *Verh. Int. Ver. Theor. Angew. Limnol.*, 15: 91—99.

Bentor, Y.K., 1961. Some geochemical aspects of the Dead Sea and the question of its age. *Geochim. Cosmochim. Acta*, 25: 563—575.

Berg, L.S., 1950. *Natural Regions of U.S.S.R.* MacMillan, New York, N.Y., 436 pp. (translation).

Chandler, D.C., 1964. The St. Lawrence Great Lakes. *Verh. Int. Ver. Theor. Angew. Limnol.*, 15: 59—75.

Denny, P., 1972. Lakes of south-western Uganda. I. Physical and chemical studies of Lake Bunyonyi. *Freshwater Biol.*, 2: 143—158.

Förstner, U., Müller, G. and Reineck, H.E., 1968. Sedimente und Sedimentgefüge des Rheindeltas im Bodensee. *Neues Jahrb. Miner. Abh.*, 109: 33—62.

Golterman, H.L., 1973. Deposition of river silts in the Rhine and Meuse Delta. *Freshwater Biol.*, 3: 267—281.

Gulati, R.D., 1972. Limnological studies on some lakes in the Netherlands. I. A limnological reconnaissance and primary production of Wijde Blik, an artificially deepened lake. *Freshwater Biol.*, 2: 37—54.

Hervey, G.A.K. and Barnes, J.A.G. (Editors), 1970. *Natural History of the Lake District.* Warne, London, 230 pp.

Hutchinson, G.E., 1957. *A Treatise on Limnology, 1. Geography, Physics, and Chemistry.* John Wiley, New York, N.Y., 1015 pp.

Jennings, J.N. and Lambert, J.M., 1951. Alluvial stratigraphy and vegetational successions in the region of the Bure Valley Broads. I. Surface features and general stratigraphy. II. Detailed vegetational—stratigraphical relationship. *J. Ecol.*, 39: 106—148.

Kelley, R.W. and Farrand, W.R., 1967. *The Glacial Lakes Around Michigan.* Geol. Surv. Mich., Lansing, Mich., Bull. 4, 23 pp.

Lambert, J.M., 1952. The past, present and future of the Norfolk Broads. *Trans. Norfolk—Norwich Nat. Soc.*, 17(4): 223—258.

Larkin, P.A., 1964. Canadian lakes. *Verh. Int. Ver. Theor. Angew. Limnol.*, 15: 76—90.

Macan, T.T., 1970. *Biological Studies of the English Lakes.* Longman, London, 260 pp.

Melack, J.M. and Kilham, P., 1972. Lake Mahega: a mesothermic, sulphato-chloride lake in Western Uganda. *Afr. J. Trop. Hydrobiol., Fish.*, 2(2): 141—150.

Müller, G., 1966. Die Sedimentbildung im Bodensee. *Naturwissenschaften*, 53: 237—248.

Müller, G., 1971. Sediments of Lake Constance. In: G. Müller (Editor), *Sedimentology of Parts of Central Europe.* Int. Sediment. Congr., 8th, Heidelberg, 1971, pp. 237—252.

Müller, G. and Gees, R.A., 1970. Distribution and thickness of Quaternary sediments in the Lake Constance Basin. *Sediment. Geol.*, 4: 81—87.

Pearsall, W.H. and Pennington, W., 1947. Ecological history of the English Lake District. *J. Ecol.*, 34: 137—148.

Powell, A., 1975. Gravel pit fisheries. *Proc. British Coarse Fish Conf., 5th, The University of Liverpool,* (in press).

Rawson, D.S., 1956a. The net plankton of Great Slave Lake. *J. Fish. Res. Board Can.,* 13(1): 53—127.

Rawson, D.S., 1956b. Algal indicators of trophic lake types. *Limnol. Oceanogr.*, 1: 18—25.

Serruya, C., 1969. Problems of sedimentation in the Lake of Geneva. *Verh. Int. Ver. Theor. Angew. Limnol.*, 17: 209—218.

Wagner, G., 1962. Zur Geschichte des Bodensees. *Jahrb. Ver. Schutze Alpenpflanzen Tiere,* 1962: 7—17.

Worthington, E.B., 1932. The lakes of Kenya and Uganda, investigated by the Cambridge Expedition, 1930—31. A paper read at the Evening Meeting of the Society, Jan. 1932. *Geogr. J.*, 79: 275—297.

CHAPTER 3

CHEMICAL COMPOSITION

3.1. INTRODUCTION; PHYSICAL PROPERTIES OF THE WATER MOLECULE

In the preceding chapter it was shown how the geology of the water shed of a lake causes differences in the chemical composition of the water. Before dealing with the compounds dissolved in water, it is necessary to examine the properties of water itself.

In a water molecule the two hydrogen—oxygen bonds form an angle (about 105°). Therefore the water molecule forms a dipole and molecular aggregates are formed. For this reason many of the physical properties of water such as melting and boiling points are much higher than one would expect from a consideration of the hydrogen compounds of the elements near to oxygen in the periodic table (Table 3.1).

In the context of limnology even more important than the abnormal melting and boiling points is the heat of vaporization which is much greater than that of related compounds and which plays an important role in the heat budget of a lake. In a lake at 50° latitude, 60 cm of water (and thus 6 l per dm^2) may evaporate during the period March—October (200 days). This is equivalent to an average heat loss of $6 \times 600/200$ kcal. dm^{-2} d^{-1} = 18 kcal. dm^{-2} d^{-1}. Thus nearly 50% of the solar energy input during this period (40—50 kcal. dm^{-2} d^{-1}) is dissipated in evaporation of water. If, in the first 100 days of this period, a 2 m deep lake warms up 20°C, the heat necessary to provide this temperature increase is only 4 kcal. dm^{-1} d^{-1}. It is proportionately greater for deeper lakes. Under tropical conditions nearly all the solar energy absorbed is lost by evaporation.

TABLE 3.1

Melting points and boiling points of hydrogen compounds of some elements near to oxygen in the periodic table

	Melting point	Boiling point
H_2O	0	100
H_2S	− 83	− 60
H_2Se	− 64	− 42
H_2Te	− 48	− 2
H_4C	−183	−161
H_3N	− 78	− 33
H_2O	0	100
HF	− 83	19.5

The formation and melting of ice has a great influence on the temperature of a lake. In spring when the surface ice melts, it absorbs heat (the latent heat of fusion of ice) from the lake water. This has a moderating influence on the rate at which the waters warm up after the winter, and the lake organisms are thus protected against sudden temperature increase. The same heat-exchange mechanism operates in reverse during winter ice-formation, when latent heat is released into the underlying waters. This has a compensating effect against sudden falls in temperature which might otherwise damage the lake orgamisms. The melting of 1 m^3 of ice prevents the warming up of 10 m^3 of water with 8°C.

Other physical consequences of the molecular structure of water are the abnormal values of its density, viscosity, and dielectric constant. The H$_2$O aggregates dissociate with increasing temperature, so these properties change markedly with changing temperature. The relation between density and temperature is well known (Fig. 3.1). The fact that water is most dense at 4°C is of the utmost importance for aquatic ecosystems, as water at this temperature will sink and not freeze during cold periods even though there may be ice at the lake surface. Overwintering of organisms can thus take place at

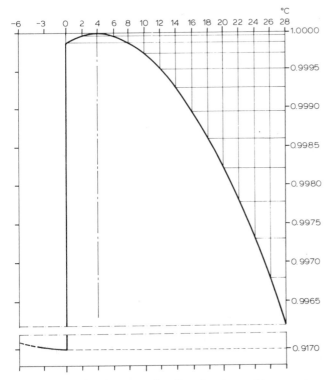

Fig. 3.1. Relation between density of water and temperature. (Redrawn after Dussart, 1966.)

the bottom of a lake, even though the upper layers are frozen. Dense water layers at 4°C are also of importance in the stratification of lakes (see Chapter 9). The relatively low density of ice is of an importance that hardly needs to be elaborated here. When surface water cools to 4°C it sinks. Ice with a density of ca. 0.92 will float, unlike solid metals, which sink down through the molten liquid.

Viscosity changes greatly with temperature:

Temperature (°C)	Viscosity (cp)	Temperature (°C)	Viscosity (cp)
0	1.8	15	1.1
4	1.6	20	1.0
8	1.4	30	0.8

The viscosity of water affects the speed with which algal cells sink. Upward movement of cells occurs either when the cell density is less than that of the water or when currents of water carry the cells upwards. In both cases the viscosity of water is important. The sinking rate of algal cells may approximate to Stoke's law, which holds for spherical particles. Most algal cells, however, deviate from this shape (ellipsoid, disk, prolate or oblate spheroid) and often have extensions for example like the "horns" of *Scenedesmus quadricauda.* The sinking rate of some of these simple shapes has been derived by Hutchinson (1957).

For evaluating the influence of temperature on sinking rate Stoke's law (for spheres) gives a first approximation:

$$v = \frac{kr^2 \Delta\rho}{\eta}$$

where v = rate of sinking (LT^{-1}); r = radius of sphere (L); $\Delta\rho$ = difference in density of sphere and medium (ML^{-3}); and η = viscosity. If a particle sinks from a warm-water layer into a colder layer, its velocity will be determined (within the temperature range 0—30°C) more by changes in water viscosity than by the change in density. This is firstly because the change in viscosity is larger than that of the density (Fig. 3.1 and the above table) and secondly because the sinking rate depends on the density difference between that of the particle and that of the surrounding medium, $\Delta\rho$. For a given temperature change within the range found in lake waters, this density difference $\Delta\rho$ changes proportionately less than the actual change in the densities themselves. Because the sinking velocity is lower at lower temperature, plankton cells which fall downwards tend to accumulate in the thermocline, the water layer between the epi- and hypolimnion (see Chapter 9). A complete treatment of the relation between sinking rate and viscosity involves considerations of the shape of the particle and water turbulence (amongst other factors) and is far from simple (see Hutchinson, 1957).

Turbulence caused by temperature differences, wind, or even fish movements, cannot be dealt with quantitatively, but it is of great importance in relation to the sinking of phytoplankton. The high viscosity of water ensures that the phytoplankton cells are carried with the water movement and are more or less prevented from sinking and thus causes them to remain dispersed throughout the whole lake depth or in the light-penetration zone. Sinking of cells may increase mineral uptake. This is because movement will prevent the formation around the plankton cells of layers with low mineral concentrations due to uptake.

Lastly, the dielectric constant of liquid water is also much higher than that of other comparable liquids, making water an excellent solvent for salts, because they will mostly dissociate into ions.

3.2. DISSOLVED COMPOUNDS

3.2.1. General composition (quantities and origin)

The elements or compounds that occur in natural waters, in dissolved state, can be allocated for convenience into the categories: major, minor, trace elements, gases, and organic compounds. Furthermore, several compounds occur in water in particular form. Rivers carry large quantities of silt (Golterman, 1975) into lakes, where, owing to lack of water movement, sedimentation will take place (see Chapter 18). The terms "major" and "minor" refer to quantities found and not to their biological significance. A summary of this quantitative classification is given in Table 3.2. Dissolved gases such as O_2, CO_2, N_2, H_2, CH_4 will be discussed in the appropriate chapters (4, 8

TABLE 3.2

Major, minor, trace elements, organic compounds, and gases normally found in natural waters (the percentages refer to relative quantities calculated from "mean" freshwater composition)

Major elements		Minor elements	Trace elements and organic compounds
Ca^{2+} (64%)	HCO_3^-	(73%) N (as NO_3^- or NH_4^+)	Fe, Cu, Co, Mo, Mn, Zn,
Mg^{2+} (17%)	SO_4^{2-}	(16%) P (as HPO_4^{2-} or $H_2PO_4^-$)	B, V (μg l^{-1})
Na^+ (16%)	Cl^-	(10%) Si (as SiO_2 or $HSiO_3^-$)	
K^+ (3%)			organic compounds such
H^+	F^-		as humic compounds, excretion products, vitamins,
(Fe^{2+})			other metabolites
(NH_4^+)			

Concentrations usually found

0.1—10 meq. l^{-1}	(< 1 mg l^{-1})	
(mean 2.4 meq. l^{-1})	(Si sometimes higher)	

and 9), whereas the microelements N, P and Si will be discussed in Chapters 5, 6 and 7.

In Table 3.2 the percentages of the major elements are calculated for a "mean" global freshwater composition (Rodhe, 1949). It is clear that on a global basis calcium bicarbonate predominates as the most abundantly occurring dissolved salt.

The sum of the cations in Table 3.2, except those placed between brackets, should equal the sum of the anions. Fe^{2+} and NH_4^+ normally occur only in anaerobic water, and in alkaline water they are not present in the ionic state. H^+ is always present of course but it contributes significantly to the ionic balance only in water where pH = 4 or less. OH^- may occur in strongly alkaline water but only together with much larger concentrations of HCO_3^- and CO_3^{2-}. It will therefore never contribute significantly to the ionic balance.

The relative quantities are calculated as though a mean freshwater composition existed. Unlike the major constituent composition of seawater, the composition of freshwater is not "standard" but is always changing, both in time and place. Because the above calculations were made before the water composition of the large African lakes was widely known, the percentages given are different from those which would be calculated now. Lake Tanganyika containing roughly 20% of the world's freshwater has 8 meq. l^{-1} cations plus 8 meq. l^{-1} anions, of which Mg = 48%, Na = 33%, K = 12%, Ca = 7%, while HCO_3^- = 89% and Cl^- = 10% (Talling and Talling, 1965). Other areas which have only recently been studied are some Australian lakes and other inland waters, especially in Queensland (Bayly and Williams, 1972; Williams and Hang Fong Wan, 1972). In many of these waters sodium dominates the cations and chloride the anions; this is so both for dilute and for quite saline waters. In other regions, however, divalent cations and bicarbonate dominate, whilst with an intermediate type the chloride content is roughly equal to the bicarbonate content. No indication is given of the probable explanation for the frequently high sodiumchloride content.

The presence of large quantities of Ca^{2+} and HCO_3^- ions is important in determining and buffering the pH of the water (see subsection 3.3.1).

The electrical conductivity of any water is roughly proportional to the concentration of its dissolved major elements. It varies enormously ranging in value between 10 and over 100 000 μS cm^{-1}.

An approximate estimate of the quantity of dissolved ionic matter in mg l^{-1} in a water sample may be made by multiplying the specific conductivity by an empirical factor varying from 0.55 to 0.9, depending on the nature of the dissolved salts. A similar estimate in meq. l^{-1} may be made by multiplying the conductance in μS cm^{-1} ($\equiv \mu mho$ cm^{-1}) by 0.01.

Talling and Talling (1965) gave examples of many waters in Africa with exceptionally low or high mineral concentrations. They divided the lakes in Africa into three classes according to their conductivity value. Class I had a conductivity < 600 μS cm^{-1}, Class II 600—6 000 μS cm^{-1}, and Class III > 6 000 μS cm^{-1}.

Some of the lowest conductivities (Class I) are found either in the Congo Basin or in lakes whose inflow drains through swampy regions. These lakes are frequently dark-coloured and contain a high concentration of organic compounds (compare with the black waters of Amazonia; Golterman, 1975). Other examples will be given in subsection 3.2.2. Class II includes large lakes such as Lake Rudolf, one of the most saline in this group, which occupies a closed basin which formerly contained a larger lake with outflow to the Nile. The salinity in these lakes is due largely to Na^+, Cl^- and HCO_3^-. Class III includes lakes in closed basins with no outflow; under these conditions salts accumulate. The highest conductivity which Talling measured was 160 000 μS cm^{-1} with alkalinities up to 1 500—2 000 meq. l^{-1} being recorded.

Cations and anions in lakewaters originate mostly from mineral sources such as erosion of rocks and soils. Goldschmidt (1937) estimated that 160 kg of rock has been eroded per cm^2 of soil surface or 600 g per kg of seawater (equivalent to a layer of 400—600 m deep over the whole earth!). Apparently the oceans are now in a state of equilibrium between input and output, but for freshwater this is certainly not yet the case; obviously long-term changes in riverwater composition are too small to be measurable for the period during which analyses are available. From a study of the chemical and physical characteristics of layers of sediments in sample profiles from lake bottoms, however, drastic differences in the erosion pattern may be inferred (Chapter 18).

Concentrations of (bi)carbonates normally exceed those of the other anions present, and in most cases the (bi) carbonate is calcium bicarbonate, although in African waters sodium and magnesium bicarbonate may reach high concentrations. If sodium bicarbonate is the main product of erosion, it will gradually precipitate any calcium from the solution owing to the low solubility of $CaCO_3$. High concentrations of soluble phosphate are then possible (Lake Rudolf and Lake Albert, section 5.1). Chloride may come from natural erosion or, near the coast, from the sea. In the latter case the chloride may be carried by the wind; thus, in the English Lake District Gorham (1958) found that the chloride concentrations in different tarns are a function of the distance from the coast to those tarns. The following relationship was found:

$$c = 0.71 \, d^{-0.33}$$

where c = concentration (meq. l^{-1}), and d = distance (km).

Gorham showed that atmospheric precipitation (both by wind and in rain) is of the greatest importance as a source of many ions especially to bogs, to upland tarns on insoluble rocks, and to more humus-rich waters. He showed that the relative proportion of ions which are derived from rain varied with the total salt content. While SO_4^{2-}, K^+, Mg^{2+} and Ca^{2+} decrease with increasing chloride concentration, Na^+ does not change as, in all waters involved, it is only of marine origin. On the other hand, K^+ also comes from the rain in the more dilute waters, but there is a steadily decreasing proportion present as total salt concentration rises.

TABLE 3.3

Chloride load in the Rhine, up- and downstream of the French potassium mines in 1955 and 1965

	1955	1965
Kembs (upstream)	8 kg sec^{-1}	9 kg sec^{-1}
Seltz (downstream)	90 kg sec^{-1}	160 kg sec^{-1}

Chloride may enter inland waters near the coast through brackish river estuaries, while in low-lying areas it may arrive even by leakage or seepage from below. Lastly, large amounts of dissolved salts may come from human activities, e.g. the river Rhine receives disastrous amounts from the French potassium mines (see Table 3.3 and section 19.1). This very high chloride load renders the downstream water unusable for agricultural purposes. International agreements which would permit pollution restrictions to be enforced across national boundaries have not yet been reached.

Fluoride is also a naturally occurring element. Wright and Mills (1967) found 10 mg l^{-1} in the Madison River in the Rocky Mountains; in some parts of Wales several mg l^{-1} are commonly found. Fluoride is eroded mainly from acidic silicate-rich rocks and from alkaline intrusive igneous rocks (Fleischer and Robinson, 1963). It also occurs in gases and waters of volcanic origin; in volcanic gases in the Valley of Ten Thousand Smokes (Alaska, U.S.A.), up to 0.03% fluoride is found so that 200 000 tonnes per year of HF enter the geochemical cycle there. Its solubility is determined by the solubility product of CaF_2, which is $4 \cdot 10^{-11}$. In waters with 2 meq. l^{-1} of Ca, only 4 mg l^{-1} of F$^-$ can dissolve. Phosphate further decreases the amount of soluble fluoride, since fluor apatite is even less soluble than CaF_2. The high concentrations of phosphate present in the Rhine limit the dissolved fluoride concentrations in it to $2 \cdot 10^{-19}$ g l^{-1}. Soluble fluoride is found in many tributaries of the Rhine, but is no longer detectable in dissolved state in the main river.

The distinction made between *major* and *minor* elements has a biological significance in addition to the chemical one. The concentrations of major elements and compounds present in lakewater tend to determine which species of organism can occur there while those of minor elements and compounds limit the relative and absolute numbers of those organisms (see limiting factors, Chapter 10). Planktonic desmids are for example normally found only in Ca-poor waters, although the distinction between calciphobe and calciphile should not be taken too rigidly. Moss (1972) attempted to establish factors affecting the distribution of algae in hard and soft waters by studying the mineral requirements of these algae. Some oligotrophic (Ca-poor) desmids were found to be unaffected by levels of calcium as high as those which actually occur in eutrophic lakes, while Ca-poor media could be used for in vitro growth both of eutrophic and oligotrophic species.

Lakes with a different calcium content will also have a different pH and bicarbonate content. The types and quantities of lake organisms present are due to a complex of factors and not simply to the concentration of one ion. Thus Provasoli et al. (1954) found that the diatom *Fragilaria capucina* Desmaz grew better when the Ca^{2+}/Mg^{2+} ratio was greater than 1 and the ratio of monovalent to divalent ions (M/D) was low, whilst some *Synura* species grew better when M/D was high and concentration of total solids was low. However, *Synura* also includes species with the opposite behaviour.

Pearsall (1923) suggested that diatoms are abundant only in calcareous waters having an M/D ratio below 1.5. Nevertheless, it is now known that in African waters, where the Ca content is low, diatoms bloom in waters which have a high M/D ratio. Thus the situation is probably much more complex than was originally supposed, and greater consideration should be given to other factors such as erosion. For example, most of the silicate supply derives from erosion and erosion will in itself affect and be correlated with the M/D ratio. M/D ratios in natural waters are therefore not independent of the silicate concentrations. Furthermore, since Ca will erode from marine limestone (see Chapter 18), there will often be more phosphate in hard than in soft waters.

The presence of chelating substances may greatly affect certain ionic ratios. Miller and Fogg (1957) demonstrated an antagonistic effect of high concentrations of Ca^{2+} ions on Mg^{2+} uptake by *Monodus*. The addition of chelating agents such as EDTA — but also of naturally occurring chelators, such as glycine (see Chapter 12) — had an enhancing effect on Mg accumulation and uptake, this being due probably to their preferential chelation of the Ca^{2+} ions. Because chelation processes occur in most natural waters, only a very cautious attempt should be made to relate data obtained from culture studies to actual field conditions. Growing algae, e.g. blue-greens, may excrete organic chelating compounds into a culture solution which originally contained only inorganic compounds (Fogg and Westlake, 1955).

An example illustrating the relative importance of K^+ or Na^+ comes from work on Russian fish ponds (Braginskii, 1961). Application of potassium phosphate caused a change in the dominance of a population from blue-green algae towards desmids. The addition of sodium phosphate caused colonies of the blue-green algae *Microcystis* and *Aphanizomenon* to disintegrate but this did not occur when potassium phosphate was added.

Many interrelations exist between the different ions. The product of the concentrations of Ca^{2+} and CO_3^{2-} can theoretically not exceed the solubility product (see subsection 3.3.1), while the same is true for Ca^{2+}, PO_4^{3-}, OH^- or F^-. Comparable interrelations exist between phosphate and iron (see subsection 9.4.4). Therefore, the effect of one or two elements should never be studied separately without reference to the whole chemical balance. This can be done by measuring the total sums of cations and of anions, and these totals (in meq. l^{-1}) should be equal (hence the use of the term "ionic balance").

TABLE 3.4

Ionic balances of several lakes (meq. l^{-1})

		H^+	(NH_4^+)	Ca^{2+}	Mg^{2+}	Na^+	K^+	(Fe)
Acidic bogs	I	0.1	0.05	0.08	0.05	0.29	0.03	—
in Drenthe	II	0.1	—	0.08	0.05	0.36	0.04	—
	III	0.1	0.1	0.08	0.05	0.22	0.07	—
Lake Vechten epilimnion		—		2.37	1.16	0.4	0.05	0.06
Lake Vechten hypolimnion			0.1	3.43	1.32	0.4	0.05	0.26
Lake of Geneva		—		2.34	0.52	0.10	0.04	—
Lake Tanganyika		—		0.5	3.6	2.5	0.9	
Dead Sea		—		650	2870	1460	162	
Lake Chad (S)		0.0		0.56	0.36	0.36	0.13	
Lake Chad (N)		0.0		2.2	3.4	4.2	2.2	
Rhine at Lobith (13—3—66)		0.1		4.24	0.86	2.74	0.15	
at Lobith (13—10—66)		0.1		4.68	1.00	5.57	0.24	
at Stein (17—10—66)		0.0		2.63	0.58	0.20	0.03	
Rhine conductivities								
at Lobith (13—3—66)		6		221	43	132	10	—
at Lobith (13—10—66)		6		248	50	268	17	—
at Stein (17—10—66)		0		140	30	10		—

[*1] Includes 0.11 NO_3^-; [*2] includes 0.14 NO_3^-; [*3] includes 0.05 NO_3^-; [*4] includes 51 meq. l^{-1} of Br^-; [*5] includes 0.03 NO_3^-.

An estimate of the accuracy of the analytically determined "ionic balance" may be made by comparing it with conductance measurements; errors or omissions in the analysis or calculations may sometimes be traced if the results do not agree very closely (examples in Table 3.4).

3.2.2. Case studies

Some examples of ionic balances are now discussed in order to illustrate the enormous variation which occurs (Table 3.4). Firstly examples of bog waters with low mineral concentrations will be considered. In acidic bogs (e.g. such as in Drenthe, The Netherlands) calcium and magnesium concentrations are rather low and are numerically equal to the acidity (H^+) which in general may be regulated by the bog moss *Sphagnum* which often dominates the plant community (Clymo, 1967). A supposed absence of sulphate in these bog waters was believed to limit the algal productivity. After close examination of details in the ionic balance, it was realised that sulphate might be present in quantities up to 0.2 meq. l^{-1} which could easily escape the relatively insensitive method of sulphate determination used. After the samples had been concentrated, sulphate was found in quantities which were sufficient to permit nearly unlimited algal growth. Sulphate is often presumed to be measurable

ΣCat	Cl^-	SO_4^{2-}	HCO_3^-	$HSiO_3^-$	ΣAn	$Cat{-}An$	
0.55	0.90					− 0.35	
0.63	0.28					+ 0.35	
0.62	0.23					+ 0.31	
4.04	0.47	0.27	2.70		3.44	0.6	
5.55	0.47	0.01	4.38	0.26	4.8	0.7	
3.00	0.07	1.00	1.82		2.92[*5]	0.08	
7.5	0.75	0.1	6.7		7.6	− 0.1	
5142	5171[*4]	188	4		5363	−221	
1.41	−	−	1.40		1.40	0.01	
12.0	−	−	10.7		10.7	1.3	
8.09	3.38	1.64	2.90	0.15	8.19[*1]	− 0.1	
11.6	6.04	2.25	2.80	0.13	11.4[*2]	0.2	
3.44	0.10	0.74	2.46	0.05	3.40[*3]	0.04	
						Calc.	*Measured*
415	250	128	122	10	516	931	690
589	450	175	118	10	760	1350	1044
180	7	60	105		172	352	300

as the difference between total cations and other major anions; this is a thoroughly unsatisfactory practice. Even lower total ion concentrations were found by Armstrong and Schindler (1971) in lakes in the Canadian Precambrian Shield. Here the total cation concentration varied between 0.15 and 0.38 meq. l^{-1}, with Ca^{2+} as the predominating ion. Specific conductance ranged between 11 and 32 μS cm^{-1}, with a mean or median of 21 μS cm^{-1}. The ratio of conductivity to total dissolved solids in meq. l^{-1} is very near to 100. Most lakes in this region are dominated by Ca^{2+} and HCO_3^- due to the presence of calcareous material in glacial drift, but the total concentrations are low as is to be expected for lakes in a granite area. A few lakes which are near the sea coast are dominated by Na^+ and Cl^-. Similarly low conductivities are found in high mountain glacier lakes, e.g. Lake Gjende, Finland (Kjensmo, 1972). Conductivity in this lake is 10 μS and total cations just under 0.1 meq. l^{-1}. Due to the presence of glacial ooze the pH of the water is relatively high (pH = 6.8—7.0), and Kjensmo believed that the ooze acts as a major buffering system in the lake.

The epilimnetic surface water of Lake Vechten (The Netherlands, see section 2.6) is a typical example of a deep lake in equilibrium with $CaCO_3$ in the sediments (Table 3.4). It is slightly supersaturated with $CaCO_3$, a common situation in many calcareous lakes (see subsection 3.3.1), although it is also possible that the lake was not in equilibrium with CO_2 in the air because of a release of CO_2 from the deeper water, the hypolimnion. In shallower calcareous lakes calcium content tends to be just below 2 meq. l^{-1}. The high values of Ca^{2+} and HCO_3^- in the hypolimnion of this lake show that, although chemical composition may be fairly uniform over large horizontal distances, it may be — perhaps usually is — much more variable over relatively small vertical distances. (The cause of this will be explained in subsection 9.4.3.)

Chemical analysis of total mineral content of the waters of Lake Vechten shows that even using the most accurate analytical methods, the sum of the cations does not always equal the sum of the anions. In this case the differences are probably due to the presence of colloidal $CaCO_3$, but in other lakes there may be different reasons for the discrepancy.

Another example of a Ca-rich lake is the Lake of Geneva (Meybeck, 1972). This type of water is typical for the Swiss alpine lakes. The chloride/sulphate ratio is atypical and results from the presence of deposits of $CaSO_4$ within the alpine watershed. Most of the lake sediments consist of a $CaCO_3$ precipitate, a common situation in calcareous lakes (see subsection 3.3.1). Meybeck's study of these sediments is discussed in Chapter 18.

In Table 3.4 data from one river, the Rhine, are included (from Rhine Report, 1966). This river originates from the same region as the Lake of Geneva and has roughly the same calcium content (Stein). Measurements of the ion content of the river Rhine more downstream (Lobith) show increases in concentrations of calcium, sodium and chloride resulting from industrial effluent. Apparently the water at Lobith must be supersaturated with free CO_2. If the water enters a lake, the supersaturation will cease once an equilibrium with air is re-established or approached. The calculated conductivity based on actual ion-content analysis values exceeds the measured values. Colloidal matter is likely to be responsible for this phenomenon; it may become partly dissolved during the analysis. If this is true then it is clear that the published values for the concentrations of dissolved components are of doubtful validity because large quantities of particulate matter are present in the water but not included in the figures given. This is particularly true for phosphate; the river Rhine brings far greater quantities of phosphate compounds into The Netherlands than the published figures would suggest. For this same reason the large quantities of heavy metals which are now known to be present in Rhine water remained undetected for many years. Presumably they were present only in the particulate fraction, whereas the earlier analyses were only of the dissolved components.

Lake Chad and the Dead Sea — although having a similar hydrological regime, i.e. inflow and no outflow with water loss due only to evaporation — show remarkable differences in mineral content.

The Dead Sea is, together with Great Salt Lake (U.S.A.), probably the most saline inland water: its salinity is 31.5%. The Dead Sea, which acts as a salt pan, with salt crystallizing from a saturated solution, receives one-third of its salts from the river Jordan and two-thirds from highly saline springs. An interesting feature of the water is its high bromide content, 5.9 g l^{-1}, which comes from relict fossil brines. The chemistry of the Dead Sea has been studied by Bentor (1961), whose analyses are included in Table 3.4 and by Nissenbaum (1970), who gave some slightly higher values. A model for the chemical evolution and behaviour of the Dead Sea has been developed by Lerman (1967).

The Dead Sea has a notoriously high salt concentration. In Lake Chad, however, the water remains fresh; the waters in the north are only 8.5 times more concentrated than those in the south, where the river Chari enters the lake. Analysis shows that for northern waters of the lake the sodium and bicarbonate concentrations exceed those in the south by a factor of 8, whereas the calcium concentration is only 4 times as great. Carmouze (1969, 1973) believed that, besides precipitation of $CaCO_3$ in the north basin, a new formation of calcium, magnesium and potassium silicates takes place in the south basin. Adsorption of these cations onto the clays which are brought into the lake in large quantities by the river Chari with a concomitant precipitation of SiO_2 cannot yet be distinguished from neogenesis of a clay-like silicate. The input of clay by the river Chari thus prevents Lake Chad resembling the Dead Sea.

Carmouze gave evidence to show that such a mechanism is clearly indicated although the additional possibility of the occurrence of an underground water outlet cannot be excluded. The existence of such an outlet might explain why the water of Lake Chad is not more concentrated.

3.3. THE CALCIUM BICARBONATE SYSTEM

3.3.1. Chemical considerations

The bicarbonate ions in the water have two important functions. In the first place they provide the main buffer system for regulating the hydrogen-ion concentration in water, while secondly they provide the carbondioxide for photosynthesis. For this reason photosynthesis has an influence on the pH of water, causing the pH to rise and carbonate ions to be formed in the light, the process being reversed in the dark.

A clear understanding of the bicarbonate system is therefore essential and is probably best obtained by considering firstly the solubility of CO_2 in water. The solubility follows Henry's law and shows a linear relationship with the partial pressure of CO_2 in the gas phase (0.03% for CO_2 = 0.23 mm Hg) and with the temperature.

Under normal conditions the solubility of CO_2 is 0.4 mg l^{-1} at 30°C, 0.5 mg l^{-1} at 20°C and 1.0 mg l^{-1} at 0°C. The solubility is independent of the pH. A small amount of CO_2 will react with water to form HCO_3^- ions:

$$CO_2 + H_2O \rightleftharpoons H_2CO_3 \rightleftharpoons H^+ + HCO_3^-$$

Larger quantities of HCO_3^- occur in mineral waters, where CO_2 is often present in excess. This excess CO_2 will increase the H^+ concentration sufficiently to dissolve carbonates:

$$CaCO_3 + H_2O + CO_2 \rightleftharpoons Ca^{2+} + 2HCO_3^-$$

The bicarbonate system can thus be described by the following reactions, which each represent an equilibrium between the ions and molecules which together regulate the system as a whole, i.e. H_2O, H^+, OH^-, CO_2, HCO_3^-, CO_3^{2-} and Ca^{2+}. The latter will be considered, for the sake of simplicity, to be the only cation present:

$$H_2O \leftrightarrow H^+ + OH^- \tag{3.1}$$
$$CO_2 + H_2O \leftrightarrow H_2CO_3 \tag{3.2a}$$
$$H_2CO_3 \leftrightarrow H^+ + HCO_3^- \tag{3.2b}$$
$$HCO_3^- \leftrightarrow H^+ + CO_3^{2-} \tag{3.3}$$
$$Ca^{2+} + CO_3^{2-} \leftrightarrow (CaCO_3)_{solid}$$

An equation may be written for each equilibrium. It is necessary that Σ cations = Σ anions, thus if no mineral cations are present:

$$[H^+] = [HCO_3^-] + [OH^-] + 2[CO_3^{2-}] \tag{3.4}$$

There are six simultaneous equations therefore and these must all be satisfied at any one time. Furthermore one may define:

$$(CO_2)_{total} = (free\ CO_2 + H_2CO_3) + HCO_3^- + CO_3^{2-} \tag{3.5}$$

The equilibrium constant for reactions (3.2a) and (3.2b) combined is called K_1, the apparent dissociation constant for the first ionisation step of carbonic acid. The dissociation constant for (3.3) is K_2. Both reactions are temperature-dependent (Table 3.5) and are also dependent on the ionic strength of the solutions.

If some CO_2 is removed from solution, for example by algae during photosynthesis or by escape into the air, reaction (3.2a) and thus (3.2b) will shift leftwards, causing the H^+ concentration to decrease. Therefore, the reactions (3.1) and (3.3) will shift to the right, causing increasing OH^- and CO_3^{2-} concentrations. Because $K_1 \gg K_2$, more HCO_3^- will react following reaction (3.2) than (3.3). The resultant increase of CO_3^{2-} is therefore less than the amount of CO_2 removed from the system. The amount of CO_3^{2-} formed can be estimated by measuring the change in pH and then using Fig. 3.2.

TABLE 3.5

Temperature dependence of the first (K_1) and second (K_2) ionisation constant of carbonic acid and of the ionisation product K_w of water ($pK = -\log K$) (see Golterman, 1971)

Temp. (°C):	0	5	10	15	20	25
pK_1	6.58	6.52	6.46	6.42	6.38	6.35
pK_2	10.63	10.56	10.49	10.43	10.38	10.33
pK_w	14.94	14.73	14.54	14.35	14.17	14.00

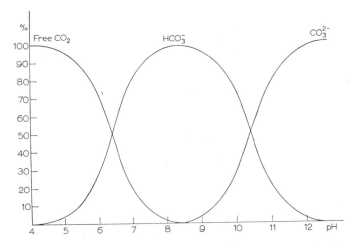

Fig. 3.2. Relation between pH and percent of total "CO_2" as free CO_2, HCO_3^- and CO_3^{2-}. (From Schmitt, 1955; and Golterman, 1971.)

Because reaction (3.3) is of little quantitative importance in most natural waters, it is useful to combine the others, (3.1) + (3.2a) + (3.2b), as:

$$HCO_3^- \rightleftharpoons OH^- + CO_2 \qquad (3.6)$$

It can be seen from (3.6) that when $NaHCO_3$ is dissolved in H_2O, the solution has a slightly alkaline reaction (pH = 8.3 for 10^{-3} M $NaHCO_3$). In this case the amount of CO_2 formed according to (3.2) is more than can dissolve in H_2O in equilibrium with CO_2 in the air, which is 0.5 mg l^{-1} of CO_2 at 20°C under atmospheric CO_2 partial pressure. Until the system is in equilibrium with air the excess CO_2 will escape. The disappearance of the CO_2 and the resultant decrease in H^+ causes the OH^- to increase.

As K_2 is much smaller than K_1, the solution becomes more strongly alkaline as when CO_2 is used during photosynthesis. The overall reaction, the combination (summation) of reactions (3.2a + b) + (3.3) may be written as:

$$2 HCO_3^- \rightleftharpoons CO_3^{2-} + H_2O + CO_2 \qquad (3.7)$$

This reaction will take place until the equilibrium between HCO_3^-, CO_3^{2-}, "equilibrium" CO_2, and pH has re-established itself. "Equilibrium" CO_2 refers to CO_2 in equilibrium with HCO_3^- and CO_3^{2-}. It is not, in general, in equilibrium with air. The equilibrium position depends on K_1 and K_2 and on the pH of the solution. Fig. 3.2 gives the percentages of HCO_3^-, CO_3^{2-}, and CO_2 present in a solution in relation to the pH. In Table 3.6 are shown the pH values and the changes of pH with time of some $NaHCO_3$ solutions, and the amounts of free CO_2, calculated from the equation:

$$pH = pK_1 + \log \frac{[HCO_3^-]}{[CO_2]}$$

TABLE 3.6

Measured pH values and calculated CO_2 concentrations in $NaHCO_3$ solutions

	After 0 h		After 24 h	
	pH	CO_2 $(mg\ l^{-1})$	pH	CO_2 $(mg\ l^{-1})$
0.001 M in unboiled H_2O	7.2	6	7.94	1.1
0.001 M in boiled H_2O	8.1	0.8	8.1	0.8
0.003 M in boiled H_2O	8.35	1.1	8.60	0.8
0.010 M in boiled H_2O	8.44	3.7	8.95	1.1

The pH of a fresh solution depends partly on the amount of CO_2 already present, and this variability is more marked in very dilute solutions of $NaHCO_3$.

In most natural waters, where Ca^{2+} is the predominating cation, the solubility product of $CaCO_3$ determines the amounts of Ca^{2+} and CO_3^{2-} (and thus the HCO_3^- concentration) which can co-exist in the solution. (The solubility product is a constant at a given temperature and is obtained as $[Ca^{2+}] \times [CO_3^{2-}]$ in a saturated solution. At $15°C$ the value is 0.99×10^{-8}, at $25°C$ it is 0.87×10^{-8}.)

Therefore there is a finite upper limit to the possible concentration of a $Ca(HCO_3)_2$ solution at a given temperature and partial CO_2 pressure. The $NaHCO_3$ system, however, has one more degree of freedom as Na_2CO_3 is very soluble. In the sodium system the limiting factor determining actual concentrations is CO_2 in equilibrium with air, whereas in the calcium system it is the solubility product of $CaCO_3$ which limits concentrations. In the absence of CO_2 the solubility of $CaCO_3$ is about 15 mg l^{-1}. In water in equilibrium with air the maximum concentration is about 0.6×10^{-3} M or 50—60 mg l^{-1}. (This amount is calculated as though $CaCO_3$ is in solution; Ca^{2+} is of course balanced by HCO_3^- ions.) For further detailed discussion see Schmitt (1955), Weber and Stumm (1963) and Stumm and Morgan (1970).

In many lakes, however, metastable conditions may persist for a long time with Ca^{2+} concentrations greatly exceeding theoretical values (see Hutchinson, 1957, p. 670).

Golterman (1973) found that the degree of supersaturation is a function of the pH. The higher the pH, the greater the supersaturation. He suggested that the presence of an electrical double layer of CO_3^{2-} or OH^- ions stabilises the colloidal particles, although Wetzel (1972) has suggested that organic matter in lakewater may also have a contributory effect. The pH dependence on the degree of supersaturation, which can occur, is then difficult to understand and especially so since the data for different lakes like Tjeukemeer (eutrophic) and Lake Vechten (oligotrophic) show the same tendency. The supersaturation can be demonstrated by dissolving $CaCO_3$ in H_2O with CO_2 gas. After aeration to remove extra CO_2, supersaturation is always found.

TABLE 3.7

Relation between $CaCO_3$ concentrations and free CO_2

$CaCO_3$ in H_2O	Equilibrium CO_2	$CaCO_3$ in H_2O	Equilibrium CO_2
1 meq. l^{-1}	0.6 mg l^{-1}	3 meq. l^{-1}	6.5 mg l^{-1}
2 meq. l^{-1}	2.5 mg l^{-1}	4 meq. l^{-1}	15.9 mg l^{-1}

The solubility of $CaCO_3$ can also be increased by the presence of an extra amount of free CO_2 which is not in equilibrium with the air. Tillmans and Heublein (1912) have given an approximate relationship between dissolved $Ca(HCO_3)_2$ and the amount of free CO_2 necessary to keep that $Ca(HCO_3)_2$ in solution (Table 3.7). This applies only to waters where pH is lower than 8.3 because the concentration of free CO_2 which can occur is negligible at pH values greater than this. These solutions are obviously not in equilibrium with the air.

The amount of free CO_2 present in excess of the equilibrium CO_2 is sometimes called "agressive" CO_2 because it is able to react with alkaline carbonates or metals.

From Table 3.7 it is possible to estimate how much $CaCO_3$ will precipitate if a river supersaturated with CO_2 enters a lake and achieves equilibrium with the air. Assuming an alkalinity in a river of 4 meq. l^{-1}, it is clear that 3 meq. l^{-1} or 3 eq. m^{-3} (or 150 g per m^3) will precipitate, because after equilibrium with air has been achieved, only 1 meq. l^{-1} will remain dissolved. If this process takes place in a water column of 10 m depth, 15 kg m^{-2} will precipitate. Entz (1959) estimated that in Lake Balaton 84 000 tonnes of $CaCO_3$ per year are deposited by this mechanism. The amount of biogenic $CaCO_3$ formed may be even greater because photosynthesis takes up larger quantities of CO_2 than would escape for physico-chemical reasons alone (see e.g. Wetzel and Otzuki, 1974). The occurrence of this effect can be observed on submerged water plants such as *Potamogeton* and *Elodea*, which are often covered with a thick layer of $CaCO_3$. *Stratiotes aloides* is reputed to accumulate so much $CaCO_3$ on its leaves during the summer that it sinks to the bottom of the water although in spring it is a free-floating plant.

Thomas (1955) used the amount of $CaCO_3$ formed to make a rough estimate of total primary production in the Swiss lakes Pfäffikersee and Greifensee.

3.3.2. Biological considerations

It has long been a controversial subject whether, during photosynthesis, aquatic plants take up CO_2 or HCO_3^- ions which are then replaced by OH^- ions formed by dissociation of water.

Because of the equilibrium between the dissolved CO_2 and the HCO_3^- ions, this has relatively little effect on the bicarbonate/carbonate system in lakes since the equilibrium conditions of all the reactions (3.1—3.4) must be met. In either case precipitation of $CaCO_3$ may occur. Ruttner (1948) demonstrated the different behaviour towards CO_2 or HCO_3^- of *Elodea*, a vascular plant, and *Fontinalis antipyretica*, a water moss. He found that both plants utilised free CO_2 at approximately the same rate (as measured by rate of O_2 production). When free CO_2 was exhausted, however, *Fontinalis* practically ceased photosynthesis while *Elodea* continued though at a decreased rate by utilising much of the bicarbonate present. This phenomenon can be demonstrated when some shoots of *Elodea* are placed in two small flasks filled with lakewater to which a few drops of the pH indicator phenolphthalein have been added; one flask is illuminated while the other is darkened. Wisps of red colour will soon appear on the upper surface of the leaves in the illuminated flask and within a few hours the whole liquid body becomes red as the pH rises. In the red bottle (red = no free CO_2) photosynthesis will proceed — though at a lower rate — and if sufficient Ca^{2+} is present, $CaCO_3$ will precipitate. The darkened flask remains colourless. If air (or CO_2) is bubbled through the coloured water, the red colouration disappears as the pH falls. The pH time curve for the experiments show the same effect as do the CO_2 time-curves. The difference in behaviour of the two plants might be explained by the operation of other mechanisms than uptake of CO_2 or HCO_3^-, e.g. by the presence or absence of the enzyme carbonic anhydrase or a direct influence of the high pH.

Steemann Nielsen (1947) also studied the problem of CO_2 and HCO_3^- uptake, working with a variety of plants: *Fontinalis antipyretica* (a moss), *Myriophyllum spicatum*, *Potamogeton lucens*, and *Cryptocoryne griffithii* (higher plants), and *Cladophora* sp. and *Ulva lactuca* (green algae). His results agreed with Ruttner's results using *Fontinalis* and showed inter alia that, with free CO_2 as C source, the rate of photosynthesis in *Myriophyllum* (but not in *Cladophora*) was independent of the ionic composition of the water, while with HCO_3^- as C source, the photosynthesis was highly dependent on the ionic composition of the water, both cations and anions having an effect. Steemann Nielsen concluded that because the ability to assimilate HCO_3^- is less widespread amongst water plants than the ability to assimilate free CO_2 (some do not assimilate HCO_3^- at all), this does not necessarily indicate that free CO_2 and not HCO_3^- is used directly during the photosynthetic process in the chloroplasts themselves. He even suggested that the opposite may be true and that impermeability of the cytoplasmic membranes to HCO_3^- ions may be the sole reason why some plants make little use of this ion. Therefore, the total ionic composition of the water, including the H^+ ions, may be important. In all studies concerning the uptake of different C sources, the influence of the pH itself is difficult to assess. Österlind (1948) and Felföldy (1960) have shown that some algae may use HCO_3^- ions quite rapidly, depending on adaptation, age and physiological conditions.

Recently Jolliffe and Tregunna (1970) demonstrated an uptake of HCO_3^- ions by some species of benthic marine algae, and not by others. Morton et al.

(1972) concluded from growth experiments that bicarbonate could be used by certain strains of *Chlorella, Anabaena* and *Microcystis*, algae which could grow in cultures with a pH value above 10. They further concluded that replenishment by diffusion of CO_2 from the atmosphere without any wind mixing can be sufficient to sustain growth rates of 2 mg dry weight per litre per day through a water depth of at least 1.7 m.

A second physiological difference which occurs between different types of organisms is the ability to utilise CO_3^{2-} at high pH values. Results of experiments using a high pH are, however, difficult to assess because HCO_3^- and CO_3^{2-} rapidly equilibrate with each other and even at pH = 11 the ratio CO_3^{2-}/HCO_3^- is still 80/20. Nevertheless, Felföldy (1960) showed that it is possible that the green algae *Chlorocloster terrestris* and *Coelastrum microporum* are capable of utilising CO_3^{2-} ions as well as HCO_3^-. During periods of intensive photosynthesis low concentrations of free CO_2 combined with relatively high pH values are normally found. If the water contains Ca ions, $CaCO_3$ may precipitate, thus shifting the equilibria towards the situation of equilibrium with the air. If equilibrium is not reached, however, or if the calcium concentration is too low to form $CaCO_3$, the low concentration of free CO_2 and vertical convective mixing allows diffusion of CO_2 from the air into the water to occur to an extent which can be quantitatively significant.

Schindler (1971) has shown recently that this diffusion can supply sufficient CO_2 to maintain a dense algal population in an experimental lake (see section 17.2). The occurrence of this diffusion thus prevents CO_2 from becoming a limiting factor for photosynthetic *yield* regardless of whether CO_2 or HCO_3^- is taken up, although the rate of photosynthesis may become reduced during periods of intensive CO_2 uptake (Schindler and Fee, 1973).

There is now sufficient evidence accumulated to show that bicarbonate will support photosynthesis under natural conditions. Whether the ion itself is taken up, or whether the enzyme carbonic anhydrase splits off the CO_2, is not yet known. It is nevertheless well established that CO_2 limitation of photosynthesis does not occur naturally, a fact which is relevant in the context of eutrophication. In some very eutrophic waters with low alkalinity the rate of photosynthesis may be limited temporarily by shortage of CO_2. but the yield is still limited ultimately by the quantities of minor elements present (see Chapter 17).

3.4. DETERMINATION OF TOTAL CO_2

Because of its importance for the study of photosynthesis, brief consideration will now be given to a method of determining the total CO_2 dissolved in the water. This can be done by titration or by calculation, if there is sufficient evidence for a particular water that the alkalinity is caused mainly by CO_3^{2-}, HCO_3^- and OH^-.

The titration method is based on the principle that the difference in volume

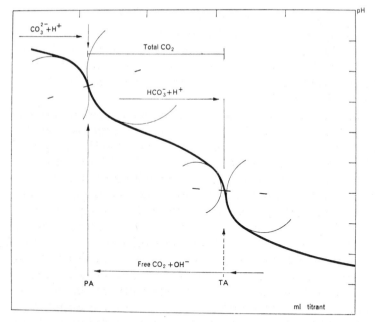

Fig. 3.3. Titration curve of Na_2CO_3 with HCl, or H_2CO_3 with NaOH. The following titrations are indicated:

$$CO_3^{2-} + H^+ \to HCO_3^-$$
$$HCO_3^- + H^+ \to H_2O + CO_2$$

free $CO_2 + OH^- \to HCO_3^-$

of the titrant between the P.A. end point (P.A. end point = phenolphthalein endpoint: pH = 8.3) and the T.A. end point (T.A. = total alkalinity: pH = 4—5) is equivalent to the total CO_2 (see Fig. 3.3). The pH of the sample is first brought to 8.3 by careful addition of CO_2-free strong acid or base. The solution is then titrated with strong acid to the T.A. end point as in a normal T.A. determination. From a knowledge of the initial pH and the total alkalinity (T.A.), it is possible to calculate the concentration of total CO_2:

$$[CO_2]_T = (C.A.)_{eq.} \ \frac{1 + \dfrac{[H^+]}{K_1} + \dfrac{K_2}{[H^+]}}{1 + \dfrac{2K_2}{[H^+]}}$$

or, representing the last part of this expression by F_{20}, the value at $20°C$, $[CO_2]_T = (C.A.)_{eq.} \ F_{20} = \{(T.A.)_{eq.} - [OH^-]\} F_{20}$, where $(C.A.)_{eq.}$ is carbonate alkalinity in eq. l^{-1} and $[CO_2]_T$ is total CO_2 in mol. l^{-1}.

For soft waters the so-called Gram titration method may be advantageous. The details are explained in Stumm and Morgan (1970). The application of this method to limnological use has been reviewed by Talling (1973).

REFERENCES

Armstrong, F.A.J. and Schindler, D.W., 1971. Preliminary chemical characterization of waters in the Experimental Lakes Area, Northwestern Ontario. *J. Fish. Res. Board Can.*, 28(2): 171—187.

Bayly, I.A.E. and Williams, W.D., 1972. The major ions of some lakes and other waters in Queensland, Australia. *Aust. J. Mar. Freshwater Res.*, 23: 121—131.

Bentor, Y.K., 1961. Some geochemical aspects of the Dead Sea and the question of its age. *Geochim. Cosmochim. Acta*, 25: 239—260.

Braginskii, L.P., 1961. O sootnoshchenii mjezdu sostawom prudnogo fitoplanktona i proyawleniem ego "potrebnosti" biogennych elementach. In: G.G. Vinberg, *Pjerechnaya produktsija morjei i vnumrennich vod.* Minsk, pp. 137—147.

Carmouze, J.-P., 1969. La salure globale et les salures spécifiques des eaux du Lac Tchad. *Cah. O.R.S.T.O.M., Sér. Hydrobiol.*, 3(2): 3—14.

Carmouze, J.-P., 1973. Régulation hydrique et saline du Lac Tchad. *Cah. O.R.S.T.O.M., Sér. Hydrobiol.*, 7(1): 17—25.

Clymo, R.S., 1967. Control of cation concentrations, and in particular of pH, in *Sphagnum* dominated communities. *IBP-Symp. Amsterdam—Nieuwersluis 1966*, pp. 273—284.

Dussart, B., 1966. *Limnologie; l'étude des eaux continentales.* Gauthier-Villars, Paris, 618 pp.

Entz, B., 1959. Chemische Charakterisierung der Gewässer in der Umgebung des Balatonsees (Plattensees) und chemische Verhältnisse des Balatonwassers. *Annal. Biol. Tihany*, 26: 131—201.

Felföldy, L.J.M., 1960. Comparative studies on photosynthesis in different *Scenedesmus* strains. *Acta Bot. Hung.*, 6(1—2): 1—13.

Fleischer, M. and Robinson, W.O., 1963. Some problems of the geochemistry of fluorine. In: D.M. Shaw (Editor), *Studies in Analytical Geochemistry.* University of Toronto Press, Toronto, Ont., 139 pp. (Special publications of the Royal Society of Canada, No. 6)

Fogg, G.E. and Westlake, D.F., 1955. The importance of extracellular products of algae in freshwater. *Verh. Int. Ver. Theor. Angew. Limnol.*, 12: 219—232.

Goldschmidt, V.M., 1937. The principles of distribution of chemical elements in minerals and rocks. *J. Chem. Soc., Lond.*, 1937: 655—673.

Golterman, H.L. (with Clymo, R.S.) (Editors), 1971. *Methods for Chemical Analysis of Fresh Waters.* IBP Handbook no. 8. Blackwell, Oxford, 166 pp.

Golterman, H.L., 1973. Vertical movement of phosphate in freshwater. In: E.J. Griffith, A. Beeton, J.M. Spencer and D.T. Mitchell (Editors), *Environmental Phosphorus Handbook.* John Wiley, New York, N.Y., pp. 509—538.

Golterman, H.L., 1975. Chemistry of running waters. In: B. Whitton (Editor), *River Ecology.* Blackwell, Oxford (in press).

Gorham, E., 1958. The influence and importance of daily weather conditions in the supply of chloride, sulphate and other ions to fresh waters from atmospheric precipitation. *Philos. Trans. R. Soc. Lond., Ser. B. Biol. Sci.*, 241(679): 147—178.

Hutchinson, G.E., 1957. *A Treatise on Limnology, 1. Geography, Physics, and Chemistry.* John Wiley, New York, N.Y., 1015 pp.

Jolliffe, E.A. and Tregunna, E.B., 1970. Studies on HCO_3^- ion uptake during photosynthesis in benthic marine algae. *Phycologia*, 9(3/4): 293—303.

Kjensmo, J., 1972. Gjende. A glacier-fed mountain lake. *Verh. Int. Ver. Theor. Angew. Limnol.*, 18: 343—348.

Lerman, A., 1967. Model of chemical evolution of a chloride lake — The Dead Sea. *Geochim. Cosmochim. Acta*, 31: 2309—2330.

Meybeck, M., 1972. Bilan hydrochimique et géochimique du Lac Léman. *Verh. Int. Ver. Theor. Angew. Limnol.*, 18: 442—453.

Miller, J.D.A. and Fogg, G.E., 1957. Studies on the growth of Xanthophyceae in pure culture. I. The mineral nutrition of *Monodus subterraneus* Petersen. *Arch. Mikrobiol.*, 28: 1—17.

Morton, S.D., Sernau, R. and Derse, P.H., 1972. Natural carbon sources, rates of replenishment, and algal growth. In: G.E. Likens (Editor), *Nutrients and Eutrophication: The Limiting Nutrient Controversy. Spec. Symp., Am. Soc. Limnol. Oceanogr.*, 1: 197—204.

Moss, B., 1972. The influence of environmental factors on the distribution of freshwater algae; An experimental study. I. Introduction and the influence of calcium concentration. *J. Ecol.*, 60: 917—932.

Nissenbaum, A., 1970. Historical article; chemical analyses of Dead Sea and Jordan river water, 1778—1830. *Israel J. Chem.*, 8: 281—287.

Österlind, S., 1948. The retarding effect of high concentrations of carbon dioxide and carbonate ions on the growth of a green alga. *Physiol. Plant.*, 1: 170—175.

Pearsall, W.H., 1923. A theory of diatom periodicity. *J. Ecol.*, 9(2): 165—182.

Provasoli, L., McLaughlin, J.J.A. and Pinter, I.J., 1954. Relative and limiting concentrations of major mineral constituents for the growth of algal flagellates. *Trans. N.Y. Acad. Sci.*, 16(8): 412—417.

Rhine Report, 1966. *Zahlentafeln der physikalisch—chemischen Untersuchungen des Rheins sowie der Mosel/Koblenz.* Internationale Kommission zum Schutze des Rheins gegen Verunreinigung.

Rodhe, W., 1949. The ionic composition of lake waters. *Verh. Int. Ver. Theor. Angew. Limnol.*, 10: 377—386.

Ruttner, F., 1948. Zur Frage der Karbonatassimilation der Wasserpflanzen. *Österr. Bot. Z.*, 95.

Schindler, D.W., 1971. Carbon, nitrogen and phosphorus and the eutrophication of freshwater lakes. *J. Phycol.*, 7(4): 321—329.

Schindler, D.W. and Fee, E.J., 1973. Diurnal variation of dissolved inorganic carbon and its use in estimating primary production and CO_2 invasion in lake 227. *J. Fish. Res. Board Can.*, 30(10): 1501—1510.

Schmitt, C., 1955. Contribution a l'étude du système chaux — carbonate de calcium, bicarbonate de calcium—acid carbonique—eau. *Ann. Ec. Nat. Sup. Méc., Nantes*, 1955: 1—135.

Steemann Nielsen, E., 1947. Photosynthesis of aquatic plants with special reference to the carbon-sources. *Dan. Bot. Ark.*, 12(8): 3—140.

Stumm, W. and Morgan, J.J., 1970. *Aquatic Chemistry; An Introduction Emphasizing Chemical Equilibria in Natural Waters.* Wiley, New York, N.Y., 583 pp.

Talling, J.F., 1973. The application of some electrochemical methods to the measurement of photosynthesis and respiration in fresh waters. *Freshwater Biol.*, 3(4): 335—362.

Talling, J.F. and Talling, I.B., 1965. The chemical composition of African lake waters. *Int. Rev. Ges. Hydrobiol.*, 50(3): 421—463.

Thomas, E.A., 1955. Sedimentation in oligotrophen und eutrophen Seen als Ausdruck der Produktivität. *Verh. Int. Ver. Theor. Angew. Limnol.*, 12: 383—393.

Tillmans, J. and Heublein, O., 1912. Über die Kohlensäuren. Kalk angreifende Kohlensäure der natürlichen Wässer. *Gesundheits-Ingenieur*, 35(34): 669—677

Weber, W.J. and Stumm, W., 1963. Mechanism of hydrogen ion buffering in natural waters. *J. Am. Water Works Assoc.*, 55(7): 1553—1578.

Wetzel, R.G., 1972. The role of carbon in hard-water marl lakes. In: G.E. Likens (Editor), *Nutrients and Eutrophication: The Limiting Nutrient Controversy. Spec. Symp., Am. Soc. Limnol. Oceanogr.*, 1: 84—97.

Wetzel, R.G. and Otzuki, A., 1974. Allochthonous organic carbon of a marl lake. *Arch. Hydrobiol.*, 73(1): 31—56.

Williams, W.D. and Hang Fong Wan, 1972. Some distinctive features of Australian inland waters. *Water Res.*, 6: 829—836.

Wright, J.C. and Mills, I.K., 1967. Productivity studies on the Madison River, Yellowstone National Park. *Limnol. Oceanogr.*, 12: 568—577.

CHAPTER 4

PRIMARY PRODUCTION IN RELATION TO PHOTOSYNTHESIS UNDER NATURAL CONDITIONS

4.1. INTRODUCTION

Photosynthesis is the biological process by which physical light energy is transformed into chemical energy. The energy is used to reduce CO_2. For this reduction the hydrogen from the water molecule is used, while the remaining oxygen is given off. The energy for this part of the process comes from the sun via photo-activated plant pigments — mainly chlorophylls. Light energy is incorporated into the pigment molecule giving an electron a higher energy content than it has in its ground state. The chlorophyll pigment is then said to be excited, and the incorporated energy is used to reduce CO_2. The first products of photosynthesis are phosphorylated sugars which are converted rapidly into other metabolites (carbohydrates, proteins, nucleic acids, etc.) if the cells are in conditions favourable for growth. Besides CO_2, H_2O, and energy, all the other essential cell constituents such as N, P, S, Fe, etc., must also be present. Some of the products formed will be used as a substrate for respiration, which occurs both in the light and in the dark. If that part which is used for respiration in light is subtracted from the total (or gross) photosynthesis, then the net photosynthesis can be evaluated. Losses also occur as a result of mineralisation of dying plants at all times. If these losses plus those due to respiration in the dark are subtracted from the net photosynthesis, a value for the net primary production (see Notation) is obtained. This quantity is of special importance in the study of food chains (Chapter 15), since it is the material available for use by the herbivores.

Some of the more important photosynthetic reactions, which will be discussed in detail in this chapter together with the environmental effects upon them, are shown in schematic form in Fig. 4.1.

Important aspects of photosynthesis from a limnological standpoint are listed below and will be considered in detail:

NOTATION

The following definitions are used:

gross photosynthesis:	total O_2 production (light plus dark bottle)
net photosynthesis:	total O_2 production minus respiration during light period
net primary productivity:	net photosynthesis minus losses due to respiration and mineralisation over 24 h
mineralisation:	oxidative breakdown of dead algal material

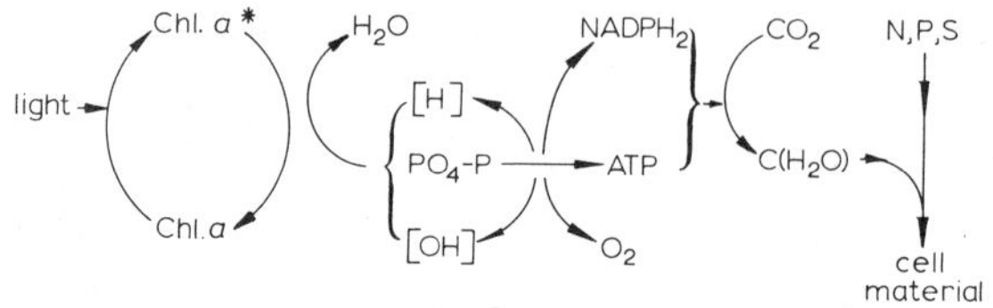

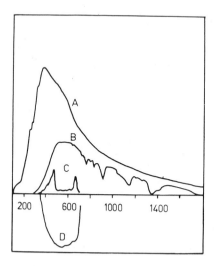

Fig. 4.1. Schematic representation of main photosynthetic processes.

(1) Physical energy supply.

(2) Relationship between photosynthesis, energy and nutrient supply.

(3) Intracellular conversions and extracellular products.

(4) Quantitative measurements of primary production.

(5) Calculation of net primary production; respiration and mineralisation losses.

(6) Variations of photosynthesis in time and space.

4.2. PHYSICAL ENERGY SUPPLY

4.2.1. Total solar energy

About half of the radiant energy of the sun between the wavelengths 200 and 1 600 nm never reaches the earth's surface because it is absorbed by the atmosphere. From Fig. 4.2 it can be seen that this is especially the case for ultraviolet light, which is absorbed principally by a very thin layer of ozone occurring at a height of several kilometres in the upper atmosphere. Energy

Fig. 4.2. Energy spectrum of radiant energy outside the atmosphere (A) and at the earth's surface (B); absorption spectrum of chlorophyll (C) and water (D) (energy and absorption in relative units). Data for D from Lake San Vicente (Tyler and Smith, 1970). Wavelengths in nm.

loss in the atmosphere is also caused by molecular absorption by other gases, especially by water vapour and CO_2, and to scattering by dust and other particles. Because the water and dust content of the atmosphere vary from day to day, accurate estimates of the amount of light that falls on a lake cannot be calculated and must be measured directly, e.g. with a solarimeter. The curves for light absorption by a chlorophyll molecule and for the light absorption by water (Fig. 4.2) indicate that only light within very limited wavelengths is likely to be used for photosynthesis. Transfer of energy from other pigments may also contribute, thus extending to some degree the range of photosynthetically utilisable wavelengths.

The amount of light that a lake can receive is a function of the elevation above sea level. The density of the atmosphere (sea level : 100% density) is 90% at 0.9 km, 80% at 1.9 km, 70% at 2.9 km and 60% at 4 km elevation. High mountain lakes receive considerably more ultraviolet light than do lakes near sea level, because the light at high altitudes has passed through less of the atmosphere. The ultraviolet light which does reach the lake is strongly absorbed by the surface layer of water.

The daily energy input is also a function of the geographical position of the lake. Table 4.1 gives a summary of approximate values as measured by Perl (1935), whose study also gives many data on diurnal changes at different latitudes.

TABLE 4.1

Estimates of total solar radiation on the 15th day of several months at various latitudes (Perl, 1935) (approximate values in cal. cm^{-2} d^{-1})

Latitude	January	March	May	July	September	November
0	500	500	500	500	500	500
30	300	500	600	600	500	300
45	150	375	600	600	400	200
60	75	250	550	600	300	50
90	0	0	500	600	25	0

For conversion into Joules per m^2, multiply the above figures by 42 000.

4.2.2. Diurnal changes

Diurnal changes of the irradiance follow a complex pattern. The main variation in irradiance is a function of solar elevation. At the equator the sun height, h, on any one day may be described by the following equation:

$$h = h_{max} \sin 2\pi \frac{t}{\tau}$$

where h = tangent of angle which the sun makes with the horizon;
h_{max} = maximal value of h;

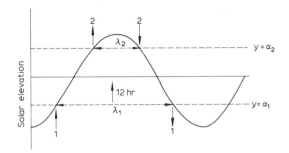

Fig. 4.3. Solar elevation above horizon during summer (*1*) or winter (*2*). Sun is seen during period λ from sunrise (↑) till sunset (↓).

τ = day length (12 h);
t = fractions of τ, starting at $t = 0\tau$ at sunrise, $t = \frac{1}{4}\tau$ at noon, $t = \frac{1}{2}\tau$ at sunset.

Depending on season and geographical position the sun is seen for a shorter or longer period than half the sine curve (Fig. 4.3).

In the summer season the sun is above the horizon during the day for a period λ_1 (above $y = \alpha_1$), whereas during winter the day length is shown by λ_2 (above $y = \alpha_2$). Clearly the value of α depends on season and latitude.

For the equation describing the relation between solar elevation and irradiance, the variable should not be τ but the variable day length λ. If then t is expressed in fractions of λ and the abscissa is shifted so that $t = 0$ at 12 h (thus $t = -\frac{1}{2}\lambda$ at sunrise and $t = \frac{1}{2}\lambda$ at sunset), the equation for the diurnal changes of irradiance I, becomes:

$$I = I_{max}\, \frac{1}{2}\left(1 + \cos\frac{2\pi t}{\lambda}\right)$$

neglecting any absorption by the atmosphere.

When the sun is low in the sky the optical path length through the atmosphere is larger, so that the irradiance after sunrise increases more steeply than shown by the sine curve in Fig. 4.3 and often approaches a straight line. The same effect causes a steep decrease before sunset. The amount of atmospheric absorption cannot be calculated as this would vary according to the length of the optical path through which the light passes. Furthermore, the absorption is different for different wavelengths and changes with such variables as the moisture content of the atmosphere, so that no exact mathematical relationship for the theoretical diurnal changes of irradiance can be given. In addition to this direct solar energy, scattered light in the sky also contributes to the energy budget. The actual amount depends so much on local conditions, such as time of the day, cloudiness, etc., that no generalised values can be given although it is possible to give an approximate value for overcast conditions.

Some of the incident solar energy falling on a lake does not enter the water but is reflected by the water surface; the amount of reflection depends on the

TABLE 4.2

Percentage reflection from water surface as a function of solar elevation

Solar elevation [*]:	5°	10°	15°	20°	30°	40°	50°
R (direct sunlight; %)	58	35	21	13	6	3.5	2
R (diffuse sky light; %)	17	15	14	13	11	9	7

[*] Angle between sun and horizon.

wind-induced disturbances of the water surface and the angle of incidence of the light. The reflected component is relatively large in early morning and evening. During these parts of the day the amount of light penetrating the lake increases (morning) or decreases (evening) proportionally more rapidly than does the total amount falling on the lake.

For a calculation of the reflection coefficient, R, the following equation may be used, neglecting any variations caused by disturbance of the water surface:

$$R = \frac{1}{2}\left(\frac{\sin^2 (i-r)}{\sin^2 (i+r)} + \frac{\tan^2 (i-r)}{\tan^2 (i+r)}\right)$$

where i = angle of incidence, and r = angle of refraction. Since $\sin i/\sin r = \mu$ (the refractive index = 1.33 for air $\rightarrow$ H_2O), the value for r can be calculated for different solar elevations.

The reflection of the diffuse radiation should also be included. Values of this are given in Table 4.2, compared with those for a cloudless sky.

For detailed studies involving energy values it seems better to use ones own measurements than to calculate the theoretical irradiance. Solarimeters which measure the total radiant energy are commercially available. With an attached recorder they provide a great deal of information, much of which is often too detailed to be useful. For most purposes values of the integrated energy flux per hour are usually sufficient. The observed reading should, however, be corrected for reflection from the lake surface as shown above.

Only part of the total light energy, roughly that between 350 and 710 nm (disregarding bacterial photosynthesis) is available for photosynthesis. Since the part between 350 and 390 nm hardly penetrates the water, the lower limit is roughly 390 nm for aquatic photosynthesis. The part between 390 and 710 nm is roughly 45—50% of the total radiant energy incident on a lake.

4.2.3. Light penetration

If light passes through a solute or solution, the proportion of light absorbed is constant per unit optical path length, or:

$$-\frac{\Delta I}{I} = \epsilon \Delta Z \qquad (4.1a)$$

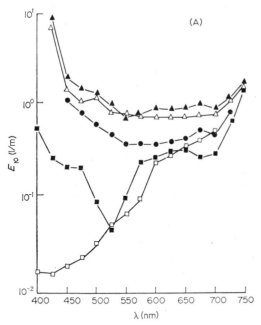

Fig. 4.4.A. Diffuse attenuation coefficient E_{10} (ln m^{-1}) for down-welling irradiance in Loch Croispol (4 August, 1970), Loch Uanagan (15 July, 1970), and Loch Leven (2 July, 1970) calculated from E_{10} for each 25-m waveband (390—750 nm). Data are included of Tyler and Smith (1967) for Crater Lake, Oregon, where down-welling irradiance was corrected for up-welling irradiance and converted data of Talling (1970) for Blelham Tarn, England. ▲, Loch Leven; △, Loch Uanagan; ■, Loch Croispol; □, Crater Lake; ●, Blelham Tarn. (From Spence et al., 1971.)

where I = incident light; ϵ = vertical attenuation (extinction) coefficient, a constant depending on the characteristics of the liquid, different for the various wavelengths; and Z = optical path length.

In its integrated form eq. 4.1a becomes:

$$I = I_0\ e^{-\epsilon Z}\ , \text{or}$$

$$\ln I = \ln I_0 - \epsilon Z \tag{4.1b}$$

where I_0 = incident light (subsurface), and I = light at depth Z.

Apparatus (e.g. spectroradiometer) is now available which permits the light extinction to be measured in situ at all wavelengths. Spence et al. (1971) using a spectroradiometer, measured the light attenuation coefficient (ϵ) in different Scottish lochs and compared their data with values given by Smith and Tyler (1967) for Crater Lake and those of Talling (1970) for Blelham Tarn (Fig. 4.4A).

These lakes represent typical cases. Loch Uanagan and Loch Leven show a fairly constant value for the extinction coefficient between 500 and 700 nm, with increasing values below and above these wavelengths. Loch Croispol

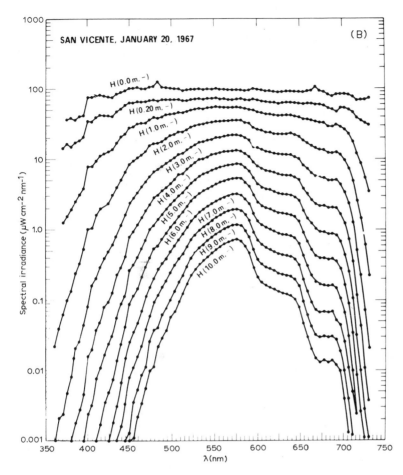

Fig. 4.4.B. Spectral irradiance (μW cm^{-2} nm^{-1}) at different depths in Lake San Vicente (20 January, 1967). (From Tyler and Smith, 1970.)

shows an incredibly sharp minimum at 525 nm (due to a passing sunbeam?) and a small plateau between 650 and 700 nm. Crater Lake, which has been described as "natural distilled water", behaves as such showing low values from 400 to 750 nm. It should be noted that the difference in light extinction, in the blue part of the spectrum, between lakewater and distilled water is usually caused mostly by dissolved organic compounds and in the red region by particulate matter.

Spence et al. (1971) demonstrated large differences in proportion of the different wavelengths of light at 1 m as compared with the values at 0 m. They discussed the possible implications for light-sensitive seeds of aquatic plant species, and for morphogenesis and zonation.

Fig. 4.4B shows light penetration at different depths in Lake San Vicente,

52

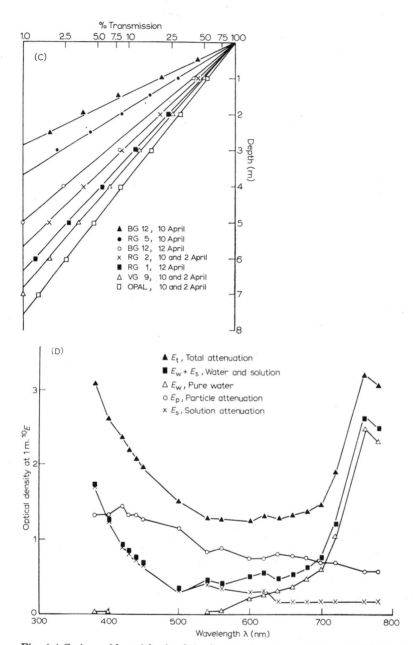

Fig. 4.4.C. A semi-logarithmic plot of percentage transmission of different wavelengths of light at various depths in Lake Bunyonyi, Uganda (Schott filter codes are indicated). (From Denny, 1972).

Fig. 4.4.D. Spectrophotometric analysis of water from Lake Bunyonyi, Uganda at 1 m depth. Settled and filtered water were both scanned and pure-water data was obtained from tables. (From Denny, 1972.)

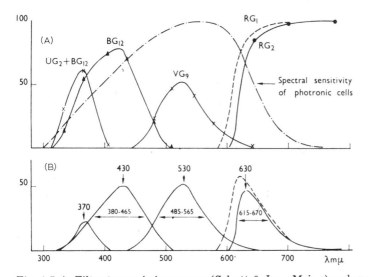

Fig. 4.5.A. Filter transmission curves (Schott & Jen., Mainz) and spectral sensitivity curve of a typical barrier layer photocell (Weston photronic cell).
B. Spectral response of the mounted photocell in air, calculated from the combination of the sensitivity curve of the photocell and the various filter transmission curves. The sensitivity centre and the 50% band width of each combination in air are indicated by arrows. (After Vollenweider, 1961.)

a lake which contains phytoplankton, zooplankton and decomposition products (Tyler and Smith, 1970).

For most purposes light-penetration measurements at 370, 430, 530 and 630 nm will be sufficient. Filters may be used for obtaining such measurements. Fig. 4.5 gives the filter transmission curves and the sensitivity curve of a Weston photocell (Vollenweider, 1961).

The extinction coefficient, ϵ, (see eq. 4.1b) is calculated using the natural logarithm of I_0/I; the depth should be expressed in metres. Vollenweider (1961) found a significant correlation between the extinction coefficients of light at different wavelengths and calculated in these cases the optical characteristics using measurements at only one single wavelength.

The amount of energy at a given depth can be evaluated from the attenuation coefficient of the most penetrating component, $\epsilon_{m.p.c.}$. Vollenweider (1961) calculated the amount from the mean of the transmission at depth Z at three wavelengths:

$$T_Z^E = 1/3\,(T_Z^{red} + T_Z^{green} + T_Z^{blue})$$

where T_Z^E is the percentage energy at depth Z. The procedure is simple, but less than Talling's calculations, although it sometimes gives reasonable estimates.

The photosynthetically available energy at depth Z during a certain exposure time is then:

$$T_Z^{400-700} = (\text{irradiance—reflection}) \times 0.5 \times \frac{T_Z^E}{100}$$

(0.5 refers to that fraction of the solar energy which is available at photosynthetically active wavelengths). The value of the attenuation coefficient depends upon two factors: light absorption can be caused both by solutes and by particles in suspension. Particles are usually algae although in some cases detritus and clay may contribute. The amount of light absorbed by each separate factor can be measured by use of appropriate laboratory apparatus. Denny (1972) compared this procedure with field observations in Lake Bunyonyi (Uganda), using optical filters (see Fig. 4.4D and C). It can be seen that the dissolved components absorbed mainly in the blue part and the particulate matter in the red part of the spectrum.

Rodhe introduced the concept of the so-called optical depth unit, $Z_{o.d.}$, which is the depth of the water in metres in which 50% of the light is absorbed. Therefore at depth $1 \times Z_{o.d.}$ $I = \frac{1}{2}I_0$, at $2 \times Z_{o.d.}$ $I = \frac{1}{4}I_0$, etc. Ten percent of I_0 will therefore be found at a depth $Z = 3.3 \, Z_{o.d.}$ and 1% at $Z = 6.6 \, Z_{o.d.}$. Furthermore $Z_{o.d.} = (\ln 2)/\epsilon$.

A first approximation for measuring light attenuation in lakewater can be made by measuring the depth at which a white disc, a so-called Secchi disc, just disappears. Vollenweider (1969) estimated from experience that this "vanishing" of the disc occurs when irradiance values are between 15% and 20% of the incident radiation. The "vanishing depth" value varies with local conditions and depends on the observer. It is therefore useful as a rough guide when describing a lake but not for detailed production studies. The value varies with turbidity but does not distinguish between algal and other suspended matter. In silty lakes like Lake Gjende (see Chapter 3), where the silt changes the optical properties of the water, Kjensmo found that a Secchi disc disappeared at depths where irradiance was more than 20% of that at the surface.

4.3. PHOTOSYNTHESIS, ENERGY AND NUTRIENT SUPPLY, AND ADAPTATION

4.3.1. Relationship between photosynthesis, energy and nutrient supply

The relationship between photosynthesis, A, and light energy, I, is well established for algal suspensions from laboratory experiments and can usually be described approximately by the following expression:

$$A = A_{\max} \frac{aI}{\sqrt{1 + (aI)^2}} \tag{4.2}$$

The equation comes from Smith (1936), and its importance for limnological studies has been recognised and emphasised by Talling (1957a, b). It shows that at lower values of I ($aI \ll 1$) the rate of photosynthesis is nearly linear with I, while at high intensities ($aI \gg 1$) photosynthesis reaches its maximal value A_{max}. The nearly linear part of the curve ($A = A_{max} \, aI$, up to $I = 1/2a$) can be extrapolated to cut the line $A = A_{max}$ at point K, where the I value is defined by Talling as I_k, of which the biological implications will be discussed below. Concerning point K where $A = A_{max}$ and $I = I_k$, it follows that:

$A_{max} = A_{max} \, aI_k$, or

$$I_k = 1/a \tag{4.3}$$

As was stated above, the near-linearity holds until I values reach $I = 1/2a$, if eq. 4.3 is used. The nearly linear range is thus from $I = 0$ up to $I = 0.5 \, I_k$.

At high irradiance, photosynthesis may be inhibited. A complete mathematical description then becomes:

$$A = A_{max} \frac{aI}{\sqrt{1 + (aI)^2}} \cdot Q\,(\alpha I) \tag{4.4}$$

The function $Q(\alpha I)$ is not known though some models of the system are described by Vollenweider (1965).

Assuming that the algae are uniformly distributed throughout the water column, the relationship between photosynthesis and depth is in principle to be found by combining eqs. 4.1 and 4.3; this is shown graphically in Fig. 4.6. At increasing depths the irradiance decreases approximately exponentially, but the photosynthetic rate remains initially through the upper layers at its saturated value A_{max}. At lower depths irradiance and photosynthesis decrease,

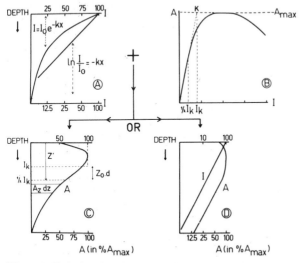

Fig. 4.6. Relationship between photosynthesis and depth. A. Relationship between irradiance and depth. B. Relationship between irradiance and photoassimilation. C. Summation of curves A and B (linear scale). D. Summation of curves A and B (log scale).

and below the depth where $I = 0.5\,I_k$ the photosynthesis decreases nearly linearly with the irradiance and thus nearly exponentially with depth.

The total photosynthesis per unit surface area of a lake can be found by integrating $A_z \cdot dZ$ between the limits $Z = 0$ and $Z = \infty$ or

$$\Sigma A_{(t)} = \int_0^\infty A_z \cdot dZ \qquad (4.5)$$

where $\Sigma A_{(t)}$ = photosynthesis at time t per m^2, and A_z = photosynthesis at depth Z per m^3.

The general solution of eq. 4.5 is:

$$\Sigma A_{(t)} = \int_0^\infty A_{max} \frac{aI}{\sqrt{1 + (aI)^2}} \cdot dZ \qquad (4.6)$$

as $dZ = -dI/\epsilon I$ and substituting $I' = aI$, so that $dI = 1/a\,(dI')$, we arrive at:

$$\Sigma A_{(t)} = \frac{A_{max}}{\epsilon} \int_{I=I_0}^{I=0} \frac{dI'}{\sqrt{1 + (I')^2}}$$

and as:

$$a = \frac{1}{I_k}, \quad \Sigma A_{(t)} = \frac{A_{max}}{\epsilon} \left[\ln \frac{I_0}{I_k} + \sqrt{1 + \left(\frac{I_0}{I_k}\right)^2} \right] \qquad (4.7)$$

for values of I_0 for which $I_0/I_k \gg 1$, eq. 4.7 becomes:

$$\Sigma A_{(t)} = \frac{A_{max}}{\epsilon} \ln \frac{2 I_0}{I_k} \qquad (4.8)$$

The agreement between eqs. 4.7 and 4.8 is good when $I_0/I_k \gg 1$ (Vollenweider, 1965). The relationship shown in eq. 4.8 was demonstrated experimentally by Talling (1957a, b) who found that:

$$\Sigma A_{(t)} = A_{max} \cdot Z' \qquad (4.9)$$

where Z' is the depth where $I = 0.5\,I_k$.

Eqs. 4.8 and 4.9 are equivalent as $0.5\,I_k = I_0\,e^{-\epsilon Z'}$, so that

$$Z' = \frac{1}{\epsilon} (\ln I_0 - \ln 0.5\,I_k) \qquad (4.10)$$

Substituting eq. 4.10 in 4.9 does indeed give eq. 4.8. The condition under which eq. 4.9 is valid, i.e. $I_0/I_k \gg 1$, is automatically met in Talling's solution as the extrapolation to I_k is only possible for I_0 values larger than I_k.

The difficulties encoutered when I_0 was less than I_k were approached by Talling (1961) by proposing:

$$f(I) = \ln \left(1 + \frac{I_0}{I_k}\right)$$

and more complex but more accurate solutions were derived by Vollenweider (1965).

In the surface layers of a lake, inhibition of photosynthesis may occur, as described by the function $Q(\alpha I)$ in eq. 4.4. In this case the photosynthesis—depth curve follows the lines shown in Fig. 4.6. Vollenweider has pointed out that an overestimation by Talling's curve in the upper water layers may be compensated by an underestimation below the points of A_{max}. This photo-inhibition may be caused by a restriction of movement of the algae, which normally are able to migrate actively either towards or away from the light. Such a restriction of algal movement may be important, especially in shallow waters where the zone of light inhibition is less than 1 m. In an extensive study Talling (1966a) has shown that apparent light inhibition in the upper layer may in fact be due to a lower concentration density of algae (in this case *Asterionella*) which tended to accumulate in deeper layers. Measurements made in situ showed that the lower total photosynthesis observed in the zone of inhibition was nevertheless maintained at the light-saturated rate. Phytoplankton from the deeper layers, however, did show a real photo-inhibition caused either by a lower concentration of algae or by an irreversible rate-depression or a reversible true inhibition (see also p. 79).

The rate of photosynthesis in the upper layers of a lake must be limited by some intrinsic limit to photosynthetic capacity with limitation being mainly determined by an external factor other than the irradiance. In most lakes this will be the supply of nutrients to algae, with nitrogen or phosphate often important in this respect. A few cases have been described where trace elements are believed to limit algal growth (Goldman, 1964), and these will be discussed in Chapter 11.

In Chapter 10 the general growth equation for algal cells is given:

$$\frac{dN}{dt} = \beta_{max} \frac{C_n}{C_1 + C_n} \cdot N$$

where N = cell number, and C_n = concentration of the growth rate-limiting nutrient, C_1 = a constant.

For a very low nutrient concentration this equation becomes:

$$\frac{dN}{dt} = \beta_{max} \frac{C_n}{C_1} \cdot N = K \cdot N \cdot C_n$$

where $K = \beta_{max}/C_1$, and this describes algal growth in relation to the concentration of the growth rate-limiting nutrient at low concentrations.

Assuming that the same kind of relationship applies in a lake and that a linear relationship exists between A_{max} and β, the equation for the total photosynthesis per unit area becomes:

$$\Sigma A_{(t)} = k \cdot \beta_{max} \cdot C_n \cdot Z' \cdot N$$

where k is a constant.

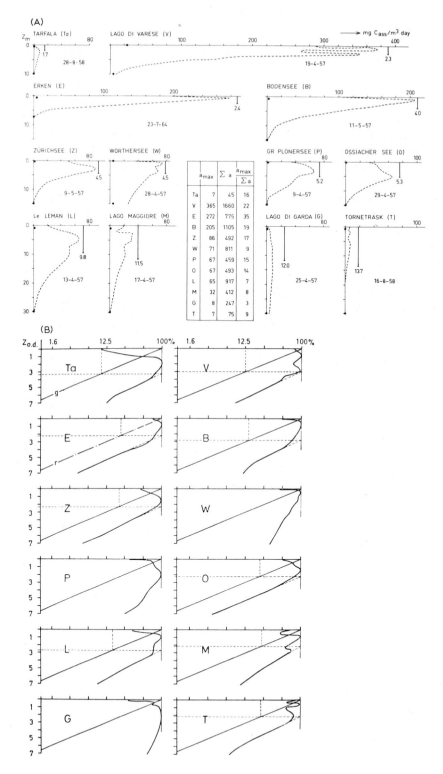

Fig. 4.7.A,B. For legend see p. 59.

As Z' (depth of $0.5\,I_k$) is inversely, but N (cell number) and C_n are directly related to nutrient supply, the relation between photosynthesis, nutrient concentration and irradiance is rather complicated. It seems likely though that in a given lake primary production will be related to the (limiting) nutrient concentration (see for example Megard's results on p. 61).

In the deeper layers it is probable that nutrient supply does not limit photosynthesis since light values will be low anyway.

Rodhe (1965) has shown (by plotting the most penetrating light component semi-logarithmically against depth with $Z_{o.d.}$ as unit) that for a large number of lakes Talling's model does in fact apply. For all the lakes which he studied a straight line was found for the light-penetration depth plot and for many lakes photosynthesis curves were found to run parallel with this light line (Fig. 4.7).

Vollenweider (1960) studied a modified formulation of Talling's model in several lakes and demonstrated a remarkable similarity between the values calculated and measured (discrepancies between 0% and 5%) but doubted the validity of the factor "Light Division Hours", a factor which Talling used to compute daily photosynthesis from hourly values, arguing that photo-

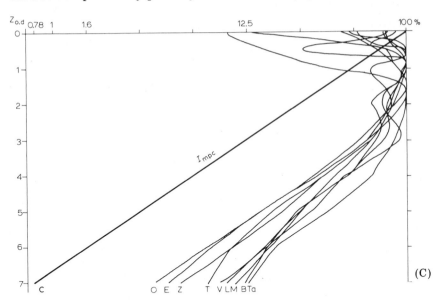

Fig. 4.7.A. Absolute photosynthetic rates illustrated in linear graphs. Dotted lines: values from 24-h exposures in light bottles with ^{14}C. Big dots: assimilation in parallel dark bottles. The Secchi-disc transparencies are also given. (From Rodhe, 1965.)
B. Relative photosynthetic rates in semi-logarithmic graphs plotted against the optical depths of the most penetrating component, $I_{m.p.c.}$ The graphical evaluation of I_k (in units of optical depth and incident $I_{m.p.c.}$) is indicated. (From Rodhe, 1965.)
C. Joint semi-logarithmic graph for the regular curves of A.

synthesis decreases during day time. (See subsection 4.7.2. for this computation.)

Vollenweider found values for I_k to be within a rather small range, 14—17 kerg sec^{-1} (1.19—1.44 cal. cm^{-2} h^{-1}). He discussed the relationship between light attenuation and phytoplankton density and the ratio $\Sigma A/A_{max}$, and showed the occurrence of a remarkably regular relationship between A_{max} and Secchi depth, S, in some lakes. His equation:

$$\frac{\Sigma A}{A_{max}} = \sim 1.75\ S^{0.7}$$

might prove useful to those who are concerned with the management of lakes, reservoirs or rivers. In a second paper, Vollenweider (1961) reported a detailed study of various correlative relationships in light attenuation.

If a lake is too large for sufficiently representative work to be carried out in situ, another solution must be found.

Sorokin (1963) measured:

(1) the photosynthetic rate of a surface sample (C_s in mg m^{-3} h^{-1} of C);

(2) the photosynthetic rate of samples taken from different depths and illuminated under equal moderate irradiance, e.g. on the deck of a ship under shade (the K_r function, indicating vertical distribution of phytoplankton);

(3) the photosynthetic rate of samples of surface water exposed at various depths expressed relative to C_s (the K_t function, which in the sea is not so variable as K_r, but this may not be so for smaller lakes).

Total production (g m^{-2} d^{-1} of C) according to Sorokin is then:

$$C_f = C_s \cdot I \cdot K_f \cdot 1000 ,$$

where K_f is a correction factor derived from K_r and K_t. On larger lakes it is usually possible to obtain ships equipped with a powerful and constant electricity supply, and if this can be arranged, light sufficient to give a saturation value could be used. Then exposures could be made on deck at different distances from the light source, the positions being chosen in relation to the light attenuation occurring in the field. Actual light values used could be measured with a photocell.

In an attempt to avoid laborious and lengthy measurements of ^{14}C profiles at a large number of stations in Lake Kinneret (Israel), which showed an unhomogeneous horizontal distribution, Rodhe (1972) measured a complete vertical profile at only one station and measured the maximal assimilation rate at optimal irradiance at other stations.

He showed that for this lake the photosynthetic gradient was apparently governed not by the $I_{m.p.c.}$ (green light, filter VG 9) but by blue light (13G12). Rodhe called this blue irradiance $I_{m.e.c.}$, the most efficient component of insolation. Maximal photosynthesis was measured at 50% of $I_{m.e.c.}$ and the Z' value was taken as the depth where $I_{m.e.c.} = 0.5\ I_k$. The model proved to be valid except during the period of *Peridinium* bloom.

Seasonal and spatial variations of $\Sigma A_{(t)}$ and chlorophyll concentration were studied by Megard (1972) in a large morphometrically complex lake, Lake Minnetonka, Minnesota (U.S.A.). Results from this study also agreed closely with Talling's model. Megard defined Z' as:

$$Z' = \frac{\ln(I_0'/0.5\,I_k)}{K_n n + K_w} = \frac{\Sigma A_{(t)}}{A_{max}}$$

where n = chlorophyll concentration; K_n = light attenuation due to n; and K_w = light attenuation due to dissolved or suspended matter in the water. He showed that with increasing chlorophyll concentration, A_{max} increased and Z' decreased. Because he found the actual relationship to be $Z' = 3 \cdot b^{-0.013n}$ instead of the theoretically predicted reciprocal, Megard suggested that daily integral photosynthesis at any locality can be calculated from measurements of n and A_{max} in samples incubated for half a day. Chlorophyll concentration increased with increasing phosphate concentrations, but a linear relationship was found to exist only when *Anabaena* species were the dominant algae. No apparent relationship was found between phosphate concentration and A_{max} during spring and autumn, but an excellent linear one occurred in the summer:

$$A_{max} = 38 \cdot 7\,P_{tot} - 0.22$$

No figures for the P loading were given, but they may be estimated from the nearby human population density to be at least $1\ \mathrm{g\ m^{-2}\ yr^{-1}}$ of PO_4-P, disregarding any further phosphate adsorbed on biogenic $CaCO_3$. Megard followed Elster's argument that specific rates of photosynthesis (mg C per mg chlorophyll a per hour) are the same both in eutrophic and oligotrophic lakes, which, according to Elster, could mean that nutrients were saturating algal growth. The argument could be reversed, however, if chlorophyll concentration is a function of P_{tot}. Apparently confusion exists between growth *rate*-limiting and growth *yield*-limiting factors (see Chapter 10).

4.3.2. Light saturation, the meaning of I_k and adaptation

In the preceding section the value I_k has been defined as that irradiance at which the saturation level of photosynthesis would be reached if the linear relationship with irradiance is extrapolated. I_k is obviously not an absolute value since it is different for different species, depends inter alia upon the history of a given cell, and is temperature-dependent. Steemann Nielsen et al. (1962) demonstrated that *Chlorella* cells grown under low illuminance (3 klux) were more efficient at low illuminance but became saturated at a lower illuminance than those grown at 30 klux (see Fig. 4.8A). The greater efficiency under the low illuminance of cells growing at a low illuminance was due to their higher chlorophyll content, and the photosynthesis per mg of chlorophyll was found to be equal in the two cultures. A similar kind of

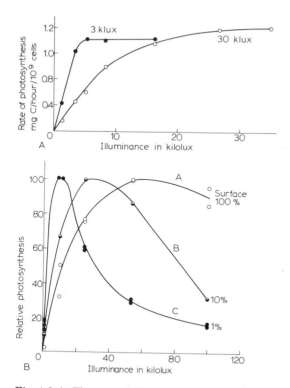

Fig. 4.8.A. The rate of photosynthesis per unit number of cells, in relation to illuminance, for *Chlorella vulgaris* grown at low (3 klux) or high (30 klux) illuminance. (From Fogg, 1968.)

B. The rate of photosynthesis, in relation to illuminance, of phytoplankton taken from depths in the Sargasso Sea to which (*A*) 100%, (*B*) 10%, and (*C*) 1% of the surface light penetrated. (From Fogg, 1968.)

adaptation was found by Ryther and Menzel (1959) for plankton in the ocean (Fig. 4.8B).

Talling (1966b) found a range between 0.8 and 2.0 mg of chlorophyll *a* per 10^9 *Asterionella* cells, the higher values being obtained in samples from deeper layers. Furthermore the onset of light saturation and photo-inhibition both occurred at lower irradiance in the deeper samples of cells. Talling pointed out that shifts in the I_k value may be an unreliable index of light adaptation because of its dependence upon fluctuations in the A_{max} values which may result from both external and internal factors such as nutrient supply, temperature, and age of population. He found that I_k increased with temperature to the same extent as A_{max} did, namely 7.7% per °C (Q_{10} = 2.1). No measurable effect of temperature upon rates at low illuminance was found, so that the initial slope of the rate—intensity curve is apparently independent of the temperature (see Talling, 1966b, fig. 8).

Furthermore he found an I_k value averaging about 2.8 klux (or 11.5 kerg cm^{-2} sec^{-1}) for *Asterionella* both from Windermere and from Blelham Tarn, with a tendency towards higher values in Windermere surface samples.

Another possible mechanism by which adaptation may operate is a decrease in the ratio of carotenoids to chlorophyll *a* with increasing depth. This higher proportion of carotenoids in the upper layers may provide protection against the higher irradiance there (see Chapter 13), but if the decrease of carotenoids with depth were due to fucoxanthin, this explanation seems unlikely since fucoxanthin itself is also effective for photosynthesis. Fucoxanthin may, of course, have a dual role both of protection and of light absorption for use in photosynthesis.

4.4. INTRACELLULAR CONVERSIONS AND EXTRACELLULAR PRODUCTS

Van Niel (1932) formulated the following general equation for photosynthesis:

$$Energy + 6\ CO_2 + 12\ H_2A \rightarrow C_6H_{12}O_6 + 12\ A + 6\ H_2O$$

where A = oxygen for green plants, and sulphur for many sulphur bacteria. With oxygen as H acceptor 686 kcal. solar energy are fixed per mole of glucose. For the reduction of CO_2 to $C(H_2O)$ 120 kcal. per mole is needed. One mole contains $6 \cdot 10^{23}$ molecules, and the same number of quanta of radiation at 900 nm and 500 nm contain only 72 kcal. and 36 kcal., respectively. A minimum of 4—5 quanta is therefore needed for each CO_2 molecule which is reduced. The energy from the photon is "trapped" in a pigment molecule which is thereby transformed into an energy-rich or excited molecule by transfer of an electron from an inner to an outer orbit, where it contains more energy. The energy value is only 40 kcal. when the photon has excited a pigment molecule, so it is clear that several electrons are necessary for the reduction of 1 molecule of CO_2.

The photons are absorbed by different pigments (such as chlorophyll, fucoxanthin and phycocyanin, which each have a different absorption spectrum) and are then transported to the "photosynthetic reaction centre". Here the excitation energy is transformed into chemical energy by means of a complex series of reactions, which eventually produce:

(1) an oxidant to oxidise the water molecule to O_2, H^+, and electrons (redox potential + 0.8 V, energy of 0.8 eV);

(2) a reductant to produce the reducing compound NADPH (nicotinamide adenine dinucleotide phosphate; redox potential —0.3 V, energy 0.3 eV (1 eV is equivalent to about 23 kcal. $mol.^{-1}$)).

Photosynthesis, like respiration, may be described as an electron transport, but in this case the electrons are transported upwards towards more negative redox potentials. When the electron "falls back", the resultant energy is incorporated into ATP molecules. It has been postulated that two different me-

64

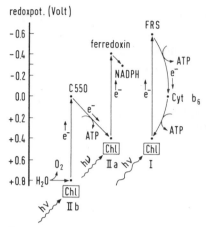

Fig. 4.9. Simplified electron-transport scheme for photosynthesis. FRS: Ferredoxin reducing substance; C550: unidentified chloroplast component. I: Cyclic photosystem I; IIa + IIb: non-cyclic photosystem II.

tabolic pathways for the ATP synthesis are involved, one being a cyclic and the other a non-cyclic photophosphorylation process. The reactions so far postulated are summarised in Fig. 4.9.

The "cyclic mechanism" in which no O_2 is produced is called Photosystem I. In the non-cyclic mechanism (Photosystem II) ATP formation is coupled with the reduction of NADP to NADPH. In green plants the hydrogen donor is H_2O, and O_2 is produced. Experimentally, however, it has been shown that in chloroplasts inhibited by DCMU, 3-(3', 4'-dichlorophenyl)-1,1 dimethylurea, then reduced ascorbic acid can be used with 2,6 dichlorophenol indophenol serving as a hydrogen carrier. Use of these dyes has made it possible to distinguish two separate photochemical reactions in one of which ATP is formed while in the other NADPH is formed (Photosystem IIa and b).

Electron carriers for the "downhill" movement of energy are vitamin K or plastoquinone and cytochromes. From a consideration of their oxidation—reduction potentials it would appear that plastoquinone or cytochrome B_6 and cytochrome f might provide for two stages of the electron transport process in Photosystem II, both of which could be coupled with ATP synthesis. The formation of ATP is thought to involve both chlorophyll and cytochrome as in cyclic photophosphorylation, but an additional accessory pigment is probably concerned in the second photochemical reaction.

In blue-green algae the pigment phycocyanin is used in the non-cyclic Photosystem II but it is not found in the heterocysts (see subsection 6.4.2). As Photosystem II is involved in O_2 production and the heterocysts are thought to maintain anaerobic conditions for N_2 fixation, it might be relevant that in the heterocyst only Photosystem I (cyclic) is believed to be active (Fay, 1969). Cox and Fay (1969) found that for *Anabaena*, if Photosystem II is inhibited by CMU (*p*-chlorophenyl-1,1 dimethylurea) no O_2 is produced but N_2 is still fixed. Non-cyclic ATP formation is thought to occur in photo-organotrophic bacteria such as the purple non-sulphur bacteria (see Chapter 16).

For every molecule of O_2 produced 3 molecules of ATP are formed. This number of ATP molecules is found to be necessary for the reduction of 1 molecule of CO_2. This reduction is performed in a complex series of reactions jointly entitled the Calvin cycle or pentose-phosphate cycle. The essential steps in the carbon reduction cycle are simplified as follows:

1 ribulose diphosphate + CO_2 → 2 phosphoglyceric acid

2 phosphoglyceric acid + $NADPH_2$ + 2 H^+ + 2 ATP $\xrightarrow{\text{reduction}}$

2 glyceraldehyde phosphate + NADP + 2 PO_4-P + 2 ADP

↳ hexose phosphate → glucose

↳ ribose phosphate + ATP → ribulose diphosphate

The light reactions may be summarised thus:

$$2\,NADP^+ + 2\,H_2O + 2\,ADP + 2P \rightarrow 2\,NADPH + 2\,H^+ + O_2 + 2\,ATP$$

$$ADP + \quad P \rightarrow \quad ATP$$

while the biosynthetic reaction is:

$$CO_2 + 2\,NADPH + 2\,H^+ + 3\,ATP \rightarrow C(H_2O) + H_2O + 2\,NADP^+ + 3\,ADP^+ + 3\,P$$

$$\text{Sum } CO_2 + 2\,H_2O \rightarrow C(H_2O) + H_2O + O_2$$

From this scheme it can be seen that 3 molecules of ATP and 1 molecule of $NADPH_2$ are in fact used. It has been found experimentally that at least 8 quanta are used for the reduction of 1 molecule of CO_2, and thus the maximal efficiency is 35%. This maximum is rarely reached under normal conditions, however.

It has now been established that the Calvin cycle operates in many types of plants, although it was originally studied in the algae *Chlorella* and *Scenedesmus* (Fogg, 1968; Rabinowitch and Govindjee, 1969). A remarkable uniformity was found (after 5 min photosynthesis) in the relative amounts of tracer among alcohol-soluble constituents of the Calvin cycle system in 27 different plants, including *Nostoc* (blue-green alga), *Porphyridium* (red alga), and many higher plants.

Even the photosynthetic bacteria seem to have essentially the same mechanism. In *Rhodospirillum rubrum* phosphoglyceraldehyde was found to be prominent among the first products of photosynthesis in experiments using $^{14}CO_2$ and H_2. In another photosynthetic bacterium, *Chlorobium thiosulphatophilum*, another carbon reduction cycle is found. It operates by reversal of the respiratory cycle, the Krebs tricarboxylic acid (TCA) cycle. This latter cycle, in which pyruvic acid is oxidised to CO_2, normally has two irreversible steps, but in *C. thiosulphatophilum* these steps are driven in a reverse direction by reduced ferredoxin, resulting in the fixation of two molecules of CO_2. The reduced ferredoxin is generated photosynthetically. The other two carboxylation steps in the Krebs cycle are normally reversible, one requiring NADPH and the other ATP. The H donor is presumably generated photochemically. One revolution of this "reversed" Krebs cycle results in the fixation of 4 molecules of CO_2 and generates 1 molecule of oxaloacetic acid. The cycle apparently occurs together with the pentose-phosphate cycle, although their relative importance in different organisms is not known. The systems may perhaps allow photo-organotrophic growth of blue-green algae (see Chapter 6) and bacteria (Chapter 16).

A third photosynthetic carbon cycle which can occur is the C_4 pathway, in which phosphopyruvate is carboxylated to oxaloacetate, after which the carboxyl group is accepted by a C_2 or C_5 acceptor molecule derived from ribulose phosphate (Hatch, 1970). The cycle has not yet been found to occur in algae, and seems to be less efficient than the Calvin cycle, unless the rate of photorespiration relative to photosynthesis in Calvin cycle plants is greater.

Some of the products of photosynthesis will be used for energy-requiring processes in the dark, and also for other processes that are dependent on chemical energy. The Krebs TCA cycle is well recognised as an energy-generating process. It seems likely, however, that some of the substances formed are lost by photorespiration. Originally it was believed that respiratory processes were independent of photosynthesis, except that the latter provided substrates for the former. At high irradiance, however, respiration may be depressed and CO_2 is produced as the result of another process. During this photorespiration, for which glycollate ($CH_2OH\ COOH$ = glycollic acid) is the substrate, energy is lost because, in many higher plants, the following exothermal reaction occurs:

$$2\ CH_2OH\ COOH + 1\tfrac{1}{2}\ O_2 + NH_3 \rightarrow CH_2NH_2COOH + 2\ CO_2 + 3\ H_2O$$
$$\text{glycollate} \qquad\qquad\qquad\quad \text{serine}$$

No ATP is formed in this reaction; a contrast with glycolytic respiration of glucose. The glycollate system may be a protective mechanism against a "short circuiting" of the reductant with O_2 under conditions of high photosynthetic O_2 production. Algae seem to be able to excrete the glycollic acid, which would give a similar protection to that provided by photorespiration in higher plants. Evidence for photorespiration in algae is scarce, if not non-existent, certainly under natural conditions. Fogg et al. working on Lake Windermere (1965) found that up to 90% of the $^{14}CO_2$ fixed was excreted as glycollic acid in conditions of high irradiance where there was a strong photo-inhibition. Fogg postulated a possible protective function for this mechanism. He was able to study glycollic acid excretion in algal cultures, since it could be induced by removing CO_2 or poisoning with INH (isonicotyl hydrazide). A possible metabolic route for the production of the glycollate is as follows:

$$\text{glucose} \to \text{ribulose diphosphate} \xrightarrow[\quad]{\displaystyle CO_2 \quad \overset{INH}{\downarrow}} \text{heptulose diphosphate} \to 2 \times C_3 \sim P$$

low ↓ CO_2

phosphoglycollate $\xrightarrow{\qquad}$ serine
glycine

↓

phosphoglycerate

If this does occur then the similarity to the supposed photorespiration route in higher plants is striking. It is surprising that very little work has been done to study the influence of N starvation on glycollic-acid production, since it is clear that the proposed pathway could be closely connected with other routes in amino-acid metabolism. In the upper layers of clear lakewater, where irradiance is high, photo-inhibited algae are often found to be growing under conditions of severe N deficiency.

Nitrogen deficiency may stimulate glycollic-acid production in two ways: firstly the products of photosynthesis cannot be converted into amino acids because a "bottle neck" forms at the transamination stage; and secondly glycollic acid itself cannot be metabolised into serine for the same reason. Fogg et al. (1965) found that in eutrophic waters organic-acid production was only about 1% of the total C fixed but in oligotrophic waters it was as high as 35%. Watt (1966) found that up to 90% of photosynthetic fixation could be in the form of organic acids at high irradiance. Nalewajko and Lean (1972), using Sephadex chromatography, found that *Chlorella*, *Anabaena*, and *Asterionella* cultures all released compounds whose molecular weight was near that of glycollate, together with other compounds of higher molecular weight. The relative quantities of the various dissolved compounds formed depended on growth rate and age of the cultures. They demonstrated that certain bacteria — both in non-anoxic cultures and in lakewater — could utilise low molecular-weight extracellular metabolites of algal origin and could form larger molecules presumably derived from the algal products.

Saunders (1972) pointed out that most studies so far reported measured the extracellular release only during periods of high solar radiation. He found that three different patterns could occur if the dark period were also included. The concentrations of dissolved compounds could be constant or they could decrease or increase at different rates according to the time of day or night. In the first case metabolically refractory compounds seemed to be involved, and in the second case the smaller compounds are probably used by bacteria while the presence of refractory compounds may apparently have decreased the rate of turnover. In the third case the compounds come from a metabolic pool in the algae, and this decreases in size during the night.

No estimations of the molecular weight of the various compounds was made.

Saunders gives an interesting discussion of the relationship between colour and optical properties of lakewater and a theoretical model is also given to express that relationship. Saunders found that Secchi-disc transparency could be expressed by:

$$Z_{s.d.} = \frac{8.69}{\alpha + K_v}$$

where α = attenuation coefficient for the $I_{m.p.c.}$ (473 nm), and K_v = attenuation coefficient for the total illuminance.

The glycollic acid and other excreted metabolites may be used by bacteria or by the algae themselves. Algae have at least two different pathways for utilisation of glycollic acid. Firstly it may be converted into serine via glycine, a process which can be inhibited by INH. Codd and Stewart (1973) found that in the light *Anabaena cylindrica* metabolised glycollic acid mainly via glyoxylate, and tartronic semialdehyde, (OH)HC-C(OH)COOH, into glycerate, while in the dark the glyoxylate from exogenous glycollic acid appeared to be metabolised mainly to malate.

The metabolic pathways in the pentose-phosphate CO_2 reduction cycle, Krebs cycle, C_4 acid cycle, and glycolysis are summarised in Appendix III. Obviously not all cells can operate all the reactions given.

Extracellular products excreted by algae consist of both intermediate and end-products of metabolism. The latter compounds often have a high molecular weight. Fogg, who reviewed the subject in 1971, called the two groups I and II. In group I Fogg put several organic acids, such as α-ketoglutaric, pyruvic and other α-keto acids, formic, acetic, and aceto-acetic acid. Carbohydrates are also excreted, e.g. pentose up to 7 mg l^{-1} by *Anabaena cylindrica*, being part of a polysaccharide of which glucose, arabinose and galactose were other constituents. Fogg suggested that these polysaccharides modify the viscosity of water in the vicinity of the cells and thus reduce the rate of sinking. Other extracellular products are amino acids, polypeptides and even proteins. The release of nitrogenous compounds such as polypeptides by blue-green algae is especially striking although these compounds are not closely associated with N_2 fixation. Other compounds, put by Fogg in group II, are volatile substances, enzymes, vitamins and other growth factors, hormones, or sexual substances. Although the production of these substances in cultures is well established, their production and ecological significance in natural habitats are doubtful.

4.5. QUANTITATIVE MEASUREMENTS OF PRIMARY PRODUCTION

4.5.1. O_2 and ^{14}C method

Algal carbon assimilation may be described as:

$$CO_2 + H_2O \rightarrow C(H_2O) + O_2$$

or taking into account the formation of proteins, etc.:

$$5\ CO_2 + 2\ H_2O + NH_3 \rightarrow C_5H_7NO_2^* + 5\ O_2 \qquad A.Q \equiv \frac{O_2}{CO_2} = 1$$

or

$$5\ CO_2 + 3\ H_2O + HNO_3 \rightarrow C_5H_7NO_2^* + 7\ O_2 \qquad A.Q. \equiv \frac{\dot{O}_2}{CO_2} = 1.4$$

Measuring algal growth means, in principle, measurement of the increase in dry weight or cell numbers with time. In almost all lakes these increases are too small to be measured during periods sufficiently short for the natural population not to have changed qualitatively. Moreover, the total particulate matter in the water includes varying amounts of material such as detritus and sediments stirred up from the bottom or freshly transported into a lake by rivers and these will make the determination of algal dry weight unreliable. In practice, therefore, the production of oxygen or the uptake of CO_2 is measured instead.

Oxygen production is usually measured using three bottles, so-called Winkler bottles, filled with lakewater. The first bottle, A, gives the initial O_2 concentration. A second, L, is placed in the light while the third bottle, D, is kept in complete darkness. The elegance and acceptability of the method lies in the fact that the measurements can be made in situ, i.e. the bottles are suspended in the lake at certain depths so that measurements are made under natural light and temperature conditions. If measurements are made in triplicate using 9 bottles, the experimental error may be reduced to 0.2% and will normally be of the order of $\pm$ 0.02 mg l^{-1} of O_2. The method becomes insensitive when the initial amount of oxygen (7—15 mg l^{-1}) is very large by comparison with that produced during the experimental time. Details of the O_2 analysis method are given in Golterman (1971b). From Fig. 4.6 it is clear that for accurate work, sufficient measurements at various depths should be made to enable the photosynthesis—depth curve to be constructed.

The usual method of sampling is to take water samples from the same depth at which the bottles will be suspended during the actual experiment. For the study of adaptation and light inhibition, it may be useful to take separate samples from one depth and to incubate these samples at different depths. Difficulties will arise if algal populations are not homogeneously dispersed through the water column. Certain physical processes could cause this but vertical heterogeneity occurs mainly because the organisms can migrate actively from one water layer to another.

Nevertheless, small differences in species composition at different depths

---

* $C_5H_7NO_2$ as mean algal composition is a fair approximation to the commonly found ratio C: dry weight = 1 : 2 and C : N = 5:1; a more reduced state may be indicated by adding more H atoms, e.g. $C_5H_9NO_2$, or H_2O molecules, e.g. $C_5H_9NO_3$.

do not necessarily influence primary production. If the chlorophyll is dispersed homogeneously through the upper water layers then for the purposes of primary production study actual species composition may be less relevant. (For patchiness see Chapter 14.)

The work of Talling in Windermere (1966a) demonstrates the great accuracy which is possible by careful use of this method.

A more sensitive method for measuring algal photosynthesis is that of Steemann Nielsen, which measures the uptake of labelled CO_2. It has the same advantages as the oxygen method in that measurements can be made in situ. Small amounts of ^{14}C-labelled bicarbonate are added to lakewater in experimental bottles and the bottles are resuspended in the lake at different depths. High sensitivity is achieved due to the fact that algae from a reasonably large volume of water can be collected on a filter, and that the initial labelled-CO_2 value is zero, although in practice the background radiation reading from the laboratory is used as the lower limit or "control value". The filters are dried and are either counted in a windowless (or thin window) gas-flow counter or combusted and counted as gas. CO_2 uptake is calculated by:

$$^{12}C \text{ assimilation} = \frac{^{14}C \text{ assimilation}}{^{14}C \text{ available}} \cdot {}^{12}C \text{ available} \cdot 1.06$$

The factor 1.06 allows for the fact that $^{14}CO_2$ is 6% less available to algae than the normal $^{12}CO_2$. The ^{14}C method usually gives no information from which to estimate losses due to respiration, excretion and mineralisation. For short exposure times (2 h) it is usually but incorrectly assumed that only net photosynthesis is measured. The usefulness of the ^{14}C method is clearly demonstrated by the work of Rodhe (section 4.7).

4.5.2. O_2 versus ^{14}C method

Several advantages and disadvantages of both methods need to be considered. The ^{14}C method is certainly the more sensitive one. Expensive counting apparatus is necessary, however, and although bottles may be incubated under field conditions, the counting apparatus must usually be in a laboratory and needs a reliable electricity supply.

Determinations of O_2, however, can be carried out in the field as accurately as in any laboratory. From ^{14}C data no respiration can be measured and therefore no net production can be calculated, whereas the oxygen method gives a good estimate of assimilation and dissimilation (see section 4.6).

The ^{14}C method shows a finite rate of dark uptake or fixation. This must not be neglected, as it may be important quantitatively both for algae and for heterotrophic bacteria if present (see Chapter 16). It is probably due — among other reactions — to the so-called Wood and Werkman reaction:

$$CH_3COCOOH + CO_2 \rightleftharpoons COOHCH_2COCOOH$$

pyruvate   oxaloacetate

This is probably very important in growing algae, because protein formation constantly withdraws oxaloacetic acid and α-ketoglutaric acid from the respiratory tricarboxylic-acid cycle, so that without a regeneration of acids this cycle would cease. The pyruvate is presumably generated from the products of photosynthesis, which will be rapidly depleted in the dark, so that this "dark fixation" is probably mainly active in and dependent on the light. As the Wood and Werkman reaction does not involve an increase in chemical energy, although normal CO_2 fixation does, it ought not to be included in the measured CO_2 uptake, but in fact there are no means of measuring the extent of this process in the light. The fact that CO_2 fixation falls rapidly after darkening is some indication of the existence of a cellular pool for carbon-dioxide acceptance. When bacteria are also present in lakewater, a CO_2 dark fixation with completely different energy aspects may occur (see Chapter 16). Such processes are often found in the hypolimnion, well below the photosynthetic zone, and their CO_2 utilisation can be rather large compared with that due to photosynthesis. The rate of bacterial CO_2 uptake does not normally decrease with time.

Morris et al. (1971) demonstrated for marine phytoplankton an increase in the ratio of light to dark fixation as algal population density increased. They believed that their findings were related to the biochemistry of the algal cells and they did not consider the possibility of bacterial CO_2 uptake which could equally well explain their results. Moreover, they found that the ammonium ion enhanced the dark fixation of CO_2 if added to nitrogen-free cultures. This would certainly seem to suggest a connection between the dark fixation and protein formation. The authors doubt the validity of the normal practice of subtracting the dark from the light fixation when measuring CO_2 uptake due to photosynthesis.

Dark fixation was measured by Gerletti (1968b) in different Italian lakes. He found a relative increase with depth and a decrease in the rate of fixation in long-term exposure samples, which he attributed to a possible depth-related bacterial population or to differing physiological states of the algae at different depths.

A further disadvantage of the ^{14}C method is that the excretion of extracellular products makes results inaccurate if a cell synthesises and then excretes organic molecules, whereas the amount of oxygen produced is not affected by this. Estimation of ^{14}C in cellular material will give no measure of any excreted extracellular organic compounds. Examples of excreted organic compounds are the glycollic-acid production under stress conditions, and polypeptide production by blue-green algae (see Chapter 6). These products can however be estimated as ^{14}C-labelled organic molecules in solution after careful removal of the excess labelled bicarbonate, although many of the results described in the literature are probably to be attributed to an incomplete removal of bicarbonate. A source of inaccuracy in the O_2 method is the assumption that O_2 utilisation is equal in dark and light bottles. Results for humic-rich lakewater

(Golterman, 1971a) have shown that when photosynthesis is inhibited, using DCMU, the light bottle has an O_2 consumption greater than that due to respiration alone (dark-bottle result), an effect probably due to the photocatalytic oxidation of certain extracellular molecules. In Lake George, however, we found a decreased O_2 consumption compared with the dark bottle, so that no generalisation is possible. Further use of DCMU for inhibiting photosynthesis in various situations may solve the problem. Finally not all the O_2 consumed in the dark is related to the oxidation of algal material, even if zooplankton is absent, since aerobic bacteria may contribute by using dissolved organic matter as a substrate (Golterman, 1971a).

4.6. CALCULATION OF NET PRODUCTION; RESPIRATION AND MINERALISATION LOSSES

If oxygen uptake can be measured sufficiently accurately in the dark, the net production may be calculated by subtracting O_2 uptake from O_2 production, both calculated per 24 hours and per unit lake area. In many eutrophic lakes this net production will be found to be small when compared with the gross photosynthesis. The apparently low figures for O_2 uptake, when measured per l per h, show a comparatively considerable increase when calculated per m^2 and per day, as appears from the following example. (It should be noted that errors are also magnified.)

In 1970 the mean gross production of Tjeukemeer (The Netherlands) was found to be 8.0 g m^{-2} d^{-1} of O_2. The mean oxygen consumption rate was 0.15 g m^{-3} 2h^{-1}, which was about 7.5% of the maximal oxygen production rate A_{max}. Oxygen production in this lake is however confined to the upper half metre, while the consumption occurs through the whole 2-m water column and throughout the whole 24-h period. The oxygen consumption rate is $2 \times 12 \times 0.15$ g m^{-3} h^{-1} = 3.6 g m^{-2} d^{-1} or 45% of the photosynthesis. Net photosynthesis can be calculated by a similar procedure, considering only the light period instead of 24 h. For many shallow productive lakes a rather low net productivity has been found, sometimes near zero (see section 15.5). In those cases where the lake has a low water retention time, such a situation is evidently impossible because some algae are washed away in the outflow. Several apparently low net production figures must therefore be attributable to experimental errors in the oxygen-uptake measurements or to a fundamentally wrong assumption, i.e. that O_2 uptake in the dark is related to oxidation of algal material. The latter is untrue in cases where much dissolved organic matter is present. Comparing O_2 uptake with decrease of cellular C and N in dark bottles we found that this decrease in the example mentioned above was equivalent to 50% of the oxygen taken up, the other 50% being used in the oxidation of the allochthonous dissolved material. A corrected value for net primary productivity was therefore in this case $8.0-0.5 \times 3.6 = 6.2$ g m^{-2} d^{-1} of O_2 instead of the 3.6 g m^{-2} d^{-1} quoted previously.

4.7. VARIATIONS OF PHOTOSYNTHESIS IN TIME AND SPACE

4.7.1. Geographical effects

In Fig. 4.7A large differences in photosynthetic rate were shown for several lakes, the range being from 7 to 365 mg m^{-3} d^{-1} of C assimilated for A_{max} and from 45 to 1660 mg m^{-2} d^{-1} of C assimilated for the integral photosynthesis (ΣA). The differences probably result from the combined effects of light penetration and nutrient supply. Lake Tarfala and Lago di Varese (Italy), although having the same light-penetration properties, have the greatest difference in primary production, apparently caused by the more abundant nutrient supply in the latter lake. The lake showing the second lowest productivity figures, Lake Torneträsk (N. Sweden, Arctic latitude), predictably has a high Secchi-disc reading, and this low primary production is apparently due to a low concentration of nutrients. It is obvious that the low light penetration in Lago di Varese is caused by the algae themselves (self-shading), whereas in Tarfala the turbidity is due to suspended clay eroded by glacier activity. Very low productivity values (5 mg m^{-3} d^{-1} of C_{ass}) caused by lack of nutrients have been recorded by Findenegg (1964, 1966) in eastern alpine lakes such as the Attersee. Rodhe (1965) showed that low primary production in the upper layers of this lake was caused by photo-inhibition.

Findenegg (1966) demonstrated that near ultraviolet light, which has a relatively high penetration in these oligotrophic lakes, increases the severity of photo-inhibition. Whether this is due specifically to the effect of short-wave radiation itself or to its contribution to the total energy which is already at an inhibitory level is a matter for discussion. Results similar to those of Findenegg were obtained by Pechlaner (1964), who measured a value of 24 mg m^{-2} d^{-1} of C after subtraction of dark uptake in the Vorderer Finster-taler See (Austria), with values for A_{max} of about 1.5 mg m^{-3} d^{-1} of C. Pechlaner measured rather low values for total phosphate (0.9—14 μg l^{-1}) and these are probably low enough to have been limiting photosynthesis in the lake.

Wetzel (1966) compared primary production in thirteen marl (Ca-rich) lakes in the glaciated Great Lakes region of the United States. Most of these waters are calcareous (2—4 meq. l^{-1} of Ca^{2+}) and have high pH and alkalinity values and excessive deposits of carbonates. Wetzel found a total range for the daily production rates of between 6 and 6 000 mg m^{-3} d^{-1} of C_{ass}, but the annual means ranged only between 200 and 700 mg m^{-3} d^{-1} (with the exception of Sylvan Lake). It is remarkable that the largest range was found for the lake with the lowest annual mean, i.e. Smith Hole.

Furthermore it should be noted that, although the annual means are rather low, values approaching the maximal production rate (see below) are often found at some times of the year. Wetzel pointed out that the extremes of the daily fluctuations of the primary production rates are less variable among the larger and more typical calcareous lakes such as Crooked Lake than among the smaller, more productive lakes such as Little Crooked Lake.

Such a general statement needs statistical investigation when more data on the nutrient budgets are available and after the influence of the seasonal effect has been taken into account. The rather low winter values tend to over-emphasise the ranges as published.

Very high production rates were found by Talling (1965) in East African lakes. With a few exceptions around 2 g m^{-2} d^{-1}, most of his values lie within the range 4—15 g m^{-2} d^{-1} of O$_2$, the highest values being obtained in Lake George. Several points arise from this study. The value A_{max} varied within 25 ± 6 mg m^{-3} of O$_2$ per hour per mg m^{-3} of chlorophyll, covering chlorophyll concentrations between 2 and 120 mg of chlorophyll a. This average value is much higher than the majority of published records for the photosynthetic capacity of phytoplankton from temperate waters. The most likely explanation for such a high value seems to be the temperature dependence of A_{max} per unit of population (Talling, 1965).

The calculated value of I_k lies between 10 and 30, with a mean value of 20 kerg cm^{-2} sec^{-1}. It seems likely that the difference between this and the value of 10 kerg cm^{-2} sec^{-1} as reported for Windermere is due to the higher temperature in the tropical lakes studied by Talling. With typical values for I' (320 kerg cm^{-2} sec^{-1}) and for I_k (20 kerg cm^{-2} sec^{-1}), formulation of ΣA (see subsection 4.3) was reduced to:

$$\Sigma A = \frac{A_{max}}{\epsilon_{min}} \cdot 2.6$$

This formulation enabled Talling to predict the *maximal* daily rate of primary production, a maximum which exists due to the self-shading effect. From observations on *Asterionella maxima* in Windermere, Talling (1965) calculated a value ϵ_s — the increment of ϵ — of 0.02 per unit of population density (ln units per mg of chlorophyll a per m^2); thus for every increase of chlorophyll concentration by 1 mg per m^2, the parameter ϵ increases by 0.02 ln units per m.

The pigment content of the euphotic zone and the ratio A_{max}/ϵ_{min} will each reach an upper limit, and using a value of ϵ_s = 0.02, Talling (1970) calculated the maximal pigment content of the euphotic zone to be about 200 mg per m^2. As A_{max} is approximately 25 mg of O$_2$ per mg per h, the maximal value for the ratio A_{max}/ϵ_{min} will be 1250 mg m^{-2} h^{-1} of O$_2$.

As the day length in the lakes studied is constant at 12 h, the conversion factor from hourly to daily production is about 9, so that the productivity values of 10—15 g m^{-2} d^{-1} as given in Talling's study are not far from the theoretically estimated maximal ones, the latter being approximated as:

$$\left(\frac{A_{max}}{\epsilon_{min}}\right)^{max} \cdot 2.6 \cdot 9 = 1250 \cdot 2.6 \cdot 9 \text{ mg m}^{-2} \text{ d}^{-1} \text{ of O}_2$$

Talling found a value of A_{max}/ϵ_{min} for Lake Victoria of 508, which is just below the maximum theoretically possible. This was consistent with the occurrence of a weak though positive correlation between plankton population changes and the seasonal variation of ϵ_{min}.

The low value for Windermere (85—89) probably indicates how important the effects of climate must be since the plankton population densities in Windermere and Lake Victoria are closely similar.

It should be realised that the high values for O_2 production rate given above refer to gross primary production. It has been stated already that due to the high algal population density rather high values for the O_2 uptake rate were found, both in light and dark, so that the net particulate productivity may be rather low (nearly zero for Lake George, see section 15.5).

In Ethiopian soda lakes, Talling et al. (1973) found values for O_2 production rate in the range 1.4—2.49 g m^{-2} h^{-1}. In one of the two lakes studied, chlorophyll concentration was $> 2\,000$ mg m^{-3}. The euphotic zone varied between 15 and 27 cm. No data for net productivity is available but it is estimated indirectly to be rather low, sometimes resulting in complete deoxygenation overnight. Talling et al. suggested that the relationship $Z_{eu} = 3.7/\epsilon_{min}$ can be exceeded either because ϵ_s can be smaller than 0.02, or because the extrapolation from thin to dense populations is not valid or because the theoretical maximum estimate of about 200 mg m^{-2} of chlorophyll a in the euphotic zone was exceeded.

4.7.2. Diurnal effects

As the irradiance changes constantly during the day, photosynthesis will also vary with the time of day.

In early morning light will be limiting even in the upper water layers and photosynthesis will decrease almost exponentially with depth. As irradiance increases, photosynthesis will increase until a maximal value of A_{max} (Fig. 4.10) is reached. Further increase can then only take place in the integral photosynthesis (per unit area) by a deepening of the layer in which light saturation occurs.

On the other hand, the layer of light inhibition may also become deeper, which will cause photosynthesis to decrease in this upper layer. Part of this loss may be compensated, however, by a higher photosynthesis than expected from the theoretical light—depth curve as shown in Fig. 4.6, below the zone of A_{max} (Vollenweider, 1965).

It is obvious from these considerations that no theoretical approach is possible for calculating the changes per day. In a few studies the rate of photosynthesis has been measured for short periods, e.g. by Talling (1957c) (Fig. 4.11).

The only meaningful method acceptable on theoretical grounds would be the summation of results of several short-term measurements obtained through-

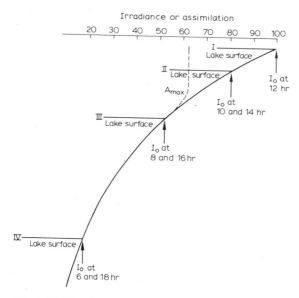

Fig. 4.10. Graphs representing diurnal changes of depth—light and depth—photosynthesis curve. By placing the lake surface at different points on the light penetration curve at the time of the day of maximal irradiance (solid line), the light penetration for different moments of the day is shown as that part of the total curve that falls below "lake surface". Photosynthesis (dotted line) shows saturation in the top layer from a certain time of the day onwards. If light is still increasing after that time, the layer of saturated photosynthesis will become deeper, but A_{max} cannot increase.
The horizontal lines I—IV represent the top of the water column (depth = 0 m) at different times of the day, while arrows indicate I_0 at the beginning of that period.

out the whole day. As such a procedure is often impractical and certainly tedious, several approximations have been made. Vollenweider (1965), in an excellent theoretical paper on the calculation of primary productivity, proposed that the day be divided into five equal periods between sunrise and sunset and that photosynthesis be measured during the second and third periods. If nutrient depletion is serious, the decrease during the afternoon is asymmetric and the contribution of the second period alone to the whole day can be expected to be 30%. If no nutrient depletion occurs, Vollenweider argues that during periods II + III, 55% of the daily production should take place. No diurnal rhythm, however, was reported by Rodhe (1958) for Lake Erken (Sweden) in one of the earliest papers published concerning the effects of time on primary productivity.

Rodhe demonstrated that an exposure time of 19 h gave lower results than the sum of 5 shorter experiments as can be seen in Table 4.3. Rodhe pointed out that the discrepancy between totals of short-term and long-term measured values is especially great in the upper zones and is probably due to photoinhibition; the algae are likely to be more inhibited during a longer exposure

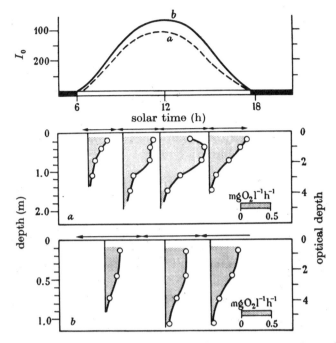

Fig. 4.11. Depth profiles of photosynthetic rate (shaded areas) for various intervals (indicated by horizontal arrows) during the day during (a) 12 December 1954, and (b) 7 October 1955. Photosynthetic rates, expressed per unit water volume, are measured horizontally from the vertical axis of each profile according with the inset scales. The upper graph shows variation of incident irradiance, I_0 (in kerg cm^{-2} sec^{-1}; spectral region 400—700 nm). (From Talling, 1957c.)

time, especially under nutrient-limiting circumstances. It can be seen that the production during period IV (415 mg m^{-2} of C) is exactly that of the production in period II times the relative light intensities, 130/160, so it seems that no diurnal effect is really discernible. The relatively lower production in period III, where a further increase in light does not cause an increase in photosynthesis, is apparently because A_{max} has already been reached and cannot increase further. Theoretically however, the primary productivity per unit area is not proportional to I_0. These characteristics should be kept in mind when trying to assess diurnal effects such as those described by Malone (1971). He found a slight diurnal rhythm in phytoplankton assimilation, the nano- and net-plankton showing maximum assimilation rates during the afternoon in tropical eutrophic seawater (NO$_3$-N: 70—168 μg l^{-1}; 25—28°C) Malone suggested that diurnal variations, which are often found in seawater, may depend upon the availability of nutrients and upon the relative importance of nano- and net plankton. The results of his experiments are, however, inconclusive in this respect.

For Dutch nutrient-rich, shallow waters, where light saturation occurs for

TABLE 4.3

Primary production at several depths during different periods in Lake Erken (values in mg m^{-3} of C per period mentioned)

	Period						Measured directly
	I(2—6 h)	II(6—10 h)	III(10—14 h)	IV(14—18 h)	V(18—21 h)	ΣI—V	
Incident light energy (kcal. cm^{-2}):	39	160	220	132	7	558	
Depth (m)							
0.2	79	139	123	131	55	527	332
1	47	149	145	141	23	505	414
2	22	114	105	86	12	339	301
3	12	65	66	36	11	190	159
4	~11	~35	~63	~21	~24	~154	~143
Total (mg m^{-3})	171	502	502	415	125	1715	1350

only a short period of the day, good results were obtained by multiplying the values obtained from a 2- or 4-h experiment by a light factor F, which is the ratio of total kcal. cm^{-2} d^{-1} to kcal. cm^{-2} per exposure time (Golterman, unpublished). Talling (1971) has shown that the diurnal ΣA assimilation curve deviates considerably from the incident light curve I_0, but follows more closely $\log I_0$ above a baseline of 0.5 kcal. cm^{-2} h^{-1} ($= I_{1/2 k}$ assuming $I_k = 1$ kcal. cm^{-2} h^{-1}). Talling suggested that the area between the $\log I_0'$ curve and the 0.5-I_k level will be the more meaningful long-term measure of effective radiation input.

Talling (1957b, 1965) defined a function of I_0', i.e.:

$$L.D. = \frac{\ln I_0' - \ln 0.5\, I_k}{\ln 2}$$

where L.D. = Light Division Hours. Talling (1957b) suggested that the factor F, by which the amount of photosynthesis occurring during the exposure time, may be converted to total daily photosynthesis values:

$$F = \frac{\ln \overline{I}_0 - \ln 0.5\, I_k}{\ln I_0' - \ln 0.5\, I_k} \cdot 0.9$$

where $\overline{I}_0$ = mean irradiance between sunrise and sunset, and I_0' = mean irradiance during exposure time.

Further difficulties in the computation of daily rates will be encountered if photo-inhibition occurs, especially during the middle of the day when most measurements are made. Fig. 4.12 with data from Marshall and Orr (1928) and from Talling (1957c) demonstrates the non-linear response caused by light saturation and photo-inhibition. The apparent occurrence of photo-inhibition in many experiments is questionable and of dubious ecological significance for the following reasons:

(1) Lack of vertical circulation throughout the experimental arrangements compared with natural movements of algae in the equivalent water column in a lake. Because of this, the artificial effect would be expected to be more serious in lakes with a shallow euphotic zone, since the algae have a smaller migration range within which they could "escape" from the inhibition zone to that of maximal photosynthesis.

(2) Influence of spectral sensitivity, as e.g. described by Findenegg (1966) for UV light, although it may be argued that inhibition is caused by an increase in total radiant energy to which the UV light contributes and not by specific inhibition by the UV as such. Halldal (1970, cited in Talling, 1971) has reviewed evidence that little selective inhibition is likely at wavelengths longer than about 320 nm and transmitted by the usual glass bottles.

(3) Decreased phytoplankton density near the water surface due to migration away from the photo-inhibition area.

(4) Lack of pronounced surface inhibition is observed in relatively long (7-day) growth-rate experiments (Lund, 1964). This may be due to com-

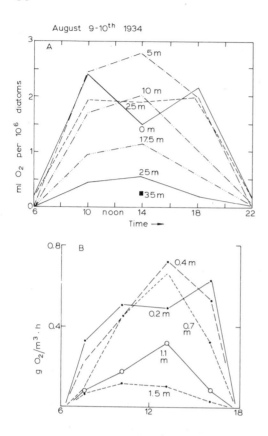

Fig. 4.12. Examples of the diurnal variation of photosynthetic rates measured at various depths in the euphotic zone. A. Photosynthesis by cultured *Coscinodiscus excentricus* in the sea off southwest England during a sunny summer day. (From Jenkin, 1937, fig. 7.) B. Photosynthesis by natural phytoplankton of a Nile reservoir south of Khartoum under cloudless tropical conditions. (From Talling, 1971.)

pensation by adaptation during the weekly exposures, and also perhaps on a daily basis, so that the experimental period is not truly representative of the whole period during which photo-inhibition may occur.

A model developed by Fee (1969) based on Vollenweider (1965) utilises a computer programme to simulate daily photosynthetic rates and photo-inhibition. Fee was forced to adopt an "idealised" ϵ value because he did not have sufficient measurements for Lake Michigan. Further relevant discussions are in Talling (1970) and Vollenweider (1970).

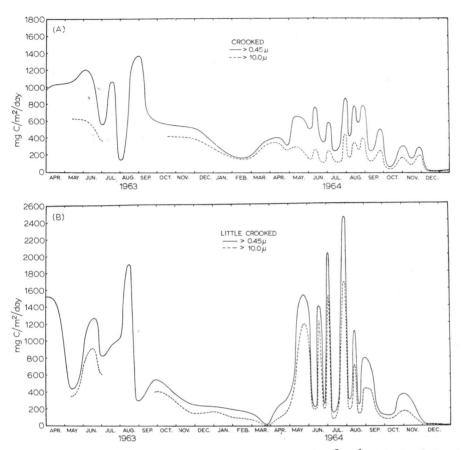

Fig. 4.13.A. Integrated in situ primary productivity (mg C m^{-2} d^{-1}), Station A, Crooked Lake, Whitley—Noble counties, Indiana. The area between the solid and broken lines represents that portion of the primary productivity resulting from the ultraplankton (< 10 μm). Sampling intervals were approximately 10—14 days in 1963 and increased to 5—7 day intervals in 1964 during the warmer periods, biweekly during the colder periods, of the year. B. Idem for Little Crooked Lake, Whitley County, Indiana. (From Wetzel, 1965.)

4.7.3. Seasonal effects

For most, if not all, non-tropical waters and probably for many tropical ones also, the amount of carbon assimilated daily per unit surface will vary according to many local factors such as light and nutrient supply, which are subject to considerable fluctuations.

In temperate regions of the Northern Hemisphere primary productivity increases considerably in February—March, the exact time depending on the disappearance of ice or of low winter temperatures, as is shown for example by Wetzel's (1965) results for Crooked and Little Crooked Lake (Fig. 4.13A and B).

Both lakes show large fluctuations and, as pointed out earlier, in Little Crooked Lake they are more pronounced than in Crooked Lake. Writers in the earlier literature described the primary productivity as being frequently high in spring and autumn with lower rates during the summer, which they called the stagnant period. Loss of nutrients in dead plants and animals which had sunk into the hypolimnion was thought to be the cause of the lower summer values. Wetzel's results and our own from the small stratifying sand pit Vechten (where most values lie between 300 and 400 mg m^{-2} d^{-1} of C_{ass} with a few higher values in August and September) do not support this idea, but it may be that in these relatively shallow lakes there is some upward diffusion of nutrients from hypo- into epilimnion (Chapter 9). It should be remembered however, as Rodhe demonstrated as early as 1958, that the daily fluctuations of photosynthetic rate can be considerable. Thus Rodhe found that the average value (mg m^{-2} d^{-1}) in Lake Erken was between 400 and 600 in 1954 and about 400 in 1955, but sudden single values reached over 1 600 and 2 000 (July—August) while spring maxima were 1 000 and 2 000 for 1954 and 1955, respectively. Rodhe therefore suggested that a fuller study of daily fluctuations should be made before data for less frequent periods are used or interpreted. Less variation and no spring maxima occur in many, if not all, shallow lakes, as can be demonstrated from values for Tjeukemeer, but even here occasional high values are found. Most values fall within 10—20% of the annual means, and the occasional high values have a disproportionately large effect upon the mean. The very low values that are found from time to time are probably caused by upwelling of sediments which temporarily decrease the light penetration. In fairly shallow lakes, where the chemical composition is regulated by the sediments and is affected by wind strength through turbulence, the amount of suspended matter in the water column is one of the most important factors controlling photosynthesis. This control is exercised by virtue of both the nutrient-supplying role of the sediments and by their effect upon light penetration.

Measurements of photosynthesis over a long time period for one place are scarce. They are nevertheless the only really satisfactory and accurate way to collect sufficient comparable material for an analysis to be made of the factors that regulate photosynthesis under natural conditions.

Although light energy is a less variable factor under tropical conditions, lakes at tropical latitudes still show seasonal effects. Talling (1966b) demonstrated a distinct seasonal periodicity for several phytoplankton species in Lake Victoria. He attributed these differences to the onset and breakdown of stratification together with variable mixing depth. For example, the August maximum of *Melosira* species was due almost entirely to the seasonal disappearance of stratification. The maximum between December and February was connected with a downward extension of the upper mixed layer, and possibly with an aftermath of a preceding bloom of *Anabaena flos aquae*. Here the changes, which result in thermal stratification, are not induced entirely by changes in solar radiation but also by seasonal increases in the southeast trade winds causing increased evaporation and thus cooling. It is remarkable that rainfall apparently had no effect on the main seasonal changes. This may be because direct rainfall on the lake accounts for approximately five-sixths of the water intake. Erosion of nutrients by heavy rainfall is therefore unimportant in the nutrient budget of this lake, but it would certainly induce seasonal effects in nutrient-limited lakes. No data are available on the input of nutrients by the main inflow, the Kagera River. Although this river replaces only 0.6% of the lake volume annually, the proportional input of nutrients might be considerably larger.

4.7.4. Local effects

Photosynthetic activity within a single lake may vary greatly due to local conditions such as morphology, water currents, local inflows of rivers, or nature of the substratum.

Gerletti (1968a) demonstrated the occurrence of a gradual decrease of primary productivity from inlet to outlet of the river Ticino along the axis of Lago Maggiore. His figures ranged from 47 to 19 mg m^{-3} of C_{ass} for a time period of 4½ h. Values higher than to be expected according to this gradient were found near a population centre and in the region of a smaller river inflow. As a result of local erosion effects, inshore production may be greater than offshore production, but much will depend on the availability of the inflowing nutrients. We noticed in Lake Edward (Uganda) a clay-rich inshore zone with a high total-phosphate concentration but with hardly any algae. Patchy algal production does of course occur in this lake (Verbeke cited in Talling, 1966b). The local inflowing river contained much suspended phosphate adsorbed on clay, but had practically no dissolved phosphate.

Megard (1972, see subsection 4.3.1) studied photosynthesis in a large complex lake. He used data from a variety of sites for a multifactorial analysis and was able to assess the relationship between light and photosynthesis, photosynthesis and phosphate concentration, etc. Such a study demonstrates that a good insight into the local variations may provide at the same time to a good collection of data which are relevant for use in a more basic treatment. In small shallow lakes relatively little local variation is to be expected.

In Tjeukemeer, we found hardly any differences between three widely spaced sampling points, unless one of these was affected by the inflow of water. Even then the differences found were hardly ever greater than 20%. Work on the larger Loch Leven (Scotland) yielded similarly homogeneous results during most of the years when it was studied (Bailey-Watts, 1974). In large shallow lakes, like Lake Chad, local differences are rather large (Lemoalle, 1973).

84

REFERENCES

Bailey-Watts, A.E., 1974. The algal plankton of Loch Leven, Kinross. *Proc. R. Soc. Edinb. B*, 1974: 135—156.
Codd, G.A. and Stewart, W.D.P., 1973. Pathways of glycollate metabolism in the blue-green alga *Anabaena cylindrica. Arch. Mikrobiol.*, 94: 11—28.
Cox, R.M. and Fay, P., 1969. Special aspects of nitrogen fixation by blue-green algae. *Proc. R. Soc. Lond., Ser. B*, 172: 357—366.
Denny, P., 1972. Lakes of south-western Uganda. I. Physical and chemical studies of Lake Bunyonyi. *Freshwater Biol.*, 2: 143—158.
Fay, P., 1969. Cell differentiation and pigment composition in *Anabaena cylindrica. Arch. Mikrobiol.*, 67: 62—70.
Fee, E., 1969. A numerical model for the estimation of photosynthetic production, integrated over time and depth, in natural waters. *Limnol. Oceanogr.*, 14(6): 906—911.
Findenegg, I., 1964. Types of planktic primary production in the lakes of the Eastern Alps as found by radioactive carbon method. *Verh. Int. Ver. Theor. Angew. Limnol.*, 15: 352—359.
Findenegg, I., 1966. Die Bedeutung kurzwelliger Strahlung für die planktische Primärproduktion in den Seen. *Verh. Int. Ver. Theor. Angew. Limnol.*, 16: 314—320.
Fogg, G.E., 1968. *Photosynthesis.* English University Press, London, 116 pp.
Fogg, G.E., 1971. Extracellular products of algae in freshwater. *Ergebnisse der Limnologie, Arch. Hydrobiol., Beih.*, 5: 1—25.
Fogg, G.E., Nalewajko, C. and Watt, W.D., 1965. Extracellular products of phytoplankton photosynthesis. *Proc. R. Soc. Lond., Ser. B*, 162: 517—534.
Gerletti, M., 1968a. Primary productivity along the axis of Lake Maggiore. *Mem. Ist. Ital. Idrobiol.*, 23: 29—47.
Gerletti, M., 1968b. Dark bottle measurements in primary productivity studies. *Mem. Ist. Ital. Idrobiol.*, 23: 197—208.
Goldman, Ch. R., 1964. Primary productivity and micro-nutrient limiting factors in some North American and New Zealand lakes. *Verh. Int. Ver. Theor. Angew. Limnol.*, 15: 365—374.
Golterman, H.L., 1971a. The determination of mineralization in correlation with the estimation of net primary production with the oxygen method and chemical inhibitors. *Freshwater Biol.*, 1(3): 249—256.
Golterman, H.L. (with Clymo, R.S.) (Editors), 1971b. *Methods for Chemical Analysis of Fresh Waters.* IBP Handbook, no. 8. Blackwell, Edinburgh, 166 pp.
Halldal, P. (Editor), 1970. *Photobiology of Microorganisms.* John Wiley, London, 479 pp.
Hatch, M.D., 1970. Chemical energy costs for CO_2 fixation by plants with differing photosynthetic pathways. In: *Prediction and Measurement of Photosynthetic Productivity. Proc. IBP/PP Tech. Meet., 1969, Třeboň.* PUDOC, Wageningen, pp. 215—220.
Jenkin, P., 1937. Oxygen production by the diatom *Coscinodiscus excentricus* Ehr. in relation to submarine illumination in the English Channel. *J. Mar. Biol. Assoc. U.K.*, 22: 301—343.
Lemoalle, J., 1973. L'énergie lumineuse et l'activité photosynthétique du phytoplancton dans le Lac Tchad. *Cah. O.R.S.T.O.M., Sér. Hydrobiol.*, 7(2): 95—116.
Lund, J.W.G., 1964. Primary production and periodicity of phytoplankton. *Verh. Int. Ver. Theor. Angew. Limnol.*, 15: 37—56.
Malone, T.C., 1971. Diurnal rhythms in netplankton and nannoplankton assimilation ratios. *Mar. Biol.*, 10: 285—289.
Marshall, S.M. and Orr, A.P., 1928. The photosynthesis of diatoms cultures in the sea. *J. Mar. Biol. Assoc. U.K.*, 15: 321—360.
Megard, R.O., 1972. Phytoplankton, photosynthesis, and phosphorus in Lake Minnetonka, Minnesota. *Limnol. Oceanogr.*, 17: 68—87.

Morris, I., Yentsch, C.M. and Yentsch, Ch.S., 1971. Relationship between light carbon dioxide fixation and dark carbon dioxide fixation by marine algae. *Limnol. Oceanogr.*, 16(6): 854—858.

Nalewajko, C. and Lean, D.R.S., 1972. Growth and excretion in planktonic algae and bacteria. *J. Phycol.*, 8: 361—366.

Pechlaner, R., 1964. Plankton production in natural lakes and hydro-electric water-basins in the alpine region of the Austrian Alps. *Verh. Int. Ver. Theor. Angew. Limnol.*, 15: 375—383.

Perl, G., 1935. Zur Kenntnis der wahren Sonnenstrahlung in verschiedenen geographischen Breiten. *Meteorol. Z.*, 52(3): 85—89.

Rabinowitch, E. and Govindjee, 1969. *Photosynthesis*. John Wiley, New York, N.Y., 273 pp.

Rodhe, W., 1958. Primärproduktion und Seetypen. *Verh. Int. Ver. Theor. Angew. Limnol.*, 13: 121—141.

Rodhe, W., 1965. Standard correlations between pelagic photosynthesis and light. *Mem. Ist. Ital. Idrobiol.*, 18 (Suppl.): 365—381.

Rodhe, W., 1972. Evaluation of primary production parameters in Lake Kinneret (Israel). *Verh. Int. Ver. Theor. Angew. Limnol.*, 18: 93—104.

Ryther, J.H. and Menzel, D.W., 1959. Light adaptation by marine phytoplankton. *Limnol. Oceanogr.*, 4: 492—497.

Ryther, J.H. and Menzel, D.W., 1960. The seasonal and geographical range of primary production in the western Sargasso Sea. *Deep-Sea Res.*, 6: 444—446.

Saunders, G.W., 1972. The kinetics of extracellular release of soluble organic matter by plankton. *Verh. Int. Ver. Theor. Angew. Limnol.*, 18: 140—146.

Smith, E.L., 1936. Photosynthesis in relation to light and carbon dioxide. *Proc. Natl. Acad. Sci. U.S.A.*, 22: 504.

Smith, R.C. and Tyler, J.E., 1967. Optical properties of clear natural water. *J. Opt. Soc. Am.*, 57(5): 589—595.

Smith, R.C., Tyler, J.E. and Goldman, Ch.R., 1973. Optical properties and color of Lake Tahoe and Crater Lake. *Limnol. Oceanogr.*, 18(2): 189—199.

Sorokin, Y.I., 1963. Primary organic production in the Atlantic Ocean. *Hydrobiologia*, 22: 306—316.

Spence, D.H.N., Campbell, R.M. and Chrystal, J., 1971. Spectral intensity in some Scottish freshwater lochs. *Freshwater Biol.*, 1(4): 321—337.

Steemann Nielsen, E., 1962. Inactivation of the photochemical mechanism in photosynthesis as a means to protect the cells against too high light intensities. *Physiol. Plant.*, 15: 161—171.

Steemann Nielsen, E., 1965. On the determination of the activity in ^{14}C ampoules for measuring primary production. *Limnol. Oceanogr.*, 10 (Suppl.): R247—R252.

Steemann Nielsen, E. and Hansen, V.Kr., 1959. Measurements with the carbon-14 technique of the respiration rates in natural populations of phytoplankton. *Deep-Sea Res.*, 5: 222—233.

Steemann Nielsen, E., Hansen, V.Kr. and Jørgensen, E.G., 1962. The adaptation to different light intensities in *Chlorella vulgaris* and the time dependence on transfer to a new light intensity. *Physiol. Plant.*, 15: 505—517.

Talling, J.F., 1957a. The phytoplankton population as a compound photosynthetic system. *New Phytol.*, 56: 133—149.

Talling, J.F., 1957b. Photosynthetic characteristics of some freshwater plankton diatoms in relation to underwater radiation. *New Phytol.*, 56: 29—50.

Talling, J.F., 1957c. Diurnal changes of stratification and photosynthesis in some tropical African waters. *Proc. R. Soc. Lond., Ser. B*, 147: 57—83.

Talling, J.F., 1961. Photosynthesis under natural conditions. *Ann. Rev. Plankt. Physiol.*, 12: 133—154.

Talling, J.F., 1965. Comparative problems of phytoplankton production and photosynthetic productivity in a tropical and a temperate lake. *Mem. Ist. Ital. Idrobiol.*, 18 (Suppl.): 399—424.

Talling, J.F., 1966a. Photosynthetic behaviour in stratified and unstratified lake populations of a planktonic diatom. *J. Ecol.*, 54: 99—127.

Talling, J.F., 1966b. The annual cycle of stratification and phytoplankton growth in Lake Victoria (East Africa). *Int. Rev. Hydrobiol.*, 51: 545—621.

Talling, J.F., 1970. Generalized and specialized features of phytoplankton as a form of photosynthetic cover. In: *Prediction and Measurement of Photosynthetic Productivity. Proc. IBP/PP Tech. Meet., 1969, Třeboň*. PUDOC, Wageningen, pp. 431—445.

Talling, J.F., 1971. The underwater light climate as a controlling factor in the production ecology of freshwater phytoplankton. *Mitt. Int. Ver. Theor. Angew. Limnol.*, 19: 214—243.

Talling, J.F., Wood, R.B., Prosser, M.V. and Baxter, R.M., 1973. The upper limit of photosynthetic productivity by phytoplankton: Evidence from Ethiopian soda lakes. *Freshwater Biol.*, 3(1): 53—76.

Tyler, J.E. and Smith, R.C., 1967. Spectroradiometric characteristics of natural light under water. *J. Opt. Soc. Am.*, 57(5): 595—601.

Tyler, J.E. and Smith, R.C., 1970. *Measurement of Spectral Irradiance Underwater*. Gordon and Breach, New York, N.Y., 103 pp.

Van Niel, C.B., 1932. On the morphology and physiology of the purple and green sulphur bacteria. *Arch. Mikrobiol.*, 3: 2—112.

Verbeke, J., 1957. Recherches écologiques sur la faune des grands lacs de l'est du Congo Belge. *Explor. Hydrobiol. Lacs Kivu, Edouard et Albert (1952—1954)*, 3(1): 177 pp. Inst. R. Nat. Belg., Brussels.

Vollenweider, R.A., 1960. Beiträge zur Kenntnis optischer Eigenschaften der Gewässer und Primärproduktion. *Mem. Ist. Ital. Idrobiol.*, 12: 201—244.

Vollenweider, R.A., 1961. Photometric studies in inland waters. 1. Relations existing in the spectral extinction of light in water. *Mem. Ist. Ital. Idrobiol.*, 13: 87—113.

Vollenweider, R.A., 1965. Calculation models of photosynthesis—depth curves and some implications regarding day rate estimates in primary production measurements. *Mem. Ist. Ital. Idrobiol.*, 18 (Suppl.): 425—457.

Vollenweider, R.A., 1969. Möglichkeiten und Grenzen elementarer Modelle der Stoffbilanz von Seen. *Arch. Hydrobiol.*, 66(1): 1—36.

Vollenweider, R.A., 1970. Models for calculating integral photosynthesis and some implications regarding structural properties of the community metabolism of aquatic systems. In: *Prediction and Measurement of Photosynthetic Productivity. Proc. IBP/PP Tech. Meet., 1969, Třeboň*. PUDOC, Wageningen, pp. 455—472.

Watt, W.D., 1966. Release of dissolved organic material from the cells of phytoplankton populations. *Proc. R. Soc. Lond.*, B, 164: 521—551.

Wetzel, R.G., 1965. Nutritional aspects of algal productivity in marl lakes with particular reference to enrichment bioassays and their interpretation. *Mem. Ist. Ital. Idrobiol.*, 18 (Suppl.): 137—157.

Wetzel, R.G., 1966. Productivity and nutrient relationship in marl lakes of Northern Indiana. *Verh. Int. Ver. Theor. Angew. Limnol.*, 16: 321—332.

CHAPTER 5

THE PHOSPHATE CYCLE*

5.1. OCCURRENCE OF PHOSPHATE IN LAKEWATER

In Chapter 4 it was shown that in the upper layers of a lake, where irradiance is high, availability of certain nutrients may be the limiting factor for algal growth. For the great majority of natural lakes in temperate regions the growth-limiting factor is phosphate (Chapter 17). Although data for tropical lakes are scarce, it seems likely that nitrogen is the limiting factor there. Evidence for this comes from Sioli (1968) for the Amazon and from Talling and Viner for lakes in East Africa, especially in the Rift Valley (see Chapters 17 and 18 and Golterman, 1975).

A rough approximation of the demand for the different nutrients can be made by the following semi-quantitative assumption:

$$5\,CO_2 + H_2O + NH_3 + 1/30\,PO_4\text{-}P \rightarrow C_5H_7NO_2P_{1/30} + 5\,O_2$$

From this reaction, which is based on P being 1% of the dry weight, a ratio of C/P (weight) = 60 can be calculated. Exceptions from this ratio exist of course (see subsection 17.4.5). From this ratio and from the measured primary production the amount of phosphate necessary for the algal growth can be estimated. As the P content varies normally between 0.5 and 1% the C/P ratio will vary between 50 and 100 (see also subsection 6.2.2).

The concentration of inorganic orthophosphate (PO_4-P) varies widely in different lake waters. Observations for Windermere indicate that 2 mg m^{-3} of PO_4-P could be considered to be a "normal" winter value. For the Swiss alpine lakes values of 5—10 mg m^{-3} seem to be normal. During the summer, algal growth may reduce the inorganic phosphate concentrations to values below those detectable by chemical analysis. High values which are found in lakes in delta regions are due to deposition of clay containing adsorbed or incorporated phosphate. In the Dutch peaty lakes the following summer values can be taken as normal: PO_4-P: 1—50 mg m^{-3} (winter values 25—50 mg m^{-3} with a possible tenfold increase if much peaty water enters the lake); Tot-P$_{diss}$:

---

* The following abbreviations of the different phosphate components will be used:
I.   Inorganic orthophosphate: PO_4-P (= $H_2PO_4^- + HPO_4^{2-} + PO_4^{3-}$)
II. Total dissolved phosphate: Tot-P$_{diss}$
III = II—I, Hydrolysable phosphate: Poly-P + Org-P$_{diss}$
IV. Particulate phosphate: Part-P (in algae, bacteria, other organisms; adsorbed on clay or humic compounds, or as pebbles and rock fragments)
V. Sum of II + IV = Tot-P = (Part-P + Tot-P$_{diss}$).

$50-100$ mg m^{-3} (partly on humic particles); and Part-P: $50-100$ mg m^{-3}.

Values of the same order of magnitude were found by Talling and Talling (1965) in lakes Albert and Edward ($100-200$ mg m^{-3}). They mention 2.6 g m^{-3} for Lake Rudolf and values between 10 and 60 g m^{-3} for some saline waters.

5.2. SOURCES OF PHOSPHATE

The main natural origin of phosphate is due to erosion, the chemical and mechanical weathering of rocks. The total amount of phosphate occurring throughout the solid crust of the earth is estimated at 10^{25} g (10^{19} tonnes). It occurs mainly as apatite, $3\,Ca_3(PO_4)_2 \cdot Ca(OH, F \text{ or } Cl)_2$, occluded in igneous rock in concentrations within the range $0.07-0.13\%$ (P). During the erosion process phosphate is mobilised, partly as dissolved inorganic phosphate, and partly adsorbed on or even into the clay particles. Before it reaches the sea, some of the phosphate may be trapped in sedimentary rock formations. Using erosion figures given by Goldschmidt (1937), it can be estimated that 10^{21} g of P are thus trapped in marine sediments, mainly adsorbed on aragonite ($CaCO_3$). These sedimentary rocks are abundant and contain a considerable quantity of phosphate; their composition is a mixture of $CaCO_3$ and forms of apatite, such as a fluor apatite and francolite, $Ca_{10}(PO_4)_6 \cdot F_2 \cdot xCaCO_3$. Phosphate trapped in this way is not permanently withdrawn from the phosphate cycle. If these marine sediments are re-exposed to the atmosphere by uplifting (orogenic activity, e.g. the Alps, see Chapter 2), weathering will start again, first allowing the $CaCO_3$ to dissolve, while later, probably due to mechanical weathering, the apatite will also dissolve and renew its journey towards the sea. This mechanical weathering can result in the deposition of phosphate pebbles in riverbeds. Mansfield (1942) calculated that $3 \cdot 10^9$ tonnes of river and land pebbles occur in the Florida streams, some consisting of up to 80% apatite. Phosphate adsorbed onto clay will often be deposited in places where rivers join the sea, making delta areas fertile.

The second most important source of phosphate is human excreta and detergents. It is estimated that 2 g of PO_4-P per person per day is excreted, partly in urine, partly in faeces. For the well-developed areas of the world it is necessary to add to this quantity another 2 g of tri-phosphate phosphorus (see Figs. 17.10 and 17.13), derived from textile-washing detergents, which will hydrolyse to orthophosphate. This mean total of about 4 g of PO_4-P per person per day is well established in several studies of lake nutrient budgets. Urbanisation and use of freshwater for receiving sewage outfalls are the main factors tending to increase the total freshwater phosphate concentration, sometimes at an alarming rate. Thomas (1968) estimated the winter phosphate content in the Lake of Zürich as follows:

Years	0—10 m	10—20 m	20—136 m	Total
1941—50	15	15	300	330
1955—59	80	60	360	500
1960—64	95	75	510	680

The consequences of this increase will be discussed in Chapter 17.

The third and last main source is agriculture and other land uses. This source includes the leaching and drainage of fertilisers and other soil nutrients and the removal of soil particles. Estimates of the phosphate content of run-off are difficult to make and few data are available. Any figures which are given should be used only as an approximate guide. Gächter found that the phosphate outwash from soils increases as the proportion of the catchment area used for agriculture increases. In the Swiss lowlands 30—40 kg km^{-2} of PO_4-P and in the lower Alps 70 kg km^{-2} can be expected per year if the catchment area is used entirely for agriculture. Phosphate loss from tree-covered areas is negligible.

Kolenbrander (1972a, b) has given a mean value of 50 kg km^{-2} per year for loss from most soils with higher values from clay and peat (200—400 kg km^{-2}). Much lower quantities (5 kg km^{-2}) are lost from forest land and from unfertilised prairie.

5.3. THE PHOSPHATE CYCLE IN LAKEWATER

5.3.1. Introduction

Inorganic phosphate is used by growing algae which are extremely efficient in removing phosphate from solution. Following the death of the algae most of the phosphate is released back into the water. For the sake of convenience a distinction can be made between an "internal" or *"metabolic"* phosphate cycle:

$$(PO_4\text{-}P)_{water} \xrightarrow{\text{primary production}} cell\text{-}PO_4 \xrightarrow{\text{mineralisation}} (PO_4\text{-}P)_{water} + Org\text{-}P_{water}$$

and an "external" phosphate cycle:

$$(PO_4\text{-}P)_{water} \longrightarrow sediments \longrightarrow (PO_4\text{-}P)_{water} + Org\text{-}P_{water}$$

The first cycle summarises biological aspects, while the second is a geochemical cycle (Fig. 5.1).

Processes in the first (biological) cycle are usually of short duration (up to a few days), though animals may use a small fraction of the phosphate for longer periods. The biochemical processes are now fairly well understood. Processes in the second cycle may be very slow, especially the solution of sediments.

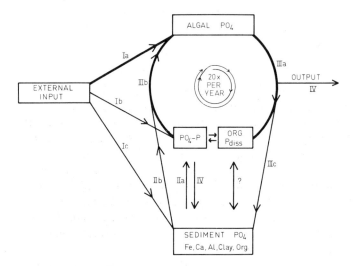

Fig. 5.1. Schematic representation of the phosphate cycle. The indicated turnover of 20 times per year may vary between 10 and 40.

5.3.2. The metabolic phosphate cycle

Thomas (1968) has demonstrated the quantitative importance of the removal of phosphate from the water by phytoplankton. In the Lake of Zürich he has found for 1960—1964 a mean annual decrease in spring from 90 to 10 tonnes of PO_4 (Fig. 5.2). A small part of this 80 tonnes is in the hypolimnion (see Fig. 17.5). Gächter (1968) has made a very detailed study of the phosphate kinetics in the Horwer Bucht in the Lake of Lucerne (Switzerland). His orthophosphate phosphorus isopleths varied from less than 1 $\mu g\ l^{-1}$ to 46 $\mu g\ l^{-1}$ in the anaerobic hypolimnion during the summer period. The phosphate concentration during the winter was homogeneous throughout the whole water column. The concentration of other soluble phosphates showed a remarkable homogeneity throughout the whole year. The particulate phosphate concentration, however, showed a few maxima at certain periods in the deeper layers, each being associated with a decrease of phosphate in the trophogenic layer, which indicates a sinking of particulate phosphates. Because the observed differences were small and particulate phosphate is measured as the difference between total phosphate minus soluble phosphate, the sinking rate cannot be reliably calculated from these figures, but one may conclude that sinking is probably not a quantitatively important process. There is considerable evidence indicating that mineralisation of the dead plankton cells probably takes place mainly in the epilimnion (see Kleerekoper, 1953; Ohle, 1965; Serruya, 1971) and that a high turnover rate of phosphate exists. Because direct demonstration of this phenomenon is extremely difficult, the daily

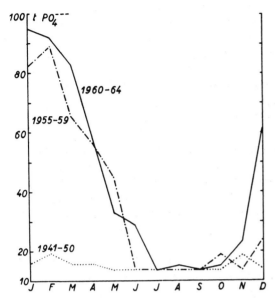

Fig. 5.2. Mean monthly concentration of PO_4-P in the Lake of Zürich (period 1960—1964).

phosphate uptake must be estimated by comparing the phosphate concentrations in the water with the rate of phosphate uptake associated with primary production. Such indirect calculations may be based upon measurements of primary production assuming a given C:P ratio (e.g. 50) or a phosphorus content of about 1% of the dry weight.

Using figures for primary production in the Lake of Zürich of 500 mg m^{-2} of C on 9 May 1957 as given by Rodhe (1965), a daily phosphorus uptake of about 10 mg m^{-2} may be calculated. With an actual concentration of about 7 mg m^{-3} of PO_4-P (Schürmann, 1964), a turnover time* for the internal phosphate cycle may be estimated as 7 to 14 days, thus indicating a rapid turnover of phosphate in the upper layer.

Gächter has made similar calculations from his data. Assuming a P:C ratio of 1:40 and assuming that the primary production during the observation period (10.00—14.00 h) was 50% of the total daily primary production, he estimated a maximum turnover time of 10 days for particulate phosphate during the summer period, increasing to 30 days during the winter. He has also calculated the turnover rates for PO_4-P and total dissolved phosphate, but these are maximal turnover times because he did not take the PO_4-P into account in the calculation for the total dissolved P and vice versa. It must be

---

* This "biological turnover time" should not be confused with Rigler's (1956) "physical turnover time", which is mainly determined by exchange processes between ^{31}P and [32]P.

pointed out that the P:C ratio may sometimes (but seldom) be as high as 1:10, which would decrease estimates of the turnover time by a factor of 4. These short turnover times add support to the increasing amount of evidence which suggests that most of the material formed by photosynthesis is mineralised in the epilimnion. It seems probable therefore that sedimentation of dead phytoplankton contributes only slightly to the sedimentation of phosphates at the bottom of the lake. The short turnover time can only be accounted for assuming a rapid liberation of phosphate from dying cells by autolysis (see Chapter 8).

Before dying cells become metabolically inactive, autolysis may cause a liberation of phosphates into the water. Golterman (1960) found that 50% of the particulate phosphate was returned into solution as PO_4-P a few hours after the onset of autolysis; this figure reached 70—80% after a few days.

At low temperatures (4°C) a considerable quantity was released as organic dissolved phosphate, which at higher temperatures hydrolyses into inorganic phosphate.

Very little is known about the biological availability of certain dissolved organic phosphate compounds which are released into the water during autolysis. Some of them, e.g. glycerophosphate, are rapidly available for algal growth (Miller and Fogg, 1957). Substances which are not available for algal metabolism may be used by bacteria since they are potential energy sources. It is only when the phosphates are not susceptible even to bacterial decomposition, such as humic-iron-phosphate complexes, that this portion of the biological phosphate may be withdrawn from the internal phosphate cycle.

Dissolved organic phosphate will be released by algae mainly during autolysis, but Johannes (1964) showed that animals such as amphipods also release organic phosphate. He also demonstrated that the organic phosphate released is rapidly hydrolysed by bacteria and by extracellular alkaline phosphatases or is partly taken up by marine diatoms. Overbeck (1968) also showed a phosphatase to be present in lake waters and suggested a bacterial origin for this enzyme. However, we found that bacteria-free cultures of *Scenedesmus obliquus* produce alkaline phosphatases* as soon as P is depleted and the logarithmic growth-rate stage is no longer maintained. In the earlier stages of growth an acid phosphatase was produced. Lien and Knutsen (1973) showed that in cultures of *Chlamydomonas* phosphatase formation is repressed if phosphate concentrations are high. They also showed a "de novo" synthesis of a different form of acid phosphatase when phosphate concentrations were reduced. The acid phosphatase was located near the cell surface. Both forms of repressed and derepressed phosphatase exhibited different enzymatic properties such as pH optimum and some kinetic constants, the derepressed enzyme showing an increased substrate affinity, which may more easily

---

* Alkaline phosphatases are enzymes that hydrolyse organic phosphate compounds; their maximal activity is reached above pH = 9. The acid phosphatases have their optimum around pH = 4.

hydrolyse external organic phosphate esters.

Bone (1971) found that *Anabaena flos aquae* produced an alkaline phosphatase when grown in a chemostat with phosphate as the growth-limiting factor. Alkaline phosphatase activity varied twentyfold, the lowest activity being found in experiments where excess phosphate was present. Berman (1970) demonstrated the occurrence of a seasonal fluctuation of alkaline phosphatase activity in Lake Kinneret (Israel) with high values recorded from the upper and central parts of the thermocline. The activities were low after the onset of homothermy at all depths. Berman attributed the phosphatases in the epilimnion to algal activity, while those in the thermocline and hypolimnion were attributed to bacterial production. He found that soluble phosphatases varied between 0 and 57% (mean 16%) of the total phosphatase activity. The biological function of these phosphatases might well be related to the fact that a large part of phosphate released during autolysis is organic.

5.3.3. The geochemical phosphate cycle

Autolytic processes limit the amount of sedimentation of particulate phosphate. Even after autolysis the dead algal cells contained sufficient phosphate for bacteria to break them down. The phosphates taken up by bacteria or released to the water by bacterial action also remain in the upper water layer and do not contribute to the sediment. In accordance with the suggestion that the short turnover time prevents much sedimentation from occurring, we could not find any significant difference between the summer and winter values for particulate phosphate using observations made over five years (1966 —1970) in the artificial sandpit Vechten, which stratifies during summer (see Annual Report Limnological Institute, 1970). Values of the phosphate concentrations in 1966 are given in Table 5.1.

These values seem to indicate that there is no great loss of particulate phosphate from the epilimnion. In this lake the primary production does not show a summer stagnation. Both facts indicate a short turnover time for the phosphate in the "internal" phosphate cycle. The turnover time of PO_4-P, estimated from the primary production figures and phosphate concentrations in the epilimnion of this sandpit is 7 to 14 days. This is within the same order of magnitude as was quoted earlier for the Lake of Zürich. In the anaerobic hypolimnion of Vechten maximal values of between 500 and 1 000 mg m^{-3} of PO_4-P were found near the bottom during the same period. As this sandpit has no water outlet or inlet and since sedimentation hardly occurs, it seems likely that its phosphate metabolism is autochthonous and that the negligible loss due to sedimentation is compensated by diffusion from the hypolimnion. It is probable that the phosphate in the hypolimnion is regulated mainly by physicochemical processes in the "external" cycle. The increase of Part-P in the hypolimnion in October may be due to precipitation of $FePO_4$, since the thermocline sinks at about this time and $Fe(OH)_3$ is precipitated (see

TABLE 5.1

Concentrations of phosphate phosphorus and organic-phosphate phosphorus in the epilimnion, and of particulate phosphorus in epi- and hypolimnion in the sandpit Vechten in 1966

Date	PO_4-P ($\mu g\ l^{-1}$)	Org-P_{diss} ($\mu g\ l^{-1}$)	Part-P	
			Epilimnion ($\mu g\ l^{-1}$)	Hypolimnion (ca. 50 cm above mud)
17- 1	12	8	7	8
14- 2	5	12	5	7
28- 2	5	2	5	6
14- 3	8	6	4	5
28- 3	14	11	6	7
12- 4	5	5	6	5
9- 5	11	0	6	11
6- 6	64	—	6	34
4- 7	1	22	3	159
15- 8	8	23	5	59
12- 9	6	7	4	79
24-10	9	0	5	110
21-11	58	0	9	7
19-12	44	—	5	7

Chapter 9). There is no obvious explanation for the high value on 4 July.

Little is known about the kinds of organic compounds which remain undecomposed, but it seems reasonable to suppose that the final products will be humic-iron-phosphate compounds or other products which are biochemically inert. Humic phosphates are always found in the NaOH extracts of sediments but very little is known about the activity of these compounds.

In addition to sedimentation of phosphates in detrital material, phosphate may also precipitate as iron or calcium phosphate if the amounts present exceed the solubility product of these compounds (see subsection 9.4.4). Apparently an equilibrium exists between the precipitate formed and the dissolved ions, and this is important when considering the influence of sediments on the chemical composition of hypolimnia (a subject discussed in Chapters 9, 17 and 18).

A final source of phosphate in lakes is that adsorbed on clay which may be either suspended or precipitated into the sediment. The clays may release phosphate, especially if the soluble PO_4-P concentration is low due to the growth of algae (Fig. 5.1: reaction IIb). Apparently a competitive relationship exists between the inorganic precipitates (Ic + IV) and algal uptake (Ia). It is not yet possible to know how much of the allochthonous phosphate which arrives in the lake is taken up by algae and how much is precipitated into the sediments, since no quantitative data exist. The sedimentary part of the phosphate is not permanently withdrawn from the cycle since it may dissolve again in certain circumstances (e.g. anaerobic hypolimnion) or it may be extracted from the sediments of shallow lakes by algae, bacteria or other organisms (IIa). Probably no phosphate will escape from sediments in very deep lakes or in stratifying lakes with an oxygenated hypolimnion (IIa + IIb = 0). In Fig. 5.1 a summary of the reactions of phosphate is given and some of these will be discussed in Chapters 9 and 18.

The kinetics of algal growth are discussed in Chapter 10. From the relationship between growth rate and phosphate concentration it can be seen that the naturally occurring phosphate concentrations limit not only the population density, but also the growth rate of most algal populations. It should be noted that with increasing phosphate concentrations and subsequent population growth, light may then become limiting due to self-shading. Even if this occurs however, the rate of phosphate uptake may not be sufficiently great to enable the algae to reach maximal division rates. This may mean that diffusion or other rate-limiting processes determining P uptake are involved and only a rather complicated model can explain the growth kinetics. The occurrence of phosphate storage in algae makes the whole situation more complex. Most algae are able to store phosphate in their cells in the form of microscopically visible poly-phosphate bodies. Under conditions of phosphate starvation these poly-phosphate reserves may be mobilised rapidly and be used to form the normal cell constituents.

In the case of shallow lakes it becomes important to know whether or not the phosphates in the sediments are available for algal metabolism (IIb). Armstrong and Harvey (1951) noted that a culture of PO_4-P-depleted marine diatoms grew in water containing $FePO_4$, but this growth probably followed the hydrolysis of the $FePO_4$ in the seawater, since they demonstrated a concurrent fall in pH from 8 to 7, at which level the growth stopped. Golterman et al. (1969) showed that $FePO_4$ could be utilised completely by *Scenedesmus* as a sole source of phosphate in a medium containing sufficient $NaHCO_3$ to maintain the pH at about 8. The growth rate was less than that which could occur with an equivalent amount of KH_2PO_4 (1 000 $\mu g \, l^{-1}$ of PO_4-P) and was similar to that which could occur with a lower concentration of KH_2PO_4 (100 $\mu g \, l^{-1}$), a value which was therefore called the "apparent" PO_4-P concentration. Furthermore, they showed that cultures provided with hydroxy-apatite produced only one-third of the growth yield which occurred with $FePO_4$, while the growth rate was equivalent to that produced by only 35 $\mu g \, l^{-1}$ of PO_4-P of KH_2PO_4. It has been shown recently by us that small crystals of hydroxy-apatite are more rapidly available to algae than are large ones; availability shows a definite correlation with crystal size.

Mud is also a source of phosphates. Muds from several different origins when added as the only source of phosphates to a Rodhe culture solution gave excellent algal growth. Of the muds which were used, the only one which was unable to support algal growth was a pure "unpolluted" clay. This clay contained considerable quantities of phosphate, but these were unavailable to the algae probably because the phosphate molecules were integrated into the clay structure. This is probably the reason why Great Slave Lake and Lake Kinneret, which receive an enormous input of PO_4-P, remain oligotrophic (see subsection 17.4.5). If clay is loaded with freshly adsorbed phosphate, most of this phosphate fraction appears to be available to algae.

Another source of unavailable phosphate is peat (Golterman, 1973a). During the winter, Tjeukemeer receives up to 1 mg l^{-1} of PO_4-P, some as soluble PO_4-P and some as phosphate adsorbed on particles. Nevertheless the algal blooms remain below 100 mg m^{-3} of chlorophyll, partly because the phosphorus adsorbed on the humus particles is not available and partly because light penetration is restricted so strongly by the brown colour and turbidity of the lakewater that light becomes limiting at any depth below 10—20 cm. It seems likely that phosphate which enters lakes from natural sources is much less available for algal growth than that coming into the water as a result of human activities (Golterman, 1973a,b).

5.4. CONCLUDING REMARKS

The role of phosphate in eutrophication is discussed in Chapter 17. One remark must be made here, however. Studying the intensive eutrophication in several lakes, it is striking that the cellular P content of these lakes, which is often about 0.1 g m^{-2}, increases only slowly even if the allochthonous input is 2—4 g m^{-2} yr^{-1}. Originally it had been suggested that possibly the incoming phosphate was sorbed directly onto the bottom material (shallow lakes) or precipitated with iron or calcium ions from the water. But looking at Fig. 5.1 with a dynamic point of view, we notice that the heavy artificial input matches the "natural input" due to internal recycling. It is striking that the natural losses may be significant, because they may occur 20 or even 40 times a year, while the loss per algal cycle may be as small as 1—5%. The formation of insoluble sedimentary P compounds does not take place directly by chemical reactions but through algal photosynthesis. Only during winter may direct chemical precipitation be important. As algal photosynthesis is the product of the total ecosystem, the phosphate uptake and sedimentation is a function of all biological and (bio)chemical processes in this ecosystem. Predictions from one lake and application to another one are, therefore, rather dangerous.

The exchange mechanism between dissolved and sediment phosphate will be further discussed in the subsections 9.4.4 and 9.5 and in Chapter 18. The influence of phosphate on the growth of algal cells in cultures is discussed in Chapter 10 and that on the growth of natural populations in Chapter 14.

REFERENCES

Annual Report of the Limnological Institute, 1970. *Verh. K. Ned. Akad. Wet., Afd. Natuur-kunde*, 2e reeks, 60(2): 74—92.

Armstrong, F.A.J. and Harvey, H.W., 1951. The cycle of phosphorus in the waters of the English Channel. *J. Mar. Biol. Assoc.*, N.S., 29: 145—162.

Berman, T., 1970. Alkaline phosphatases and phosphorus availability in Lake Kinneret. *Limnol. Oceanogr.*, 15(5): 663—674.

Bone, D.H., 1971. Relationship between phosphates and alkaline phosphatase of *Anabaena flos aquae* in continuous culture. *Arch. Mikrobiol.*, 80: 147—153.

Gächter, R., 1968. Phosphorhaushalt und planktische Primärproduktion im Vierwaldstätter-see (Horwer Bucht). *Schweiz. Z. Hydrol.*, 30(1): 1—66.

Goldschmidt, V.M., 1937. The principles of distribution of chemical elements in minerals and rocks. *J. Chem. Soc., Lond.*, 1937: 655—673.

Golterman, H.L., 1960. Studies on the cycle of elements in fresh water. *Acta Bot. Neerl.*, 9: 1—58.

Golterman, H.L., 1968. Investigations on photosynthesis and mineralization in experimental ponds. *Mitt. Int. Ver. Theor. Angew. Limnol.*, 14: 25—30.

Golterman, H.L., 1973a. Natural phosphate sources in relation to phosphate budgets: a contribution to the understanding of eutrophication. *Water Res.*, 7: 3—17.

Golterman, H.L., 1973b. Vertical movement of phosphate in freshwater. In: E.J. Griffith, A. Beeton, J.M. Spencer and D.T. Mitchell (Editors), *Environmental Phosphorus Handbook*. John Wiley, London, pp. 509—538.

Golterman, H.L., 1975. Chemistry of running waters. In: B. Whitton (Editor), *River Ecology*. Blackwell, Oxford (in press).

Golterman, H.L., Bakels, C.C. and Jakobs-Möglin, J., 1969. Availability of mud phosphates for the growth of algae. *Verh. Int. Ver. Theor. Angew. Limnol.*, 17: 467—479.

Johannes, R.E., 1964. Uptake and release of phosphorus by a benthic marine amphipod. *Limnol. Oceanogr.*, 9: 235—242.

Kleerekoper, H., 1953. The mineralization of plankton. *J. Fish. Res. Board Can.*, 10(5): 283—291.

Kolenbrander, G.J., 1972a. Does leaching of fertilizers affect the quality of ground water at the waterworks? *Stikstof*, no. 15: 8—15.

Kolenbrander, G.J., 1972b. The eutrophication of surface water by agriculture and the urban population. *Stikstof*, no. 15: 56—67.

Lien, T. and Knutsen, G., 1973. Synchronous cultures of *Chlamydomonas reinhardti*: Properties and regulation of repressible phosphatases. *Physiol. Plant.*, 28: 291—298.

Mansfield, G.R., 1942. Phosphate resources of Florida. *Bull. U.S. Geol. Surv.*, no. 934: 82 pp.

Miller, J.D.A. and Fogg, G.E., 1957. Studies on the growth of Xanthophyceae in pure culture. I. The mineral nutrition of *Monodus subterraneus* Petersen. *Arch. Mikrobiol.*, 28: 1—17.

Ohle, W., 1965. Die Eutrophierung der Seen und die radikale Umstellung ihres Stoffhaushaltes. In: *Limnologisymposium Finland*, 1964, pp. 10—24.

Overbeck, J., 1968. Prinzipielles zum Vorkommen der Bakterien im See. *Mitt. Int. Ver. Theor. Angew. Limnol.*, 14: 134—144.

Rigler, F.H., 1956. A tracer study of the phosphorus cycle in lake water. *Ecology*, 37(3): 550—562.

Rodhe, W., 1965. Standard correlations between pelagic photosynthesis and light. *Mem. Ist. Ital. Idrobiol.*, 18(Suppl.): 365—381.

98

Schürmann, J., 1964. Untersuchungen über organische Stoffe im Wasser des Zürichsees. *Vierteljahresschr. Naturforsch. Ges. Zürich*, 109: 409—460.

Serruya, C., 1971. Lake Kinneret. *Limnol. Oceanogr.*, 16(3): 510—521.

Sioli, H., 1968. Principal biotopes of primary production in the waters of Amazonia. *Proc. Symp. Rec. Adv. Trop. Ecol., 1968*, pp. 591—600.

Talling, J.F. and Talling, I.B., 1965. The chemical composition of African lake waters. *Int. Rev. Ges. Hydrobiol. Hydrogr.*, 50(3): 421—463.

Thomas, E.A., 1968. Die Phosphattrophierung des Zürichsees und anderer Schweizer Seen. *Mitt. Int. Ver. Theor. Angew. Limnol.*, 14: 231—242.

Viner, A.B., 1970. The chemistry of the water of Lake George, Uganda. *Verh. Int. Ver. Theor. Angew. Limnol.*, 17: 289—296.

CHAPTER 6

NITROGEN CYCLE AND BLUE-GREEN ALGAE

6.1. INTRODUCTION

Unlike the other minor elements or compounds, nitrogen occurs in molecules at several different oxidation states in lakewater. The following forms are quantitatively of importance:

(1) N_2, dissolved in water in equilibrium with the nitrogen in the atmosphere (about 15 mg per litre at $20°C$).

(2) NH_3, in equilibrium with NH_4^+ and NH_4OH, which is produced from organically bound nitrogen, either from allochthonous or from autochthonous sources.

(3) NO_2^-; small quantities ($0.1-10$ mg m^{-3}) of nitrite are a normal constituent of most lakewaters. Larger quantities occur under certain conditions. They are formed during the oxidation of ammonia, but will normally be oxidised to NO_3^-, except in anaerobic conditions.

(4) NO_3^-, the only stable form of a nitrogen compound, unless taken up by algae and bacteria.

(5) Organic N, either in dissolved state or as particulate nitrogen. Stages in the oxidation—reduction series are (reductions as shown):

Oxidation stage of N: $NO_3^- \xrightarrow{2e} NO_2^- \xrightarrow{3e} N \xrightarrow{3e} NH_3 \longrightarrow$ organic N

Electric charge N atom: +5   +3   0   −3   −3

The concentrations of all nitrogen compounds vary considerably in different waters, the range being between hardly detectable traces and several mg l^{-1}. The occurrence of hydroxylamine is very seldom mentioned, but Baxter et al. (1973) reported its presence in small quantities in an Ethiopian lake and gave some circumstantial evidence suggesting that it has a short turnover time. Frinck (1967) attempted to account for the amounts of nitrogen in various compartments in a small eutrophic lake, Bantam Lake (Connecticut, U.S.A.). He found 51% in the sediment top layer, 3% in aquatic weeds, 33% in algae, and 13% of total N in the lakewater.

Several species of bacteria which can convert one form of nitrogen compound into another are known. The oxidation steps produce energy for the bacteria; the bacteriological aspects of some of these processes will be discussed in more detail in Chapter 16.

Nitrogen enters water from very different sources. The nitrogen content of igneous rocks is only 46 g tonne^{-1} (46 p.p.m.), although that of certain sedimentary rocks may be ten times higher. Larger quantities come from erosion of both natural and artificially fertilised soils (some figures for this

100

outwash are given in Chapter 17). This source leads sometimes to the occurrence of constantly high concentrations in some rivers, values of up to 10 mg l^{-1} being not uncommon (see Golterman, 1975). Other sources are rainfall and dust precipitation, and sewage. The subject has been reviewed by Feth (1966).

6.2. INTERCONVERSIONS OF THE DIFFERENT FORMS OF NITROGEN

6.2.1. Oxidation of ammonia

As early as 1890 Winogradsky isolated two species of bacteria which oxidise NH_3 in a two-stage process:

$$NH_3 \xrightarrow{\textit{Nitrosomonas sp.}} NO_2^- \xrightarrow{\textit{Nitrobacter sp.}} NO_3^-$$

Each of these stages yields energy which is used by the bacteria for reduction of CO_2. If ammonium salts are added to lakewater, several days usually elapse before the bacterial populations are sufficiently large to produce a measureable oxidation rate. At low temperatures and in the presence of organic compounds this rate is often low as can be seen from the changes of the concentrations of NH_3 and NO_3^- in Dutch polder lakes (Fig. 6.1).

During the autumn and early part of the winter, water entering those lakes has ammonia concentrations of up to 2 mg l^{-1} of NH_3-N. The concentration remains high until the temperature begins to rise. Polder water contains much organic matter which may partly inhibit oxidation of the NH_3; nevertheless, if the lakewater temperature is raised to $10°C$ either in the field or in the laboratory, the oxidation begins after a lag of a few days, indicating that the low temperature in the lake is partly responsible for the low rate of ammonia oxidation. It is possible that the "lag phase" before oxidation of NH_4^+ begins may be due to the preliminary oxidation of certain inhibitory substances by other species of bacteria.

It can be seen from Fig. 6.1. that not all the added NH_3 is recovered as NO_3^-.

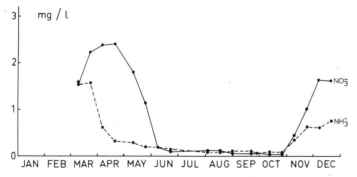

Fig. 6.1. Concentration of NH_3-N and NO_3^--N in Tjeukemeer. (From Golterman, 1973a).

Because only a part of the difference will have been converted into algal and bacterial cell material a deficit is found. Similar losses are also described from shallow artificially managed fish ponds and from deeper lakes. At least under anaerobic circumstances, losses may be caused by bacterial denitrification, when nitrogen is lost as the gas N_2 (see subsection 6.2.3.).

Gode and Overbeck (1972) examined ammonia oxidation by autotrophic and heterotrophic bacteria. The latter, which cannot use the oxidation energy for their growth and which need an organic carbon supply, seemed to occur in larger quantities than the autotrophic ones. The details of their activity and the biochemical pathways are however still obscure.

6.2.2. Synthesis of cellular nitrogen

Most algae have a relatively high protein content, often exceeding 50% of their dry weight. Their nitrogen requirements are therefore quite considerable and are normally supplied by NO_3^- and NH_3, both of which forms are of common occurrence. Assuming a mean algal composition $C_5H_7NO_2$, the following reactions describe the conversion of soluble inorganic compounds into cellular matter:

$$5\ CO_2 + 2\ H_2O + NH_3 \longrightarrow C_5H_7NO_2 + 5\ O_2$$

$$5\ CO_2 + 3\ H_2O + HNO_3 \longrightarrow C_5H_7NO_2 + 7\ O_2$$

The formula $C_5H_7NO_2$ does not mean that there are sufficient data to suggest that the mean "global" algal composition can be calculated. It merely describes the empirical observation that the C:N ratio is often 5—6 and that the protein content is about 50%. The formula implies the total nitrogen content to be 13% of the dry weight. Allowing 3% for non-amino-acid nitrogen, the remaining 10% or so must represent amino-acid nitrogen, and with the commonly used conversion factor 6.25 this 10% means roughly 60% protein for algal cells. If the algae contain more fat than normal — some diatoms in a state of starvation and in synchronised algae (see section 10.6) — the relative hydrogen content may be increased. The formula is useful insofar as it enables rough calculations to be made concerning nutrient budgets, turnover times of different algal nutrients and the assimilation quotient.

The assimilation quotient reflects the oxidation state of the nitrogen source (A.Q. = O_2/CO_2). Thus it is approximately 1.0 during NH_3 and 1.4 during NO_3^- uptake.

If algae are given both NH_3 and NO_3^-, they appear to prefer NH_3 as nitrogen source (Fogg, 1966b; Golterman, 1967b). When an algal culture is given NH_4NO_3, the pH falls at first, as a result of the NH_4^+ uptake, but it returns later to its original value due to NO_3^- uptake. Cultures growing on KNO_3 become alkaline.

The preference for NH_3 before NO_3^- as nitrogen source can be attributed to the inhibition of the enzyme nitrate reductase by NH_3 (Syrett and Morris, 1963; Herrera et al., 1972; and Thacker and Syrett, 1972a, b). Thacker and Syrett demonstrated that *Chlamydomonas reinhardtii*, like *Chlorella* in earlier studies, did not assimilate nitrate or ammonia unless a suitable source of carbon was provided, either by photosynthesis or in darkness by acetate or accumulated carbon reserves in nitrogen-starved cells. Ammonium inhibition seems to be caused more by products of ammonium assimilation or depletion of carbon sources than by the ammonium itself. After depletion of the NH_3 had occurred, the nitrate-reductase activity was restored again after a lag phase of about 2 h. Herrera et al. (1972) showed that ammonium inactivates the nitrate reductase primarily by raising the cellular level of reducing power (—NADH), which in turn determines the rate of reversible reduction of the enzyme. Inactivation by NADH in cell-free extracts could be prevented by NO_3^-.

Besides organic nitrogen in particulate form, there are always dissolved organic compounds present in lakewater. These will often be in a refractory form with rather low turnover rates. The concentration of these dissolved compounds may often be as high as 0.5 mg l^{-1} of nitrogen. Manny (1972) followed the seasonal changes of dissolved organic nitrogen in six Michigan lakes, but could not discern a distinct seasonal pattern. In Lawrence Lake fluctuations occurred between 0.1 and 0.25 mg l^{-1} and no correlation with the dissolved organic carbon could be detected. Manny discussed several possible interactions of dissolved organic nitrogen with trace elements (chelation) and with clays and other suspended particles but gave no experimental evidence to support his suggestions. Some of the dissolved organic nitrogen may be present as soluble fulvic acids (see Chapter 13).

6.2.3. Denitrification

Some microorganisms, especially of the genera *Pseudomonas, Bacillus* and *Micrococcus* which can grow aerobically, can also grow anaerobically when they use substances other than O_2 as the final electron acceptor.

Some *Bacillus* species metabolise anaerobically, converting nitrate to ammonia, although their presence is not yet reported for many lakes or other surface waters. Baier (1936), however, demonstrated reduction of NO_3^- to NH_3 in various ponds in Kiel in some experiments, whereas in other experiments N_2 and NO_2^- were produced. Baier reported the presence of bacteria comparable to *Thiobacillus denitrificans*, which would suggest the occurrence of the reaction:

$$5\,S + 6\,NO_3^- + 2\,H_2O \rightarrow 5\,SO_4^{2-} + 3\,N_2 + 4\,H^+$$

Strains of *Pseudomonas* spp. which can convert NO_3^- to N_2 are well-known; they are used in sewage-water treatment for removing nitrogen. The same process occurs in anaerobic hypolimnia even before oxygen is completely removed.

Chen et al. (1972) showed in a field investigation that when NO_3-N added to Lake Mendota sediments was converted to organic and NH_4-N fractions, only 37% was recovered after four days while the remaining 63% was thought to be lost through denitrification. When NO_3^- was added to lake sediments in vitro, up to 90% of the added NO_3-N disappeared from the sample within 2 h. These processes must therefore be taken into account in nutrient budgets, especially in seepage lakes receiving nitrate in groundwaters. Brezonik and Lee (1968) reported a denitrification loss of 11% of the annual nitrogen input in Lake Mendota.

For a long time denitrification was considered not to occur in well-oxygenated shallow lakes since the process is anaerobic. Jannasch and Pritchard (1972), however, found that if these *Pseudomonas* bacteria occur in water containing much suspended material, then denitrification may take place, probably because a micro-anaerobic zone may develop around the particles. Whether or not this denitrification is the cause of the nitrogen losses described in subsection 6.2.1 remains to be discovered.

6.3. MINERALISATION OF CELLULAR NITROGEN

The first extensive studies concerning the mineralisation of cellular nitrogen were carried out by Waksmann et al. (1937) with marine algae, and by Von Brand et al. (1937—1942) with freshwater algae. Waksmann killed diatoms by suspending them in seawater in the dark and concluded from the resultant bacterial growth that nitrogen was liberated from the diatoms. Von Brand, using the same technique, found a 100% liberation of NH_3, which subsequently oxidised to NO_2^- and then to NO_3^- (Fig. 6.2).

Von Brand described his results as showing a quantitative conversion, but in most of his experiments an important increase in total inorganic nitrogen was found compared with the amounts of organic cellular nitrogen originally present. Von Brand assumed that this increase was due to oxidation of dissolved

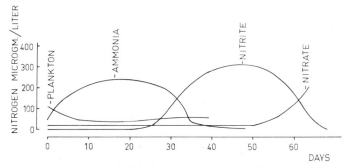

Fig. 6.2. The decomposition of nitrogenous organic matter in mixed plankton, showing the appearance of soluble nitrogen compounds in the water in which it is suspended (see Von Brand et al., 1937—1942).

organic-nitrogen compounds, but this seems unlikely as in his later experiments the dissolved organic fraction increased rather than decreased. It is perhaps more likely that the amount of particulate nitrogen at zero time was underestimated. His experiments are nevertheless important because they provided the first real evidence for the occurrence of rapid bacterial mineralisation.

The time scale which Von Brand used in static laboratory experiments should not however be extended to apply to a natural situation. The rate of nitrite formation in his experiments was quite slow. This may be due to the long period required to build up a sufficiently large population of the relevant bacteria but this may also be partly due to inhibition by organic matter as described in subsection 6.2.1. In natural conditions NH_3 is released every day, so the *Nitrosomonas* population is presumably constantly active and the non-cellular organic-matter concentration is much lower than in cultures where dead algae are introduced suddenly. Possible survival of these bacteria on organic compounds is discussed in subsection 16.2.6. Continuous culture could simulate the natural situation more closely.

Golterman (1960, 1964, 1968, 1973b) could never detect a 100% conversion of particulate nitrogen to ammonia but found values between 50 and 75% after 10—15 days. After this time only traces of nitrite were found and about the same amount of nitrate as ammonia (partly unpublished results). He suggested (see also Chapter 8) that the incomplete conversion to ammonia could be due to the fact that the bacteria use one part of the algal protein to build up their own proteins, while a second part is used for the energy demands of the bacteria, the nitrogen of this latter part alone being released as ammonia. Furthermore the bacterial nitrogen cannot in practice be separated from the residual algal nitrogen (not all algal nitrogen is digestable by bacteria) so that Von Brand's values, indicating a complete disappearance of particulate nitrogen, seem unlikely.

That part of the particulate nitrogen which cannot be hydrolysed or used by bacteria will eventually sediment. Very little is known about the processes which occur in the sediments. In some cases N_2 formation has been demonstrated (it is often present in natural gas), but probably the main product formed is ammonia. This might be liberated directly (as has been shown for hypolimnetic waters) or adsorbed physically and released later (Chapter 9). Part of the refractory nitrogen occurs in humic compounds (Chapter 12).

6.4. NITROGEN FIXATION AND BLUE-GREEN ALGAE

6.4.1. Occurrence of N_2 fixation

Nitrogen fixation is the process by which organisms reduce N_2 gas to the level of ammonia. A review of the earliest demonstrations of the occurrence of N_2 fixation by free-living organisms in water is given by Fogg and Horne (1967), from whom the following is taken. Mortimer (1939) compared the annual in- and output of nitrogen in Windermere which he found to be 326 and 318 tonnes, respectively. A similar budget for Lake Mendota gave an input of 156 tonnes and an output of 40 tonnes per year (Rohlich and Lea, 1949). The loss of nitrogen is probably caused by sedimentation and subsequent denitrification. Goering and Neess (1964) demonstrated the occurrence of N_2 fixation in Lake Mendota. Hutchinson (1941), using bottles of Linsley Pond water, found an increase of 0.5 g m^{-3} of N in 10 days at a time when there was a dense population of the blue-green alga *Anabaena circinalis*.

Real progress was made when Neess et al. (1962) developed a method in which $^{15}N_2$ was used, not unlike the ^{14}C method for measuring primary production. The difference is that ^{15}N is not radioactive and so an expensive mass-spectrometer is essential. The water sample to be analysed is placed in a special 1-l bottle leaving 40 ml of air above the water. The N_2 in the sample is then removed by flushing with a mixture of oxygen (25%), argon (74.96%) and carbon dioxide (0.04%). The bottle is then replaced in the water to be studied, quite often for a period of 24 h. $^{15}N/^{14}N$ (atom excess) ratios are then determined from which the amount of N_2 fixed can be calculated if the amount of particulate nitrogen is known. Stewart (1966) and Neess et al. (1962) found fixation in seawater under normal conditions in the range of 0.05—0.15 mg m^{-3} of N_2 per hour, with maximal values of 2, these being an order of magnitude greater than those for ammonia uptake. Using the same technique Horne and Fogg (1970) found total fixation for Windermere to be 10 tonnes per year. Fogg pointed out that although this quantity is probably not important as a significant factor of the N budget for the whole lake, it may be important for some water layers at particular times. The fixation appeared to be dependent on both light and the presence of blue-green algae.

In 1967, Stewart et al. studied blue-green algal activity both in lakewater and in cultures, using a technique based on the fact that acetylene can be reduced by the algal N-fixing enzyme complex; the ethylene formed is then measured by gas chromatography. It has been demonstrated that the reaction:

$$CH{\equiv}CH + H_2 \rightarrow CH_2 = CH_2$$

acetylene   ethylene

can be used to obtain a quantitative measurement of the N_2-fixing process since the N-fixing reaction in cells is:

$$N \equiv N + 3 H_2 \rightarrow 2 NH_3$$

although the ratio of N_2 fixation to C_2H_2 reduction varies somewhat with organism, history and environment.

Similarities between the two reactions are shown by the following facts:

(1) Reduction of acetylene like that of N_2 needs ATP and a reducing agent.

(2) Purification of the N-fixing enzyme complex enhances the acetylene-reducing capacity.

(3) N_2 fixation and acetylene reduction are both inhibited or inactivated at the same rate by the same agents.

(4) Acetylene (C_2H_2) inhibits N_2 reduction and probably acts as a competitive H acceptor.

Its sensitivity and cheapness make this method more suitable than the ^{15}N method for a wide range of investigations and especially for cultures and cell-free extracts. Comparison between older biochemical work using anaerobic bacteria and experiments with blue-green algae is now possible. It is well known that the facultative anaerobic bacteria *Azotobacter* and *Clostridium* can fix N_2 but with a low efficiency. Stewart (1969) found a fixation of 10—20 mg of N_2 for *Azotobacter* and 2.27 mg of N_2 for *Clostridium* both per g carbohydrate oxidised. For lakes this quantity would be of no importance.

The fixation efficiency increases with decreasing O_2 concentration reaching a maximum at about 0.001 atm. Apparently there is competition between O_2 and N_2 as final hydrogen acceptor.

As early as 1937 N_2 fixation was demonstrated in Myxophyceae (Cyanophyceae), commonly called blue-green algae, especially in the filamentous heterocyst-forming orders Stigonematinales and Nostocales. The following genera of the order Nostocales are of importance in aquatic habitats, especially in eutrophic waters:

Anabaena (*flos aquae* and *spiroides*)   *Oscillatoria* (e.g. *agardhii*)
Nostoc   *Anacystis* (or *Aphanocapsa*)
Aphanizomenon

The blue-green algae have a nitrogenase enzyme very similar to that occurring in the N-fixing bacteria, with iron and molybdenum in the prosthetic group. Nitrogenase activity has been shown to be positively correlated with the presence of heterocysts. Fixation is in the order of 0.4—10 μg of N_2 per mg of particulate nitrogen per day.

For Tjeukemeer, Horne found a fixation of 5 μg per mg of N per day, in the autumn, a considerable contribution to the N balance. For Lake George a value of 33% of the total nitrogen input is given (Viner and Horne, 1971). It seems likely that in eutrophic lakes N_2 fixation makes an important contribution especially in layers where the inorganic nitrogen compounds are exhausted, but sufficient data to make a generalisation are not yet available. If we assume a 10% N and a 1.5% chlorophyll content for algal cells, the N_2 fixation can be expressed as 2.8—70 μg of N per mg of chlorophyll per day. Assuming a respiration rate of 24 mg of O_2 expressed per mg of chlorophyll per day, we can calculate that N_2 is fixed with an efficiency of 3 mg of N_2 per g carbohydrate which compares favourably with the efficiencies in the N-fixing anaerobic bacteria, especially if it is realised that the protein-nitrogen component measured usually includes non-blue-green algal protein as well.

6.4.2. Site of N_2 fixation in blue-green algal cells

Stewart (1969) suggested that since nitrogenase activity is correlated with the number of heterocysts, a count of those could perhaps be used as a rough estimate of N_2 fixation in field studies. Heterocysts are large pale green cells with thick cell walls and are formed by division from vegetative cells. The normal blue-green cells are different from those of other algae in that they contain no visible nucleus or chloroplasts. The chlorophyll is homogeneously dispersed through the whole cell in a manner similar to that found in many coloured bacteria (procaryotic cell structure). In addition to chlorophyll and yellow pigments blue-green algae contain phycocyanin, a pigment that is photosynthetically active (see Chapter 13).

The arguments that the heterocyst is the site of N_2 fixation are summarised by Fay et al. (1968) as follows:

(1) Only heterocyst-containing species of blue-green algae are able to fix N_2; other members of the group are unable to do so at least under aerobic conditions.

(2) NH_3 or NO_3^- inhibits both N_2 fixation and heterocyst formation.

(3) New heterocysts are formed in filaments between two old ones, which might be due to a NH_3 gradient.

(4) Spore formation is induced by NH_3 or *near* heterocysts if no NH_3 is available, but N_2 fixation does take place, which suggests a transport of NH_3 from the heterocysts to surrounding cells.

(5) In isolated heterocysts respiration creates anaerobic conditions which are essential for the nitrogenase activity.

Furthermore, Bone (1971) has shown that the presence of nitrate decreases the amount of nitrogenase activity per heterocyst in a chemostat culture of *Anabaena flos aquae*, probably as the result of ammonia formation which may inhibit N_2 fixation. The activity of the nitrogenase is decreased by a factor of 6 or 9, so that counting of heterocysts in a natural algal population gives no indication of N-fixing capacity in the field if NO_3^- is present.

Ohmori and Hattori (1972) however found neither ammonia, nor nitrate suppressed immediately nitrogenase activity in *A. cylindrica*. When ammonia was added to the culture medium, the nitrogenase activity decreased, becoming negligible within 24 h. Variation in enzyme activity during growth in the presence of nitrate was relatively small. The concentrations of ammonia used are however from an ecological point of view rather high (10^{-3}—10^{-2} M). *A. cylindrica* and *A. flos aquae* can thus take up nitrogen concurrently from both N_2 and nitrate, provided that the two are simultaneously supplied.

Fay (1969) has shown that although heterocysts do contain chlorophyll, they lack phycocyanin, which is part of photosystem II (the part that produces the O_2) and which normal cells contain. Furthermore, Fay (1970) showed that the action spectrum of acetylene reduction corresponds with the light absorption of chlorophyll a, indicating the involvement of photosystem I only.

As this is the only system present in the heterocysts and it performs cyclic photophosphorylation without O_2 production, it may be that in the heterocyst anaerobic conditions may occur while ATP formation is possible through photosystem I. Absence of photosystem II in the heterocysts of the bluegreen alga *Anabaena* was demonstrated by Donze et al. (1972). Cox and Fay (1969) showed that even when CO_2 uptake was inhibited by CMU (chlorophenyl dimethyl urea), N_2 fixation continued, being especially great in Nstarved cells. CMU inhibits photosystem II specifically and this system is absent from heterocysts.

Bothe and Loos (1972), using uncouplers of the photosynthetic electrontransport chain, found that the direct role of photosynthesis in nitrogen fixation is restricted to that of supplying energy by cyclic photophosphorylation, and thus without oxygen production. The reductant — probably ferredoxin — is reduced independently from the photosynthetic electron transport. It probably involves pyruvate as the reducible substrate through a pyruvate clastic reaction and not the glucose-6-phosphate, glucose-6-phosphate dehydrogenase, $NADP^+$, NADPH-ferredoxin oxido-reductase pathway.

Thomas and David (1972) also showed the heterocyst to be the site of N_2 fixation, at least during active aerobic growth. Photosystem II pigments reappear in old heterocysts and N_2 fixation capacity is then greatly reduced.

Heterocyst development in *Anabaena cylindrica* was studied by Kulasooriya et al. (1972), who found that it correlated with the N_2-fixing capacity. Both processes reached a steady state when the C : N ratio increased from 4.5 : 1 to 8 : 1 if the cells were originally cultured with NH_3. Anaerobic incubation enhanced heterocyst production as well as nitrogenase activity. Weare and Benemann (1973) performed experiments using a blender to disrupt filaments and showed that only heterocysts attached at both ends to vegetative cells appeared to be functional in N_2 fixation. These results add further support to the view that heterocysts are the site of N_2 fixation. Weare and Benemann suggested that the vegetative cells supply a $C(H_2O)$ molecule, from which NADPH is produced which, together with nitrogenase and with ATP derived from photosystem I, reduces the N_2. (See further note on p. 119.)

Oxygen production by the normal blue-green cells inhibits nitrogenase activity although this enzyme is present in normal vegetative cells (Haystead et al., 1970). Under normal conditions therefore, the nitrogenase in the vegetative cells is presumably not active. Nitrogen fixation does occur in nonheterocyst-containing filamentous algae and in unicellular blue-green algae, but only in micro-aerophilic conditions and not in the air (Stewart, 1971; Fogg, 1971b). Van Gorkom and Donze (1971) reported that in *Anabaena* under aerobic conditions nitrogen fixation is confined to the heterocysts, whereas in anaerobic conditions the vegetative cells fix nitrogen as well. Stewart (1973) believed that there is conflicting evidence whether or not the vegetative cells have an active nitrogenase under micro-aerobic circumstances.

6.5. PHYSIOLOGICAL PROPERTIES OF BLUE-GREEN ALGAE WHICH MAY STIMULATE THEIR BLOOM

Whenever there is an enormous increase in algal population in eutrophic inland waters, species of the blue-green group are usually involved, and these species dominate the flora. This great population increase is called an "algal bloom".

In addition to the N-fixing abilities of blue-green algae, several other physiological differences between them and other groups may separately or jointly be the cause of the "bloom" phenomenon, which is often a considerable nuisance.

6.5.1. Growth in dim light

Rather low light saturation values (I_k) have been reported for blue-green algae. Baker et al. (1969) found light saturation for *Oscillatoria agardhii* under field conditions at only 10% of the maximal insolation, while high photosynthetic rates were found even at 4%. Strong inhibition was found in the surface layers, where light values exceeded 30% of the total insolation. In a study of the ecology and physiology of *Oscillatoria*, Zimmermann (1969) found an optimal temperature of 14°C and an I_k value of 1 000—1 500 lux. Due to adaptation processes, the optimum of the culture which was kept at 22°C shifted to 29°C. Zimmermann's extensive studies showed that either the maximal densities are reached in nature under conditions far from those giving optimal growth in cultures or that a rapid adaptation to laboratory conditions may take place. His laboratory experiments were performed in a relatively rich culture solution, and it is perhaps unwise to assume that the results are directly applicable to field conditions.

Besides having low I_k values, several species can grow facultatively photoheterotrophically although most blue-green algae are obligately photoautotrophic (see also section 16.1). Van Baalen et al. (1971) found that two species of blue-green algae, *Agmenellum quadruplicatum* and *Lyngbya lagerheimii*, could utilise glucose at an irradiance which was too low to support autotrophic growth. In dim light glucose made a greater contribution to cell constituents (amino acids) than at high irradiance. The increase in dry weight of his cultures was also higher with glucose in dim light than with normal photosynthesis at high irradiance. Additions of acetate, formate, pyruvate and succinate had no apparent effect, however, and this makes the heterotrophic growth with glucose less likely to be of great ecological significance. Most unicellular blue-green algae are obligately photoautotrophic. Two metabolic features — the low permeability (e.g. lack of glucose permease) and the absence of the tricarboxylic acid (TCA) or Krebs cycle — may account for this. Pelroy et al. (1972) and Rippka (1972) found that the oxidation of glucose in *Aphanocapsa* (= *Anacystis*) strain 6714, which can grow facultatively chemo-

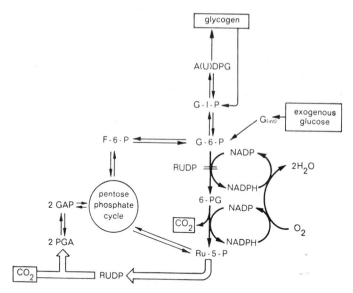

Fig. 6.3. Metabolic map of the pentose-phosphate cycle (Pelroy et al., 1972).

heterotrophically, follows the pentose-phosphate pathway as the principal if not the only route; this resembles the pathways which occur in the filamentous blue-green algae such as *Anabaena variabilis*. The necessary enzymes such as glucose permease, hexokinase and phosphatases were all found to occur in this strain. The obligately photoautotrophic strain A 6308 is different from 6714 only in its lack of the glucose permease, and in neither strain does the TCA cycle occur. Strain A 6308 can grow photoheterotrophically with glucose and DCMU but not in the dark. It seems that in the light no permeability barrier was present, which suggests that the importance of the light lies in its suppression of the permeability barrier, not as a source of energy. From Pelroy's metabolic map (Fig. 6.3) for these blue-green algae, it can be seen that externally supplied glucose may be converted into glucose-6-phosphate, which may either be converted into 6-phosphogluconic acid or enter the pentose-phosphate cycle. The enzyme for the first reaction is inhibited by ribulose-diphosphate (RUDP), a product of the Calvin cycle, so that in strong light glucose can not be oxidised to 6-phosphogluconic acid. DCMU, which inhibits RUDP formation, removes the RUDP block. In dim light the RUDP block does not exist either so that external glucose may then be used.

Pelroy pointed out that it is only in the blue-green algae that NADP is used for electron transport between oxidisable substrate and respiratory chain, whereas in the normal glucose oxidation NAD is used. NADP-linked dehydrogenases mediate oxido-reductive steps in biosynthetic pathways, whereas NAD-linked enzymes function in dissimilatory (catabolic) pathways.

Utilisation of exogenous organic substances by blue-green algae under light conditions (photoheterotrophic growth) is reasonably accounted for and described by a pathway such as that shown in Fig. 6.3. The pathway has been incorporated in the scheme of Appendix III which shows most alternative carbon cycles known to operate in plant cells. (Heterotrophy is further discussed in Chapter 16.)

Pearce and Carr (1967) have shown that in *Anabaena variabilis* acetic acid can contribute about 20% of the cellular carbon in light; the extent of acetate incorporation by *A. variabilis* did not change when the irradiance was reduced but stopped in the dark or when photosynthesis was inhibited by DCMU. Acetate did not alter the rate of growth or respiration of *A. variabilis* nor of *Anacystis nidulans*, nor were the enzymes of the TCA cycle stimulated. Incorporation of ^{14}C-labelled acetate showed that the cells were not impermeable to acetate. The incorporated acetate competed with respiration via the TCA cycle. Pearce and Carr (1969) showed that in darkness ^{14}C glucose was metabolised by *Anabaena variabilis* via the pentose-phosphate pathway and contributed up to 46% of the total dry weight. Enzyme activity was not increased and Pearce and Carr suggested that the failure to do so may account for the inability of *A. variabilis* to increase its growth rate in the presence of exogenous substrates. Sugars which are photoassimilated include sucrose and fructose.

Ingram et al. (1973) studied photoheterotrophic growth of the unicellular blue-green alga *Agmenellum quadruplicatum*, by supplying it with various reduced organic compounds. This species was unable to utilise any substances in the dark but could utilise glycerol with high efficiency when light was present for growth. This high efficiency of substrate utilisation, the action spectrum for growth on glycerol and the types of pigments in the cell all suggest that both photosystems I and II are probably involved during the photoheterotrophic growth of these organisms (compare purple non-sulphur bacteria, in which photosystem II is responsible for the high efficiency of heterotrophic growth; see subsection 16.4.3).

Both the low I_k value and facultative photoheterotrophy may help to explain why blue-green algae are often found below the thermocline (see Fig. 6.4 and subsection 16.4.2) (Gorlenko and Kuznetsov, 1972). In such regions the irradiance is low and the mineralisation processes result in release of organic substances or molecules, some of which may form a substrate for heterotrophic nutrition of the blue-green algae.

Saunders (1972) presented evidence to show a heterotrophic uptake of glucose and acetate by *Oscillatoria agardhii* and other *Oscillatoria* species at substrate concentrations known to occur naturally. Other species of blue-green algae, e.g. *Aphanizomenon flos aquae*, were apparently unable to utilise glucose or acetate. No comparison was made with photoheterotrophic uptake which might occur in the lake at the depth from which samples were taken.

112

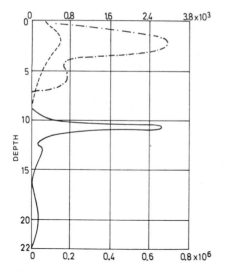

Fig. 6.4. Number per ml of diatoms (------), *Oscillatoria* (———), and *Microcystis* (—·—) in Lake Kononjer (U.S.S.R.). (From Gorlenko and Kuznetsov, 1972.)

Saunders observed that *O. agardhii* occurred in winter in Frains Lake (U.S.A.) in deep water, where the illuminance was $< 10^{-3}$ lux.

When anaerobic conditions developed in the hypolimnion in late spring, *O. agardhii* disappeared in the lowest water layers and migrated upwards as anaerobiosis intensified. This species was even found in H_2S-containing waters. Saunders suggested that when the algae reach shallower waters photo-lithotrophy develops, the heterotrophic mechanism being repressed (see below and Chapter 16). This considerable flexibility of assimilation methods might be related to the somewhat bacterial character of the blue-green algae; it certainly confers upon them the ability to grow in two different niches of the lake habitat.

The ecological significance of photoheterotrophic growth is however still obscure. Concentrations of glucose and acetate in lakewaters remain low, even if a bloom of algae occurs. Blue-green algae cannot compete successfully for organic substances with most heterotrophic bacteria and, in any case these substances may be the result of, and not the primary cause of, algal blooms.

A few other blue-green algae, especially species of *Nostoc*, have also been shown to be able to grow heterotrophically in the dark. This has been cited so often to explain the predominance of all Cyanophyta that its importance has probably been greatly overemphasised.

Stanier (1973) explained the observation that only a few types of organic compounds, i.e. glucose, fructose and one or two disaccharides, can support dark growth by the fact that the pentose-phosphate pathway is the sole energy-yielding dissimilatory pathway in blue-green algae. Since this is so, only compounds which are easily converted into glucose-6-phosphate would be expected to be used. He attributes the fact that only a few types of blue-green algae can use glucose to the absence of glucose permease in most blue-green algal species. Obligate photoautotrophy in blue-green algae closely resembles that of *Thiobacilli* (see Chapter 16). Metabolism of blue-green algae has been reviewed by Holm Hansen (1968), Smith (1973) and Stanier (1973). Recent reviews on heterotrophic and photoheterotrophic growth are published by Carr and Whitton (1973) and by Fogg et al. (1973).

6.5.2. Presence of gas vacuoles

Various species of blue-green algae contain so-called gas vacuoles which can be seen under the light-microscope as highly refractive bodies with an irregular outline. They are not merely passive bubbles produced by some process but have a definite structure. The units are surrounded by self-erecting three-dimensional membranes of about 2—3 nm thickness and have a diameter of about 70 nm and a length of 100—300 nm. Since the membranes are freely permeable to gases, the spaces must be initiated in the protoplasm and then be filled passively with gas. They may be destroyed by sudden pressure if insufficient time is given for the gas to equilibrate.

Walsby (1969) found that they are produced more abundantly in *Anabaena flos aquae* at low than at high irradiance. The presence of gas vacuoles enhances the buoyancy of the cells. Thus Klebahn (1922) measured an increase of specific gravity from 1.0065 to 1.0085 after destruction of the gas vacuoles, whereafter the algae sank. He destroyed the gas vacuoles easily by applying a sharp blow to the cork of a completely filled blue-green algal culture bottle. The algae sank quickly to form a sediment.

Dinsdale and Walsby (1972) found that the turgor pressure required to collapse gas vacuoles in cells of *Anabaena flos aquae* varied over the range of 265 to 459 kN m^{-2}. When the cell suspension, which had been pretreated for 3 days below 50 lux, was transferred to 10 000 lux, the turgor pressure of the cells rose from 330 to 396 kN m^{-2}, causing the disappearance of gas vacuoles with a critical pressure of less than 350 kN m^{-2}. The cell turgor increase, which took place at a rate of up to 94 kN m^{-2} h^{-1}, appeared to be the result of the accumulation of photosynthate. The turgor rise on transfer to high irradiance can account for collapse of sufficient of the gas vacuoles to destroy buoyancy, but it does not always account for all gas vacuoles which are lost. Other factors might be for example a change in pH.

Waaland et al. (1971) suggested that in *Nostoc muscorum* gas vacuoles whose formation is induced on transfer to high irradiance provide light-shielding of the underlying photosynthetic apparatus.

By varying the number of these gas vacuoles the blue-green algae are able to migrate between the zone of optimal light conditions and the dark, more nutrient-rich, layers. The influence of low and high irradiance on the abundance of the gas vacuoles favours this movement. The migration movements increase the supply of nutrients to the plants, since micro-layers of nutrient-poor water may otherwise develop around stationary algal cells. It is also possible that cells may hover at depths where only dim light penetrates and where the heterotrophic photoassimilation described earlier may take place, although there is no real proof for the occurrence of this phenomenon in nature.

Very little evidence for the occurrence of diurnal blue-green algal movements exists although diurnal migrations are frequently referred to in the literature. Reynolds (1972), studying a natural population of *Anabaena circinalis*, found that the gas-vacuole volume is determined primarily by the rate of increase of the alga. It decreased from 6% to 1% of the cell volume if the doubling time decreased from > 50 to about 4 days. Reynolds also showed that in cultures the vacuole volume was influenced by cell turgor pressure, as on transfer to favourable light conditions the buoyancy — measured as flotation rate — was reduced to 40%. After the population was returned to darkness, the effect was reversed. Reynolds' method of studying the cell turgor pressure was to apply pressure with compressed N_2 while the cells were suspended in different concentrations of sucrose. This method is quite ingenious and can easily be used even in field conditions.

A convenient field apparatus to measure the turgor pressure in the field has been designed by Walsby (1973). It is based on the change of turbidity which occurs when the gas vacuoles are destroyed.

Reynolds suggested that buoyancy of algae in the densely populated surface layers of a lake is caused by an inhibition of photosynthesis, resulting in an increased relative vacuole volume. This vacuole volume may be increased further in darkness or when growth is arrested by a limiting factor (P concentration decreased from 1.6% to 0.5%). The bloom would then be the result of a redistribution of an existing population. Finally it is noteworthy to report Reynolds' figures for the ratio heterocysts to vegetative cells. It was generally greater than 1 : 1 000 in May, with a maximal value of 1 : 13 000 on 18 May when inorganic nitrogen was still present, but changed to 1 : 100—300 in August during the second period of blooming.

The structure and possible ecological significance of gas vacuoles have been discussed in detail by Walsby (1971, 1972) who has also reviewed blue-green algae in more general terms (Walsby, 1970). Other relevant papers are those of Fogg (1969, 1971b), Fogg and Walsby (1971), Stewart (1971), and Fogg et al. (1973).

6.5.3. Organic substances including toxins

Blue-green algae often occur simultaneously with high concentrations of organic matter in lakewater. These high concentrations are often due either to large volumes of untreated household sewage entering the lake or to effluent from certain processing industries, mainly those dealing with food or agricultural products. Pearsall (1932) in the English Lake District found a correlation between blue-green algal blooms and the organic nitrogen concentration that had occurred in the previous month, but this correlation does not always hold.

Results of work by Singh (1955), Brook (1959) and Vance (1965) supported a general impression that organic pollution and myxophycean blooms are related, although most planktonic algae are less successful than bacteria in competing for organic substances. The problem seems to be that the presence of organic matter may be partly an effect of, and not the cause of, the algal bloom, since many blue-green algae excrete considerable amounts of organic compounds. The majority of waste products originally in the water would normally be broken down quickly, however, and in none of the studies of Pearsall, Singh, Brook, or Vance it was shown that the organic compounds were the original ones rather than extracellular products of the algae themselves.

Fogg (1952, 1966a and 1971b) has found that on occasion more than 50% of the assimilated nitrogen can be excreted, mostly as poly-amino acids. These compounds are end-products of algal metabolism, however, and unlike the production of glycollate, they cannot be considered as substances that can promote growth. It is unlikely that these compounds can act as a substrate either for photoassimilation or for heterotrophic growth, at least in their original form.

Many of the excreted products (e.g. for *Microcystis*) are potent toxins, which may cause death of animals (Hughes et al., 1958; Gorham, 1964; Shilo, 1964, 1967; Heany, 1971; and the impressive review list of Schwimmer and Schwimmer, 1968, in which several cases of mass deaths of fish, cattle and waterfowl have been described). Some of these toxins have been identified as cyclic polypeptides (mol. weight about 1 200). Gorham et al. (1964) showed that the ability to produce toxins is not necessarily a property of the whole species but only of certain strains. As blooms may contain an association of several strains only some of which may produce toxins, it is easy to understand why not all blooms are equally toxic or why some are not toxic at all. It should also be realised that a toxin for one animal species will not automatically be so for another species. Heany (1971) described samples of *Microcystis aeruginosa* from reservoirs of the London Metropolitan Water Board which were toxic to mice but produced no abnormal mortalities of fish or birds. Hughes et al. (1958) isolated a unialgal culture of *M. aeruginosa* which produced toxic symptoms similar to those described for natural blooms. Two distinct factors were present in the culture, one causing a slow death and the other causing a fast death (in mice). The fast death is probably the result of an endotoxin which is detected when cells leak out their contents either by autolysis or by disintegration.

Although the toxicity of the excretion products of the blue-green algae is well documented — certainly for cattle and waterfowl —, its ecological significance is difficult to establish. There is no good evidence to suggest an influence of blue-green algae on other algae, and descriptions of situations purporting to show this should mostly be attributed to other causes, probably for example to high pH values (see Chapter 14).

Supposed competitive effects of blue-green algae against other species are quite often mentioned hypothetically, but are rarely shown to occur (e.g. Vance, 1965). If they do in fact occur, it is more likely that a competition for nutrients is involved rather than a poisonous inhibitor. Nutrient supply and possible depletion should always be assessed and controlled in experiments studying competition.

Production of toxins is not restricted to blue-green algae. Shilo, who reviewed the literature (1964) and suggested a model for describing the formation and mode of action of algal toxins (1967), showed that an ichthyotoxin was produced by the flagellate *Prymnesium parvum*. He showed that a large number of physiological and environmental factors were involved concurrently in the processes of toxin formation and excretion. Synthesis of the toxins was greatest during the late stages of the logarithmic phase of growth. In the stationary growth phase P depletion enhanced toxin production even before it affected growth. Activation of the excreted products was a second essential step, depending perhaps on presence of suitable chelators, favourable pH, etc.

Light was essential for toxin production, although light in the 400—510 nm and UV ranges inactivated them slowly. Other factors inactivating the toxins were their adsorption on colloids and changes of the pH. Thus not only biosynthesis but also extracellular stability of the toxin were affected by environmental conditions.

Blue-green algae seem to be less grazed upon by zooplankton or even by fish than other algae are, which could result in their dominance. This inhibition of grazing might be caused by their toxins, but evidence to show such an effect is scarce. (Grazing is discussed in more detail in Chapter 15.) The question of whether or not toxin production by blue-green algae does stimulate their blooms under *natural* conditions is still open to doubt. The direct influence of blue-green algal blooms on fish is also difficult to prove. The blooms in which fish deaths occur are often very dense and the resultant O_2 depletion should therefore be taken into account. Enormous fish deaths in Lake George, for example, seemed to be related more to the anaerobic conditions caused by excessive respiration and mineralisation rates than to any other factors (Ganf and Viner, 1973).

Such anaerobic conditions may also favour development of certain species of bacteria including *Clostridium botulinum*, which is extremely toxic to many organisms. It is known to occur in several Dutch canals and ditches and it is stimulated both by anaerobic conditions and artificial heating (thermal pollution).

6.5.4. Concluding remarks

Low soluble phosphate and nitrogen concentrations are sometimes reported to accompany blue-green algal blooms (e.g. Reynolds, 1972). These may be more the effects of the bloom rather than their cause. There is no evidence that the blue-green algal cells have a lower than average N or P content, and it is not correct to consider in this respect only the remaining inorganic P and N compounds in solution. From the shallow Dutch lakes there is evidence that particulate phosphate and nitrogen concentration are as high during blooms of blue-green algae as they are during any other type of algal bloom.

It is likely that the real cause is a combination of several factors. Quantitative effects of algal blooms on grazing are difficult to ascertain and more work on this is desirable. Predation as a factor controlling the size of algal populations will be discussed in the Chapters 14 and 15.

It is clear that although local remedies are possible the only real basis for control is to decrease the nutrient input, especially of phosphate, into lakes and rivers. As a possible controlling factor disturbance of stratification should be mentioned since this may interrupt migration of algae. This method was introduced by the London Metropolitan Water Board as a general means of algal control but it seems in fact to be especially effective against blue-green algae (Fogg, 1969; Ridley, 1969).

6.6. THE N CYCLE

A schematic cycle can be composed in a similar form to that already described for summarising phosphate processes (Fig. 6.5).

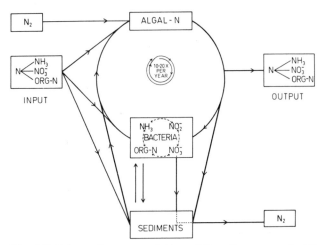

Fig. 6.5. Schematic representation of the nitrogen cycle. The indicated turnover of 20 times per year may vary between 10 and 40.

Inorganic sources of N for algae are NH_3, NO_3^- and also N_2 in cases where N_2 fixation occurs. The uptake of inorganic nitrogen is dependent on C sources, either from active photosynthesis, from heterotrophic growth, or from carbon products accumulated during photosynthesis under N starvation.

Algal nitrogen content will be about 10% of the dry weight and after the death of the algae a rapid mineralisation will occur, mostly by bacterial action (see section 6.3 and Chapter 8). By this process nitrogen is liberated mainly as NH_3, although small amounts of organic-nitrogen compounds will escape to the water. Probably a larger proportion of nitrogen than of phosphate will be converted into refractory compounds. Humic substances contain about 4—5% of N. In dissolved organic matter a C:N ratio of 5—10 is found; a value which is close to that occurring in living material. Some of these refractory compounds will sediment and mineralisation will occur slowly. Nitrogen may be lost from sediments by denitrification as N_2. Very little is known of the turnover time of nitrogen; it is more difficult to estimate than that of phosphate. Besides the accumulation of nitrogen compounds as refractory material in the sediments, adsorption and subsequent liberation of ammonia occurs, although in shallow lakes the mud will not remove as much ammonia as it does phosphate (Golterman, 1967a; and section 9.5). Although most if not all the reactions in the N cycle are well known, there are as yet insufficient data on which to base a fully quantitative model.

Thomann et al. (1971) suggested a mathematical model for nitrification processes and algal cycles in estuaries. Analysis of the nitrification phenomenon was carried out using a generalised steady-state multidimensional feedback model. Any configuration of reaction systems seemed to be permissible provided that first-order reaction kinetics were adhered to and that linearity was maintained. For more complex algal growth situations, non-linear equations were necessary to describe the phytoplankton dynamics. In the equations features such as growth, death and predation were described as functions of factors such as temperature, irradiance and nutrient concentrations. The equations could only be used with the aid of a sophisticated computer and applied only to situations where the initial cellular mass was large relative to the initial substrate concentrations. Throughout the whole study nitrogen was assumed to be the growth-limiting factor and the influence of phosphate was never considered.

Filter-feeding rates for zooplankton were measured and included, but many other data were taken from the literature. Use of a computer makes it possible to test different trial models until a good fit is obtained between predicted and actual values. The model which provides the best conformity between predicted and actual values is not necessarily the most correct, and since not all the possible nitrogen reactions were included in the scheme, the significance of the results obtained is difficult to evaluate.

Note: Winkenbach and Wolk (1973) found that heterocysts of *Anabaena cylindrica* had higher activities (per mg of soluble protein) of glucose-6-phosphate dehydrogenase, 6-phosphogluconate dehydrogenase and hexokinase and lower activities of ribulose diphosphate carboxylase and glyceraldehyde-3-phosphate dehydrogenase than whole filaments. Heterocysts contained 74—80% of the total filament activity of glucose-6-phosphate dehydrogenase. Both facts suggest that sugars, formed photosynthetically in vegetative cells, are a major source of electrons for reduction processes in the heterocysts.

REFERENCES

Baier, C.R., 1936. Studien zur Hydrobakteriologie stehender Binnengewässer. *Arch. Hydrobiol.*, 29: 183—264.

Baker, A.L., Brook, A.J. and Klemer, A.R., 1969. Some photosynthetic characteristics of a naturally occurring population of *Oscillatoria agardhii* Gomont. *Limnol. Oceanogr.*, 14: 327—333.

Baxter, R.M., Wood, R.B. and Prosser, A.V., 1973. The probable occurrence of hydroxylamine in the water of an Ethiopian lake. *Limnol. Oceanogr.*, 18(3): 470—472.

Bone, D.H., 1971. Nitrogenase activity and nitrogen assimilation in *Anabaena flos aquae* growing in continuous culture. *Arch. Mikrobiol.*, 80: 234—241.

Bothe, H. and Loos, E., 1972. Effect of far red light and inhibitors on nitrogen fixation and photosynthesis in the blue-green alga *Anabaena cylindrica*. *Arch. Mikrobiol.*, 86: 241—254.

Brezonik, P.L. and Lee, G.F., 1968. Denitrification as a nitrogen sink in Lake Mendota, Wis. *Env. Sci. Techn.*, 2: 120—125.

Brook, A.J., 1959. The waterbloom problem. *Proc. Soc. Water Treat. Exam.*, 8: 133—137.

Carr, N.G. and Whitton, B.A. (Editors), 1973. *The Biology of Blue-Green Algae*. Blackwell, Oxford, 676 pp.

Chen, R.L., Keeney, D.R., Graetz, D.A. and Holding, A.J., 1972. Denitrification and nitrate reduction in Wisconsin lake sediments. *J. Env. Qual.*, 1(2): 158—162.

Cox, R.M. and Fay, P., 1969. Special aspects of nitrogen fixation by blue-green algae. *Proc. R. Soc. Lond., Ser. B*, 172: 357—366.

Dinsdale, M.T. and Walsby, A.E., 1972. The interrelations of cell turgor pressure, gasvacuolation and buoyancy in a blue-green alga. *J. Exp. Bot.*, 23: 560—570.

Donze, M., Haveman, J. and Schiereck, P., 1972. Absence of photosystem 2 in heterocysts of the blue-green alga *Anabaena*. *Biochim. Biophys. Acta*, 256: 157—161.

Fay, P., 1969. Cell differentiation and pigment composition in *Anabaena cylindrica*. *Arch. Mikrobiol.*, 67: 62—70.

Fay, P., 1970. Photostimulation of nitrogen fixation in *Anabaena cylindrica*. *Biochim. Biophys. Acta*, 216: 353—356.

Fay, P., Stewart, W.D.P., Walsby, A.E. and Fogg, G.E., 1968. Is the heterocyst the site of nitrogen fixation in blue-green algae? *Nature*, 220(5169): 810—812.

Feth, J.H., 1966. Nitrogen compounds in natural water — A review. *Water Resour. Res.*, 2(1): 41—58.

Fogg, G.E., 1952. The production of extracellular nitrogenous substances by a blue-green alga. *Proc. R. Soc. Lond., Ser. B*, 139: 372—397.

Fogg, G.E., 1961. Recent advances in our knowledge of nitrogen fixation by blue-green algae. In: *Symposium on Algology*. Indian Counc. Agric. Res., New Delhi.

Fogg, G.E., 1962. Photosynthesis and nitrogen fixation in blue-green algae. *Vortr. Gesamtgeb. Bot., N.F.*, 1: 89—91.

Fogg, G.E., 1966a. The extracellular products of algae. *Oceanogr. Mar. Biol. Ann. Rev.*, 4: 195—212.

Fogg, G.E., 1966b. *Algal Cultures and Phytoplankton Ecology.* Athlone Press, London, 126 pp.
Fogg, G.E., 1969. The Leeuwenhoek Lecture, 1968. The physiology of an algal nuisance. *Proc. R. Soc. Lond., Ser. B,* 173: 175—189.
Fogg, G.E., 1971a. Extracellular products of algae in freshwater. *Arch. Hydrobiol., Ergebn. Limnol., Beih.,* 5: 1—25.
Fogg, G.E., 1971b. Nitrogen fixation in lakes. In: T.A. Lie and E.G. Mulder (Editors), *Biological Nitrogen Fixation in Natural and Agricultural Habitats. Plant Soil,* Spec. Vol.: 393—401.
Fogg, G.E. and Horne, A.J., 1967. The determination of nitrogen fixation in aquatic environments. *Proc. IBP-Symp. Amsterdam—Nieuwersluis, 1966,* pp. 115—120.
Fogg, G.E. and Walsby, A.E., 1971. Buoyancy regulation and the growth of planktonic blue-green algae. *Mitt. Int. Ver. Theor. Angew. Limnol.,* 19: 182—188.
Fogg, G.E., Stewart, W.D.P., Fay, P. and Walsby, A.E. (Editors), 1973. *The Blue-Green Algae.* Academic Press, London, 459 pp.
Frinck, C.R., 1967. Nutrient budget: Rational analysis of eutrophication in a Connecticut lake. *Env. Sci. Techn.,* 1: 425—428.
Ganf, G.G. and Viner, A.B., 1973. Ecological stability in a shallow equatorial lake (Lake George, Uganda). *Proc. R. Soc. Lond., Ser. B,* 184: 321—346.
Gode, P. and Overbeck, J., 1972. Untersuchungen zur heterotrophen Nitrifikation im See. *Z. Allg. Mikrobiol.,* 12(7): 567—574.
Goering, J.J. and Neess, J.C., 1964. Nitrogen fixation in two Wisconsin lakes. *Limnol. Oceanogr.,* 9: 530—539.
Golterman, H.L., 1960. Studies on the cycle of elements in fresh water. *Acta Bot. Neerl.,* 9: 1—58.
Golterman, H.L., 1964. Mineralization of algae under sterile condition or by bacterial breakdown. *Verh. Int. Ver. Theor. Angew. Limnol.,* 15: 544—548.
Golterman, H.L., 1967a. Influence of the mud on the chemistry of water in relation to productivity. *Proc. IBP-Symp. Amsterdam—Nieuwersluis, 1966,* pp. 297—313.
Golterman, H.L., 1967b. Tetraethylsilicate as a "molybdate unreactive" silicon source for diatom cultures. *Proc. IBP-Symp. Amsterdam-Nieuwersluis, 1966,* pp. 56—62.
Golterman, H.L., 1968. Investigation on photosynthesis and mineralization in experimental ponds. *Mitt. Int. Ver. Theor. Angew. Limnol.,* 14: 25—30.
Golterman, H.L., 1973a. Deposition of river silts in the Rhine and Meuse Delta. *Freshwater Biol.,* 3: 267—281.
Golterman, H.L., 1973b. The role of phytoplankton in detritus formation. *Proc. IBP—UNESCO Symp., Pallanza (Italy), 1972. Mem. Ist. Idrobiol.,* 29 (Suppl.): 89—103.
Golterman, H.L., 1975. Chemistry of running waters. In: B. Whitton (Editor), *River Ecology.* Blackwell, Oxford (in press).
Gorham, E., 1964. Molybdenum, manganese and iron in lake muds. *Verh. Int. Ver. Theor. Angew. Limnol.,* 15: 330—332.
Gorham, P.R., McLachlan, J., Hammer, U.T. and Kim, W.K., 1964. Isolation and culture of toxic strains of *Anabaena flos-aquae* (Lyngb.) de Bréb. *Verh. Int. Ver. Theor. Angew. Limnol.,* 15: 796—804.
Gorlenko, W.M. and Kuznetsov, S.I., 1972. Über die photosynthesierende Bakterien des Kononjer Sees. *Arch. Hydrobiol.,* 70(1): 1—13.
Haystead, A., Robinson, R. and Stewart, W.D.P., 1970. Nitrogenase activity in extracts of heterocystous and non-heteroeystous blue-green algae. *Arch. Mikrobiol.,* 74: 235—243.
Heany, S.I., 1971. The toxicity of *Microcystis aeruginosa* Kutz from some English Reservoirs. *Water Treat. Exam.,* 20: 235—244.
Herrera, J., Paneque, A., Maldonado, J.M. et al., 1972. Regulation by ammonia of nitrate reductase synthesis and activity in *Chlamydomonas reinhardti. Biochem. Biophys. Res. Comm.,* 48(4): 996—1003.

Holm Hansen, O., 1968. Ecology, physiology, and biochemistry of blue-green algae. *Ann. Rev. Microbiol.*, 22: 47—70.

Horne, A.J. and Fogg, G.E., 1970. Nitrogen fixation in some English lakes. *Proc. R. Soc., Lond. Ser. B*, 175: 351—366.

Hughes, E.O., Gorham, P.R and Zehnder, A., 1958. Toxicity of an unialgal culture of *Microcystis aeruginosa*. *Can. J. Microbiol.*, 4: 225—236.

Hutchinson, G.E., 1941. Limnological studies in Connecticut. IV. The mechanism of intermediary metabolism in stratified lakes. *Ecol. Monogr.*, 11: 21—60.

Ingram, L.O., Van Baalen, C. and Calder, J.A., 1973. Role of reduced exogenous organic compounds in the physiology of the blue-green bacteria (algae): Photoheterotrophic growth of an "autotrophic" blue-green bacterium. *J. Bacteriol.*, 114(2): 701—705.

Jannasch, H.W. and Pritchard, P.H., 1972. The role of inert particulate matter in the activity of aquatic microorganisms. *Mem. Ist. Ital. Idrobiol.*, 29 (Suppl.): 289—308.

Klebahn, H., 1922. Neue Untersuchungen über die Gasvakuolen. *Jahrb. Wiss. Bot.*, 61: 535.

Kulasooriya, S.A., Lang, N.J. and Fay, P., 1972. The heterocysts of blue-green algae. III. Differentiation and nitrogenase activity. *Proc. R. Soc. Lond., Ser. B*, 181: 199—209.

Manny, B.A., 1972. Seasonal changes in organic nitrogen content of net- and nanophytoplankton in two hardwater lakes. *Arch. Hydrobiol.*, 71(1): 103—123.

Mortimer, C.H., 1939. The nitrogen balance of large bodies of water. *Off. Circ. Br. Waterworks Assoc.*, 21: 1—10.

Neess, J.C., Dugdale, R.C. and Goering, J.J. et al., 1962. Use of nitrogen-15 for measurement of rates in the nitrogen cycle. In: *Radioecology; Proc. Nat. Symp. Radioecology, 1st, Colorado State University, 1961*, pp. 481—484.

Ohmori, M. and Hattori, A., 1972. Effect of nitrate on nitrogen-fixation by blue-green alga *Anabaena cylindrica*. *Plant Cell Physiol.*, 13: 589—599.

Pearce, J. and Carr, N.G., 1967. The metabolism of acetate by blue-green algae, *Anabaena variabilis* and *Anacystis nidulans*. *J. Gen. Microbiol.*, 49: 301—313.

Pearce, J. and Carr, N.G., 1969. The incorporation and metabolism of glucose by *Anabaena variabilis*. *J. Gen. Microbiol.*, 54: 451—462.

Pearsall, W.H., 1932. Phytoplankton in the English lakes. II. The composition of the phytoplankton in relation to dissolved substances. *J. Ecol.*, 20: 239—262.

Pelroy, R., Rippka, R. and Stanier, R.Y., 1972. Metabolism of glucose by unicellular blue-green algae. *Arch. Mikrobiol.*, 87: 303—322.

Reynolds, C.S., 1972. Growth, gas vacuolation and buoyancy in a natural population of a planktonic blue-green alga. *Freshwater Biol.*, 2(2): 87—106.

Ridley, J.E., 1969. Artificial destratification of waterworks compoundments for the control of algal blooms. *Br. Phycol. J.*, 4(2): 215.

Rippka, R., 1972. Photoheterotrophy and chemoheterotrophy among unicellular blue-green algae. *Arch. Mikrobiol.*, 87: 93—98.

Rohlich, G.A. and Lea, W.L., 1949. The origin of plant nutrients in Lake Mendota. *Rep. Univ. Wisc. Lake Investigations Committee*.

Saunders, G.W., 1972. Potential heterotrophy in a natural population of *Oscillatoria agardhii* var. *Isothrix Skuja*. *Limnol. Oceanogr.*, 17(5): 704—711.

Schwimmer, M. and Schwimmer, D., 1968. Medical aspects of phycology. In: D.F. Jackson (Editor), *Algae, Man, and the Environment*. Syracuse Univ. Press, Syracuse, N.Y., pp. 279—358.

Shilo, M., 1964. Review on toxigenic algae. *Verh. Int. Ver. Theor. Angew. Limnol.*, 15: 782—795.

Shilo, M., 1967. Formation and mode of action of algal toxins. *Bact. Rev.*, 31: 180—193.

Singh, R.N., 1955. Limnological relation of Indian inland waters with special reference to waterblooms. *Verh. Int. Ver. Theor. Angew. Limnol.*, 12: 831—836.

Smith, A.J., 1973. Synthesis of metabolic intermediates. In: N.G. Carr and B.A. Whitton (Editors), *Biology of Blue-Green Algae.* Blackwell, Oxford, pp. 1—38.

Smith, A.J., London J. and Stanier, R.Y., 1967. Biochemical basis of obligate autotrophy in blue-green algae and thiobacilli. *J. Bacteriol.,* 94: 972—983.

Stanier, R.Y., 1973. Autotrophy and heterotrophy in unicellular blue-green algae. In: N.G. Carr and B.A. Whitton (Editors), *Biology of Blue-Green Algae.* Blackwell, Oxford, pp. 501—518.

Stewart, W.D.P., 1966. *Nitrogen Fixation in Plants.* Athlone Press, University of London, London, 168 pp.

Stewart, W.D.P., 1969. Biological and ecological aspects of nitrogen fixation by free-living micro-organisms. *Proc. R. Soc. Lond., Ser. B,* 172: 367—388.

Stewart, W.D.P., 1971. Physiological studies on nitrogen-fixing blue-green algae. *Plant Soil, Spec. Vol.,* pp. 377—391.

Stewart, W.D.P., 1973. Nitrogen fixation by photosynthetic microorganisms. *Ann. Rev. Microbiol.,* 27: 283—316.

Stewart, W.D.P., Fitzgerald, G.P. and Burris, R.H., 1967. In situ studies on N_2 fixation using the acetylene reduction technique. *Proc. Natl. Acad. Sci. U.S.A.,* 58(5): 2071—2078.

Syrett, P.J. and Morris, I., 1963. The inhibition of nitrate assimilation by ammonium in *Chlorella. Biochim. Biophys. Acta,* 67: 566.

Thacker, A. and Syrett, P.J., 1972a. The assimilation of nitrate and ammonium by *Chlamydomonas reinhardti. New Phytol.,* 71: 423—433.

Thacker, A. and Syrett, P.J., 1972b. Disappearance of nitrate reductase activity from *Chlamydomonas reinhardti. New Phytol.,* 71: 435—441.

Thomann, R.V., O'Connor, D.J. and Di Toro, D.M., 1971. Modeling of the nitrogen and algal cycles in estuaries. *Proc. Int. Water. Res. Conf., 5th,* III-9/1—III-9/14.

Thomas, J. and David, K.A.V., 1972. Site of nitrogenase activity in the blue-green alga *Anabaena* sp. *Nature, New Biol.,* 238: 219—221.

Van Baalen, Ch., Hoare, D.S. and Brandt, E., 1971. Heterotrophic growth of blue-green algae in dim light. *J. Bacteriol.,* 105(3): 685—688.

Vance, B.D., 1965. Composition and succession of cyanophycean water blooms. *J. Phycol.,* 1: 81—86.

Van Gorkom, H.J. and Donze, M., 1971. Localisation of nitrogen fixation in *Anabaena. Nature,* 234: 231—232.

Viner, A.B. and Horne, A.J., 1971. Nitrogen fixation and its significance in tropical Lake George, Uganda. *Nature,* 232: 417—418.

Von Brand, Th., Rakestraw, N.W. and Renn, Ch.E., 1937. The experimental decomposition and regeneration of nitrogenous organic matter in sea water. *Biol. Bull.,* 72: 165—175.

Von Brand, Th., Rakestraw, N.W. and Renn, Ch.E., 1939. Further experiments on the decomposition and regeneration of nitrogenous organic matter in sea water. *Biol. Bull.,* 77: 285—296.

Von Brand, Th., Rakestraw, N.W. and Renn, Ch.E., 1940. Decomposition and regeneration of nitrogenous organic matter in sea water. III. Influence of temperature and source and condition of water. *Biol. Bull.,* 79(2): 231—236.

Von Brand, Th., Rakestraw, N.W. and Renn, Ch.E., 1941. Decomposition and regeneration of nitrogenous organic matter in sea water. IV. Interrelationship of various stages; influence of concentration and nature of particulate matter. *Biol. Bull.,* 81: 63—69.

Von Brand, Th., Rakestraw, N.W. and Zabor, J.W., 1942. Decomposition and regeneration of nitrogenous organic matter in sea water. V. Factors influencing the length of the cycle; observations upon the gaseous and dissolved organic nitrogen. *Biol. Bull.,* 83: 273—282.

Waaland, J.R., Waaland, S.D. and Branton, D., 1971. Gas vacuoles. Light shielding in blue-green algae. *J. Cell Biol.*, 48: 212—215.

Waksmann, S.A., Stokes, J. and Butler, R., 1937. Relation of bacteria to diatoms in sea water. *J. Mar. Biol. Assoc. U.K.*, 22: 359.

Walsby, A.E., 1969. The permeability of the blue-green alga gas-vacuole membrane to gas. *Proc. R. Soc. Lond., Ser. B*, 173: 235.

Walsby, A.E., 1970. The nuisance algae: Curiosities in the biology of planktonic blue-green algae. *Water Treat. Exam.*, 19: 359—373.

Walsby, A.E., 1971. The pressure relationships of gas vacuoles. *Proc. R. Soc. Lond., Ser. B*, 178: 301—326.

Walsby, A.E., 1972. Structure and function of gas vacuoles. *Bacteriol. Rev.*, 36: 1—32.

Walsby, A.E., 1973. A portable apparatus for measuring relative gas vacuolation, the strength of gas vacuoles, and turgor pressure in planktonic blue-green algae and bacteria. *Limnol. Oceanogr.*, 18(4): 653—658.

Weare, N.M. and Benemann, J.R., 1973. Nitrogen fixation by *Anabaena cylindrica*. I. Localization of nitrogen fixation in the heterocysts. *Arch. Mikrobiol.*, 90: 323—332.

Winkenbach, F. and Wolk, C.P., 1973. Activities of enzymes of the oxidative and the reductive pentose phosphate pathways in heterocysts of blue-green algae. *Plant Physiol.*, 52: 480—483.

Zimmermann, U., 1969. Oekologische und physiologische Untersuchungen an der planktischen Blaualge *Oscillatoria rubescens* D.C. unter Berücksichtigung von Licht und Temperatur. *Schweiz. Z. Hydrol.*, 31: 1—58.

CHAPTER 7

SILICON AND ITS COMPOUNDS — CHEMISTRY, BIOCHEMISTRY AND PHYSIOLOGY OF UPTAKE BY DIATOMS

7.1. INTRODUCTION

SiO_2 in various rocks or minerals forms 50% of the earth's crust, and erosion causes large quantities of silicates to occur in natural waters. In volcanic regions porous igneous rocks are eroded particularly easily, and thus very high concentrations can be found in rivers draining them. Kobayashi (1967) gives 30 mg l^{-1} of SiO_2 as the mean value for rivers in the Philippines and 19 mg l^{-1} for Japanese rivers.

Livingstone (1963) gives 13 mg l^{-1} for the world's average. Several rivers transport even larger amounts of SiO_2 in the form of suspended clays and other erosion products (Golterman, 1975). The concentration of soluble SiO_2 in lakewater is much lower than these values for rivers and is usually no more than a few milligrams per litre. Kobayashi recorded that in Lake Biwa 40 000 tonnes of SiO_2 precipitate annually although the soluble content is only a few milligrams, as it is in most Japanese lakes (Horie, 1969).

The biological importance of silicate in lakewater is due mainly to the fact that it is used in considerable quantities by diatoms for their cell walls (Table 7.1). *Asterionella* contains about 140 μg of SiO_2 per 10^6 cells compared with 0.5–1 μg of PO_4-P (Lund, 1950). Silicate concentrations in lakewaters can therefore often act as a yield-limiting factor for diatom populations. During diatom growth the silicate concentration may fall to less than 0.1 mg per litre, making further growth impossible (see Chapter 14). Since silicates in lakes and rivers come from erosion of rocks and soils, the first stage in the production of soluble silicates is an incongruent dissolution of aluminium silicates, e.g.:

$$KAl\text{-silicate} + H_2CO_3 + H_2O \rightarrow HCO_3^- + Si(OH)_4 + Al\text{-silicate} + K^+$$

i.e., a primary mineral is converted into a secondary mineral, which finally dissolves further. The breakdown process is accompanied by a release of cations and silica. The water becomes alkaline due to the interaction of H_2CO_3 as proton donor, and the solid residue has a higher acidity than that of the original aluminium silicate. Erosion will be discussed in more detail in Chapter 18.

Another source of silica is sedimentary sandstone. Johnson and Johnson (1972) found silica concentrations of both ground and surface waters from the Triassic Hawkesbury Sandstone (N.S. Wales, Australia) to fall in the range between 1.5 and 6 mg l^{-1} of SiO_2-Si. The silica content did not depend on

TABLE 7.1

Amount of silica in planktonic diatoms from different places (For original references see Lund, 1965)

Species	No. of samples	SiO_2 (μg per 10^6 cells)		
		Mean	Range	
Asterionella formosa	5	139	97—171	Germany
A. formosa	28	137	100—175	England
A. formosa	2	153	140—165	England
A. formosa	1	178		U.S.A.
A. formosa	3	50	43— 59	(culture, Moed 1973)
Cyclotella bodanica	1	ca. 32 000		Germany
C. comta	1	750		Germany
C. glomerata	1	220		Germany
C. socialis	2	1 000	850—1 150	Germany
Diatoma elongatum	1	250		Germany
Fragilaria crotonensis	4	189	170—215	Germany
F. crotonensis	2	188	189—190	England
Melosira granulata	1	130		Germany
M. granulata	1	620		Sudan
M. islandica subsp. helvetica	1	280		Germany
M. italica	1	320		Germany
M. italica subsp. subarctica	3	237	165—274	England
Synedra acus var. angustissima	2	1 175	1 100—1 250	Germany
S. acus var. radians	1	286		England
Stephanodiscus astraea	1	8 500		Germany
S. hantzschii	3	41	32— 47	England
S. hantzschii var. pusillus	1	100		Germany
Tabellaria flocculosa var. asterionelloides	5	310	250—420	England
T. flocculosa var. asterionelloides	2	395	370—420	Germany/ Switzerland

the total ionic content but on the dissolution rate. During the first few days after rainfall when new material was leached, the dissolution rate was high but it became much lower after some weeks when an equilibrium concentration caused by inhibiting surface reactions on the mineral grains was established. The sandstone consisted mainly of quartz (70%) with some kaolinite, feldspar and mica (see section 18.2).

7.2. INORGANIC CHEMISTRY

The chemistry of the silicate system resembles that of carbonates since Si is placed immediately below C in the periodic table of elements. The following equilibria occur in lake waters (Stumm and Morgan, 1970):

$$(SiO_2)_s(\text{quartz}) + 2H_2O \; \rightleftharpoons \; Si(OH)_4 \qquad\qquad\qquad\qquad\quad \log K = -\;3.7$$
$$(SiO_2)\text{amorphous} + H_2O \; \rightleftharpoons \; Si(OH)_4 \qquad\qquad\qquad\quad \log K = -\;2.7$$
$$Si(OH)_4 \qquad\qquad\qquad \rightleftharpoons \; H^+ + SiO(OH)_3^- \qquad\qquad \log K = -\;9.46$$
$$SiO(OH)_3^- \qquad\qquad\; \rightleftharpoons \; H^+ + SiO_2(OH)_2^{2-} \qquad\quad \log K = -12.56$$
$$4\,Si(OH)_4 \qquad\qquad\; \rightleftharpoons \; Si_4O_6(OH)_6^{2-} + 2H^+ + 4H_2O \quad \log K = -12.57$$

Fig. 7.1 gives a diagram of the solubility of the different ionic species and is constructed from the equilibria above. As in the case of CO_2, the solubility of $Si(OH)_4$ is independent of the pH; Stumm and Morgan give the solubility as $10^{-2.4}$ mole per litre, which is a very high value and near to that which Tessenow (1966) has used, 60—80 mg l^{-1} at 0°C and 100—140 mg l^{-1} at 25°C. The low solubility of quartz is supposed to be due to the low solubilisation rate resulting from the smoothness of the crystal surface. When preparing solutions of SiO_2 by hydrolysing tetraethyl silicate (TES = $Si(OC_2H_5)_4$) we could not prepare solutions of SiO_2 of concentrations above 5—10 mg l^{-1} of SiO_2. The higher values given above are probably due to the occurrence of colloidal suspensions, which may be obtained quite easily by acidifying Na_2SiO_3. This causes the formation of colloidal SiO_2 solutions, which are apparently more strongly stabilised than those prepared from TES.

The SiO_2 or $Si(OH)_4$ dissolved in the water is in equilibrium with the $HSiO_3^-$ (or $SiO(OH)_3^-$) ion. This equilibrium depends on the pH of the solution and on the concentrations of cations. Tessenow pointed out that con-

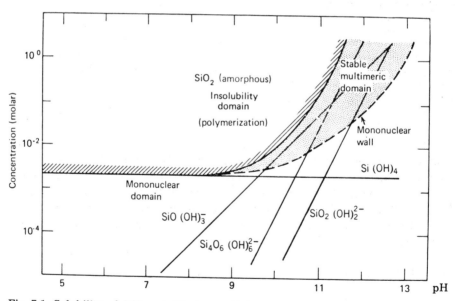

Fig. 7.1. Solubility of different silicate species in equilibrium with amorphous silica. The line surrounding the shaded area indicates maximum soluble silica. No polymeric silica will be found in natural water. (From Stumm and Morgan, 1970.)

centrations found in natural waters are always below the saturation value. His highest values were found in interstitial waters and may easily have been colloidal as many cations were present.

The silicate in the colloidal solutions prepared with TES, like that in river waters containing high silicate concentrations, will not react with the usual molybdate reagent for determining silicate. However, diatoms can use such waters as a source of silicate. This phenomenon, which is in contrast to what normally happens, e.g. for phosphate and nitrogen, was demonstrated by Golterman (1967) using solutions of TES which functioned as a Si source for his diatom cultures. Stable SiO_2 solutions of up to 10 mg l^{-1} could be obtained in which diatoms grew excellently but in which silicate was not detectable with the molybdate reagent. Just as with certain SiO_2-containing river waters, the TES solutions become molybdate-reactive after dilution, showing that the colloidal SiO_2 is in equilibrium with the soluble fraction (Kobayashi, 1967). TES proved to be an excellent source of SiO_2 for diatom cultures since no problems arose from increasing pH values during growth, in contrast to what happens with Na_2SiO_3.

The rate of silicate uptake from Na_2SiO_3 solutions was equal to or less than that from TES suspensions in distilled water. It seems likely therefore that it is SiO_2 which is taken up and not $HSiO_3^-$, although the problem remains as difficult to interpret as that of $CO_2-HCO_3^-$ uptake (see also Lewin, 1962). Most silicate will be present in natural waters as SiO_2, because HCO_3^- is the main buffering ion in lakewater, and because H_2SiO_3 is a weaker acid than H_2CO_3, as can be seen from the K_1 and K_2 values of H_2CO_3 and H_2SiO_3.

	H_2CO_3	H_2SiO_3
K_1	$4.3 \cdot 10^{-7}$	$2 \cdot 10^{-10}$
K_2	$5.6 \cdot 10^{-11}$	$1 \cdot 10^{-12}$

When comparing these values, it should be remembered that in the K_1 value for bicarbonate the equilibrium with CO_2 is taken into account, but this is not the case for SiO_2. The difference between K_1 and K_2 for the silicate system is therefore nearer to the "normal" difference for dibasic acids, i.e. a factor of 100 instead of 10 000. Furthermore, because the silicate concentrations are lower in a HCO_3^--buffered lake the silicate equilibrium is regulated by the pH which is in turn regulated by the carbonate system. Using the equilibrium constant K_1, it can be calculated that at a pH = 8, $HSiO_3^-$ is about 2% of the SiO_2, while at lower pH values the proportion is even lower. Only in the alkaline soda lakes, such as those of East Africa, may the percentage of $HSiO_3^-$ be much higher, for example about 65% at pH = 10.

7.3. BIOCHEMISTRY AND PHYSIOLOGY

Diatom cell walls are composed largely of SiO_2 (5—60% of the cell dry weight, depending on species), which indeed often forms the greatest part of the dry weight of the cells. The silica is in the form of a hydrated amorphous gel $SiO_2 \cdot nH_2O$ and contains about 10% of H_2O. The cell walls also contain several percent Al_2O_3 and Fe_2O_3. These compounds may have an integral structural function or they may be only adsorbed.

The silica frustule is encased in an organic coating of proteins and sugars. Hecky et al. (1973) found that the cell-wall protein is enriched in serine, threonine and glycine and is depleted in acidic sulphur-containing and aromatic amino acids when compared to cellular protein. The sugars of the cell-wall carbohydrates are quite variable and seem to vary with species. Three freshwater species studied had relatively little fucose and more glucose compared with three estuarine species.

Hecky et al. suggested that the inner surface of the silicalemma consists of the special protein template, of which the hydroxyl ions undergo a condensation with silicic acid. This thin template may be linked to a structural carbohydrate layer at the other, the environmental side of the diatom cell wall. This carbohydrate may have a protective function against dissolution of the silicate. Both protein and carbohydrate of the cell wall form a small proportion of the cell protein and carbohydrates.

Each diatom cell or frustule consists of two halves, one of which fits on the other like the two halves of a petri dish. The sides of the box are called the girdle; the top and bottom are the valves (Fig. 7.2). The microscopic appearance of the diatom will depend on whether it is viewed with valve or with girdle uppermost. The cell walls are very resistant to breakdown after the death of the cell, so that a vast deposit of remains of past and present marine

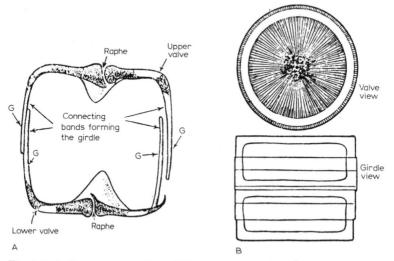

Fig. 7.2. A. Transverse section of *Navicula* (*G* = girdle). B. *Melosira* cell (schematic). (Redrawn after Clegg, 1965.)

and freshwater species is built up on the bottom of present or previously existing waterbodies. This indestructibility has in the past led to the mistaken idea that no dissolution of diatomaceous SiO_2 occurred at all. (The solubilisation of this SiO_2 will be discussed in Chapter 8.) The skeletons are often exquisitely ornamented, mostly on the valve side, with numerous fine transverse lines and depressions, and were often used for testing the resolution of optical microscopes. The patterns can best be observed after removing the living matter by treatment with acid and are important for species identification. The detailed patterns are just as easily recognised on material from fossil sediments as from fresh cultures.

Diatoms reproduce by cell division. The valves are pushed apart by the growing contents of the cell, which then divides into two parts each retaining one of the original valves. A new valve is then formed on the naked side. This may happen even before they are completely separated. After a while this often leads to a gradual decrease in cell size. At intervals another method of reproduction is thus necessary. Such a method is by fusion of two cells which form either one or two auxospores, which will then develop into larger new cells.

Lauritis et al. (1968) described cell-wall formation in *Nitzschia alba*. In this species the silica is deposited within a membrane system, the silicalemma, which develops progressively until the wall is complete. They described how, in contrast to some other diatoms (e.g. *Navicula pelliculosa* and *Cylindrotheca fusiformis*), the raphe is deposited last. In all these diatoms, however, the initial silicon deposition vesicle is formed in the central region of the dividing cell. Lauritis et al. were not able to give a clear description of details of the genesis and interrelationships of membranes involved in division and silicon deposition. It appeared that the two daughter-cell protoplasts are formed within the parent frustule by invagination of the plasmalemma. The membranes are, however, indistinguishable since they are closely adpressed to the silicon once that is deposited. Silicon uptake depends on prior nuclear division, which can be induced artificially by such compounds as colchicine. Coombs et al. (1968a, b), studying the effect of colchicine on growth, viability and fine structure of *Navicula pelliculosa*, suggested that the nucleus, the Golgi complex and microtubules all had a role to play in the deposition of silicon during wall formation.

Silica plays a major role in the periodicity of diatom growth (see Chapter 14). One of the most striking events is the abrupt death of the population, either in nature or in cultures, as soon as silicon supply is depleted. Although many algal cultures will not die abruptly when for example nitrogen or phosphate supply is depleted, a diatom culture will rapidly deteriorate if deprived of silicate especially when left in bright light. In dim light it may survive much longer. Thus, the question arises whether silicate metabolism is interlinked with the normal cell metabolism, so that a shortage of silicate inhibits the other cellular processes. There are indeed many indications that silicate uptake is related to cellular metabolism and that silicon is not simply deposited around the cells as $CaCO_3$ may be in some organisms.

Lewin (1954, 1955) studied silicate uptake by placing Si-depleted cells of *Navicula pelliculosa* in the dark in a nitrogen-free medium. Although no growth could take place without nitrogen, he found that the amount of bound silica doubled. The Si uptake process was shown to be linked with aerobic respiration since it could be inhibited by respiration inhibitors such as CN^-, F^-, iodo acetate and others. Si uptake stopped if the cells were deprived of their endogenous respiration substrate, but could be restored again by the addition of glucose or acetate. Respiration rate increased following the addition of 2,4-dinitrophenol, while at the same time Si uptake was inhibited, which indicates forcibly that the energy for Si uptake comes from ATP. Respiration and Si uptake could also be uncoupled by washing the cells repeatedly with distilled water. Si uptake could be restored by adding solutions of glutathione, cysteine, or methionine. Lewin therefore suggested that SH groups on the outside of the cell membrane may be active. Low concentrations of cadmium (10^{-3} M) also depressed respiration and Si uptake. Lewin suggested that the cadmium formed a chelate with the SH groups. The effect of washing could not be repeated using cultures of other species of diatoms (Golterman, unpublished).

Coombs et al. (1967) found a decrease in the ratio of ATP to organic carbon during Si uptake in synchronised cultures of *Cylindrotheca fusiformis*. Werner (1966) studied physiological changes which occurred after Si depletion in *Cyclotella cryptica*. If Si-rich cells were placed in the light in a silica-free medium, there was an initial protein and carbohydrate synthesis after which protein synthesis became blocked. Later the synthesis of chlorophyll, DNA, RNA and xanthophyll ceased in that order. A shift occurred from carbohydrate towards lipid accumulation. Cells deprived of nitrogen or phosphate had been previously shown to store carbohydrate. The chlorophyll became even more difficult to extract, indicating a protein denaturation. If the cells were placed in the dark, the lipid content remained constant and carbohydrates were converted into protein, but this process stopped when silica accumulation in the cells ceased. Cells put in a medium without silicate and originally in darkness accumulated lipids when illuminated. Silica deficiency also affected the R.Q. which decreased in *Navicula pelliculosa* from 0.8—0.9 to 0.3 after four days of starvation in darkness, indicating the use of lipids for respiration. This may relate directly to silicate deficiency or it may merely reflect the depletion of carbohydrate and use of lipid as respiratory substrate. Werner claimed to have purified a protein—silica complex in which the ratio of proteins to silica remained constant after repeated precipitations. As the complex was precipitated every time with 66% saturated $(NH_4)_2SO_4$ (pH = 5.6) and was dissolved with borax (pH = 7.6), it is quite possible that both compounds are precipitated and dissolved without necessarily being bound to each other, or it may be that the silica is adsorbed passively by the proteins.

The existence of a metabolic protein—silica complex (different from the cell-wall template protein) has not yet been described.

Healey et al. (1967) argued that, although the relation between silica uptake and cell metabolism seems well established, the two processes are not necessarily interdependent. It is possible that the silica-starved biprotoplastic cell may switch to a slower metabolism as occurs in resting cells. Furthermore, the accumulation of fats instead of carbohydrates means an accumulation of more energy, which would normally be used for silicate uptake. The amount of SiO_2 involved is large. If as little as one ATP molecule is needed for the uptake of one molecule of SiO_2, the depletion of silica might cause drastic changes in metabolism, without necessarily being directly coupled.

Moed et al. (1975) found biprotoplastic cells of *Diatoma elongatum* during its decline period following silica depletion in Tjeukemeer. The percentage of biprotoplastic cells decreased and that of dead cells increased during a two-week period when the cells remained in darkness in the sediments. Apparently the biprotoplastic cells die rapidly even in darkness.

Sullivan and Volcani (1973a, b) found that in silicic acid-starved cultures of *Cylindrotheca fusiformis* DNA synthesis stopped. This was owing not to a lack of energy in the form of ATP or precursors but to lower DNA-polymerase activity. An immediate and specific increase in the activity of this enzyme occurred upon the addition of silicic acid to a silica-starved culture. Stimulation of the enzyme activity in vitro could not be demonstrated. The DNA-polymerase levels fluctuate during the division cycle, however, so an indirect relation between this enzyme activity and general metabolism may be suggested. Induced changes of this general metabolism, e.g. de novo protein synthesis due to Si starvation, are then reflected indirectly in the DNA-polymerase activity.

Besides the normal pigments, the living cells of diatoms contain two specific pigments: fucoxanthin, which can transfer light energy to chlorophyll, and diadinoxanthin, which is not active in light transport (see Chapter 13). These pigments give the diatoms their yellow-brownish colour.

Changes occur also in the pigments of the diatom cells. These changes may be relevant to the survival of diatoms during the normal seasonal periodicity (see Chapter 14). The synthesis of the two energy-transporting pigments chlorophyll and fucoxanthin ceases during Si depletion. Synthesis of diadinoxanthin during Si depletion decreases in darkness but is maintained in the light: a fact which suggests its possible role as a protective pigment. It seems possible therefore that diatoms which encounter increasing irradiance during their growth are protected against too much light under normal conditions but not during Si starvation. In temperate lakes diatom growth is often limited to early spring because it is only then that the winter period of Si replenishment has produced a sufficient concentration threshold for growth. The apparent adaptation of diatoms to low temperatures would thus be merely a reflection of silicate concentrations and light regime and not the inherent

132

cause of their spring bloom. Diatom growth is certainly not limited to temperate regions with low spring temperatures. Talling (1957) described the occurrence of large populations of diatoms in Lake Albert and Lake Edward (East Africa), but data on periodicity and pigment composition (diadino-xanthin) are scarce. Periodicity under these circumstances is however more related to inwash of nutrients by rain than to changes of irradiance.

REFERENCES

Clegg, H., 1965. *The Freshwater Life of the British Isles.* Warne, London, 2nd ed., 352 pp.

Coombs, J., Halicki, P.J., Holm-Hansen, O. and Volcani, B.E., 1967. Studies on the biochemistry and fine structure of silica shell formation in diatoms. Changes in concentration of nucleoside triphosphates during synchronized division of *Cylindrotheca fusiformis* Reimann and Lewin. *Exp. Cell Res.,* 47: 302—314.

Coombs, J., Lauritis, J.A., Darley, W.D. and Volcani, B.E., 1968a. Studies on the biochemistry and fine structure of silica shell formation in diatoms. V. Effects of colchicine on wall formation in *Navicula pelliculosa* (Bréb.) Hilse. *Z. Pflanzenphysiol.,* 59: 124—152.

Coombs, J., Lauritis, J.A., Darley, W.D. and Volcani, B.E., 1968b. Studies on the biochemistry and fine structure of silica shell formation in diatoms. VI. Fine structure of colchicine-induced polyploids of *Navicula pelliculosa* (Bréb) Hilse. *Z. Pflanzenphysiol.,* 59: 274—284.

Golterman, H.L., 1967. Tetraethylsilicate as a "molybdate unreactive" silicon source for diatom cultures. *Proc. IBP-Symp. Amsterdam—Nieuwersluis, 1966,* pp. 56—62.

Golterman, H.L., 1975. Chemistry of running waters. In: B. Whitton (Editor), *River Ecology.* Blackwell, Oxford (in press).

Healey, F.P., Coombs, J. and Volcani, B.E., 1967. Changes in pigment content of the diatom *Navicula pelliculosa* (Bréb.) Hilse in silicon-starvation synchrony. *Arch. Mikrobiol.,* 59: 131—142.

Hecky, R.E., Mopper, K., Kilham, P. and Degens, E.T., 1973. The amino acid and sugar composition of diatom-cell walls. *Mar. Biol.,* 19(4): 323—331.

Horie, S., 1969. Late Pleistocene limnetic history of Japanese ancient lakes Biwa, Yogo, Suwa, and Kizaki. *Mitt. Int. Ver. Theor. Angew. Limnol.,* 17: 436—445.

Johnson, M. and Johnson, W.D., 1972. Occurrence of silica in the natural waters of the Huntley—Robertson district, southern New South Wales. *Aust. J. Mar. Freshwater Res.,* 23: 105—119.

Kobayashi, J., 1967. Silica in fresh waters and estuaries. *Proc. IBP-Symp. Amsterdam—Nieuwersluis, 1966,* pp. 41—55.

Lauritis, J.A., Coombs, J. and Volcani, B.E., 1968. Studies on the biochemistry and fine structure of silica shell formation in diatoms, IV. Fine structure of the apochloritic diatom *Nitzschia alba* Lewin and Lewin. *Arch. Mikrobiol.,* 62: 1—16.

Lewin, J.C., 1954. Silicon metabolism in diatoms. I. Evidence for the role of reduced sulfur compounds in silicon utilization. *J. Gen. Physiol.,* 37: 589—599.

Lewin, J.C., 1955. Silicon metabolism in diatoms. III. Respiration and silicon uptake in *Navicula pelliculosa. J. Gen. Physiol.,* 39: 1—10.

Lewin, R.A. (Editor), 1962. *Physiology and Biochemistry of Algae.* Academic Press, New York, N.Y., 929 pp.

Livingstone, D.A., 1963. Chemical composition of rivers and lakes. In: *Data of Geochemistry. Geol. Surv. Prof. Pap.,* 440, pp.G1—G64.

Lund, J.W.G., 1950. Studies on *Asterionella formosa* Hass. 2. Nutrient depletion and the spring maximum. Part 1: Observations on Windermere, Esthwaite Water and Blelham Tarn. *J. Ecol.*, 38(1): 1—14.

Lund, J.W.G., 1965. The ecology of the freshwater phytoplankton. *Biol. Rev.*, 40: 231—293.

Moed, J.R., 1973. Effect of combined action of light and silicon depletion on *Asterionella formosa* Hass. *Verh. Int. Ver. Theor. Angew. Limnol.*, 18 (part 3): 1367—1374.

Moed, J.R., Hoogveld, H. and Apeldoorn, W., 1975. Dominant diatoms in Tjeukemeer. II. Silica depletion. *Freshwater Biol.* (in press).

Stumm, W. and Morgan, J.J., 1970. *Aquatic Chemistry; An Introduction Emphasizing Chemical Equilibria in Natural Waters.* John Wiley, New York, N.Y., 583 pp.

Sullivan, C.W. and Volcani, B.E., 1973a. Role of silicon in diatom metabolism. II. Endogenous nucleoside triphosphate pools during silicic acid starvation of synchronised *Cylindrotheca fusiformis. Biochim. Biophys. Acta*, 308: 205—211.

Sullivan, C.W. and Volcani, B.E., 1973b. Role of silicon in diatom metabolism. III. The effects of silicic acid on DNA polymerase, TMP kinase and DNA synthesis in *Cylindrotheca fusiformis. Biochim. Biophys. Acta*, 308: 212—229.

Talling, J.F., 1957. Diurnal changes of stratification and photosynthesis in some tropical African waters. *Proc. R. Soc. Lond., Ser. B*, 147: 57—83.

Tessenow, U., 1966. Untersuchungen über den Kieselhaushalt der Binnengewässer. *Arch. Hydrobiol., Suppl.*, 32(1): 1—136.

Werner, D., 1966. Die Kieselsäure im Stoffwechsel von *Cyclotella cryptica* Reimann, Lewin and Guillard. *Arch. Mikrobiol.*, 55: 278—308.

CHAPTER 8

RECYCLING OF NUTRIENTS AND MINERALISATION

8.1. INTRODUCTION

If continued availability of nutrients for subsequent generations of organisms is to be maintained, death of the organism must necessarily be followed by decomposition. The process of decomposition is often called mineralisation, since most of the inorganic nutrients bound in the cells are reoxidised to inorganic nutrients without need of an external energy source.

In aquatic habitats phytoplankton is decomposed mainly by zooplankton and bacteria. Zooplankton acts mainly on living cells; this will be considered in Chapter 15. Bacterial decomposition may take place aerobically or anaerobically and will mineralise algal cells mostly after their death. Only partial mineralisation takes place under anaerobic conditions, and some energy remains in the anaerobic end-products. In stratifying lakes (see Chapter 9) the processes of aerobic and anaerobic decomposition are separated both in time and space. In shallow lakes, especially those where wind sometimes disturbs the sediments, aerobic breakdown is the main process. The study of mineralisation in lakewater in situ is difficult because many processes occur simultaneously. Much work has therefore been done with cultures, especially on the recycling of phosphates, nitrogen and silicate. Discussion begins with these experiments and a consideration of mineralisation processes in lakes themselves follows in the second part of this chapter. Johannes (1968) doubts the importance of bacterial recycling of nutrients in aquatic environments. Direct evidence for this is indeed scarce. For too long it has been taken for granted that the patterns found in soils also occur in water. Johannes stressed the importance of animals in the recycling process, but as it is clear that animals account for only a small part of the community oxygen uptake (see section 15.5), it is obvious that they do not mineralise the main part of the algal material. Johannes correctly stated that in many studies on mineralisation no efforts were made to exclude microorganisms other than bacteria, e.g. protozoa. The qualitative role of protozoa has yet to be evaluated; their place in food chains has been almost completely neglected (see Chapter 15). Their quantitative influence in laboratory experiments is however often overestimated, because they reach much higher densities if dead material is added to lakewater than they do in nature (see e.g. Waksmann et al., 1937).

8.2. CULTURE EXPERIMENTS

The lifespan of individual algae and even of most algal populations is relatively short compared with that of most higher plants. Unfavourable physical chemical conditions both cause the populations to die (see Chapter 14) and bacterial attack will soon follow. Healthy and growing algal cells are not usually attacked (Waksmann et al., 1937; Niewolak, 1971); introduced bacteria will often die. It seems that the normal proteolytic bacteria have no way of penetrating the living cell wall or membranes. Just before death, autolysis makes the cells susceptible to bacterial attack. During autolysis normal cell enzymes — such as phosphatases for example — may start to operate "in reverse" and to hydrolyse instead of to synthesise. The cells become permeable and bacterial enzymes may enter; soluble compounds in the algal cells will leach into the surrounding solution. After autolysis the main bacterial attack will follow and will result in solubilisation and oxidation of those products which were not previously dissolved during autolysis. The process of autolysis has been studied by many workers including Waksmann et al. (1937), Steiner (1938a, b), Hoffmann and Reinhardt (1952), Hoffmann (1953, 1956), Harvey (1955), and Golterman (1960, 1973).

Golterman induced autolysis with heat, UV radiation, or $CHCl_3$, and found that in all these cases the resultant processes, which are related to mineralisation, were the same.

The main phenomena during autolysis are (Fig. 8.1): (1) a slight decrease of dry weight of about 15% (Golterman, 1973); (2) a rapid liberation of up to 80% of cell phosphate in two or three days; and (3) liberation of 15—20% of total cell nitrogen as soluble nitrogen compounds.

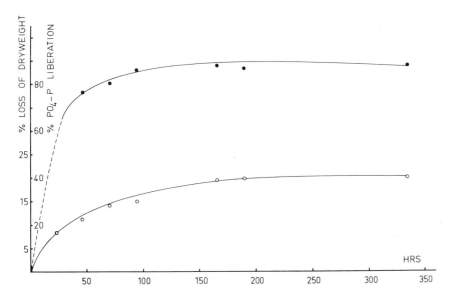

Fig. 8.1. Losses of dry weight (○——○) and phosphate (●——●) of *Scenedesmus* cells during autolysis with $CHCl_3$.

Most of the phosphate appears in solution as inorganic phosphate, but if autolysis is induced at $4°C$, a considerable quantity appears as organic phosphate. This difference may be due to phosphatases being inactive at $4°C$. Phosphate liberation is partly an enzymatic process and is especially active on those cell constituents which are extractable from live cells with alcohol and cold trichloracetic acid. Most soluble cell constituents will leach out. The phosphate fraction bound on nucleic acids or proteins is retained. Those parts of the cell remaining after autolysis still contain sufficient phosphate to support bacterial growth. This fraction, though small, is sufficient for the demands of the bacteria, because a small biomass of bacteria can oxidise a large mass of algal material. This may be the answer to Johannes' question: From which source can bacteria meet their demands for nutrients such as nitrogen and phosphate?

"Leached cells" suspended in lakewater undergo a rapid oxidative breakdown. This breakdown is not caused by protozoa; they were excluded by heating the water for 1 h at $45°C$ (e.g. Golterman, 1973). Proteolytic bacteria start growing and 50—75% of the cellular nitrogen is converted into ammonia (Fig. 8.2). Algal cells have a high protein content, part of which is used for bacterial proteins, but a considerable amount is needed simply as a source of energy by the bacteria.

$$\text{Algal protein} \underset{\searrow \text{ bacterial aminoacid}}{\overset{\nearrow \text{ } CO_2 + NH_3 + \text{energy} \nearrow}{}} \rightarrow \text{bacterial growth}$$

Golterman recovered some 20—30% of the original algal dry weight as bacteria plus algal detritus. The rest was used as energy source with a concomitant production of ammonia. Not all the nitrogen that is not released as ammonia is used by the bacteria. Refractory compounds containing nitrogen will remain, but it is not technically possible to separate these from the bacterial cells.

A large part of the energy incorporated in algal cellular material is thus necessarily used during decomposition. The breakdown process is therefore never limited by its own energy demands. Nitrogen, an important nutrient for algal growth, is automatically returned to the nutrient cycle (see also section 6.3).

From enrichment cultures Golterman isolated a strain of *Pseudomonas*, that is capable of converting cellular proteins directly into ammonia (Fig. 8.2). The bacteria released extracellular proteolytic enzymes into their culture solution. Using the bacterium with aspartic acid as a substrate, about 50% of the aspartic nitrogen was converted into ammonia. If growth was stopped, by inhibiting all synthetic reactions with $5 \cdot 10^{-5}$ M DNP (dinitrophenol), 71% of the aspartic acid was converted. This occurred because no NH_3 could be used for formation of bacterial cell material. Even after four days, bacterial activity was not inactivated by DNP. This fact will later be used in lake studies. The mechanism of the action of DNP is complicated but can

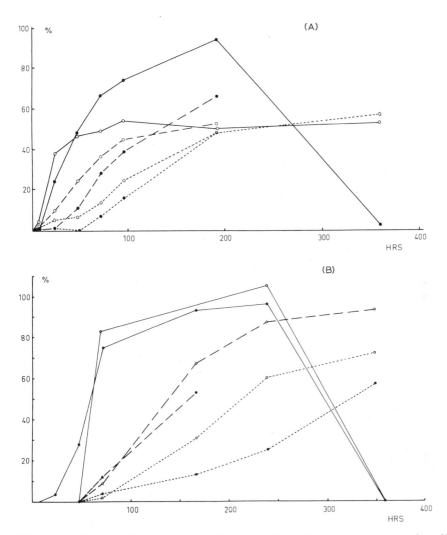

Fig. 8.2. A. Release of ammonia from "leached" *Scenedesmus* cells during mineralisation in lake water (●——●) or in cultures of *Pseudomonas boreopolis* (○——○) at three temperatures: ———, 30°C; — — —, 20°C; - - - - -, 10°C. B. Ammonia produced from "leached" algal cells after suspension in lakewater at different temperatures: ○—— ○, no treatment; ●——●, lakewater heated to 65°C for 1 h; ———, 30°C; — — —, 20°C; - - - - -, 10°C.

shortly be described by stating that it inhibits all processes which use ATP, but it has no influence on the energy-producing processes.

Similar results were found by Forree et al. (1970), who found that regeneration of nitrogen and phosphate depended on the extent of organic decomposition and on the initial nutrient content. High initial quantities of nutrients usually corresponded with young cultures growing under abundant nutrient conditions, and a greater degree of regeneration was observed. These results are essentially the same as Golterman's finding that polyphosphates, which are the form in which phosphate is stored under abundant nutrient conditions, are regenerated rapidly. There will therefore be less regeneration of phosphate from cells grown in natural lake conditions.

Golterman (1960) found the mineralisation of SiO_2 to be a slow non-enzymatic process. From dead *Stephanodiscus hantzschii* cells only 20—30% of the silicate was liberated after some weeks. Adding $NaHCO_3$ to the decaying suspensions increased the percentage of silicate that came into solution but not the rate of this process. Jørgensen (1955) found the liberation of silicate to be dependent on pH and species. For *Nitzschia* at pH = 10 the maximal amount of silicate found in solution was 20% after 85 days, whereas at the same pH for *Thalassiosira* 97% dissolved after only 37 days. Moed (1974) found that Si-depleted *Asterionella* cells released 50% silicate within 20 days after the number of cells had started to decrease.

The decomposition of phytoplankton is different from that of macro-vegetation, such as reeds, which contain large quantities of cellulose and other polysaccharides. Bacteria attacking these substrates will obtain their nitrogen from dissolved compounds, so that in this case the decomposition of the dead plant may lead to an increased particulate-nitrogen concentration. This contrasts with the decomposition of algal material (of high protein content) resulting in accumulation of NH_3 in solution. Thus the decomposition of the macro-vegetation may be limited by nitrogen, whilst the decomposition of algal material is not. Since macro-vegetation only occurs in the shallow parts of lakes, the ammonia adsorbed on lake mud may be important to bacteria involved in decomposition of non-algal material.

8.3. MINERALISATION UNDER NATURAL CONDITIONS IN LAKES

8.3.1. Oxidative mineralisation

Mineralisation (in this chapter meaning bacterial mineralisation) of algal material can be measured as O_2 consumption in darkness. In shallow eutrophic lakes daily O_2 consumption per unit area may be larger than O_2 production. For short periods it may be much larger (see section 4.6). The mean annual values for O_2 uptake in Tjeukemeer varied between 3 and 7 g m^{-2} d^{-1}, while photosynthesis ranged from 5.7 to 8 g m^{-2} d^{-1} and in some years net production was negative. Similar situations occur in Lake George, in Loch Leven (see section 15.5) and in Lake Kinneret (Serruya and Serruya, 1972). Oxygen consumption — after removing animals — is due to: (1) algal respiration, which leads to the decrease of cellular carbon; (2) mineralisation, which leads to a decrease of cellular carbon and nitrogen (and other elements); and (3) oxidation of dissolved organic compounds, which does not lead to a decrease of cellular products. From these facts it would seem that the problem of measuring mineralisation might be solved by measuring the decrease of cellular (particulate) carbon and nitrogen. As this decrease may be small, short incubation periods will not give reliable results, whereas bacterial growth introduces large errors during long incubation periods. Golterman (1971) therefore used dinitrophenol to keep the bacterial population stable. In cultures it has been shown (see section 8.2) that proteolytic processes are not inhibited by DNP. In these conditions he found that the proportional decrease of particulate carbon was roughly equal to the decrease of particulate nitrogen and was lower than the simultaneous O_2 uptake. This indicates the occurrence of process (3), the oxidation of dissolved organic matter, in addition to the others. The mineralisation appeared to be strongly temperature-dependent and was scarcely detectable at $3°C$, which would suggest that the mineralisation of the particulate material is caused by bacterial processes. About 5—10% was mineralised per day, the higher value being observed in the tropical Lake George. Such values would mean a life expectancy for the algal populations of 20—10 days. These experimental results in Lake George were confirmed by Ganf (see section 15.5). It is probable that these processes also occur in the epilimnion after the death of algal cells, although it has been originally assumed that mineralisation occurs mainly in the hypolimnion (see subsection 9.4.1). The sinking rate of the algae will determine the time during which the bacterial processes can occur in the epilimnion. Because this rate is not known, predictions about the extent of mineralisation in the epilimnion cannot be made from these experiments. Ohle (1962, 1965) collected sinking detritus in a sediment trap, and estimated the mineralisation in the epilimnion to be up to 95%. This high value may however be due to increased mineralisation within the trap itself as a result of increased bacterial substrate concentration. Furthermore, the presence of the trap may itself increase the amount sedimenting by reducing water movements which would otherwise keep the algae floating. If live algae are trapped in this way they may be killed earlier than would happen normally, causing an apparently high mineralisation rate. Other disadvantages of sediment traps are discussed in section 17.1.

Owing to the increased viscosity in the metalimnion (or thermocline, Chapters 3 and 9), the sinking rate becomes slower. Studies of oxygen consumption in the metalimnion (see Ambühl's (1969) experiments, Fig. 9.5) showed that mineralisation can occur there too. Furthermore, Collins (1963) found the number of heterotrophic bacteria in the metalimnion to be greater than in the epilimnion. High oxygen consumption was found by Serruya (1972) in the metalimnic layer and by Serruya and Serruya (1972) in the epilimnion of Lake Kinneret (Israel); the rate of O_2 consumption in the metalimnion was so great that denitrification had begun and Fe^{2+} and Mn^{2+} accumulated there. Serruya suggested that the conditions in the metalimnion are favourable to the development of sulphur bacteria and blue-green algae, which are often found in these layers (sections 6.5 and 16.4.2). The high rate of mineralisation in the epi- and metalimnion is in accordance with the observed short turnover times found for phosphate. Gächter (1968; see Chapter 5) found a turnover time for phosphate of about 5—10 days. Mineralisation of phosphate is difficult to demonstrate in the field due to rapid uptake by other organisms. It has, however, been demonstrated by Otto and Benndorf (1971) in experiments in which natural populations were left sedimenting in polyethylene enclosures. Otto and Benndorf found that the *Fragilaria* population sank more rapidly after the population began to decrease in numbers. At the same time the phosphate content per cell fell sharply from 4 to about 0.3 μg P per 10^6 cells. Dry weight of the decaying cell decreased much less.

Recently Fuhs (1973) described an improved device, with a reliable closure mechanism, for the collection of sedimenting matter. From the amounts of biomass and phosphate collected in the trap, Fuhs calculated that phosphate recycled 10—11 times during the observation period (about 200 days). He had to assume, however, that the sediments had been produced in the trophogenic zone and under an area equal to the opening of the trap and that mineralisation did not occur in the trap, although grazing ciliates were found there. Bacterial growth in the trap was not measured at all. The effect of continued mineralisation in the trap and input of other sediments (for example eroded particles from inflowing rivers) may partly compensate each other.

8.3.2. Anaerobic mineralisation: the [H] donor and the carbon sources

The final stages of decomposition of the sunken detritus take place at the sediment surface. Only decay-resistant substances such as cellulose and chitin-like substances will reach the lake bottom sediments. Some of these materials can serve as [H] donors in the reduction of sulphate and iron (see Chapter 9). Sulphate reduction is carried out by *Desulphovibrio*, which is unable to oxidise acetate because it lacks some of the enzymes necessary for the TCA (Krebs) cycle (see subsection 16.2.7). The following reaction may therefore occur:

$$H_2SO_4 + \text{organic matter} -[H] \rightarrow H_2S + H_2O + CH_3COOH$$

Some of the organic molecules may thus be converted into acetate, but other compounds entering the biochemical pathways of *Desulphovibrio* through different cycles may easily be used more completely, exactly as happens for example with glucose and acetate in blue-green algae (see sections 6.5.1 and 16.4). The production of acetate may thus be stoicheiometrically much smaller than sulphate reduction. Reduction of iron by organic compounds (see subsection 9.4.2) will also not lead to CO_2 as final oxidation product of the C skeleton. Therefore, as a result of these two reactions, the sunken organic matter is divided into a carbon source (e.g. acetic acid) and a hydrogen donor. These products may be utilised separately in further reactions. By the action of some of the anaerobic bacteria energy is thus fixed in compounds such as H_2S or Fe^{2+} and this may later be returned if O_2 becomes available for the oxidation of H_2S or Fe^{2+} (chemolithotrophic bacteria, see section 16.2).

Allochthonous input of organic matter may supplement the primary production as a carbon source, although the allochthonous material may often be more refractory than the algal cells themselves (see Chapter 13). In interstitial water of mud in Lake Vechten we found 300—500 mg l^{-1} of oxidisable organic matter. This is such a high value that its turnover time must be very long (low rate of decomposition and low rate of production).

The carbon sources will be reduced partly to methane. Three processes at least may operate:

(1) CH_3COOH (acetic acid) $\rightarrow$ $CO_2 + CH_4$

an internal oxidation reduction reaction which uses no [H] donor;

(2) $2CH_3CH_2OH$ (ethylalcohol) $+ CO_2 \rightarrow CH_3COOH$ (acetic acid) $+ CH_4$

where alcohol is a [H] donor and the oxidation product is subsequently metabolised as in reaction (1);

(3) $H_2 + CO_2 \rightarrow CH_4$

a reaction that is clearly competitive with SO_4^{2-} and Fe^{3+} reduction*. An important parameter is therefore the original SO_4^{2-} concentration. Reaction (3) may take place only after the SO_4 concentration reaches a very low level. Indeed Ohle (1968) found a lag phase between maximal primary production and maximal methane production.

Pretorius (1972) demonstrated that in continuous cultures acetate alone was unable to maintain a methanogenic enrichment culture. Formate could provide the necessary energy for the growth of the acetate-decomposing methanogenic bacteria. Formate was probably not utilised directly but was first decomposed to form H_2 and CO_2. Rod-shaped bacteria rather than Sarcinae were responsible for the decomposition of acetate.

Methane will be derived in part from the organic matter produced in the lake and perhaps in part from allochthonous organic matter. The process takes place only in the mud. Here saturation levels of CH_4 are frequently exceeded and bubbles are formed. As early as 1911, Birge and Juday found considerable methane concentrations in hypolimnetic waters, for example 38.5 ml l^{-1} in Garvin Lake. Deuser et al. (1973) found 50 km^3 of methane in Lake Kivu and suggested that most of the methane was formed by bacteria from abiogenetic CO_2 and H_2. In Lake Mendota a production in 100 days of 28 ml CH_4 per litre of mud has been measured (Allgeier et al., 1932). This is equivalent to 2.8 l of CH_4 per m^2 (if one assumes an active depth of 1 dm). Methane bubbles, when formed, will begin to rise into the hypolimnion

---

* Possible competition for the [H] donor between sulphate and iron reduction — both being competitive with methane formation — is further discussed in Chapter 9.4.2.

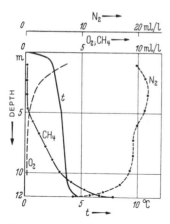

Fig. 8.3. Vertical distribution of temperature, oxygen, nitrogen, and methane in Lake Beloje, March 7, 1938. (From Kuznetsov, 1959.)

where they will be oxidised by the methane-oxidising bacteria. It is clear that only a few percent of the carbon fixed during primary production can be transformed into methane. Fig. 8.3 shows the winter distribution of O_2, N_2 and CH_4 in Lake Beloje, a small eutrophic lake near Moscow. There is little difference in the amounts of nitrogen, except for a decrease just above the bottom brought about by nitrogen-fixing bacteria. In contrast, the CH_4 curve shows a maximum at the surface of the ooze with a rapid decline in the free water because of diffusion and oxidation by oxidising bacteria (Overbeck and Ohle, 1964). The simultaneous presence in eutrophic lakes of CH_4 and H_2 in the lower part of the hypolimnion and O_2 in the upper layers of the hypolimnion is favourable for the growth of methane- and hydrogen-oxidising bacteria (see subsections 16.2.3 and 16.2.5).

Kuznetsov (1959), and Anagnostidis and Overbeck (1966) showed that the disappearance of O_2 in the hypolimnion is related to the activity of these bacteria. Anagnostidis and Overbeck calculated that 70% of the oxygen consumption in the Pluss See (Germany) during certain parts of the year may be caused by biological oxidation of methane. During the oxidation the carbon from the methane molecule will be oxidised to CO_2 in solution; only the methane which escapes as bubbles is a carbon loss from the lake.

Whether or not CH_4 itself can serve as [H] donor for sulphate reduction is still a matter for discussion.

Methane-forming bacteria are often remarkably substrate specific. Examples of some typical species are:

(1) Non spore-forming rods, *Methanobacterium* (e.g. *Methanobacterium formicicum*, which can oxidise only H_2, CO and HCOOH, and *Methanobacterium suboxydans*, which can oxidise only the three higher fatty acids, butyric, valeric and caproic acid).

143

(2) Spore-forming rods, *Methanobacillus.*
(3) Micrococci, *Methanococcus (mazei).*
(4) Sarcinae, *Methanosarcina (methanica).*
Other methane-forming bacteria are *M. sohngenii,* which is important in lakes because it can produce methane even at 5°C. Another species, *Methanobacterium omelianskii,* is active only at higher temperatures. Some doubt exists, however, about the purity and substrate specificity and demand of these bacteria.

Although most of the quantitatively significant processes occurring in the hypolimnion are now fairly well documented, little is yet understood about the factors which control their interrelationships and determine which reactions will actually occur. This is especially the case for the competition between methane formation and sulphate and iron reduction.

REFERENCES

Allgeier, R.J., Peterson, W.H., Juday, C. and Birge, E.A., 1932. The anaerobic fermentation of lake deposits. *Int. Rev. Ges. Hydrobiol. Hydrogr.,* 26: 444—461.
Ambühl, H., 1969. Die neueste Entwicklung des Vierwaldstättersees (Lake of Lucerne). *Verh. Int. Ver. Theor. Angew. Limnol.,* 17: 219—230.
Anagnostidis, K. and Overbeck, J., 1966. Methanoxydierer und hypolimnische Schwefelbakterien. Studie zur ökologischen Biocönotik der Gewässermikroorganismen. *Ber. Dtsch. Bot. Ges.,* 79(3): 163—174.
Birge, E.A. and Juday, C., 1911. The inland lakes of Wisconsin. The dissolved gases and their biological significance. *Bull. Wis. Geol. Nat. Hist. Surv.,* 22: 259 pp.
Collins, V., 1963. The distribution and ecology of bacteria in freshwater. *Proc. Soc. Water Treat. Exam.,* 12: 40—73.
Deuser, W.G., Degens, E.T., Harvey, G.R. and Rubin, M., 1973. Methane in Lake Kivu: New data bearing on its origin. *Science,* 181: 51—53.
Forree, E.G., Jewell, W.J. and McCarty, P.L., 1970. The extent of nitrogen and phosphorus regeneration from decomposing algae. In: S.H. Jenkins (Editor), *Adv. Water Pollut. Res.,* 2: III—27/1—27/15.
Fuhs, G.W., 1973. Improved device for the collection of sedimenting matter. *Limnol. Oceanogr.,* 18(6): 989—993.
Gächter, R., 1968. Phosphorhaushalt und planktische Primärproduktion im Vierwaldstättersee (Horwer Bucht). *Schweiz. Z. Hydrol.,* 30(1): 1—66.
Golterman, H.L., 1960. Studies on the cycle of elements in fresh water. *Acta Bot. Neerl.,* 9: 1—58.
Golterman, H.L., 1971. The determination of mineralization in correlation with the estimation of net primary production with the oxygen method and chemical inhibitors. *Freshwater Biol.,* 1(3): 249—256.
Golterman, H.L., 1973. Vertical movement of phosphate in freshwater. In: E.J. Griffith, A. Beeton, J.M. Spencer and D.T. Mitchell (Editors), *Environmental Phosphorus Handbook.* John Wiley, New York, N.Y., pp. 509—538.
Harvey, H.W., 1955. *Chemistry and Fertility of Sea Water.* Cambridge University Press, Cambridge, 240 pp.
Hoffmann, C., 1953. Weitere Beiträge zur Kenntnis der Remineralisierung des Phosphors bei Meeresalgen. *Planta,* 42: 156—176.

144

Hoffmann, C., 1956. Untersuchungen über die Remineralisation des Phosphors im Plankton. *Kieler Meeresforsch.*, 12(1): 25—36.

Hoffmann, C. and Reinhardt, M., 1952. Zur Frage der Remineralisation des Phosphors bei Benthosalgen. *Kieler Meeresforsch.*, 8(2): 135—144.

Johannes, R.E., 1968. Nutrient regeneration in lakes and oceans. In: M.R. Droop and E.J. Ferguson Wood (Editors), *Advances in Microbiology of the Sea*, 1. Academic Press, London, pp. 203—213.

Jørgensen, E.G., 1955. Solubility of the silica in diatoms. *Physiol. Plant.*, 8: 846—851.

Kuznetsov, S.I., 1959. Microbiological characteristics of the Volga reservoirs. *Verh. Inst. Biol. Stuwmeren*, 1: 69—81 (in Russian).

Moed, J.R., 1974. Effect of combined action of light and silicon depletion on *Asterionella formosa* Hass. *Verh. Int. Ver. Theor. Angew. Limnol.*, 18 (part 3): 1367—1374.

Niewolak, S., 1971. The influence of living and dead cells of *Chlorella vulgaris* and *Scenedesmus obliquus* on aquatic microorganisms. *Polsk. Archwm Hydrobiol.*, 18(1): 43—54.

Ohle, W., 1962. Der Stoffhaushalt der Seen als Grundlage einer allgemeinen Stoffwechseldynamik der Gewässer. *Kieler Meeresforsch.*, 18(3): 107—120.

Ohle, W., 1964. Interstitiallösungen der Sedimente, Nährstoffgehalt des Wassers und Primärproduktion des Phytoplankton in Seen. *Helgol. Wiss. Meeresunters.*, 10(1—4): 411—429.

Ohle, W., 1965. Primärproduktion des Phytoplanktons und Bioaktivität der Seen, Methoden und Ergebnissen. *Limnologissymposium, Finland*, 1964, pp. 24—43.

Ohle, W., 1968. Chemische und mikrobiologische Aspekte des biogenen Stoffhaushaltes der Binnengewässer. *Mitt. Int. Ver. Theor. Angew. Limnol.*, 14: 123—133.

Otto, G. and Benndorf, J., 1971. Über den Einfluss des physiologischen Zustandes sedimentierender Phytoplankter auf die Abbauvorgänge während der Sedimentation. *Limnologica*, 8(2): 365—370.

Overbeck, J. and Ohle, W., 1964. Contribution to the biology of methane oxidizing bacteria. *Verh. Int. Ver. Theor. Angew. Limnol.*, 15: 535—543.

Pretorius, W.A., 1972. The effect of formate on the growth of acetate utilizing methanogenic bacteria. *Water Res.*, 6: 1213—1217.

Serruya, C., 1971. Lake Kinneret: The nutrient chemistry of the sediment. *Limnol. Oceanogr.*, 16(3): 510—521.

Serruya, C., 1972. Metalimnic layer in Lake Kinneret, Israel. *Hydrobiologia*, 40(3): 355—359.

Serruya, C. and Serruya, S., 1972. Oxygen content in Lake Kinneret: physical and biological influences. *Verh. Int. Ver. Theor. Angew. Limnol.*, 18: 580—587.

Steiner, M., 1938a. Zur Kenntnis des Phosphatkreislaufes in Seen. *Naturwissenschaften*, 26(44): 723—724.

Steiner, M., 1938b. Untersuchungen über den natürlichen Kreislauf des Phosphors. *Angew. Chem.*, 51: 839.

Waksmann, S.A., Stokes, J.L. and Butler, M.R., 1937. Relation of bacteria to diatoms in sea water. *J. Mar. Biol. Assoc. U.K.*, 22: 359—373.

CHAPTER 9

STRATIFICATION IN DEEP LAKES

9.1. INTRODUCTION; PHYSICAL ASPECTS

Most deep lakes in temperate regions show a so-called stratification. Vertical mixing is incomplete, so that two or more distinct layers occur during at least part of the year. In late autumn and in winter (if no ice is formed) the lakewater is mixed homogeneously; the lake is then described as having a complete circulation. In spring increasing amounts of solar radiation heat the upper layers more than the lower layers. Since the density of the warmer surface water is less than that of the cooler water beneath, it forms a separate layer — the epilimnion — which "floats" on the cold layer, the hypolimnion.

Between epi- and hypolimnion is a transition layer, called thermocline or metalimnion. An ideal case is shown diagrammatically in Fig. 9.1. During autumn, the density of the epilimnion increases as the temperature falls and the two layers may mix again, causing the lake to be in complete circulation. Autumn storms often effect this mixing before the temperatures of the two layers have become equal. Winter stratification can occur if ice forms, the layers now having an inverse temperature difference. If no ice forms, the lakewater will remain completely mixed during the winter. Lakes which have both summer and winter stratification, with autumn and spring mixing are called dimictic. Monomictic lakes (one annual mixing period) are subdivided into two classes: the cold monomictic lakes with a summer circulation below 4°C (polar situation), and the warm monomictic lakes with a winter circulation above 4°C (tropical situation).

Another factor determining whether or not stratification will occur is the ratio between the depth of water layer which can be mixed by wind action

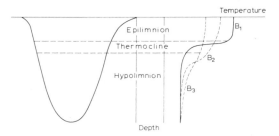

Fig. 9.1. Profile through stratified lake with temperature—depth curve. The broken lines (B_2, B_3) represent temperature—depth curves when the epilimnion is cooling in autumn. The boundaries of the thermocline (here only indicated for B_1) will gradually sink.

146

and the total depth. The depth of wind mixing depends upon the surface area and the degree of exposure to wind, which will depend upon the extent of mountains, trees, and on local topography, etc. If many storms occur during the spring warming-up period, the heat will be distributed through a much deeper water column than normal, thus diminishing the likelihood of stratification. Once stratification has become well established, the stability of the system is so great that no storm can disturb the overall pattern, although the depth of the epilimnion may vary.

The stability depends upon the density difference between warm and cold water. For example, at 8°C and at 20°C the densities are 0.999849 and 0.998203, so that 1 m³ is 1.6 kg heavier at 8°C than at 20°C. The work needed to replace 1 m³ of this hypolimnetic water at 8°C with 1 m³ of the warmer water 10 m above it is 16 kg(f) m (= 160 Nm). It is usual to express results per square metre. Multiplying this figure by the lake area gives the work needed to overturn the whole lake. The stability can be calculated by estimating the volume of the different layers and their centres of gravity. The stability of the lake is then expressed as the amount of work necessary to lift the centre of gravity of epi- plus hypolimnion to that of the mixed homogeneous lake, times the weight of the lake (volume times density). For accurate calculations the temperature of the water after mixing should be measured (or calculated) and the density can then be predicted. A better estimate can be made by calculating the amount of work (per m²) necessary to lift the hypolimnetic column through a distance of half the epilimnetic depth, while the amount of "free" work resulting from sinking of the equivalent epilimnetic water must be subtracted. The amount of work (per unit area) is therefore:

$$W = \tfrac{1}{2}(d_2 - d_1)g(Z-h)h$$

where h = depth of epilimnion; Z = depth of the lake; d_2 = density of hypolimnion; d_1 = density of epilimnion; and g = acceleration due to gravity.

In spring, when the stratification is developing, the difference between d_2 and d_1 is small due to the low temperatures (Table 9.1) and thus it is clear that wind-generated currents still transport heat downwards. It should be realised that the rate of change of density with temperature increases with temperature (Table 9.1). Because of this the stability of a tropical lake with

TABLE 9.1

Difference of water density for 1°C change of temperature at different temperatures

$\Delta T(°C)$	Δ density	$\Delta T(°C)$	Δ density
0— 1	0.00006	14—15	0.00015
3— 4	0.000008	19—20	0.00020
4— 5	0.000008	24—25	0.00025
9—10	0.00008	29—30	0.00030

a $5°$ temperature difference, e.g. between $30°C$ and $25°C$, is comparable to
that of a temperate lake with a temperature difference of $12°$ between, say,
$20°C$ and $8°C$.

Thermal stratification is thus the combined result of solar heating and
wind influence at the surface. Even after stratification, heat may be transported for example by eddy diffusion. The same mechanism is responsible
for the transport of various dissolved substances, although water movement
will increase the influence of diffusion considerably (the so-called eddy diffusion). For the heat transported by eddy diffusion the relationship:

$$\frac{dT}{dt} = A \frac{d^2T}{dz^2} \tag{9.1}$$

may be used, where T = temperature; t = time; z = depth below the surface;
and A = eddy diffusion coefficient.

Lerman and Stiller (1969) computed A for Lake Kinneret from temperature
—depth profiles after estimation of the rate of temperature change, dT/dt,
and the second derivative with respect to depth, d^2T/dz^2, following the method of Hutchinson (1957). They found high values for A above the thermocline ($1-20$ cm^2 sec^{-1}), low values below the thermocline ($15-30$ m:
$5 \cdot 10^{-2}$ cm^2 sec^{-1}) and higher values below 30 m ($30 \cdot 10^{-2}$ cm^2 sec^{-1}). The
errors in this computation of A may be as high as 20—50%. If heat is being
absorbed in the epilimnion and convective currents occur, the equation of
heat transfer becomes:

$$\frac{dT}{dt} = A \frac{d^2T}{dz^2} - \frac{d(VT)}{dz} + Q(z, t) \tag{9.2}$$

where $Q(z, t)$ = heat source function; and V = convective velocity in z direction (Lerman and Stiller, 1969).

Li (1973) found the vertical eddy diffusion coefficient in the Lake of
Zürich to be a strong function of both depth and season. Maximum values
were found in April, varying from 0.7 cm^2 sec^{-1} (5 m) to 4.3 cm^2 sec^{-1}
(50 m). During summer the coefficient decreased to low values, for example
in September to $2 \cdot 10^{-2}$ cm^2 sec^{-1} at 10 m and $20 \cdot 10^{-2}$ at 50 m.

Even in shallow tropical lakes quite a stable thermal stratification may develop, but since the amount of work necessary for mixing depends on the
length of the path through which the water must be moved (depth of lake),
this stability is more easily destroyed than in deep cold lakes. In the large
and deep tropical lakes an annual pattern of stratification occurs (Talling,
1969). Talling suggested that at higher latitudes the overall controlling mechanism is the variable input of solar energy, and that near the equator the
wind regime and humidity regulate the seasonal cooling and mixing process.
Lakes Tanganyika and Nyasa (= Malawi) do therefore have a "cool" mixed
epilimnion (ca. 200 m) only in July, whilst in Lake Victoria then complete

148

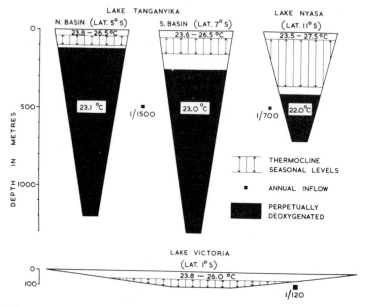

Fig. 9.2. Profiles or vertical sections of some African lakes. (From Beauchamp, 1964.)

mixing occurs. The stratification pattern in these three lakes is well shown by Beauchamp (1964), from whom Fig. 9.2 is taken.

Thermal stratification is usually shown in a time—depth diagram, in which lines connecting points of equal temperature (isotherms) are drawn. Periods of mixing are easily recognisable from the vertical lines, whereas stratification is indicated by a series of horizontal lines (Fig. 9.3).

9.2. MEROMICTIC LAKES

Meromictic lakes are lakes in which stratification occurs throughout the whole year. This may be owing to physical processes (see below for Lake Tanganyika) or to chemical processes. Water containing dissolved material has a greater density than pure water at the same temperature. Therefore, if dissolved salts in the hypolimnion become more concentrated during stratification, the stability increases and autumnal turnover may not occur. Once this process has begun, the dissolved-salt concentration may increase further, and the longer the period of stratification is maintained, the more stable the system becomes.

Findenegg (1937), who contributed much to the knowledge of meromixis, used the terms "static meromixis" if the origin of the solutes is geological, and "dynamic meromixis" if the dissolved material derives from decomposition of sediments, mainly of dead algal material, which may give rise to in-

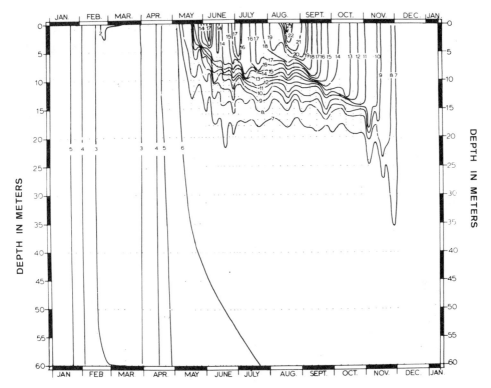

Fig. 9.3. Temperature in Windermere, North Basin (1947). Points of equal temperature are connected. (From Lund et al., 1963.)

creasing concentrations of SiO_2 and $Ca(HCO_3)_2$. Increase of SiO_2 concentration may however result also from erosion and solution.

Hutchinson (1957) calls the dissolving sediment process biogenic, and he uses the term crenogenic to describe situations where saline springs bring dense water into a lake. It is probable that in many meromictic lakes solutes have several different origins. The influence of the reducing power (lower redox potential) of the hypolimnion on solubilisation of products of erosion should not be overlooked.

The lower layer which does not mix through the whole lake is often called the *monimolimnion*; the upper part is called the *mixolimnion* and that between the *chemocline*. The three layers defined by concentration differences are comparable with hypo-, epi- and metalimnion, respectively. The names for the layers defined by temperature or chemical concentration are often interchanged although the layers are not necessarily coincident. In many cases one needs a term to describe the layer of changing temperature or concentration gradient or both. It would be convenient to keep thermocline and chemocline for restricted use, where temperature or concentration are intended specifically, and to use metalimnion in a general sense.

Meromictic lakes are in greatest danger of overturning during autumn, when the temperature difference is least and temperatures are near 4°C where the density differences are smallest. In December in Lac le Pavin, a crater lake in Auvergne (France), Pelletier (1968) found the temperature to be nearly 6°C in the upper 20-m layer (Fig. 9.4). Between 20 m and 60 m the temperature was 4°C, while below this layer the temperature increased to 5.5°C. For this layer we deduced from the measured conductivity an increase in dissolved salt content of about 200 mg l^{-1}, of which $CaCO_3$, Fe^{2+} and SiO_2 were the main constituents. The increase in density due to such a concentration of salts would be 0.00014, which is considerably greater than the difference caused by the temperature difference (4—6°C). Concentrations of Na^+ and K^+ also increased with depth. Pelletier suggested that the main origin of the chemical stratification is biological but that the accumulation of compounds from erosion must also be considered. Non-biological meromixis can arise from NaCl from seawater or Na_2SO_4 and other salts from dissolution of rock.

Kjensmo (1967, 1968) has described several soft-water lakes where variation in iron content was the primary factor rendering the lakes meromictic. The iron concentration in Lake Store Aaklungen (Norway) exceeded 300 mg l^{-1}. Nevertheless, this mechanism may not be wholly abiogenic, since the power to reduce Fe^{3+} to Fe^{2+} certainly originates from the decay of dead algal material.

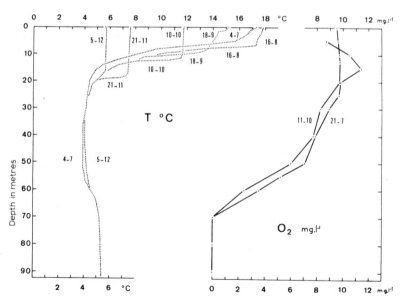

Fig. 9.4. Vertical temperature (4 July and 5 December, 1963) and oxygen distribution (21 July and 11 October, 1963) in Lac le Pavin. (From Pelletier, 1968.)

Lakes Tanganyika and Malawi are also meromictic, but there is no differential concentration of dissolved salts increasing stability. Meromixis in these cases is due to the great depth of the lakes and to the annual cycle of physical events.

A special case is Lake Kivu, where the conductivity increases with depth from nearly 1 400 to well over 2 300. The associated density increase is probably equivalent to that of a 1% NaCl solution (1.005 at 20° C) and this lake is thus meromictic because of its differential salt concentration.

9.3. DESTRATIFICATION

At the end of the warm season stratification gradually disappears, the epilimnion becomes cooler, and the thermocline sinks gradually towards the bottom. This can be seen in Fig. 9.3 where the isotherms show a downwards slope after the horizontal period. When the thermocline sinks, the upper layers of the hypolimnion mix with the epilimnion and since the chemical differences between the upper hypo- and epilimnion waters are not great, few changes occur in the epilimnion. In lakes of moderate depth even the lowest waters eventually become included in the circulation, but effects of their contribution to the greatly increased volume of mixed epilimnion are negligible. If the hypolimnion is not very deep, is anaerobic, and has a high nutrient concentration, important changes in the epilimnion may follow an abrupt mixing period.

For very deep lakes it is probably true to say that when the temperature of the epilimnion reaches that of the hypolimnion they mix. In these very deep lakes the main detectable effects following mixing are a short partial O_2 depletion and slight nutrient release; the hypolimnion differs much less from the epilimnion than in shallower stratifying lakes. The duration of the gradual mixing process depends on climate (autumn storms, winter temperature) and on the depth of the lake. If the lake is deep, sufficient heat is available to prevent freezing and no ice will cover the lake, which will remain mixed throughout the whole winter. When ice does form, and inverse stratification may develop. The stability will be low as the density differences at near freezing temperatures are small. Some wind may therefore break and melt the ice, and the whole lake will become cooler, because the ice has been melted by the warmer water from below.

The influence of wind on water movement is complicated. Wind pushes the surface waters and generates a current whose flow rate depends on the wind speed. As the current reaches the shore it bends downwards and travels in the opposite direction.

Both currents interact causing turbulence by which heat transfer may occur (eddy diffusion). Furthermore, local temperature differences may cause turbulence, especially if a shadow of a cloud moves over the lake and solar radiation is not evenly spread across the lake or when the shadow of mountains falls unevenly. Another source of turbulence is evaporation which causes cooling of the top layers.

Thus many processes can occur causing an extensive movement of water and resulting in energy transport. Currents can also occur in the hypolimnion even if no river flows into the lake. Using aluminium vanes, attached below a cork and flag and suspended at a known depth in the hypolimnion of the small stratified sandpit Lake Vechten, we were able to show strong horizontal currents several metres below the thermocline. These were caused by movements in the epilimnion and perhaps by heat turbulence. In these sandpits there is no movement due to inflowing water or to the rotation of the earth.

Mortimer (1969) has described how the thermocline may start oscillating in stormy times especially when storms or winds arise periodically and when the oscillation period of the thermocline resonates with the wind period. In Lake Kinneret Serruya et al. (1969) found oscillations, with an amplitude of 13.5 m, which were monitored by temperature measurements. The thermocline may ultimately oscillate over a distance of several metres and water from the hypolimnion may "escape" into the epilimnion causing a partial destratification. This is a special case of a common situation in which there is an initial shift of the epilimnion downwind and of the hypolimnion upwind causing a gradient on the lake surface. As a result a surface wind-driven current and a gradient-driven return current running upwind along the top of the thermocline may cause considerable internal mixing in the hypolimnion. Descriptions of movement in stratified lakes and thermoclines and of the dynamics of the autumn overturn are published by Mortimer (1952, 1955, 1961).

9.4. CHEMICAL STRATIFICATION RESULTING FROM PHYSICAL STRATIFICATION

9.4.1. O_2 consumption and CO_2 production

The primary stratification reactions

When the physical stratification is well established, chemical changes in the hypolimnion follow. Due to the sinking of organic matter and the presence of heterotrophic bacteria, O_2 will be consumed and CO_2 will be produced. The organic matter comes mostly from dead phytoplankton which sinks since it no longer has the "power" of upward migration. It is not yet known how much of the phytoplankton is mineralised in the hypolimnion. Early authors believed that the major part would be oxidised below the thermoclines (see e.g. Lund et al., 1963). Later Ohle (1962, 1964, 1968) suggested that as much as 90% of the dead phytoplankton could be oxidised in the epilimnion, but little experimental evidence for this exists. (The problem is discussed in Chapter 8.) Hargrave (1972a) calculated that the O_2 uptake in the sediments of Lake Esrom (Denmark, see page 155) equalled about one-third of the phytoplankton production. Values between 50 and 90% may be tentatively suggested. In a study of the Lake of Lucerne, Ambühl (1969) measured the oxygen deficit by calculating the differences between the March/April and the October concentration values. He showed that the oxygen-loss curves are differently shaped in time and space for waters in several parts of the lake. The curves at his four sampling stations show remarkable differences not related to the expected differences in primary production. The curves show two distinct maxima, one at about 20 m, just below or even in the thermocline, and the other at the contact zone of ooze and water. Usage of O_2 below the thermocline is even larger than this, due to diffusion from above (Li, 1973). The oxygen-loss curves for station I between 1961 and 1967 do not show an increase in oxygen deficit in spite of the progressive eutrophication of the lake during this period (Fig. 9.5).

The degree of oxygen depletion in the hypolimnion depends on the total amount of oxygen present, which varies with lake depth and with the amount of sinking dead organic material which is indirectly related to the primary

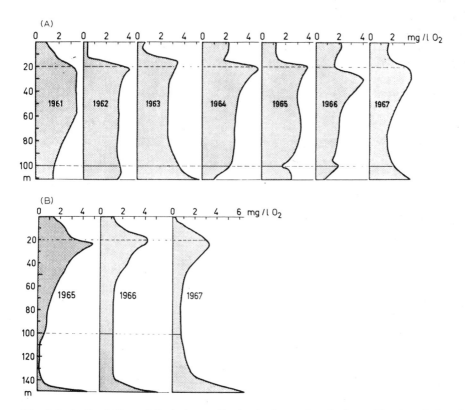

Fig. 9.5. A. Depletion of O_2 between the beginning and end of stratification; Lake of Lucerne, 1961—1967, station I. B. 1961—1967, station II. (From Ambühl, 1969.)

production. The influence of depth is well demonstrated in Windermere, where the deeper North Basin shows higher oxygen values than the shallower South Basin (Lund et al., 1963). The influence of primary production generally is less clear than that of depth.

Very oligotrophic lakes show no oxygen depletion at all (e.g. Lake of Constance, before 1940), whereas very eutrophic lakes (Lake of Zürich until 1970) have a completely anaerobic part near the bottom (Thomas, 1971; see Chapter 17). Mineralisation in the epilimnion may prevent the hypolimnion from becoming anaerobic by decreasing the amount of sinking dead material. The carbon from the sinking dead material is partially converted into CO_2, but there is not always a constant ratio between CO_2 produced and O_2 consumed in the hypolimnion. Several heterotrophic bacteria do not convert their substrates completely to CO_2. Pseudomonads may excrete several different organic acids in considerable quantities. Furthermore when the oxygen is depleted, fermentation processes may occur and these may yield organic compounds. Finally a part of the carbon may even be reduced to methane. On the other hand, the CO_2 produced will lead to a lower pH value which may cause dissolution of $CaCO_3$. The ratio between O_2 and CO_2 is therefore not a simple one. The disappearance of O_2 will lead to a decreased redox potential, while the production of CO_2 leads to a decrease of the pH. Both processes induce a series of events, called secondary hypolimnetic reactions, which will be discussed separately.

Oxygen uptake

Lake sediments take up oxygen as a result of chemical and bacteriological decomposition processes. A distinction between these processes can be made by the addition of poisons such as mercuric chloride, formalin or toluene.

The rate of oxygen uptake of natural sediments can easily be measured with the method of Knowless et al. (1962), even when no sophisticated equipment is available.

Edwards and Rolley (1965) measured an uptake of $0.1-0.2$ g m^{-2} h^{-1} of O_2 ($2.4-4.8$ g m^{-2} d^{-1}). If this oxygen came from a 1 m deep water layer, 50% of the dissolved O_2 had been consumed by the mud. Edwards found no significant change in O_2 consumption with depth in the mud between 4 and 17 cm, contrary to the findings of Baity (1938) and of Fair et al. (1941a, b, c). Baity found that the relation between O_2 demand (Y in mg l^{-1}) and sludge depth (X, in cm) is given by:

$$Y = 2700 \, X^{0.485} \tag{9.3}$$

In all three cases the composition of the mud and also the accumulation times were probably different.

Edwards (1958) stressed the importance of invertebrates in affecting the physical, chemical and microbiological conditions within the benthal deposits in which oxygen demands were measured.

Fair et al. (1941b, c) studied some other factors influencing the O_2 uptake of river mud and have given some mathematical equations concerning the effects of temperature. They also studied (1941a, c) the decrease in potential oxygen demand of the decomposable matter. This decrease was caused by anaerobic decomposition, which becomes more important when the sediments grow in depth, although quantitatively the aerobic processes still remain the most important. Although their work on river muds is connected with sewage and sludge purification, comparable processes are bound to occur at the bottom of most eutrophic waters. A high productivity in the water will therefore always lead to the formation of a sediment with a high oxygen uptake, but the depth of the hypolimnion through which the sinking detritus falls has a modifying influence.

Hargrave (1972a) compared the oxygen deficit in the hypolimnion of Lake Esrom with the sediment oxygen uptake. This uptake was high (about 40 ml m^{-2} h^{-1} of O_2) in June — a month after the peak in primary production — reflecting the time lag before sedimentation of the phytoplankton. The O_2 uptake rate decreased during summer owing to O_2 depletion in the hypolimnion. In October the rate peaked again following oxygenation of the hypolimnion. The O_2 uptake calculated from O_2 depletion in the hypolimnetic water agreed closely with that measured with cores. Hargrave therefore suggested that all O_2 was taken up by the sediments: markedly different from Ambühl's conclusions. Hargrave found no biological O_2 demand in the water in 6 h, but it is possible that longer exposure times would have shown measurable O_2 uptake. The difference from Ambühl's results may also be explained by the great difference in depth. Ambühl measured O_2 uptake just under the thermocline at about 25 m, which is about the same depth as the bottom of Lake Esrom (max. depth 22 m). In a second paper Hargrave (1972b) found a much higher O_2 uptake during incubation of Lake Esrom mud in the laboratory at 20°C. The increase as compared with the mud in situ is partly owed to shaking, partly to the higher temperature, and partly to higher O_2 supply. In these circumstances, Hargrave found a simple relationship between surface area, organic content and O_2 uptake of the sediments and thus with size and metabolism of the communities of bacteria associated with the particles. Hargrave suggested that microorganisms colonising various surfaces may have similar rates of metabolism and that in aerobic habitats colonisation and succession may occur to the point where all the O_2 supply is fully used. Edberg and Hofsten (1973) compared oxygen uptake by soft bottom sediments in situ and in cores in the laboratory. Oxygen uptake rates in the field were between 0.3 and 3.0 g m^{-2} d^{-1}. In the cores the rates were always lower although sometimes only slightly so. This difference might be due to the different ratio between mud and overlying water in the two devices and to the induced hydrodynamic differences.

156

9.4.2. The oxidation–reduction (redox) sequences

If a solution has oxidative or reducing capacity, this capacity can be described by the so-called redox potential, which is the equivalent free energy change per mole of electrons for a given reduction in a *reversible system.*

It is possible to define an electron activity [pE = −log e⁻] (just as it is possible to define the pH), although the solution does not contain free electrons (or protons). The pE is large and positive in strongly oxidising solutions (low electron activity) as is pH in alkaline solutions. The redox potential can be measured with metal rods placed in the solution.

According to Nernst the potential difference of a metal rod in a solution of its ions depends on the tendency of the metal to dissolve and the concentration C of the metal ions in solution:

$$E = \frac{RT}{nF} \ln \frac{K}{C} \tag{9.4}$$

where R = molar gas constant; T = absolute temperature; n = numbers of electrons involved per atom; F = the Faraday; and K = a constant. When the concentration exceeds a certain value (saturation value) the metal tends to precipitate on the metallic rod. Two solutions with the same redox potential may have completely different reducing powers. A concentrated solution of a weak reductant may have the same redox potential as a dilute solution of a strong reductant. A metallic rod in a solution of its salt is called an "indicator electrode". It is not possible to measure the potential of only one electrode. Two rods must be placed in two solutions of concentrations C_1 and C_2 and connected by a salt bridge. The potential between the two rods will be:

$$E = E_1 - E_2 = \frac{RT}{nF} \ln \frac{C_2}{C_1} \tag{9.5}$$

The potential of an electrode is measured by comparing it with the potential of an arbitrary standard electrode, the "reference electrode"; the normal hydrogen electrode was originally used but now it is more common to use a calomel one. The type of reference electrode used should always be specified, as should be the reference zero: hydrogen or calomel.

The hydrogen zero is preferable. The E.M.F. between two Pt rods in two separate solutions (A and B) is:

$$E = \frac{RT}{nF} \left(\ln \frac{[Ox]_A}{[Red]_A} + \ln \frac{[Ox]_B}{[Red]_B} \right) \tag{9.6}$$

provided that the oxidation–reduction reactions are reversible. If distilled water is saturated with O_2, the redox potential is about 500–600 mV (relative to the calomel reference electrode). It is said that the liquid has oxidative power since it will have a tendency to take up electrons. Water from a hypolimnion which is deprived of O_2 will show a lower redox potential, down to 100 mV; the solution has a strong reducing capacity — it will try to give off electrons.

Below the mud—water interface the redox potential often falls sharply, values of −100 mV being not infrequently found. But the disappearance of oxygen does not of itself give the water a reducing power. For a reduction to occur electrons are necessary. For a *specific* reaction the quantity of electrons available controls the amount of matter which can be reduced, whilst the redox potential is an indication of how powerful the reduction can be, and this will govern which types of compounds can be reduced. The disappearance of oxygen itself will allow the existence of compounds which otherwise would be oxidised. These compounds, when present, give the water reducing capacity. Among these substances are Mn^{2+}, Fe^{2+}, H_2S and several organic compounds. Because the organic compounds in hypolimnetic waters are often not reversibly oxidisable and reducible, the redox potential in a lake as measured with a platinum and calomel electrode is quite different from the thermodynamically well-defined redox potential. The redox potentials of some organic compounds are sensitive to the pH of the solution. For purposes of comparison the potentials of pH-sensitive systems are calculated to E_7, that is the potential the system would have at pH = 7. The calculation is made by assuming a rise of one pH unit to be the equivalent of a decrease of 58 mV. This adjusted value is not the potential which actually operates in the system.

Potentials measured with Pt electrodes in different (non reversible) lake-waters or mud systems should therefore be compared only very cautiously and should not be called redox potentials. It has been suggested that they be called "apparent redox reaction". The sharp decrease of redox potential as measured with Pt electrodes at the sediment—water interface may have a limited meaning. Because measurements of the "apparent redox reaction" give a good relative indication of the sequence of events in anaerobic systems, limnologists will certainly continue to use these. Nowadays it is possible to make continuous measurements (Effenberger, 1967; Machan and Ott, 1972). The latter authors questioned the significance of single isolated measurements of short duration and reported that during cyclic events comparable values were recorded for a series of measurements involving more than one cycle; they considered such results to indicate that electrode poisoning does not occur even after a long period.

As an example of a typical redox sequence in mud—water systems, Mortimer's results from Esthwaite Water will be discussed here. Mortimer (1941, 1942) made extensive studies of the changes of the (apparent) redox reactions in the mud and overlying water during circulation and stratification in samples as little disturbed as possible. He also made measurements in artificially established ooze—water systems, with electrodes mounted on a frame at fixed height in the columns. In oxygenated water above the mud of natural

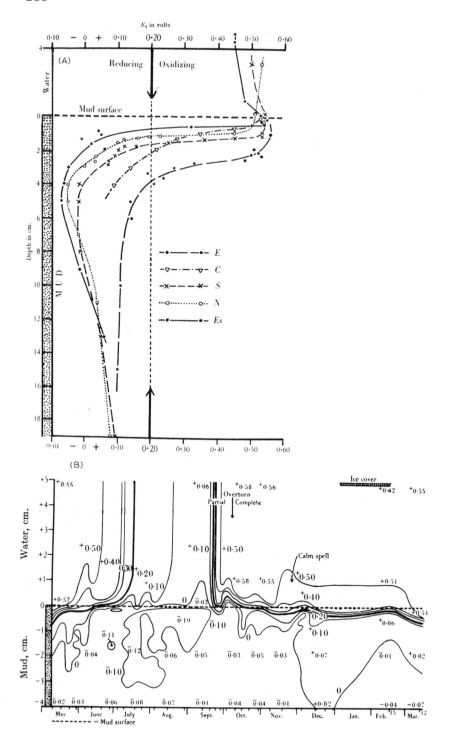

E_7 in volts

(A)

Reducing   Oxidizing

Mud surface

Water

Depth in cm.

M U D

- - ● - - E
- - ▽ - - C
- - ✕ - - S
······ ○ ····· N
── ● ── Es

(B)

Water, cm.

Mud, cm.

Ice cover

Overturn

Partial | Complete

Calm spell

------ = Mud surface

May   June   July   Aug.   Sept.   Oct.   Nov.   Dec.   Jan.   Feb.   Mar.

systems (oligotrophic lakes or eutrophic lakes during circulation), redox potentials ($E_7 = \underline{E}$ at pH = 7) of about 600 mV were found, which is the normal value for free oxygenated water. Within the sediments E_7 declined rapidly and reached a minimum at about 5 cm below the surface. At lower depths the redox potential sometimes increased, perhaps indicating diminished metabolic activity. Fig. 9.6A shows the courses of the E_7 curves in the sediments of four lakes of different trophic level at times of circulation and stratification.

If the oxygen in the hypolimnion declines, the redox potentials in mud and overlying water decrease and the horizontal surfaces of equal E_7 values migrate upward towards the surface of the sediments (Fig. 9.6B, Fig. 7).

As long as high values prevail at the surface of the sediments, $Fe(OH)_3$ will be found; the mud is brownish. As soon as the surface value falls below 200 mV, a series of reduction reactions set in, which Mortimer followed both in the lakes and in artificial tanks. The most important reactions are the reductions of nitrate, nitrite, ferric iron and sulphate. It is not certain whether the nitrite is formed from nitrate. The ammonia is certainly not produced from nitrite. These reactions were demonstrated both in the field and in experimental tanks. The reductions began when the "apparent redox reaction" fell below a certain value and were completed at a somewhat lower value. The results are given in Table 9.2, which includes both the "apparent redox reaction" and coincident O_2 concentration.

Reductions may of course have started at values of the "apparent redox reaction", higher than those at which reduced compounds were first detected. This is because the reduced compounds had to diffuse out of the mud, and by the time they were detected, the "apparent redox reaction" of the mud was lower still.

The "apparent redox reactions" of Table 9.2 are not those of the redox reactions 1—4: 1, 2 and 4 are bacteriological processes. The reactions 1 and 2 are partly discussed in Chapter 6, as they occur in the N cycle. The sequence of the bacteriological reactions may be explained by the changed course of bacterial respiration. As long as O_2 is available, the bacteria will use O_2 as electron acceptor for their respiration. Below O_2 concentrations of about 4 mg l^{-1}, NO_3^- will partly replace the O_2, with formation of N_2. When O_2

Fig. 9.6. A. Typical winter distribution of redox potential (E_7 in volts) in the surface mud cores from the deep regions of various lakes. Ennerdale Water (E), 40 m; Crummock Water (C), 40.8 m; Windermere, South Basin (S), 31 m; Windermere, North Basin (N), 65 m; and Esthwaite Water (Es), 14 m. (Mortimer, 1941, 1942.)

Fig. 9.6. B. Esthwaite Water, 1940. Depth—time diagram of the distribution of redox potential (E, in volts) above and below the mud surface. The water at this point is 14 m deep. (Mortimer, 1941, 1942.)

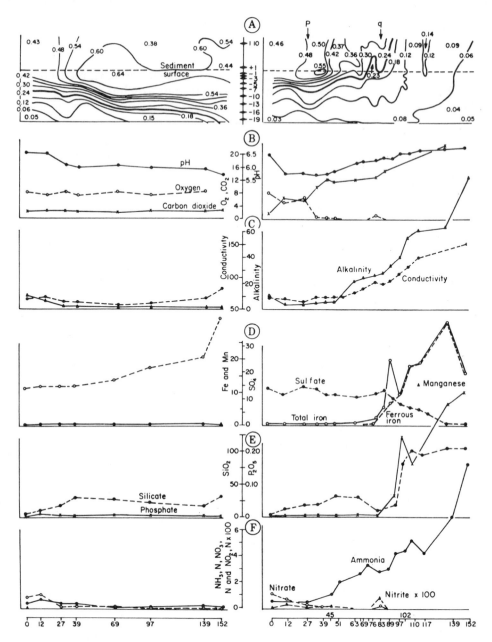

Fig. 9.7. Comparative measurements over an interval of 152 days in two experimental sediment—water systems. Left-hand series: the aerated tank which remained exposed to the atmosphere. Right-hand series: the anoxic tank which was sealed off from the atmosphere. The following measurements were made: A, distribution of redox potential (E_7) across the sediment—water interface; B, pH and concentration in the water of O_2 and CO_2; C, alkalinity (as $CaCO_3$) and electrical conductivity ($K_{18}^2 \times 10^{-6}$); D, iron (total and ferrous as Fe) and sulfate (SO_4); E, phosphate (P_2O_5) and silicate (SiO_2); F, inorganic forms of combined nitrogen (nitrate, nitrite $\times$ 100, and ammonia, all as N). Concentrations in mg l^{-1}. (From Mortimer, 1941.)

TABLE 9.2

Reduction of some compounds and apparent redox potentials and prevalent O_2 concentrations

Reduction reaction	Apparent redox potential (mV)	Prevalent O_2 concentration (mg l^{-1})
1. $NO_3^- \rightarrow NO_2^-$	450—400	4
2. $NO_2^- \rightarrow NH_3$	400—350	0.4
3. $Fe^{3+} \rightarrow Fe^{2+}$	300—200	0.1
4. $SO_4^{2-} \rightarrow S^{2-}$	100— 60	0.0

Note: The lower potentials in the given range are those below which Mortimer could not detect the oxidised phase.

has been completely exhausted, sulphate will be used as electron acceptor for the respiration of the strictly anaerobic *Desulphovibrio desulphuricans.* (The bacteria will be considered in Chapter 16.) Much work has been done on the chemistry of reactions 3 and 4 (Einsele, 1936—1938; Ohle, 1937, 1938).

According to Einsele the reaction that occurs first when the "apparent redox reaction" falls is the reduction of sulphate by the bacterium *Desulphovibrio desulphuricans*:

$$H_2SO_4 + 8[H] \rightarrow H_2S + 4H_2O \tag{9.6}$$

The hydrogen [H] comes from the organic compounds — the [H] donor — which are being formed during the anaerobic decomposition of organic matter.

The reduction of $Fe(OH)_3$ in the mud would then be according to the reaction:

$$2Fe(OH)_3 + S^{2-} \rightarrow 2Fe^{2+} + S + 6(OH)^- \tag{9.7}$$

The Fe^{2+} appears in the water of the hypolimnion and is supposed to be balanced by HCO_3^- ions. The exact nature of this process is unknown; the reaction:

$$FeS + 2CO_2 + 2H_2O \rightarrow Fe^{2+} + 2HCO_3^- + H_2S \tag{9.8}$$

as proposed by Ohle (1937) and by Einsele seems improbable as can be calculated from the pK's of H_2CO_3 and H_2S and the prevailing environmental conditions.

The maximum solubility of Fe^{2+} can be calculated according to Stumm and Morgan (1970, p. 184):

$$(FeCO_3)_s \rightleftharpoons Fe^{2+} + CO_3^{2-} \qquad \log K_s \approx -10.4$$
$$H^+ + CO_3^{2-} \rightleftharpoons HCO_3^- \qquad \log K_2 \approx -10.1$$

$$(FeCO_3)_s + H^+ \leftrightarrow Fe^{2+} + HCO_3^- \qquad \log K_s^\star \approx -0.3$$

therefore $\log [Fe^{2+}] = \log K_s^* - pH - \log [HCO_3^-]$.

In the hypolimnion of Ca-rich lakes where an alkalinity of $6 \cdot 10^{-3}$ eq. l^{-1} may be found at pH = 7, we obtain $\log [Fe^{2+}] = -5.1$, which means that the maximum solubility can be only 0.5 mg l^{-1}. In softer water, with an alkalinity of 10^{-4} eq. l^{-1} $\log [Fe^{2+}] = -3.3$ or $5 \cdot 10^{-4} M \approx 25$ mg l^{-1}. The actual measured values are higher than this. It would seem that the Fe^{2+} in the hypolimnion must be in colloidal solution, at least in waters with an alkalinity above 1 meq. l^{-1}. Kjensmo (1967) supposed the iron in Lake Store Aaklungen to be in solution as iron bicarbonate, but calculations from his conductivity values indicate that irregularities were occurring, which suggests that iron is not present entirely as Fe ions. Unfortunately his ionic analyses are incomplete, so this idea cannot be tested.

The reduction of Fe^{3+} by H_2S has been studied extensively by Einsele (1936), who found that the reaction is not stoicheiometric. In his experiments an eightfold excess of H_2S was necessary to cause all the iron to appear in solution.

It is not clear why the first formed Fe^{2+} ions are not immediately precipitated by the remaining H_2S. The excess amount required can be reduced by adding the H_2S very slowly to the Fe^{2+} solution. An excess is still needed and will indeed partially precipitate the iron as FeS. Although there are no indications in the literature, Golterman (1967) has suggested a direct reaction between the Fe^{3+} and some [H] donors (see Chapter 8). Many organic acids (oxalic acid and others) can chemically reduce Fe^{3+} especially in the light. If this were so, the reduction of Fe^{3+} would not be a sequential reaction following on the SO_4^{2-} reduction but a competitive one. This is consistent with Mortimer's results. He found that the Fe^{3+} reduction started at 0.1 mg l^{-1} of O_2 and the SO_4^{2-} reduction at 0.0 mg l^{-1}. Further indications for a direct reduction by organic matter may be found in the work of Tessenow (1972), although Tessenow did not measure a possible sulphate reduction.

When the lake stratification is destroyed, the "dissolved" Fe^{2+} is oxidised by O_2. As the pH is usually greater than 6.0, the Fe^{3+} ions formed precipitate as hydrated $Fe(OH)_3$, which is the starting point of reaction (9.7). When phosphates are present, these are either co-precipitated as $FePO_4$ or, more likely, are adsorbed onto $Fe(OH)_3$ (see subsection 9.5).

Using very small platinum-tipped redox probes, Hargrave (1972b) studied oxidation—reduction potentials in undisturbed sediment cores from Lake Esrom, and confirmed many of Mortimer's results. Hargrave gave much attention to stabilisation and to the possibility of poisoning of the electrodes. Just as in Mortimer's experiments, reducing conditions rose to the surface of Lake Esrom mud when O_2 was depleted. If O_2 was present the oxidised crust was about 0.5—1 cm thick. It is remarkable to note how closely the O_2 concentrations and concomitant E_h values parallel those found by Mortimer. The O_2 uptake by mud at various depths was inversely related to E_h. E_h and

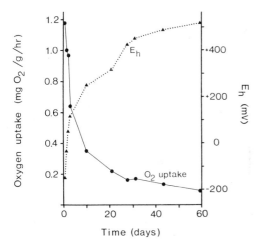

Fig. 9.8. Changes in E_h potential and oxygen uptake in subsurface profundal mud from Lake Esrom during aeration (Hargrave, 1972b).

square root O_2 uptake were also inversely related, suggesting that intensity and capacity of O_2 consumption are related (Fig. 9.8). Formalin reduced O_2 uptake to 20% at the surface of undisturbed cores. Presumably 80% was biological uptake and the rest chemical. Below 1 cm all oxygen was consumed chemically.

9.4.3. The pH sequence

As a result of the increase of CO_2 content in the hypolimnion the pH will decrease. Mortimer found pH values in Esthwaite Water around pH = 6. In calcium-rich lakes CO_2 may dissolve $CaCO_3$ which either occurs in the mud or which sediments due to biogenic or abiogenic decalcification of the epilimnion. Thus the alkalinity will increase:

$$CaCO_3 + CO_2 + H_2O \rightarrow Ca^{2+} + 2HCO_3^- \tag{9.10}$$

Two molecules of HCO_3^- are not necessarily formed for every molecule of CO_2 because the pH shift must be taken into account (see subsection 3.3.1) and considerable amounts of free CO_2 will remain in solution. In the Vechten sandpit we found an alkalinity value in the hypolimnion of about 6 meq. l^{-1}, at pH = 7, while the epilimnion contained 2 meq. l^{-1} at pH = 8. As at pH = 7 the CO_2 will be 20% of HCO_3^-, the CO_2 will have a concentration of $0.2 \cdot 6$ meq. l^{-1} = 1.2 meq. l^{-1} = 52 mg l^{-1} of $(CO_2)_{free}$. Therefore the water contains a hundred times more free CO_2 than in the situation in which the water is in equilibrium with the air. Few data are available on the vertical distribution of Ca^{2+}, HCO_3^- and pH in hypolimnia. Mortimer has studied the maximum concentrations occurring just above the mud. He found that in experimental tanks more HCO_3^- accumulated in anaerobic than in aerobic conditions.

Hutchinson's (1957) conclusion that HCO_3^- diffuses more easily from anaerobic than from aerobic mud is probably incorrect because in Mortimer's experiments the aerobic tank was in equilibrium with air so that the CO_2 produced could escape, but the anaerobic tank was sealed from the air. Quantitative experiments in which redox conditions are changed independently of the pH and the CO_2–HCO_3^- system are clearly necessary. As long as these processes continue to be so imperfectly understood, it is probably worthless to calculate the respiratory quotient of a lake from the changes of O_2 and CO_2 concentrations of the hypolimnion. In calcium-rich waters calcium carbonate may precipitate as the pH rises in the epilimnion during photosynthesis. The precipitate will sink until it reaches the layers at the bottom, with lower pH values resulting from CO_2 release, where it will dissolve again. The process may take place near the mud and cannot therefore be distinguished from direct dissolution of calcium carbonate from the mud. Because calcium carbonate precipitates may adsorb phosphate, the calcium metabolism may control that of the phosphate (see below).

9.4.4. Secondary chemical stratification

As a result of the O_2 and CO_2 reactions, many other reactions will occur due either to the lower "apparent redox reaction" or to the increased acidity. In most hypolimnia there will be increasing concentrations of phosphate, ammonia and silicate as well as of Fe^{2+}, Ca^{2+} and HCO_3^- as reducing conditions establish. The cause of the ferrous-iron appearance has been described above. At the end of stratification, when $Fe(OH)_3$ forms, the phosphate may be adsorbed onto this compound. In many hypolimnia for about a week or so a brownish precipitate of $Fe(OH)_3$ may be found. The filtered precipitate has been found to contain considerable amounts of phosphate and so by this mechanism most phosphate is trapped in the hypolimnion. It seems more likely that "iron phosphate" is a precipitate of iron hydroxide with adsorbed phosphate than that it is true $FePO_4$, although in waters with a chemically suitable composition $FePO_4$ may occur. For the sake of simplicity the formula "$FePO_4$" will still be used. The phosphate is thought to be liberated from the iron precipitate after reduction and to be released in soluble form simultaneously with the ferrous ions. When soluble phosphate is found in the hypolimnion, it will diffuse towards the epilimnion by eddy diffusion, which may be stimulated by migrating animals causing currents in the water. Whether the animals transport phosphate due to their metabolism is still an open question. Tessenow (1964) described how *Chironomus* larvae by their pumping movements increased four times the silicate diffusion from interstitial waters into the hypolimnion. It is doubtful if the extrapolation from laboratory experiments to natural population densities is justified because in natural situations other water movements also occur and the influence of animals in the laboratory may thus be overestimated.

The simultaneous appearance of Fe^{2+} and phosphate in the hypolimnion is well established. A study of iron- and phosphate-concentration changes over a period during the development of the hypolimnion reveals a striking resemblance between the two concentration—time curves, with marked correlation of any irregularities. Nevertheless, no proof exists that the soluble phosphate in the hypolimnion was originally bound onto $Fe(OH)_3$. Golterman (1973) has argued that if phosphate is being continuously released from the mud, it may become swept away upwards during the mixing period. But as soon as stratification occurs, the water above the mud becomes stationary and the phosphate concentration starts to increase. If observed irregularities in the Fe^{2+} time curve are due to special water movements, those same irregularities would be expected to be found in the phosphate curve. Evidence that the hydrated $FePO_4$ is not so easily reduced as has been commonly supposed comes from the work of Thomas (1965), who proved that $FePO_4$ in activated sludge does not pass into solution during the anaerobic digestion stage. If this is true, the phosphate released from lake mud should come from compounds different from the phosphate compounds accumulating in the mud, because the formation of iron phosphate in mud is well established. Indications of a slow but constant release of phosphate from the sediments are to be found in the work of Tessenow (1972), who studied silicate, iron and phosphate release in anaerobic and aerobic tanks. Tessenow stressed the importance of the high concentrations in the interstitial waters and believed that iron phosphate is not reduced and thus is not released from the oxidised crust. A great part of the phosphate released in Tessenow's experiments came from the diatoms that he let sediment on the ooze and that apparently died there.

Under natural circumstances most of the plankton phosphate would have been released in the water layers far above the sediments. It is remarkable that in Tessenow's aerobic experiment the ooze released phosphate between 1 and 6 cm but bound phosphate in the upper centimetres, while in the anaerobic experiment phosphate was released from the top 4 cm. Iron behaved similarly. Tessenow did not try to measure "$FePO_4$" in the sediments; measurements which would have given interesting results. Data on the fate of the organic material are lacking.

A second mechanism governs phosphate concentrations in calcium-rich waters. Phosphate may precipitate with $CaCO_3$, either being adsorbed on or included as hydroxy apatite, $3\,Ca_3(PO_4)_2 \cdot Ca(OH_2)_2$, which may form during photosynthesis in the epilimnion. Theoretical aspects of the system have been discussed by Golterman (1973); experimental proof of its occurrence in marl lakes is given by Otsuki and Wetzel (1972), who used radioisotopes. When the precipitate sinks towards the hypolimnion the low pH near the mud may cause the phosphate to dissolve again although this may be a slow process.

The solubility of calcium phosphate changes markedly with pH. In the equation for the solubility product the PO_4^{3-} concentration occurs to the sixth power and the PO_4^{3-} proportion of the total inorganic phosphate ($H_2PO_4^- + HPO_4^{2-} + PO_4^{3-}$) is strongly pH-dependent:

$$PO_4^{3-} = \frac{1}{1 + 2.8 \cdot 10^{12} \cdot [H^+] + 1.4 \cdot 10^{19} \cdot [H^+]^2} \times PO_4\text{-}P \qquad (9.11)$$

The solubility product of $3\,Ca_3(PO_4)_2 \cdot Ca(OH)_2$ is:

$$[Ca^{2+}]^{10} \, [PO_4^{3-}]^6 \, [OH^-]^2 = \sim 10^{-100} \qquad (9.12)$$

For the Lake of Geneva PO_4-P concentration has been shown gradually to increase in time up to a certain point and thereafter to decrease abruptly. The product of Ca^{2+}, PO_4^{3-} and OH^- ion concentrations at the period of abrupt change was exactly the value for the solubility product of hydroxy apatite, which suggests that in this lake phosphate was indeed controlled by the calcium concentration (cited in Golterman, 1973). Hydroxy apatite may also be formed secondarily in the mud after sedimentation of other insoluble phosphates. Thus all phosphates in the deeper sediments of Lake Erie appear to be in this form (Williams and Mayer, 1972).

Silicate in the hypolimnion will also increase and the silicate time curve resembles the iron curve in the same way as does the phosphate curve. Early workers suggested that silicates in the hypolimnion might resemble phosphates in being coupled with the iron system. But since iron concentrations (in mg l^{-1}) of Lake Vechten are often little more than twice those of SiO_2-Si, this supposition seems unlikely to be true. Dependence of SiO_2 concentrations on the redox reactions seems also to be unlikely.

If clay is present, a weathering reaction such as:

$$NaAlSi_3O_8 + CO_2 + 5\tfrac{1}{2}H_2O \rightleftharpoons Na^+ + HCO_3^- + 2H_4SiO_4 + \tfrac{1}{2}Al_2SiO_2O_5(OH)_4$$
$$\text{kaolinite}$$
$$(9.13)$$

may occur (Chapter 18). When studying the SiO_2 concentrations for some years in the hypolimnion of one lake, the reproducibility of these values is striking and must be explained by assuming a constant release in time or a stoicheiometric reaction with the increased CO_2. After the autumn overturn most of the silicate will be found in the epilimnion. During the next period of diatom growth it becomes incorporated in algal cell walls and will precipitate towards the bottom as the cells die. If mineralisation of these cell walls occurs to any great extent (see Chapter 7), SiO_2 might be continually given off from the mud in a manner similar to that tentatively suggested for phosphate. This would provide an explanation for the similarity between the silicate and iron concentration—time curves.

Solubilisation may be dependent on CO_2 where silicates derive from clay, or on the HCO_3^- concentration where they come from mineralisation of diatoms. This would explain why in Mortimer's experiments the silicate concentrations increased less in the aerobic than in the anaerobic tanks. In Tessenow's experiments no difference was found between aerobic and anaerobic experiments. In his experiments aerobic and anaerobic HCO_3^- concentrations were probably equal as both experiments were being flushed through with air or N_2. It is striking that the silicate concentration remained rather low (1.5 mg l^{-1}); in his earlier work he reported much higher values. Apparently after the addition of diatoms no extra silicate was released.

Under reducing conditions manganese, like iron, is released from the mud. It is not known whether manganese occurs in the mud as the tri- or tetravalent oxide (Mn_2O_3 or MnO_2, both in hydrated form). Since the E_0 values of both forms are higher than that for iron, one might expect to detect manganous ions at redox potentials too high for those allowing ferrous ions to dissolve, if one disregards possible influences of concentrations of organic compounds, etc. Mn^{2+} can in fact be found in the upper layers of the hypolimnion. Theoretically it seems likely that it will be oxidised there and settle as MnO_2, which on its way back would oxidise the Fe^{2+} diffusing upwards. The decrease of Mn^{2+} in Linsley Pond (Hutchinson, 1957) in the 5—7 m level while it was increasing in the deeper layers and at the same time (between 22 August and 16 September, 1938) the concentration of Fe^{2+} in the whole hypolimnion was increasing, could thus be explained. The explanation is the more plausible because the thermocline was sinking during this period, causing more oxygen to diffuse downwards. This mechanism could also explain the fact that although the ratio of Fe to Mn is 19 to 1 (by weight) in the sediments, it is much smaller in the hypolimnetic water above the sediments. The seasonal variation of manganese concentrations in Lake Mendota (U.S.A.) has been described by Delfino and Lee (1971), who suggested that besides dissolved oxygen the pH may also be an important factor controlling the Mn metabolism.

Iron and manganese may also come into solution from sources other than the bottom sediments. Serruya (1972) found very distinct maxima of both dissolved iron and manganese just below or even in the metalimnion, where reducing conditions prevailed owing to sinking dead algal cells (see Chapter 8). Deeper in the hypolimnion lower concentrations of both iron and manganese occurred, indicating that in the metalimnion sedimenting iron and manganese probably derived from the inflow streams were solubilised by reduction.

Lastly the NH_3 concentration increases in the hypolimnion, particularly just above the mud. Very little information on its origin is available, but the following three sources are possible:

(1) Release from decomposing detritus. Since most of the nitrogen in phytoplankton is protein nitrogen, most of which will be mineralised in the epilimnion, this source seems unlikely to be quantitatively significant because only the more refractory non-protein nitrogen could contribute and part of it will be converted into humic material.

(2) Anaerobic denitrification of nitrate. The occurrence of bacteria which reduce NO_3^- to NH_3 has already been mentioned, besides those which reduce NO_3^- to N_2 (Chapter 6), but the extent of their metabolic activities in the hypolimnion has not yet been evaluated.

(3) Adsorption of NH_4^+ on sediments. H. Verdouw (unpublished) has shown that equilibria exist between NH_4^+ adsorbed onto mud and that dissolved in water. He found indications that more NH_4^+ is adsorbed under oxidising than under reducing conditions, suggesting the involvement of iron once again. Nevertheless, it is not clear how a certain amount of ammonia can circulate in the hypolimnion without escaping into the upper levels during the autumn overturn, since it seems unlikely that the re-oxidised sediments can re-adsorb the ammonia from layers a few metres higher up in the hypolimnion. It seems improbable that there should be an "ammonia trap" similar to that suggested for phosphate.

An annual cyclic input is insufficient to explain the phenomenon because the difficulties mentioned earlier under point (1) would still operate. Mortimer's result for his anaerobic tank showed that ammonia concentrations began to rise before those of iron but their most rapid increase occurred later when the redox potential was about 100 mV. A constant release over the whole year, becoming apparent only after stratification, should then be balanced by a constant input. It is likely that all three events may contribute to the phenomenon, the relative importance of each mechanism probably varying from lake to lake.

9.5. ADSORPTION ON SEDIMENTS

Adsorption of phosphate on $Fe(OH)_3$ and $CaCO_3$ has already been discussed, but so far little mention has been made of adsorption on clay minerals, which occur in most lake sediments and originate from eroded material transported into the lake by the river(s) (see Chapter 18).

Although clays are commonly recognised as cation exchangers, they do also adsorb large quantities of phosphate. Golterman (1973) reviewed phosphate-binding processes on sediments and discussed possible mechanisms. He cited values of adsorbed phosphate (per 100 g) on kaolinite of 31.2 meq. at pH = 7.2 and 88.2 meq. at pH = 8.2, while for montmorillonite values were 22 meq. at pH = 6.8 increasing to 100 meq. at pH = 2.8.

Livingstone and Boykin (1962) studied phosphate adsorption on sediments of Linsley Pond. The mineral part of the sediments could not have been apatite since it was shown to be insoluble in acids; thus it seems likely that phosphate was adsorbed by clays. Similar indications were reported by Mac-Pherson et al. (1958), who found that dried mud and its ash showed a maximum adsorption between pH = 5 and pH = 7.

Experiments with $^{32}PO_4$ enable one to disentangle the processes involved. A clear distinction must be made in such experiments between the *adsorption process*:

$$X \sim (Y)_3 + PO_4^{3-} \rightleftharpoons X \sim PO_4 + 3Y^- \tag{9.14}$$

where Y stands for OH^- or other such ions and X for Al^{3+} or Fe^{3+} in the clay, and the isotopic *exchange process*:

$$X^{31}PO_4 + {}^{32}PO_4^{3-} \rightleftharpoons X^{32}PO_4 + {}^{31}PO_4^{3-} \tag{9.15}$$

A careful study was made by Olsen (1958, 1964), who equilibrated a labelled phosphate solution with a coarse sediment from 3 m deep water and with a fine-grained deep-water sediment, both in oxidised and reduced form. For all four cases the gross adsorbed quantity could be described by the same mathematical equations (see Fig. 9.9):

$$A = K \cdot C^v \tag{9.16}$$

where A = amount of PO_4-P on dry mud; C = concentration of PO_4-P in water; K and v are constants.

The relation for the exchange process was found to be:

$$b = K_b \cdot C^{-V_b} \tag{9.17}$$

so that the net adsorbed quantity was:

$$a = A - b = K \cdot C^v - K_b \cdot C^{-V_b} \tag{9.18}$$

Olsen showed that the exchange of phosphate is a rapid process and that the uptake of phosphate from the water by algae will be followed by a release of phosphate from the sediments with different rate constants for oxidised

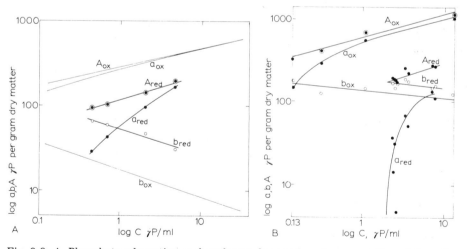

Fig. 9.9. A. Phosphate adsorption and exchange for a reduced coarse sediment from 3 m in Furesø (Denmark). For comparison the results from experiments with the same sediment in oxidised state are indicated. For further details see text. (From Olsen, 1964.) B. As A, but for a very fine-grained deep-water sediment from 35 m in Furesø.

and reduced sediments. Comparison of the two equations showed that the oxidised sediments adsorb larger quantities than the reduced ones. Therefore reduction of the clay would be expected to result in the release of phosphate, and this is found to be so in the hypolimnion of most lakes. In reduced state the sediment from deeper water showed no adsorption of phosphate (and thus only release) when the concentration of phosphate in the water was below the relatively high value of 2 mg l^{-1}. Our knowledge of the relation between iron in the clay and the influence on this of the redox potential is fragmentary. It may be that Olsen's reduced "deep" sediment had lost all its iron and that two adsorption mechanisms are operative. The main difficulty is that iron may be both part of the clay mineral itself and may also be adsorbed as a cation. These fractions would behave differently when reduced, the adsorbed atoms being more labile than those which are part of the clay structure.

It seems probable that the so-called "organic phosphate" in sediments — calculated as the difference between total minus (acid + alkaline extractable) fraction — is mostly clay-bound phosphate (Golterman, 1973). Organic phosphate probably exists only as humic-iron-phosphate. This is an iron-phosphate bound on organic matter. True organic phosphates have a C-O-P group which is more labile than that of the postulated sediment "organic phosphorus", and they should therefore hydrolyse rapidly.

9.6. TRANSPORT FROM HYPOLIMNION TO EPILIMNION

Summarising the preceding subsections, the following changes of concentrations can be found in the hypolimnion:

Increase	Increase	Decrease
$CO_2 + HCO_3^-$	PO_4-P	O_2
Ca^{2+}	NH_4-N	NO_3^-
Fe^{2+}	SiO_2-Si	SO_4^{2-}
Mn^{2+}	H_2S	

As concentrations change, diffusion rates change too because they are controlled by the concentration gradient. The concentration gradient in the hypolimnion will in turn depend on the release of these compounds from the mud, where in the interstitial water the concentrations may be higher than in the hypolimnetic water. Transport from interstitial water to hypolimnetic water depends partly on diffusion, partly on hydrodynamics and perhaps partly on the animals living in the mud.

Quantitative data are scarce. From the bottom upwards in still water the concentration decreases approximately exponentially. Reduced compounds may be detectable for only a few metres above the sediments. The total amount of reduced compounds that can be produced is a function of the reducing power derived from the sinking detritus — a subject discussed in Chapter 8. The upward diffusion of Fe^{2+} and Mn^{2+} will be limited by the diffusion of O_2 downwards, causing the two metals to be re-oxidised and thus to precipitate. A special interaction of Mn^{2+} in the electron transport between O_2 and Fe^{2+} has been suggested. The minor constituents nitrogen and silica are not restricted by a chemical barrier in their upward movement. This may influence the nutrient budget of the epilimnion. In Chapter 17 a formula is given to estimate the minimum phosphate flux. Migration of animals (e.g. diurnal migration of *Chaoborus*) and movement of fishes will increase the rate of upward movement and the same will be true of water currents, even if the flow is nearly horizontal. It is, however, not possible to describe these influences quantitatively as no equations can be presented. Daily changes may be important.

Another method of transport can be the "escape" of hypolimnetic waters through an oscillating thermocline. Only the upper layer of the hypolimnion can escape, and the concentration changes are small in this layer being furthest from the bottom. In many cases the concentration of most substances, except O_2 in the upper layer of the hypolimnion, is nearly the same as in the epilimnion. Large amounts of accumulated matter may come into the epilimnion during turnover. Phosphate will however largely be trapped with the sedimenting $Fe(OH)_3$, and the ammonia will quickly be oxidised to nitrate unless it is trapped by an exchange reaction.

REFERENCES

Ambühl, H., 1969. Die neueste Entwicklung des Vierwaldstättersees (Lake of Lucerne). *Verh. Int. Ver. Theor. Angew. Limnol.*, 17: 219—230.

Baity, H.G., 1938. Some factors affecting the aerobic decomposition of sewage sludge deposits. *Sewage Works J.*, 10: 539—568.

Beauchamp, R.S.A., 1964. The Rift Valley lakes of Africa. *Verh. Int. Ver. Theor. Angew. Limnol.*, 15: 91—99.

Delfino, J.J. and Lee, G.F., 1971. Variation of manganese, dissolved oxygen and related chemical parameters in the bottom waters of Lake Mendota, Wisconsin. *Water Res.*, 5: 1207—1217.

Edberg, N. and Hofsten, B.V., 1973. Oxygen uptake of bottom sediments studied *in situ* and in the laboratory. *Water Res.*, 7: 1285—1294.

Edwards, R.W., 1958. The effect of larvae of *Chironomus riparius* Meigen on the redox potentials of settled activated sludge. *Ann. Appl. Biol.*, 46(3): 457—464.

Edwards, R.W. and Rolley, H.L.J., 1965. Oxygen consumption of river muds. *J. Ecol.*, 53(1): 1—19.

Effenberger, M., 1967. A simple flow cell for the continuous determination of oxidation—reduction potential. *Proc. IBP-Symp. Amsterdam—Nieuwersluis, 1966*, pp. 123—126.

Einsele, W., 1936. Über die Beziehungen des Eisenkreislaufs zum Phosphatkreislauf im eutrophen See. *Arch. Hydrobiol.*, 29: 664—686.

Einsele, W., 1937. Physikalisch—chemische Betrachtung einiger Probleme des limnischen Mangan- und Eisenkreislaufs. *Verh. Int. Ver. Theor. Angew. Limnol.*, 8: 69—84.

Einsele, W., 1938. Über chemische und kolloidchemische Vorgänge in Eisen- und Phosphatsystemen unter limnochemischen und limnogeologischen Gesichtspunkten. *Arch. Hydrobiol.*, 33: 361—387.

Fair, G.M., Moore, E.W. and Thomas, H.A., 1941a. The natural purification of river muds and pollutional sediments. *Sewage Works J.*, 13(2): 270—307.

Fair, G.M., Moore, E.W. and Thomas, H.A., 1941b. The natural purification of river muds and pollutional sediments. *Sewage Works J.*, 14(4): 765—779.

Fair, G.M., Moore, E.W. and Thomas, H.A., 1941c. The natural purification of river muds and pollutional sediments. *Sewage Works J.*, 13(6): 1209—1228.

Findenegg, I., 1937. Holomiktische und meromiktische Seen. *Int. Rev. Ges. Hydrobiol. Hydrogr.*, 35: 586—610.

Golterman, H.L., 1967. Influence of the mud on the chemistry of water in relation to productivity. *Proc. IBP-Symp. Amsterdam—Nieuwersluis*, 1966, pp. 297—313.

Golterman, H.L., 1973. Natural phosphate sources in relation to phosphate budgets: A contribution to the understanding of eutrophication. *Water Res.*, 7: 3—17.

Hargrave, B.T., 1972a. A comparison of sediment oxygen uptake, hypolimnetic oxygen deficit and primary production in Lake Esrom, Denmark. *Verh. Int. Ver. Theor. Angew. Limnol.*, 18: 134—139.

Hargrave, B.T., 1972b. Aerobic decomposition of sediment and detritus as a function of particle surface area and organic content. *Limnol. Oceanogr.*, 18: 583—596.

Hutchinson, G.E., 1957. A Treatise on Limnology. 1. Geography, Physics, and Chemistry. John Wiley, New York, N.Y., 1015 pp.

Kjensmo, J., 1967. The development and some main features of "iron-meromictic" soft water lakes. *Arch. Hydrobiol., Suppl.*, 32(2): 137—312.

Kjensmo, J., 1968. Iron as the primary factor rendering lakes meromictic, and related problems. *Mitt. Int. Ver. Theor. Angew. Limnol.*, nr. 14: 83—93.

Knowless, G., Edwards, R.W. and Briggs, R., 1962. Polarographic measurement of the rate of respiration of natural sediments. *Limnol. Oceanogr.*, 7(4): 481—483.

Lerman, A. and Stiller, M., 1969. Vertical eddy diffusion in Lake Tiberias. *Verh. Int. Ver. Theor. Angew. Limnol.*, 17: 323—333.

Li, Y.-H., 1973. Vertical eddy diffusion coefficient in Lake Zürich. *Schweiz. Z. Hydrol.*, 35(1): 1—7.

Livingstone, D.A. and Boykin, J.C., 1962. Vertical distribution of phosphorus in Linsley Pond mud. *Limnol. Oceanogr.*, 7: 57—62.

Lund, J.W.G., Mackereth, F.J.H. and Mortimer, C.H., 1963. Changes in depth and time of certain chemical and physical conditions and of the standing crop of *Asterionella formosa* Hass. in the North Basin of Windermere in 1947. *Philos. Trans. Lond., Ser. B*, 246: 255—290.

Machan, R. and Ott, J., 1972. Problems and methods of continuous *in situ* measurements of redox potentials in marine sediments. *Limnol. Oceanogr.*, 17(4): 622—626.

MacPherson, L.B., Sinclair, N.R. and Hayes, F.R., 1958. Lake water and sediment. III. The effect of pH on the partition of inorganic phosphate between water and oxidized mud or its ash. *Limnol. Oceanogr.*, 3: 318—326.

Mortimer, C.H., 1941. The exchange of dissolved substances between mud and water in lakes. 1. *J. Ecol.*, 29: 280—329.

Mortimer, C.H., 1942. The exchange of dissolved substances between mud and water in lakes. 2. *J. Ecol.*, 30: 147—201.

Mortimer, C.H., 1952. Water movements in stratified lakes deduced from observations in Windermere and model experiments. *Union Geod. Geophys., Bruxelles, Assoc. Int. Hydrol., C.R. Rapp.*, 3: 335—349.

Mortimer, C.H., 1955. The dynamics of the autumn overturn in a lake. *Union Geod. Geophys., Rome, Assoc. Int. Hydrol., C.R. Rapp.*, 3: 15—24.

Mortimer, C.H., 1961. Motion in thermoclines. *Verh. Int. Ver. Theor. Angew. Limnol.*, 14: 79—83.

Mortimer, C.H., 1969. Physical factors with bearing on eutrophication in lakes in general and large lakes in particular. In: *Eutrophication: Causes, Consequences, Correctives.* Proc. Symp. held at the University of Wisconsin, Madison, 1967. Washington, Natl. Acad. Sci., pp. 340—371.

Ohle, W., 1937. Kolloidgele als Nährstoffregulatoren der Gewässer. *Naturwissenschaften,* 25: 471—474.

Ohle, W., 1938. Die Bedeutung der Austauschvorgänge zwischen Schlamm und Wasser für den Stoffkreislauf der Gewässer. *Jahrb. Wasser,* 13: 87—97.

Ohle, W., 1962. Der Stoffhaushalt der Seen als Grundlage einer allgemeinen Stoffwechseldynamik der Gewässer. *Kieler Meeresforsch.,* 18(3): 107—120.

Ohle, W., 1964. Kolloidkomplexe als Kationen- und Anionenaustauscher in Binnengewässern. *Jahrb. Wasser,* 30: 50—64.

Ohle, W., 1968. Chemische und mikrobiologische Aspekte des biogenen Stoffhaushaltes der Binnengewässer. *Mitt. Int. Ver. Theor. Angew. Limnol.,* nr. 14: 122—133.

Olsen, S., 1958. Phosphate adsorption and isotopic exchange in lake muds. Experiments with P-32 (Preliminary report). *Verh. Int. Ver. Theor. Angew. Limnol.,* 13: 915—922.

Olsen, S., 1964. Phosphate equilibrium between reduced sediments and water, laboratory experiments with radioactive phosphorus. *Verh. Int. Ver. Theor. Angew. Limnol.,* 15: 333—341.

Otsuki, A. and Wetzel, R.G., 1972. Coprecipitation of phosphate with carbonates in a marl lake. *Limnol. Oceanogr.,* 17: 763—767.

Pelletier, J.-P., 1968. Un lac méromictique, le Pavin (Auvergne). *Ann. Station Biol. Besse-en-Chandesse,* 3.

Serruya, C., 1972. Metalimnic layer in Lake Kinneret, Israel. *Hydrobiologia,* 40(3): 355—359.

Serruya, C., Serruya, S. and Berman, T., 1969. Preliminary observations of the hydromechanics, nutrient cycles and eutrophication status of Lake Kinneret (Lake Tiberias). *Verh. Int. Ver. Theor. Angew. Limnol.,* 17: 342—351.

Stumm, W. and Morgan, J.J., 1970. *Aquatic Chemistry; An Introduction Emphasizing Chemical Equilibria in Natural Waters.* John Wiley, New York, N.Y., 583 pp.

Talling, J.F., 1969. The incidence of vertical mixing, and some biological and chemical consequences in tropical African lakes. *Verh. Int. Ver. Theor. Angew. Limnol.,* 17: 998—1012.

Tessenow, U., 1964. Experimentaluntersuchungen zur Kieselsäurerückführung aus dem Schlamm der Seen durch Chironomidenlarven (*Plumosus*-Gruppe). *Arch. Hydrobiol.,* 60(4): 497—504.

Tessenow, U., 1972. Lösungs-, Diffusions- und Sorptionsprozesse in der Oberschicht von Seesedimenten. 1. Ein Langzeitexperiment unter aeroben und anaeroben Bedingungen im Fliessgleichgewicht. *Arch. Hydrobiol., Suppl.,* 38(4): 353—398.

Thomas, E.A., 1965. Phosphat-Elimination in der Belebtschlammanlage von Männedorf und Phosphat-Fixation in See- und Klärschlamm. *Vierteljahresschr. Naturforsch. Ges. Zürich,* 110: 419—434.

Thomas, E.A., 1971. Oligotrophierung des Zürichsees. *Vierteljahresschr. Naturforsch. Ges. Zürich,* 116(1): 165—179.

Williams, J.D.H. and Mayer, T., 1972. Effects of sediment diagenesis and regeneration of phosphorus with special reference to lakes Erie and Ontario. In: H.E. Allen and J.R. Kramer (Editors), *Nutrients in Natural Waters.* Wiley-Interscience, New York, N.Y., pp. 281—315.

CHAPTER 10

ALGAL CULTURES AND GROWTH EQUATIONS

10.1. INTRODUCTION

To get a fuller understanding of conditions affecting algal growth in nature it is often necessary to use laboratory cultures for experimental studies. Examples of such studies are those on the influence on the growth rate of growth-limiting substances, physiological features of the algal cells that control their growth, and the assessment of equations used to describe algal growth. The last is of importance in the evaluation of mathematical models for predation and competition.

Two different approaches are usually required:

(1) When studying the growth of a given algal species (e.g. a diatom) in a specific lake, one will try to isolate that species and to determine its growth characteristics.

(2) When a certain physiological feature must be studied one can use an algal species which shows the characteristic feature. A study of glycollic-acid production for instance should, to begin with, be made with a strain that excretes this compound in considerable quantities so that the favourable conditions can be determined. The knowledge thus got may later be applied to natural populations.

For both problems the choice of suitable algal culture solutions is essential. For physiological work suitable culture solutions are taken to be those which allow both rapid growth of cells and high population density. It is assumed implicitly that the process studied — e.g. the establishment of a photosynthetic pathway — is not dependent on the population density. For ecological work, however, one will try to prepare and to work with a culture solution of a composition comparable with that of natural lakewater. For most lakewaters this would mean population densities well below 10^7 cells per litre, and in most cases there would therefore be insufficient material available to measure anything at all, so one has to compromise. The normal range for ecological work is often between 10^6 and 10^9 cells per litre, whereas algal physiologists often use an inoculum of 10^9 cells per litre and will then have final cell densities of well over 10^{12} cells per litre. (The figures given here are an indication of mean values and depend greatly of course on the size of the individual cells.)

Isolation of cells required for culture can in principle be made in two different ways. The first method consists of "solidifying" lakewater with agar-agar in a petri dish, after which the plates are inoculated with lakewater. After incubation in the light (artificial or, preferably, near a north window) several colonies will appear. These may be isolated using a sterile needle and transferred into a small volume (1 ml) of culture solution. The second method is to isolate the desired cell with a glass micropipette under a microscope with low magnification and to wash the cell a few times in a drop of sterile culture solution on a microscope slide and finally to transfer it to a small culture tube containing culture solution[*].

10.2. THE CULTURE SOLUTION

One of the first attempts to culture algae was made by Pringsheim (1946), who sterilised water with different types of soil in culture tubes. He then inoculated the tubes with algae. His book — which summarises earlier developments in this field as well — is still a very useful aid for the student of algal growth. Later the need for more reproducible solutions arose and pure cultures were required with no particles remaining from foreign bodies such as the soil, and no precipitates. Natural water is quite often unsatisfactory for *sustained* growth, but enriched with nitrate, phosphate and possibly silicate it may prove useful in the process of establishing a well-defined medium.

As far as possible a culture medium should resemble the natural environment. Chu (1942) developed a successful medium of this kind and his culture solution no. 10 is still used. His work was followed by the now classical paper of Rodhe (1948), who systematically varied the concentrations of all the constituents. Rodhe's paper should be carefully studied by students of algal growth and ecology, as it is still one of the best examples of a synthesis of algal-growth physiology and ecology.

Both Chu's no. 10 and Rodhe's no. 8 resemble natural eutrophic water. The cations in Rodhe's solution are in proportions which resemble the mean freshwater composition:

	Rodhe's no. 8	World average
Ca	63%	65%
Mg	13%	17%
Na	25%	15%
K	1%	3%

For the anions this is not true, because Rodhe's solution supplies nitrogen only as nitrate, and pH and alkalinity have to be adjusted to certain values. When nitrate is utilised it becomes replaced gradually by bicarbonate, so that the pH increases and bicarbonate becomes the dominant anion (as in natural waters). It is essential to realise that for ecological work PO_4-P concentration

---

[*] If a particular species is required, the easiest procedure is to write to the curator of one of the excellent algal culture collections, e.g. those in Cambridge or Göttingen. These collections contain a wide variety of species of protozoa and other unicellular organisms, often bacteria-free. The curators may also give advice about culture solutions and conditions. One can then slowly modify the solution towards that which is required for experimental purposes.

should not exceed 1 mg l^{-1}. At this concentration it will not limit the algal growth rate (see below). In algae the ratio N : P averages about 10, so that the nitrogen concentration in the culture solution should be about 10 mg l^{-1}.

Table 10.1 gives the composition of two culture solutions suitable for freshwater algae.

In some cases it is useful to make modifications. If a constant pH is required (Rodhe's no. 8 may increase up to pH = 11), NH_4NO_3 can replace $Ca(NO_3)_2$, at least for those algae for which NH_3 is not toxic. This includes most species which will grow in large dense cultures. Calcium may then be given as $CaCl_2$ or $CaCO_3$, of which 50 mg l^{-1} will dissolve and will give a "natural" alkalinity. Na_2SiO_3, which will cause the pH to increase considerably, might be replaced by tetraethyl silicate (Golterman, 1967). In this case $NaHCO_3$ may be used to adjust the alkalinity. Fe-citrate can be replaced by the more stable iron compound Fe-EDTA. The pH may be adjusted, after sterilisation. This may be necessary because Fe-EDTA and KH_2PO_4 give acid solutions. The adjustment is made with the calculated amount of sterile $NaHCO_3$. The amount needed is calculated from the results of a small-scale trial. Under aeration $CaCO_3$ will dissolve in quantities up to 50 mg l^{-1} and gives slightly alkaline solutions. CO_2 should be provided by bubbling sterile air through the culture. Enrichment with excess CO_2 either gives low pH values or makes unecologically high concentrations of $NaHCO_3$ necessary in order to keep the pH within

TABLE 10.1

Concentration of elements in two culture solution, with concentrations of available salts

| Element or Compound | Rodhe's culture solution no. 8 | | | Chu's culture solution no. 10 | | |
	Conc. (mg l^{-1})	given as	Conc. (mg l^{-1})	Conc. (mg l^{-1})	given as	Conc. (mg l^{-1})
Ca	15	$Ca(NO_3)_2 \cdot 4H_2O$ [1]	90	9.7	$Ca(NO_3)_2$	40
N	10			6.8		
SO_4	4	$MgSO_4 \cdot 7H_2O$	10	9.7	$MgSO_4 \cdot 7H_2O$	25
Mg	1			2.5		
K	2.2	K_2HPO_4	5	4.5	K_2HPO_4	(5 or) 10
P	0.9			1.8		
Na	8	$Na_2SiO_3 \cdot 9H_2O$ [1]	50	9.5	Na_2SiO_3	25
Si	5.5			6.6		
Fe	0.18	Fe-citrate + citric acid	1+1	0.18	$FeCl_3$	8—10
Mn	0.01	$MnSO_4$	0.03			
HCO_3^-	[2]	$NaHCO_3$		[2]	Na_2CO_3	20

[1] Rodhe used the anhydrous salts.
[2] Depends on pH.

the normal range. Culture solutions with an acid pH value are more difficult; they cannot be buffered because most weak acids are organic or toxic. NH_4Cl can be used as N source because the pH will tend to decrease after NH_3 uptake. For cultures of *Chlorella* and *Scenedesmus* the normal laboratory grade chemicals can be used. Rodhe found that addition of Hoagland's "A—Z" trace-element solution or of Mn^{2+} had no observable effect. Presumably laboratory grade chemicals contain sufficient trace elements. Nevertheless, Mn^{2+} is usually added to provide for algae with a higher Mn^{2+} requirement than that of the species which Rodhe used (*Ankistrodesmus falcatus*). If very pure chemicals must be used, trace elements should be supplied separately. Metals such as copper, which could be toxic in high concentrations, are rendered harmless by the addition of EDTA. When adding Fe-EDTA to calcium-rich cultures it must be remembered that a proportion of the EDTA will chelate with calcium, so that iron may nevertheless precipitate and be unavailable to some algae (see Chapter 11).

For cultures of certain species of algae soil extract must still be used. Sometimes it can be successfully replaced by asparagine or (purified) yeast autolysate (Golterman, 1960). Yeast autolysate is easier to work with and can be made more reproducible in quality and quantity than soil extract.

Normally the cultures are shaken to prevent sedimentation, but bubbling with air may serve the same purpose. For some algae however this should not be done. *Nitzschia* starts to grow better after inoculation if air is not bubbled for the first few days. As soon as growth is visible, bubbling with air does no harm. Fogg and Than-Tun (1960) found that the growth of *Anabaena* doubled when cultures were shaken 90 instead of 65 times per minute and ceased entirely when cultures were shaken 140 times per minute.

If a culture of a species is difficult to obtain, it will be necessary to test for unusual requirements, for example for vitamins or trace elements. For some algae — mainly marine species — vitamins mostly from the B group do seem to be necessary; vitamin B_{12} in particular is often used for flagellates.

The ultimate goal of the algologist is to obtain pure cultures free from other organisms (especially bacteria). Such cultures are called axenic, and are not easily obtained. Several diatoms grow less well or cease to grow at all if one tries to eliminate the bacteria. Fogg (1966) suggested that this is possibly due to the supply of vitamin B_{12} by the bacteria. This suggestion has not been supported and the positive effect of the bacteria can not be replaced by B_{12}. For a long time it was believed that *Asterionella* could not be grown without bacteria but Lund (unpublished) has managed to get some axenic cultures. It may be that the efforts to remove bacteria were equally unfavourable to the diatoms; most diatoms are just as sensitive as bacteria to such treatments as antibiotics of UV light.

It seems probable that many failures to get thriving growth have a physical cause owing to differences between cultural and natural conditions. Tungsten filament bulbs may overheat a culture. Fluorescent tubes generate less heat but the spectral composition of their light is very different from that occurring in natural aquatic habitats.

Many algae, but not all, prefer a certain period of darkness. This may be arranged by use of lights wired to an automatic time switch. If cultures of less than 10 l are needed, they may be grown in the light from a north window. Temperature should be kept within a normal range, usually 5—20°C, and if growth characteristics are to be measured temperatures should be kept constant. Irradiance and temperature may both be within the range where each of them is suitable if applied separately, but may nevertheless be lethal in combination, causing failures which are not easily recognised.

10.3. GROWTH EQUATIONS

If growth of an algal cell is not limited by an external factor such as nutrient or light supply or if these external factors can be kept constant during growth, an exponential growth of the population will occur. One single original cell gives two, two cells give four cells, etc. This arises from a more general description of the growth of the population: the rate of growth is proportional to the number of cells at any particular time. Thus:

$$\frac{dN}{dt} = \beta N \tag{10.1a}$$

This equation can be integrated and then becomes:

$$\int \frac{dN}{N} = \beta \int dt, \quad \text{so} \quad \ln N_{t=1} = \ln N_{t=0} + \beta t, \quad \text{or} \quad N_t = N_0 \, e^{\beta t} \tag{10.1b}$$

where N_t = number of cells at time t; N_0 = number of cells at time zero; t = time; and β = growth rate constant (dimension t^{-1}, units usually d^{-1}).

The time interval in which the cell number doubles is called the "doubling" time, t_d. Thus:

$$N_{t=2} = 2 N_{t=1}$$

It follows that:

$$2 = e^{\beta t_d}, \quad \text{so that} \quad t_d = \frac{\ln 2}{\beta}.$$

If for example t_d is one day, $\beta = 0.7$.

In a limited volume of culture solution external factors such as nutrient concentrations will not remain constant. This may be achieved however by a constant dilution of the culture in a so-called continuous culture (section 10.5).

If there are no nutrient-limiting factors, growth will depend on cell properties only, at given optimal light and temperature values. Mathematically this can be written as:

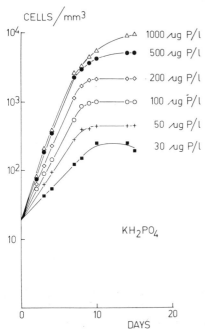

179

CELLS/mm³

1000 µg P/l
500 µg P/l
200 µg P/l
100 µg P/l
50 µg P/l
30 µg P/l

KH₂PO₄

DAYS

Fig. 10.1. Growth of *Scenedesmus* sp. as a function of phosphate concentration. (From Golterman et al., 1969.)

$$\left(\frac{dN}{dt} \cdot \frac{1}{N} = \beta_{max} \right)_{\text{light, temp.}} \qquad (10.2)$$

where N = cell number; and β_{max} = maximal growth constant (in optimal light, temperature and nutrient conditions).

If growth is plotted on semi-logarithmic graph paper, the result should be linear with time (Fig. 10.1). The tangent of the angle between this line and the time axis is β. The value of β depends on light, temperature and nutrient concentrations amongst other factors. The relationship with nutrient concentration may be described by an empirical equation as e.g. of Monod (1950):

$$\beta = \beta_{max} \frac{C_n}{C_n + C_1} \qquad (10.3)$$

where β_{max} = maximum value of β; C_n = concentration of growth rate-limiting nutrient; and C_1 = a constant (dimension of concentration), which is equal to the concentration of the limiting substance when $\beta = 0.5\,\beta_{max}$. Typical examples of such curves are given in Figs. 10.1 and 10.2. β is a constant only at given (constant) light and temperature values. (The relation between growth and irradiance has been discussed in subsection 4.3.1.) If light is not saturating, growth will result in an increase in cell numbers and as growth continues the cells shade one another more and more, so that subsequently the growth rate will decrease.

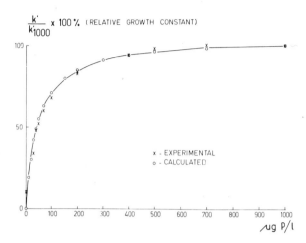

Fig. 10.2. Relative growth constants from Fig. 10.1.

In Fig. 10.1 the concepts of growth-limiting factor are represented graphically. Liebig's effect (the yield-limiting effect) is shown by the different density of cells in the cultures with different initial phosphate concentrations. It may be expressed as yield per mg of PO_4-P. Nowadays the results of such experiments are also interpreted in terms of the concept of growth *rate*-limiting factor. At PO_4-P values up to 500 μg l^{-1}, the angle between the growth curve and the x-axis becomes larger, i.e. the growth rate increases (Fig. 10.2). In eutrophication studies confusion often arises between "yield"- and "rate"-limiting concepts.

Kinetics of phosphate uptake and growth of *Scenedesmus* in a continuous culture were studied by Rhee (1973). He found a curve similar to that of Fig. 10.2 but with a lower apparent half-saturation concentration for growth, K_m, than that of Fig. 10.2 and a half-saturation constant of phosphate uptake larger than K_m by an order of magnitude. Rhee measured, however, only the external phosphate (phosphate in the filtrate), while in Fig. 10.2 the actual concentrations of the different cultures are given at zero time, or the total phosphate concentration during the growth of *Scenedesmus*. During growth the external phosphate (meaning in Rhee's paper phosphate in the filtrate) of Fig. 10.1 was of course much lower. The uptake of phosphate, measured as uptake of $^{32}PO_4$-P without correction for exchange, was correlated positively with the phosphate in solution and negatively with polyphosphates stored in the cells. Growth rate furthermore depended on cellular phosphate concentration, with no growth being possible if the cellular phosphate was below a certain critical level (1 200 μg μm$^{-3} \cdot 10^{-12}$, or about 1% of the dry weight expressed as PO_4-P).

Paasche (1973) found that the relation between growth and silicate uptake of five species of marine diatoms could be described by the same Michaelis-

Menten type of hyperbola, provided that a correction was made for the presence of a quantity of silicate that could not be used. The mean values of the half-saturation constants were in the range of 25—100 μg l^{-1} of SiO_2-Si; the concentrations that could not be used were in the range of about 10—40 μg l^{-1} of SiO_2-Si. Paasche suggested that these low silicate concentrations may exert a selective influence on species successions, but he did not discuss the possibility that in nature phosphate or nitrate may be limiting the growth rate.

It is easier to study the effect of growth rate-limiting factors in a continuous culture; more such studies will be discussed in the section which deals with this method (see section 10.5).

Eq. 10.2 holds true for any given value of β, even if a growth rate-limiting factor is present provided that this concentration and thus β can be kept constant.

Monod's equation for β as a function of the concentration of the limiting nutrient concentration is only one of several possible descriptions. An exponential relationship is sometimes used. Verduin (1964) used the Baule-Mitscherlich equation:

$$\beta = \beta_{max} \, (1-e^{-0.7x/h}) \, (1-e^{0.7y/h}) \, (1-e^{-0.7z/h}) \tag{10.4}$$

where x, y and z are concentrations of growth-limiting nutrients; and $0.7/h$ = factor introduced to facilitate computation. Verduin gave no explanation for his preference for this equation to that used by Monod. In plotting results from continuous cultures the exponential curve is sometimes said to fit better than that of Monod. Nevertheless, the Monod equation is more in accordance with modern biochemical ideas.

Growth of the population eventually falls below an exponential relationship, and finally stops. Two possible situations can occur: either the algal cell concentration remains constant, or it decreases (if more cells die than are produced). The first situation, which is often found in cultures of *Chlorella* and *Scenedesmus*, can be described by:

$$\frac{dN}{dt} = \beta \frac{K-N}{K} \cdot N = \beta N - \frac{\beta N^2}{K} \tag{10.5}$$

where K = maximum which N may attain (10^{10} cells per mg l^{-1} of PO_4-P in the case of Fig. 10.1). An equation often used in bacteriology may also be used to describe the phase of declining growth rates:

$$\frac{dN}{dt} = \beta N - \gamma N^2$$

where γ = decline coefficient. Since $\gamma \ll \beta$, the term βN is initially the largest one, while eventually γN^2 will equal it. It is obvious that eq. 10.5 is similar to eq. 10.4 with $\beta/K = \gamma$. If two algae are competing for one limiting nutrient, a general set of equations for competition can be used:

$$\frac{dN_1}{dt} = \beta_1 N_1 \frac{K_1 - N_1 - \alpha N_2}{K_1}$$

$$\frac{dN_2}{dt} = \beta_2 N_2 \frac{K_2 - N_2 - \gamma N_1}{K_2}$$

where α and γ are interaction constants. Using these equations, Slobodkin (1961) showed that grazing may change the outcome of algal competition.

Before the inoculum starts growing exponentially, there is often a so-called lag phase during which no increase in numbers occurs. Spencer (1954)

TABLE 10.2

Relative growth constants, β (in $\log_{10}$ day units) and mean doubling times (in h) of various planktonic and non-planktonic algae grown in continuous light approximately saturating for photosynthesis (Fogg, 1966)

Species	β	G (h)	Temp. (°C)	Reference
Chlorophyceae				
Chlorella pyrenoidosa	0.12	60.2	10	Fogg and Belcher, 1961
planktonic strain	0.37	19.6	20	Fogg and Belcher, 1961
Chlorella pyrenoidosa	0.93	7.75	25	Sorokin, 1959
Emerson strain	0.90	8.0	25	Sorokin, 1959
7-11-05 strain	2.78	2.6	39	Sorokin, 1959
Xanthophyceae				
Monodus subterraneus	0.074	97.7	15	Fogg et al., 1959
	0.191	37.8	20	Fogg et al., 1959
	0.297	24.3	25	Fogg et al., 1959
	0.169	42.7	30	Fogg et al., 1959
Chrysophyceae				
Isochrysis galbana	0.24	30.2	20	Kain and Fogg, 1960
Cricosphaera (Syracosphaera) carterae	0.36	20.1	18	Parsons et al., 1961
Bacillariophyceae				
Asterionella formosa	0.75	9.6	20	Lund, 1949
Asterionella japonica	0.52	13.9	20—25	Kain and Fogg, 1960
Phaeodactylum tricornutum	0.72	10.0	25	Spencer, 1954
Skeletonema costatum	0.55	13.1	18	Parsons et al., 1961
Coscinodiscus sp.	0.20	30.0	18	Parsons et al., 1961
Dinophyceae				
Amphidinium carteri	0.82	8.8	18	Parsons et al., 1961
Prorocentrum micans	0.13	55.5	20	Kain and Fogg, 1960
Ceratium tripos	0.087	82.8	20	Nordli, 1957
Myxophyceae				
Anabaena cylindrica	0.68	10.6	25	Fogg, 1949
Anacystis nidulans	3.55	2.0	41	Kratz and Myers, 1955

working with *Nitzschia closterium* and Fogg (1944) with *Anabaena cylindrica* showed that the lag phase did not occur if the inoculum was taken from a culture which was growing exponentially. Inocula taken from cultures which had a long stationary phase showed a long lag phase. In fact the duration of the preceding stationary phase was reflected in the length of the lag phase. The lag phase might be the period during which synthesis of special enzymes occurs. Some species of algae show a lag phase even though the inoculum came from an exponentially growing culture, provided that only a small inoculum is taken. In this case adaptation to a different pH value and HCO_3^- concentration may be the cause of the temporary cessation of growth. Spencer (1954) found that if cells from a P-deficient culture of *Phaeodactylum* were transferred to a high-P culture medium, the lag phase was long, but if they were transferred to a low-P culture medium, the lag phase was shorter. Effects such as these may be part of the explanation for the difficulties experienced when trying to establish the first colonies from a new isolate. The addition of glycollic acid (1 mg l^{-1}) shortens the lag phase (Sen and Fogg, 1966) and this may be important in natural conditions since considerable quantities of this acid seem to be excreted by algae in some conditions of stress. Glucose and various other organic acids did not have this shortening effect. Fogg (1966) reviewed relative growth constants for several species of algae (Table 10.2).

It seems that the smaller the alga, the larger the value of β and the shorter the doubling time. This might be explained by the larger surface:volume ratio of the smaller algae since for them diffusion of nutrients would be expected to be more efficient than for larger algae (see also section 14.2). Influence of temperature is quite clear in cultures of *Monodus subterraneus*; the Q_{10} is usually between 2 and 4 until high temperatures are reached.

10.4. BATCH CULTURES

One of the simplest cultures is the batch culture. The inoculum is introduced into a limited volume of nutrient solution. The culture vessel used is often an Erlenmeyer flask or a Fernbach flask. The Fernbach flask has a larger ratio of surface to volume (Fig. 10.3). The neck of both types of flask can be plugged with cotton wool and both can be sterilised by autoclaving. The volume of solution used varies from a few ml to 10 l. It is usually not practicable to sterilise more than 10 l of solution. Large volumes of solutions need to be bubbled with air but smaller flasks receive sufficient air by diffusion through the cotton wool plug. Cells are most easily harvested from large volumes by stopping air bubbling overnight. Next morning most of the culture solution can be decanted. A few species of algae do not settle out and cannot be harvested in this way. The culture may be further concentrated by centrifugation or filtration or both.

In the earlier stages the growth rate is exponential, but if light is not satu-

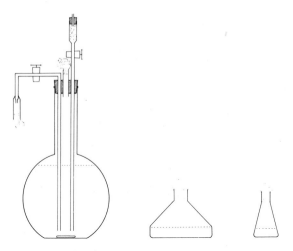

Fig. 10.3. Culture flasks for batch cultures.

rating the growth rate will decrease due to self-shading. This is an unavoidable disadvantage of culture in a limited volume (batch culture). Nutrient supply will also decrease as growth continues, so phosphate or nitrate concentrations will eventually become limiting. In most cultures larger than 1 or 2 l, CO_2 will often be limiting and the growth rate will then depend on the rate at which air can be flushed through the culture. The air may be enriched with extra CO_2, but, in that case, one has to accept either unecologically low pH values or, if the pH is raised by adding bicarbonate, unecologically high concentrations of bicarbonate. Meaningful measurements of growth rate can therefore be made only in the earlier stages of the culture, when cell concentrations are low. If sufficient material is to be collected to measure anything at all, large volumes are then necessary and these **must** be centrifuged. Because of all these difficulties it has become necessary to develop a new technique. This is the continuous-culture method, in which the culture parameters remain constant. This effect is achieved by continuous addition of fresh solution and continuous harvesting or both.

Batch cultures are useful for growing large quantities of algal cells and for simulation of natural algal growth.

10.5. CONTINUOUS CULTURES

10.5.1. Theoretical considerations

If a fresh supply of nutrient solution is added to a culture and at the same time the same quantity of the culture is withdrawn, a steady state will be reached. This technique is called continuous culture.

In most cases the supply is controlled by a valve or pump and the withdrawal takes place automatically. If the replacement rate is F (volume time^{-1}) and the dilution rate is D, then:

$$F = \frac{V_s}{t}$$

and

$$D = \frac{F}{V_c} = \frac{V_s}{V_c} \cdot \frac{1}{t}$$

where V_s = supply volume; V_c = volume of culture; and t = time. The dimension of D is t^{-1}; the same as that of the growth constant β.

The change in cell numbers due to D is:

$$\frac{dN}{dt} = -DN$$

so that the total change (caused by growth and dilution) is:

$$\frac{dN}{dt} = \beta N - DN$$

If the dilution rate is controlled so that N is constant, then:

$$0 = \frac{dN}{dt} = \beta N - DN$$

and thus:

$$\beta = D$$

If therefore the cell number is kept constant, the growth rate can be calculated from measurements of V_s and V_c alone. This continuous culture can be set up as a chemostat or as a turbidostat. In the chemostat the dilution rate is so low that one nutrient is depleted and thus the concentration of this nutrient con-

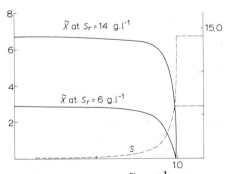

Fig. 10.4. Cell density ($\tilde{X}$, g l^{-1}, left-hand y-axis) and concentration of limiting nutrient (S, g l^{-1}, right-hand y-axis) in steady state in continuous culture in relation to dilution rate. (From Jannasch, 1962.)

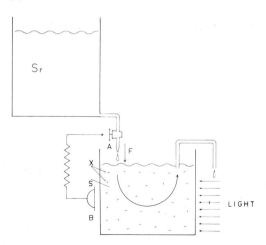

Fig. 10.5. Continuous culture (adapted from Jannasch, 1962). *A* is a tap or pump operated via photocell *B*.

trols the cell density. In the turbidostat no nutrient is depleted but the cell density is kept constant at a lower level (Fig. 10.4).

When heterotrophic bacteria are grown in continuous cultures, they get their energy from organic compounds in the inflowing liquid. If the concentration of the energy source, rather than its rate of flow, is used to limit the cell density, the system is called a chemostat. The cell density does not depend on D, which must be smaller than the maximum growth rate under the experimental circumstances (Fig. 10.4). As the substrate concentration in the culture flask will be low (because the substrate is used by the bacteria), β will be smaller than β_{max} and will adjust *itself* to be equal to $D(D < \beta_{max})$. This contrasts with the turbidostat in which D is adjusted to equal β. The chemostat system can in principle be applied to algae with a limited nutrient source, provided that energy as light is supplied in excess. If we increase D and keep the cell density constant, the substrate will not be exhausted and thus β will increase and can be studied as a function of a substrate or nutrient concentration. The addition of fresh solution to keep the cell density (N) constant can be controlled automatically. Light from a stabilised lamp passes through the culture (or outflow) and falls on a photocell. When the culture density increases slightly, the photocell output falls, and this fall operates a relay controlling the valve or pump which allows fresh solution to enter the culture (Fig. 10.5).

One of the great advantages of the turbidostat is that the culture can be kept constant at different cell-density levels with the same dilution rate (Fig. 10.6). By starting the dilution $D = \beta$ at different times (indicated by the arrows) of the exponential growth curve, the cell population may be kept constant at density *a*, *b* or *c*, provided that the nutrient concentration is not limiting the growth (although light may be limiting).

Because β is constant in this case throughout the phase of exponential

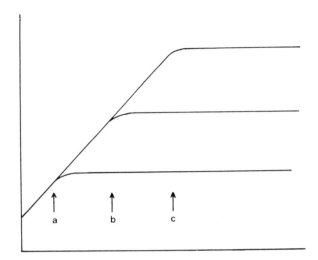

Fig. 10.6. Time course of cell densities in a turbidostat in which cells are growing with the growth rate β. If dilution at rate $D = \beta$ begins at one of the three arrows, the levels a, b or c will be maintained. Because β is constant throughout the phase of exponential growth, the dilution rate remains the same for cases a, b and c.

growth, the dilution rate is the same independent of the time at which it is started. For short experiments the rate supply can be set with a glass stop-cock, the rate being adjusted to get a constant cell density in the outflowing fluid. It is of course easier to keep the cell density constant by measuring the turbidity and adjusting D automatically. When $D \leqslant \beta_{max}$ the cell density can be kept constant at a preset value. If D is larger than β_{max} the algae will be washed out, and the density will eventually decrease to zero (Fig. 10.7). If in a turbidostat D is less than β_{max} then some external factors such as light or temperature must be limiting.

The substrate concentration can be calculated in the same way as cell numbers. Thus:

$$\frac{dS_1}{dt} = DS_0 - DS_1 \frac{\beta N_1}{Y}$$

where S_0 = substrate concentration of inflow, S_1 = substrate concentration of outflow; and $\beta N_1 / Y$ = utilisation, because Y = degree of uptake = $N_1/(S_0 - S_1)$. The difference $S_0 - S_1$ is the amount of substrate incorporated. In the steady state:

$$\frac{dN}{dt} \text{ and } \frac{dS_1}{dt} = 0$$

so that:

188

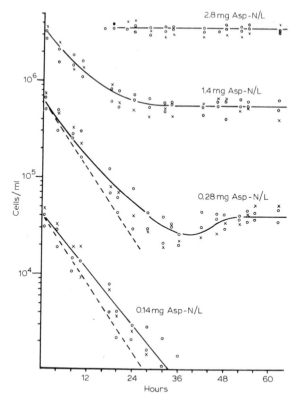

Fig. 10.7. Establishment of steady state (constant cell concentration $\tilde{X}$) after stepwise lowering of the concentration of the growth rate-limiting substrate concentration (asparagine). (From Jannasch, 1962.)

$$S_1 \approx \frac{D}{\beta_{max} - D} C_1$$

and:

$$N_1 = Y(S_0 - S_1) = Y\left(S_0 - C_1 \frac{D}{\beta_{max} - D}\right)$$

Cells in a continuous culture all pass through the stages in their life cycle at different moments. At any one moment the culture therefore contains cells in all physiological stages. If these particular stages are to be the subject of a study, it is better to synchronise the different phases of the life cycle, e.g. with alternating light and dark periods, combined if necessary with serial dilution. The latter is necessary to keep the culture both growing continuously and of a manageable population density and thus to avoid the disadvantages of a batch culture.

10.5.2. Experimental results using continuous-culture methods

Fuhs (1969) studied the effect of phosphate concentration on the growth rate of two diatoms, *Cyclotella nana* and *Thalassiosira fluviatilis*, in chemo- and turbidostats. Vitamin B_{12} and all nutrients, except phosphate, were present in excess. Phosphate was in limiting concentrations of 50 or 100 $\mu g\ l^{-1}$. In the chemostat phosphate was not detectable in the outflow, so nearly all phosphate had been taken up by the cells. Fuhs obtained typical saturation curves by plotting the number of divisions per day against phosphate content per cell. He found maximum values for β of about 1.5 per day at 5 500 lux at 18—22°C. At lower illuminance or temperatures he found values for β of only 0.5. Fuhs also graphed β/β_{max} against P content per cell as a multiple of the minimum P content per cell (Fig. 10.8).

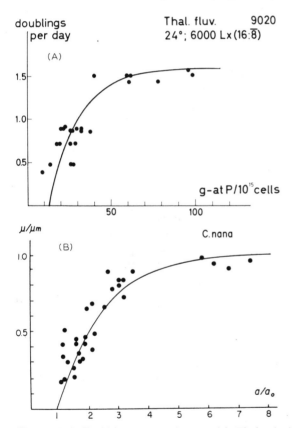

Fig. 10.8.A. Turbidostat experiment with *Thalassiosira fluviatilis* in light cycle; 50 $\mu g\ l^{-1}$ of PO_4-P. B. Summary of chemostat experiments with *Cyclotella nana*. Each point derived from arithmetic mean of all cell counts from a steady-state period. Ordinate represents growth rate as fraction of observed maximum, abscissa represents P content in multiples of minimum P content per cell. (From Fuhs, 1969.)

The growth rate of *Cyclotella nana* increased as the P concentration per cell rose, reaching a maximum at six times the minimum concentration. For this diatom the minimum concentration is 30 mg of PO_4-P in 10^{12} cells. It is remarkable that below this value apparently no growth occurred. Fuhs (1969) worked only with 50 and 100 $\mu g\, l^{-1}$ of PO_4-P which he said are limiting values but he did not give the actual cell counts per litre. Fuhs stated that his results do not fit Monod's equation (10.3) for limiting factor growth rate but that they do obey Mitscherlich's equation (10.4). He gave no figures to support these statements however. It is interesting to note that he reported that a change in the rate of P supply and the resultant change in bound P per cell affects both growth rate and the carbon content of the cells. Balanced growth, defined as a state of constant rate of multiplication during which all properties of the organism (including chemical composition) remain constant, was therefore not obtained, although the number of cells was kept constant. The difficulty is likely to be a recurring one in chemostat work with algae.

An extensive and careful study on growth and phosphate demand of *Nitzschia actinastroides* was made by Müller (1972), using both batch cultures and a chemostat of 2- or 4-l capacity. He showed that the medium used was not growth rate-limiting but that the yield in batch cultures was limited by the Si concentration. The value β_{max} was found to be 0.083 ± 0.0007 (h^{-1}) or 1.9 (d^{-1}). From this, a value of $t_d = 0.4$ days can be calculated. The production (mg) of dry weight per mg PO_4-P was 105; C_1 was measured by adjusting $D = 0.5\,\beta_{max}$. Since:

$$S_1 \approx C_1\, D/\beta_{max} - D$$

C_1 can be approximated from S_1, i.e. the concentration of PO_4-P in the outflow. Müller found a value for $C_1 = 0.44\,\mu g\, l^{-1}$ of PO_4-P. The relationship between growth rate and incorporated phosphate is identical to that between growth rate and dissolved phosphate. The curve does not go through the origin, but $\beta = 0$ is found at a low phosphate concentration, which may therefore be considered to be the minimum phosphate value.

Müller found the relation between growth rate and phosphate concentration, for values of β ranging between 0.0287 and 0.0651, to be:

$$\beta = \beta_{max}\, \frac{P_i - P_{min}}{C_1 + (P_i - P_{min})}$$

where P_i = phosphate concentration in the cells. The curve of best fit gave a P_{min} value of 8.8 μg PO_4-P per 10^8 cells. The P content per 10^8 cells at $\beta = \frac{1}{2}\beta_{max}$ was 18.9 μg. A linear correlation was found between P_1 (PO_4-P conc. outflow; $\mu g\, l^{-1}$) and P_i ($\mu g\ 10^{-8}$ cells):

$$P_i = 8.77 + 19.55\, P_1 \quad (r = 0.993;\ n = 6)$$

Obviously the concept of a "growth-limiting substrate concentration" derived from bacterial-growth studies needs to be modified for algal cultures, because the concentration of phosphate inside the cells is much more important than that in the solution. For this reason Müller's approximation for C_1 is much lower than those found by Rodhe or Golterman for cultures of *Asterionella* and *Scenedesmus*, respectively. Although some differences between species are to be expected, those between Rodhe's and Golterman's values on the one hand and Müller's values on the other are too large to be left unexplained. Müller's value for P_1 should not be compared with Rodhe's and Golterman's values, because they gave the PO_4-P concentration at the start of the culture, while Müller's PO_4-P value is recorded in the outflow only (P_1 value). Although Rodhe's and Golterman's approach is different in principle from Müller's (and Rhee's, see section 10.3), it seems likely that Golterman's and Rodhe's values should be compared with Müller's $P_1 + P_i$. Furthermore, Müller found that the P content and chlorophyll *a* content of cells grown under P-limiting conditions are directly proportional to the phosphate concentration in the outflow. Phosphate storage occurred only if growth was limited by factors other than phosphate concentration. Therefore the yield coefficient is constant only if related to the amount of chlorophyll *a* produced. The N content increased with increasing growth rate and increasing storage of phosphate but the silicate concentration was not affected.

Shelev et al. (1971) measured the growth rate of *Chlorella pyrenoidosa* as a function of nitrogen concentration. They claim that their results fitted the exponential curve better than the Monod curve, although this is not obvious from the published figures. Maximum growth rate increased substantially with temperature and it was accompanied by increasing nitrogen requirement. On the other hand, the efficiency of nitrogen uptake increased with temperature. The fact that they found that a mixed natural culture of algae showed the same specific growth rate at a much lower nitrogen concentration than *Chlorella pyrenoidosa* illustrates the likelihood of adaptation of laboratory cultures to high nutrient concentrations. This should act as a caveat when one is trying to relate experimental findings to natural conditions, and is of particular importance for the problem of algal bioassays (see section 10.7).

10.6. SYNCHRONOUS CULTURES

In continuous cultures, cells may be found in any part of the life cycle at any moment. For the study of the physiology of algae it is sometimes necessary to have all cells at the same stage, and this is now possible following the pioneering work of Tamiya and co-workers (Tamiya et al., 1953, 1961; Tamiya, 1963, 1966), Spektorova et al. (1968) and Lorenzen (1970). In such synchronised cultures the properties of the whole population are fairly similar to those of the individual cells.

In synchronised cultures cell division can be organised to take place "at a given moment", although of course that moment is always of finite duration becar٫ ٫ of the variability of individual cells. Tamiya et al. (1953) classified

192

different stages of cell development of *Chlorella ellipsoidea* and distinguished
"dark" and "light" cells, which they could separate by differential centrifugation. "Dark" cells are small and will be produced from "light" cells if these
have grown sufficiently to divide, even if they are kept in darkness. The auxospores thus formed — "dark" cells — have a high photosynthetic and low respiratory activity in contrast with the "light" cells, which can only be produced
from the "dark" cells if these are illuminated (Fig. 10.9 and 10.10).

"Dark" cells will accumulate in dark conditions, so intermittent illumination
(a dark—light rhythm) combined with serial dilution is the main method used
nowadays to produce a synchronised algal culture.

Later Tamiya made a more detailed classification into four different light
stages ($L_1 \ldots L_4$), two dark stages (D_1, D_2) and a transient stage ($D \sim L$)
(Fig. 10.9). There are concomitant changes in photosynthetic activity and in
the elementary composition of the cells. Nihei et al. (1954) found that "dark"
cells contained 39.3% of C and 37.5% of O, while light cells contained 48.3%
of C and 29.2% of O. The photosynthetic quotient ($\Delta O_2 / -\Delta CO_2$) of "dark"
cells was about 1, but rose to over 3, indicating fat production by the time
the formation of "light" cells was completed. This change in metabolism was
not found by Sorokin (1957) and Lorenzen (1959), who used a continued al-
ternation of light and dark periods, but without the additional mechanical
separation used by Tamiya. It is arguable that this in itself could be the

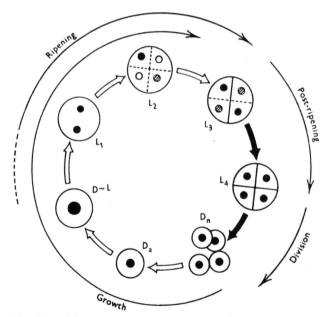

Fig. 10.9. Schematic representation of the life cycle of *Chlorella ellipsoidea* (Tamiya et
al., 1961). The white arrows indicate the light-dependent processes, while the black arrows show the transformations occurring independently of light.

cause of the difference. Perhaps the fact that the two groups of workers used a different species or strain of *Chlorella* may be relevant. The recent work of Tamiya and his group (Tamiya, 1963; Hase, 1962) deals with the biochemical aspects of cell division. From an ecological point of view it may be important that deficiency of nitrate, phosphate, potassium and magnesium will stop further divisions after the first production of daughter cells.

Lien and Knutsen (1973b) working with synchronised cultures of *Chlamydomonas reinhardtii*, showed that phosphorus deficiency led to a shortening of the G_1 (first growth) phase with an earlier start of DNA synthesis but a lower yield of DNA together with a wider dispersion in time of nuclear division and sporulation. In the phosphate-starved culture, protein accumulated at a rate varying parallel to that of RNA. Lien and Knutsen (1973a) suggested that phosphate might affect timing of cell division through the depression of phosphatases (see Chapter 5), although simple shortage of phosphate could just as easily explain the phenomenon. In natural conditions, where phosphate is usually in short supply, the lack of phosphate would therefore strongly inhibit any tendency to synchrony.

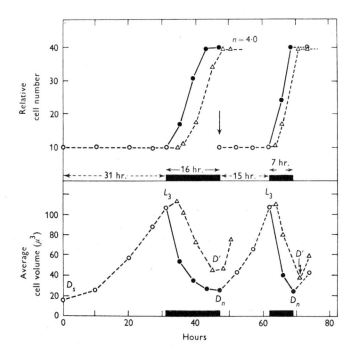

Fig. 10.10. Changes of relative cell number and average cell volume in the four cycles illustrated in Fig. 10.9. Dashed and solid curves show the processes occurring in the light and in the dark respectively. Black bars along the abscissa indicate the dark periods. At the end of each dark period the culture was diluted to the original cell concentration (indicated by vertical arrows). (Tamiya et al., 1961.)

The work done with *Chlorella* seems to be applicable more generally. Šetlík et al. (1972), working with *Scenedesmus quadricauda*, used a similar description of the life cycle for this alga and extended their study to include the nuclear division cycles. They found that the auxospore division pattern was characterised by several consecutive nuclear cycles which together constituted one cell cycle. In the nuclear cycle the sequence DNA replication, nuclear segregation and protoplast fission occurred. The sequence was initiated when either the volume or the RNA content of the cell exceeded a certain value; it then continued to completion without an external supply of energy if sufficient cell reserves were available. The number of daughter cells produced and the duration of the cycle depended on both irradiance and temperature. The attainment of the critical cell volume depended on photosynthesis but the other events in the nuclear cycle were dependent only on temperature.

Extrapolation of the results from synchronous cultures to natural conditions should be made only with the greatest caution. The increase of division stages ("dark" cells) that is sometimes observed overnight in lakes is no indication of a synchronous state but may be caused simply by these stages lasting longer during the night. Furthermore, the work with cultures has always been done with rather concentrated culture solutions and at high temperatures. As soon as a nutrient becomes growth-limiting, the synchronous state apparently ceases. It may be concluded therefore that although work with synchronous cultures may lead to a better understanding of algal physiology it is unlikely to help to explain the problems of algal growth in natural conditions.

10.7. ALGAL BIOASSAYS

If the growth of an organism (especially its yield) depends on one limiting factor, the yield can be used to determine the concentration of that factor. Bacteria have been widely used for this purpose. It is often not necessary to measure the actual yield itself because a suitable metabolic product may be used instead. Lactic-acid bacteria can easily be used to determine most of the vitamins of the B group, and with these one needs only to measure the lactic acid produced. A calibration curve is first made relating the amount of acid produced with known standard quantities of the vitamin to be assayed. The same principle has been used by Carlucci, who used algae with requirements for vitamins to measure concentrations of these vitamins in seawater and, more recently, in freshwater (Carlucci and Bowes, 1972). Nowadays bioassays are widely applied to assess the (potential) fertility of waters, especially in relation to studies on eutrophication of waters. The growth of the test algae in the water samples is a measure of the fertility of that water. Not all waters will support the growth of all algae, however, and this limits the use of the method of course. A potential fertility indicated by good growth of the test algae does not necessarily mean that high yields will actually be realised under natural conditions. Similarly, poor growth of test algae may result from factors which are unfavourable for that particular species, but it does not necessarily mean that the water is infertile for all species. As a means for comparison between different lakes however, the test may have some value. Bioassay can also be used for measuring growth after addition or even removal of a specific factor. A well-described standardised procedure has been published by Lund et al. (1971). They used *Asterionella formosa* for tests on water from British lakes. They filtered the water through a glass-fibre filter and inoculated it with *Asterionella formosa* washed free of culture solution. Lund showed that autoclaving the water reduced its fertility, while boiling the test water might increase its fertility, but this was not always so. Storing samples in polythene bottles sometimes decreased the fertility, probably as a result of bacterial growth on the bottle surface. It might be argued however, that removal by filtration of any algae present in a sample removes potential fertility since recycling provides an important supply of nutrients. Forsberg (1972) used *Selenastrum capricornutum* to monitor the fertility of sewage effluents. The same organism was used by Skulberg (1967) in bioassay studies of some Norwegian waters.

Skulberg also filtered his water samples before carrying out the bioassay. He demonstrated that river water entering a lake showed a higher potential fertility than the water at the outlet. Apparently this was so because the algae in the lake used some of the nutrients and were removed from the water by filtration. In the lake itself the nutrients contained in the algae would continue to contribute to the fertility after mineralisation. Skulberg found that *Selenastrum capricornutum* has several superior qualities as a laboratory organism: it is solitary, easy to identify, and grows well in cultures. Forsberg succeeded in reducing the incubation time to only three days. He was planning to develop a mixture of test organisms which will include *Euglena*, *Scenedesmus*, *Ankistrodesmus* and *Pandorina*. Shelev et al. (1971) used a continuous-flow turbidostat to assay algal growth, and demonstrated that natural populations showed a considerable growth rate at nitrogen levels where the test alga did not produce measurable growth. The problem of adaptation (or selection) of the test alga to higher nutrient concentrations in culture seems to be one of the difficulties that still has to be overcome. Goldman (1965) made a more rapid and sensitive bioassay by measuring the increase of ^{14}C uptake of a natural population after adding several micro-nutrients. How far these short-duration measurements of growth rate do describe the potential fertility or lack thereof, is still a matter for further investigation and argument. Cain and Trainor (1973), trying to overcome difficulties due to the short duration, suggested measuring the growth rate β of a test organism added to the water to be tested. The cells would be transferred daily for five days to freshwater in order to keep cell densities constant. The method seems worth further investigation, but the choice of one organism has all the disadvantages already discussed.

In the U.S.A. an "Algal Assay Procedure" is now widely used to stu· ιy the extent of eutrophication. The limitations of the method are particularly serious in this case. If a lake shows an increased algal concentration, perhaps due to phosphate input, another element, for example iron, may become depleted. If one tests the water after removal of the algal cells, iron will apparently be the growth-limiting factor in the filtrate. A recommendation that iron be removed from the lake may erroneously be made, but it is obvious that the design of the experiment itself is at fault, and this may lead to much waste of time and effort. The A.A.P. bottle test uses inoculation with 10^6 cells per litre of *Selenastrum capricornutum*, $50 \cdot 10^3$ cells per litre of *Microcystis aeruginosa*, or $50 \cdot 10^6$ cells per litre of *Anabaena flos aquae*. These densities seem to be so high that they are more appropriate for testing sewage effluents than for testing natural lakes. This bioassay seems therefore to be more useful for measuring fertility or toxicity of sewage than for making predictions about the eutrophic state of a lake. A serious disadvantage of the bioassay is that it gives results related to maximum algal crop but not related to total growth because no recycling takes place during the assay. In earlier chapters it has been explained that elements such as phosphorus and nitrogen may be used several times per growing season. In bioassays there is no time for this to happen. Furthermore, the bioassay gives information about the capacity for growth at a certain moment only. Continuous input of nutrients as happens with input of sewage waters is completely overlooked. Bioassays are useful therefore only in such cases as those described by Lund (see above and Chapter 17) where algal growth is limited in its yield mainly by the winter concentrations. Bioassay may be useful in estimating the availability of nutrients for algae. This is especially so for non-soluble nutrients. Fitzgerald (1970) used *Selenastrum* to test several phosphorus, nitrogen and iron sources. He found that many "insoluble" compounds would support growth. These included iron pyrites as iron source, iron phosphate both as iron and phosphate source and marble as a carbon source. Fitzgerald found no growth using mud as phosphate source. Golterman (1973) suggested that the algae must be in physical contact with the mud. If the algae and mud were separated by a dialysis tube, no growth occurred.

Growth of algae on particulate nutrients may be restricted by the particle size. Thus the availability of phosphate in apatite precipitates was controlled by the grain size, only the microcrystals being available (Chapter 5). In shallow lakes algal growth may therefore not only be limited by the nutrient concentration in the water but also by the concentration in the sediments.

REFERENCES

Cain, J.R. and Trainor, F.R., 1973. A bioassay compromise. *Phycologia*, 12(3/4): 227—232.

Carlucci, A.F. and Bowes, P.M., 1972. Determination of vitamin B_{12}, thiamine, and biotin in Lake Tahoe waters using modified marine bioassay techniques. *Limnol. Oceanogr.*, 17(5): 774—776.

Chu, S.P., 1942. The influence of the mineral composition of the medium on the growth of planktonic algae. Part I: Methods and culture media. *J. Ecol.*, 30: 284—325.

Fitzgerald, G.P., 1970. Evaluations of the availability of sources of nitrogen and phosphorus for algae. *J. Phycol.*, 6: 239—247.

Fogg, G.E., 1944. Growth and heterocyst production in *Anabaena cylindrica* Lemm. *New Phytol.*, 43: 164—175.

Fogg, G.E., 1949. Growth and heterocyst production in *Anabaena cylindrica* Lemm. Part II: In relation to carbon and nitrogen metabolism. *Ann. Bot.*, *N.S.*, 13: 241—259.

Fogg, G.E., 1966. *Algal Cultures and Phytoplankton Ecology*. University of Wisconsin Press, Madison, Wis., 126 pp.

Fogg, G.E. and Belcher, J.H., 1961. Pigments from the bottom deposits of an English lake. *New Phytol.*, 60: 129—142.

Fogg, G.E. and Than-Tun, 1960. Interrelations of photosynthesis and assimilation of elementary nitrogen in a blue-green alga. *Proc. R. Soc. Lond.*, *Ser. B*, 153: 111—127.

Fogg, G.E., Smith, W.E.E. and Miller, J.D.A., 1959. An apparatus for the culture of algae under controlled conditions. *J. Biochem. Microbiol. Techn. Eng.*, 1: 59—76.

Forsberg, C., 1972. Algal assay procedure. *J. Water Pollut. Control Fed.*, 44(8): 1623—1628.

Fuhs, G.W., 1969. Phosphorus content and rate of growth in the diatoms *Cyclotella nana* and *Thalassiosira fluviatilis*. *J. Phycol.*, 5: 312—321.

Goldman, Ch.R., 1965. Micronutrients limiting factors and their detection in natural phytoplankton populations. *Mem. Ist. Ital. Idrobiol.*, 18(Suppl.): 121—135.

Golterman, H.L., 1960. Studies on the cycle of elements in fresh water. *Acta Bot. Neerl.*, 9: 1—58.

Golterman, H.L., 1967. Tetraethylsilicate as a "molybdate unreactive" silicon source for diatom cultures. *Proc. IBP-Symp. Amsterdam—Nieuwersluis, 1966*, pp. 56—62.

Golterman, H.L., 1973. Vertical movements of phosphate in freshwater. In: E.J. Griffith, A. Beeton, J.M. Spencer and D.T. Mitchell (Editors), *Environmental Phosphorus Handbook*. John Wiley, New York, N.Y., pp. 509—538.

Golterman, H.L., Bakels, C.C. and Jakobs-Möglin, J., 1969. Availability of mud phosphates for the growth of algae. *Verh. Int. Ver. Theor. Angew. Limnol.*, 17: 467—479.

Hase, E., 1962. Cell division. In: R.A. Lewin (Editor), *Physiology and Biochemistry of Algae*. Academic Press, New York, N.Y., pp. 617—624.

Jannasch, H.W., 1962. Die kontinuierliche Kultur in der experimentellen Ökologie mariner Mikroorganismen. *Kieler Meeresforsch.*, 18(3): 67—73.

Kain, J.M. and Fogg, G.E., 1960. Studies on the growth of marine phytoplankton. III. *Prorocentrum micans* Ehrenberg. *J. Mar. Biol. Assoc. U.K.*, 39: 33—50.

Kratz, W.A. and Myers, J., 1955. Photosynthesis and respiration of three blue-green algae. *Plant Physiol.*, 30: 275—280.

Lien, T. and Knutsen, G., 1973a. Synchronous cultures of *Chlamydomonas reinhardti*: Properties and regulation of repressible phosphatases. *Physiol. Plant.*, 28: 291—298.

Lien, T. and Knutsen, G., 1973b. Phosphate as a control factor in cell division of *Chlamydomonas reinhardti*, studied in synchronous culture. *Exp. Cell Res.*, 78: 79—88.

Lorenzen, H., 1959. Die photosynthetische Sauerstoffproduktion wachsender *Chlorella* bei langfristig intermittierender Belichtung. *Flora*, 147: 382—404.

Lorenzen, H., 1970. Synchronous cultures. In: P. Halldal (Editor), *Photobiology of Microorganisms*. Interscience, New York, N.Y., pp. 187—212.

Lund, J.W.G., 1949. Studies on *Asterionella*. I. The origin and nature of the cells producing seasonal maxima. *J. Ecol.*, 37: 389—419.

Lund, J.W.G., Jaworski, G.H.M. and Bucka, H., 1971. A technique for bioassay of freshwater, with special reference to algal ecology. *Acta Hydrobiol.*, 13(3): 235—249.

Monod, J., 1950. La technique de culture continue; théorie et applications. *Ann. Inst. Pasteur*, 79: 390—401.

198

Müller, H., 1972. Wachstum und Phosphatbedarf von *Nitzschia actinastroides* (Lemm.) v. Goor in statischer und homokontinuierlicher Kultur unter Phosphatlimitierung. *Arch. Hydrobiol., Suppl.*, 38(4): 399—484.

Nihei, T., Sasa, T., Miyachi, S., Suzuki, K. and Tamiya, H., 1954. Change of photosynthetic activity of *Chlorella* cells during the course of their normal life cycle. *Arch. Mikrobiol.*, 21: 155—164.

Nordli, E., 1957. Experimental studies on the ecology of *Ceratia. Acta Oecol. Scand.*, 8: 200—265.

Paasche, E., 1973. Silicon and the ecology of marine plankton diatoms. II. Silicate-uptake kinetics in five diatom species. *Mar. Biol.*, 19(3): 262—269.

Parsons, T., Stephens, K. and Strickland, J.D.H., 1961. On the chemical composition of eleven species of marine phytoplankters. *J. Fish. Res. Board Can.*, 18: 1001—1016.

Pringsheim, E.G., 1946. *Pure Cultures of Algae; Their Preparation and Maintenance.* Cambridge University Press, Cambridge, 119 pp.

Rhee, G.-Y., 1973. A continuous culture study of phosphate uptake, growth rate and polyphosphate in *Scenedesmus* sp. *J. Phycol.*, 9: 495—506.

Rodhe, W., 1948. Environmental requirements of freshwater plankton algae. *Symb. Bot. Upps.*, 10: 1—149.

Sen, N. and Fogg, G.E., 1966. Effects of glycollate on the growth of a planktonic *Chlorella. J. Exp. Bot.*, 17(51): 417—425.

Šetlík, I., et al., 1972. The coupling of synthetic and reproduction processes in *Scenedesmus quadricauda. Arch. Hydrobiol., Suppl. 41 (Algological Studies)* 7: 172—213.

Shelev, G., Oswald, W.J. and Golueke, C.C., 1971. Assaying algal growth with respect to nitrate concentration by a continuous flow turbidostat. In: *Proc. Int. Water Pollut. Res. Conf., 5th, 1970,* pp.III-25/1—25/9.

Skulberg, O.M., 1967. Algal cultures as a means to assess the fertilizing influence of pollution. In: *Advances in Water Pollution Research. Proc. Int. Water Pollut. Res. Conf., 3rd,* 1 113.

Slobodkin, L.B., 1961. *Growth and Regulation of Animal Populations.* Holt, Rinehart and Winston, New York, N.Y., 184 pp.

Sorokin, C., 1957. Changes in photosynthetic activity in the course of cell development in *Chlorella. Physiol. Plant.*, 10: 659—666.

Sorokin, C., 1959. Tabular comparative data for the low- and high-temperature strains of *Chlorella. Nature*, 184: 613—614.

Spektorova, L.V., Kleshnin, A.F. and Spektorov, K.S., 1968. Issledovanie dinamiki processa fotosinteza na intaktnykh suspenziyakh synkhronnykh kultur Khlorelly. *Fiziol. Rast.*, 15: 26—33.

Spencer, C.P., 1954. Studies on the culture of a marine diatom. *J. Mar. Biol. Assoc. U.K.*, 33: 265—290.

Tamiya, H., 1963. Cell differentiation in *Chlorella. Symp. Soc. Exp. Biol.*, 17: 188—214.

Tamiya, H., 1966. Synchronous cultures of algae. *Ann. Rev. Plant Physiol.*, 17: 1—26.

Tamiya, H., et al., 1953. Correlation between photosynthesis and light-independent metabolism in the growth of *Chlorella. Biochim. Biophys. Acta*, 12: 23.

Tamiya, H., et al., 1961. Mode of nuclear division in synchronous cultures of *Chlorella*: Comparison of various methods of synchronization. *Plant Cell Physiol.*, 2: 383—403.

Verduin, J., 1964. Principles of primary productivity. Photosynthesis under completely natural conditions. In: D.F. Jackson (Editor), *Algae and Man. Proc. NATO Adv. Study Inst.* Plenum Press, New York, N.Y., pp. 211—238.

CHAPTER 11

TRACE ELEMENTS

11.1. INTRODUCTION

In addition to elements such as C, H, O, N, P, S, K, Ca and Mg, most organisms also need several elements in very small quantities. These elements are called trace elements. Fe, Mn, Cu, Co and Mo are generally regarded as essential for most organisms, whilst V, B and Zn are reported as being essential in some cases at least. Most of these trace elements have a function in an enzyme or in an active group in an enzyme. A list summarising some of these associations is given in Table 11.1.

TABLE 11.1

Metabolic associations of several trace elements

Trace elements	Associated with
Mn	photosystem II
	peptidase
Cu	cytohaemine (cytochrome a)
	plastocyanin for photoreduction (photosystem I)
Co	peptidase, N_2 fixation
	vitamin B_{12}
Mo	nitrate reductase (= molybdo-flavoproteine)
	nitrogenase
Fe	cytochromes, ferredoxin
	flavoproteins, nitrogenase
	nitrate reductase
Zn	peptidase, lactic dehydrogenase
	carbonic acid anhydrase
Mg	chlorophyll

The diet of most animals seems to provide adequate quantities of all necessary trace elements. This chapter therefore deals with the importance of trace elements in algal growth and the relative availability of those elements most commonly acknowledged to be essential for most species.

Most of these elements are very insoluble within the pH range which occurs in most unpolluted lakes. The only exceptions are iron and manganese, which, in anaerobic conditions, are present in considerable quantities as Fe^{2+} and Mn^{2+} (see subsection 9.4.2). In oxygenated water Fe^{2+} will be oxidised to

200

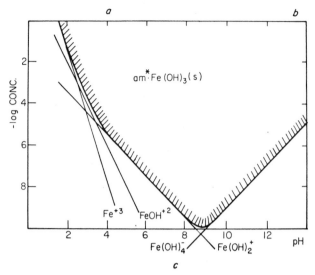

Fig. 11.1. Solubility of amorphous $Fe(OH)_3$. The possible occurrence of polynuclear complexes, for example $Fe_2(OH)_2^{4+}$, has been ignored. (From Stumm and Morgan, 1970.)

Fe^{3+}; the solubility of $Fe(OH)_3$ is given in Fig. 11.1. Thus the solubility in the pH range 7—8 is about 10^{-9} mole l^{-1} or a fraction of a microgram. This is only a theoretical value for one compound however; in most lakes iron will occur additionally both as colloidal $Fe(OH)_3$ and chelated as an organic complex with humic compounds. Furthermore in many lakes iron-containing silts occur. The quantity of iron in them is hardly detectable with common analytical methods, but is nevertheless probably of a significance at least equal to that of the soluble fraction.

With the exception of iron, it is difficult to demonstrate a definite individual trace-element requirement in algae. When Rodhe (1948) added trace elements to cultures of *Ankistrodesmus*, *Asterionella* and *Scenedesmus*, he detected no stimulation of growth, even with manganese, although the demand for this element is often supposed to be relatively large. The reason for such negative results is probably that most laboratory chemicals, even the most pure ones (analytical grade), contain sufficient of these elements to prevent limitation of growth by them. Most laboratory iron compounds contain traces of manganese for instance. Nevertheless, it is general practice to add to every algal culture solution a mixture of "recognised trace elements". In modern culture techniques most trace elements, especially iron, are kept in solution by adding the chelating agent EDTA (ethylenediamine tetraacetic acid):

A somewhat weaker complex is that of iron with NTA (nitrilotriacetic acid: $N\text{-}(CH_2COOH)_3$). In natural conditions iron-humate complexes probably serve the same purpose.

11.2. CHELATION OF TRACE ELEMENTS

The combination of a metal cation with an anion containing free pairs of electrons is called a coordination complex. The cation is called the central atom and the anion the ligand. Complexes with more than one ligand group are called chelates. A common feature of most chelates is the formation of a ring arrangement. Iron forms complexes with compounds such as glycine, oxalate, citrate and EDTA:

Iron glycinate

Metal chelate of ethylenediamine tetraacetate ligand

Such a metal complex has a different coordination number from that of the metal ion. Most metal cations have an even coordination number, normally 2, 4 or 6 and occasionally 8, and this number determines the particular special structure of the chelate which they may form. The most commonly used chelate is Fe-EDTA, and it is in this form that iron is provided in algal cultures. Ferric iron forms a fairly strong complex with EDTA:

$$Fe^{3+} + EDTA^{4-} \rightleftharpoons (Fe\text{-}EDTA)^- \qquad\qquad \log K_{FeY} = 25.1$$

where K_{FeY} is the stability constant of iron with the ligand Y (in this case EDTA).

The stability of this complex may be compared with the ferric hydroxides formed in alkaline solutions. Because iron is a weak base, cations such as $Fe(OH)_n$ are also involved as well as Fe^{3+}. The logarithm of the ratio of total iron to cationic iron can be written as:

$$\log \frac{[Fe_T]}{[Fe^{3+}] + [Fe(OH)_n]} = \log \frac{[Fe_T]}{[Fe^1]} = pFe^1 - pFe_T$$

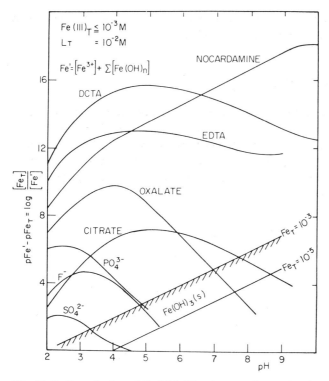

Fig. 11.2. Complexing of Fe(III). The degree of complex formation is expressed in terms of ΔpFe for various ligands (10^{-2} M). Mono-, di- and tri-dentate ligands (10^{-2} M) are not able to keep 10^{-3} M Fe(III) in solution at higher pH values. (DCTA is 1,2-diaminocyclohexane tetraacetate. Nocardamine is a trihydroxamate.) (From Stumm and Morgan, 1970.)

where $[Fe^1] = [Fe^{3+}]$ + all cations such as $Fe(OH)_n$, and $[Fe_T]$ = cationic plus chelated iron.

The stability of some iron complexes is given in Fig. 11.2.

The competitive effects of H^+ at low pH values and of OH^- at high pH values result in the degree of complex formation being dependent on the pH.

It can be seen that $10^{-2} M$ solutions of sulphate, fluoride and phosphate can keep $10^{-3} M$ ferric iron in solution up to pH values of 3.3, 4.7 and 4.8, respectively. The organic complexes oxalate and citrate are more powerful, keeping iron soluble up to pH 6.9 and 7.6. In culture solutions Fe-citrate is better than inorganic iron only between pH 4.7 and 7.6. EDTA is so much more stable a complex, that Fe-EDTA may be added to culture solutions before sterilising. But high temperature enhances the formation of hydroxides so that the Fe-citrate complex must be sterilised separately and added after cooling.

When calculating metal-complex stability, it is necessary to allow for the influence of pH on the relative quantities of the different ionic forms. Thus the Fe-NTA complex can be described as:

(1) $NTA^{3-} + H^+ \rightleftharpoons HNTA^{2-}$ $\log K = 10.3$
(2) $HNTA^{2-} + H^+ \rightleftharpoons H_2NTA^-$ $\log K = 13.3$
(3) $H_2NTA^- + H^+ \rightleftharpoons H_3NTA$ $\log K = 14.9$
(4) $Fe^{3+} + NTA^{3-} \rightleftharpoons FeNTA$ $\log K = 15.9$
(5) $2\ Fe^{3+} + 2\ NTA^{3-} + 2\ H_2O \rightleftharpoons (Fe(OH)NTA)_2^{2-} + 2\ H^+$ $\log K = 25.8$
(6) $Fe^{3+} + 3\ OH^- \rightarrow Fe(OH)_3$

Most of the NTA will thus follow the 5th reaction to the right unless the pH
is excessively high or low.

When Ca^{2+} is present in the solution it reduces the formation of the iron
complex, although the Ca-EDTA complex is much weaker than that of iron:

$$Ca^{2+} + EDTA^{4-} \rightleftharpoons Ca\text{-}EDTA^{2-} \qquad\qquad \log K_{CaY} = 10.7$$

Therefore the complex formation of calcium, iron and EDTA could be written
as:

$$Fe^{3+} + CaY^{2-} \rightleftharpoons FeY^- + Ca^{2+} \qquad\qquad \log K_{FeY}/K_{CaY} = 14.4$$

Although Ca-EDTA is thus a much weaker complex than the Fe-EDTA complex, EDTA when added to lakewater or calcium-rich culture solutions will
partly combine with the calcium present because the calcium concentration
is normally three orders of magnitude higher than that of iron. Also the EDTA
from Fe-EDTA added to lakewater will partly combine with the calcium. The
ionic iron which is thus released by the presence of calcium may be in excess
of the solubility of $Fe(OH)_3$ and may then precipitate causing a new quantity
of ionic iron to be formed, etc. This is particularly likely to happen during
sterilisation of culture solutions.

Fig. 11.3 shows that in the presence of Ca^{2+}, above about pH = 8.5, even
ten times as much EDTA as Fe^{3+} will not prevent precipitation of the iron as
$Fe(OH)_3$. This occurs to an even greater extent in culture solutions if iron
and EDTA are provided in stoicheiometric quantities (for example as the
commercially available Fe-EDTA disodium salt).

Many different ligands occur in natural waters as do competing ions other
than calcium (e.g. Mg^{2+}), the quantities of ligands involved usually ranging
between 0.1 and 1 mg l^{-1}. No complete model is yet available to describe
such systems but the problem is discussed in an excellent theoretical review
by Stumm and Morgan (1970). It is an open question to what extent such
complexes play a role in nature, where one finds colloidal metal-oxide particles too. These particles may be stabilised by peptisation, the process by
which a colloidal particle is kept in solution. It may occur when the colloidal
particle is surrounded by small quantities of another compound which may
form an electric double layer around the particle. This layer prevents the
particles aggregating into larger units which would then precipitate. To keep
the same amount of Fe in solution by peptisation needs much less organic

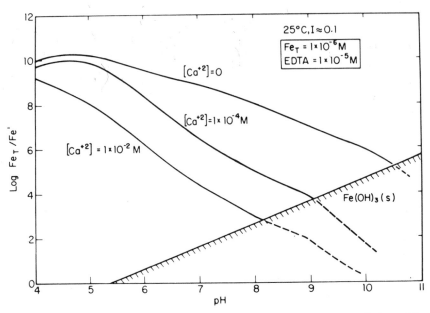

Fig. 11.3. Effects of total Ca concentration upon Fe(III)-EDTA equilibrium. (From Stumm and Morgan, 1970.)

matter than to keep it in solution by chelation. For a complex such as iron glycinate 3 mg of glycinate can chelate only 1 mg of iron, but the amount peptised by 3 mg of glycinate is several times larger. Shapiro (1967a) used yellow organic acids as peptising agents for Fe(OH)$_3$. He found an "iron holding capacity" of 40 μmoles of iron (= 2 mg) per mg dry organic matter. This is more than twice the maximum amount of iron which could be account- ed for by chelation. It is quite possible that the addition of more iron than can be chelated causes a precipitate which, after forming an unstable Fe(OH)$_3$ sol, is then stabilised by organic matter. This will not happen in nature, how- ever, because the organic chelator is usually present in excess. Shapiro gave no details of how he filtered off the iron, so that one cannot decide whether the iron was colloidal or not. We found that a "colloidal" solution of 1 mg l^{-1} of Fe as Fe(OH)$_3$ would pass through a 0.45-μm filter, but not through a 0.1-μm filter. Siegel (1971) states that this is also true of Shapiro's iron.

Christman (1967) showed that data used as proof of a non-chelation model could, with some speculative assumptions, be used to support a chela- tion mechanism equally well. Christman and Minear (1971) and Siegel (1971) believed that interaction between iron and coloured organic matter can be explained in terms of simple solution chemistry, without resort to the idea of peptised sols. The organic compounds most involved in these processes in natural waters are humic or fulvic acids (see Chapter 12), even in colourless waters. Because most other trace elements will react similarly to iron, it is likely that these compounds are responsible for keeping all trace elements in solution.

11.3. AVAILABILITY OF IRON

The question: "How much iron (and other trace elements) is available for algal uptake?" is difficult to answer. As early as 1937 Harvey considered the problem and concluded that iron was more easily available to marine diatoms from an $Fe(OH)_3$ suspension than from an inorganic ferric salt in true solution. At present there seems little doubt that Fe^{3+} ions are available, and Harvey's results may perhaps be attributed to precipitation in his solutions. Age, hydration and coarseness of such a precipitate will certainly influence availability. Goldberg (1952) reported that *Asterionella japonica* can utilise only particulate or colloidal iron and that chelates with citrate, ascorbate and humate are not available. Since Pringsheim's (1946) work, however, it has been commonly believed that the salutary effect of soil extract on algal growth is mainly due to the humic chelates in such extracts. On the other hand, the presence of chelators can have a negative effect on iron availability. If the iron is very strongly chelated, it is unavailable to algae. After adding $0.01\,M\,l^{-1}$ NTA to Rodhe's culture solution (see section 10.2) we could obtain no growth of *Scenedesmus*, but when $0.01\,M\,l^{-1}$ of Ca^{2+} was also added growth resumed. On the other hand, Golterman (1960) could obtain no growth of certain diatoms if only Fe-EDTA was added but good growth occurred if extra EDTA was added in amounts sufficient to chelate the Ca^{2+} as well. The relative amounts of EDTA, iron and calcium seem therefore to be critical.

Rodhe (1948) found that the iron available in Lake Erken water to *Scenedesmus quadricauda* corresponded with the fraction of iron that reacted with phenanthroline. Shapiro (1967b) studied iron availability from chloride, citrate and EDTA complexes and found large differences for different species. His negative results for citrate could be explained easily by supposing incomplete complex formation. Unfortunately, Shapiro gave no information on calcium content and pH of his cultures. There is no doubt that *Asterionella formosa* can grow both with iron citrate or the EDTA salt. Shapiro distinguished the iron fractions in lakewater by measuring the colour with thiocyanate at different pH values. He called this diagram a "ferrigram" and showed that the fraction which reacted at pH = 1.5 corresponded approximately with the iron available for *Microcystis* growth. However, he also found that other algae behaved differently. We noticed that *Microcystis* could use iron from hard water only after addition of NTA but that *Skeletonema* could use it directly. The "ferrigrams" are difficult to reproduce; apparently some steps — perhaps the speed of reagent additions — are critical.

11.4. TRACE ELEMENTS AS FACTORS LIMITING ALGAL GROWTH

Goldman made an extensive study of the effect of trace elements on algal

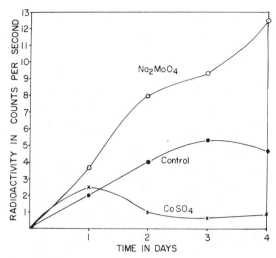

Fig. 11.4. Response of Castle Lake's natural phytoplankton population to the addition of 100 ppb molybdenum and 5 ppb cobalt, as measured by ^{14}C assimilation. The hard-glass culture containers were maintained in the lake under natural conditions of light and temperature. (From Goldman, 1965.)

growth and reviewed the subject in 1965. Goldman's method is essentially to detect an increase in growth rate following the addition of a particular element. Growth rate is measured by $^{14}CO_2$ uptake during a short period. Exposures are mostly made in situ. One of the most difficult problems is to avoid contamination: glassware, chemical reagents and culture containers must be extremely clean.

Molybdenum is necessary for the formation of the enzyme nitrate reductase and is involved in the formation of the enzymes necessary for nitrogen fixation (nitrogen hydrogenase). Molybdenum-deficient *Scenedesmus* cells show a depressed chlorophyll synthesis (Arnon, 1956). Goldman (1965) found increased $^{14}CO_2$ uptake in Castle Lake (U.S.A.) in June 1959 after addition of Na_2MoO_4 to containers of lakewater suspended in situ (see Fig. 11.4). On October 19 no stimulation of $^{14}CO_2$ uptake was found and on October 20 and 21 there was a slight stimulation. Goldman concluded that light inhibition was an important factor (it was the cause of the uptake being lower on the 21st than on the 20th), so one might argue that the real effect of molybdenum was to prevent light inhibition. As the total biomass increase is not given by Goldman, other explanations — depressed respiration or mineralisation — are possible though. To investigate total lake response to molybdenum fertilisation Goldman added 6.27 kg of molybdenum to the epilimnion in July 1963. The molybdenum remained in solution throughout the year with measurable depletion occurring only through outflow.

Because primary and secondary productivity normally vary greatly from year to year it is difficult to interpret the results of algal-growth measurements in this experiment. Primary productivity increased slightly, after an initial decrease. If this decrease caused a "sparing effect" of some other limiting factors, the subsequent increase cannot be attributed with certainty to molybdenum fertilisation. The numbers of copepods increased after the addition of molybdenum, but only during the period of increased primary productivity.

The amount of molybdenum added was equivalent to an addition of 3.14 μg per cm^2 of lake sediments; they contained 11.5 to 28.4 mg of molybdenum per litre. Gorham and Swaine (1965) reported an average of 3.5 ppm molybdenum in Windermere sediments, with up to 10 ppm in the oxidised crust. Goldman (1964, 1965) reported that twelve other lakes in northern California and a number of lakes in South Island, New Zealand, responded to molybdenum addition. Several odd features of Goldman's (1964) results remain unexplained. In Picayane and Upper Cliff lakes addition of molybdenum alone gave stimulation, while the trace-element mixture containing molybdenum did not. Goldman's explanation, that other trace elements may be present in inhibiting concentrations, makes the situation even more confused.

Dumont (1972) reported that additions smaller than those originally supplied in Goldman's studies (100 μg l^{-1}) were even more effective. In winter about 25 mg m^{-3} and in summer 5 mg m^{-3} was optimal. Dumont studied the biological cycle of molybdenum in an eutrophic shallow pond (Lake Donk) and found that a quick liberation from sediments occurred in spring with gradual decrease during the course of the summer. He believed that winter depletion was due to sediment-binding. Summer peak values were high — up to 12—14 mg m^{-3}; but the molybdenum cycle closely followed the phosphate cycle — both showed an increase in spring — and one could argue that in both cases release followed reduction of the sediments. The phosphate would be related to the iron cycle and the molybdenum to the manganese cycle (molybdenum easily coprecipitates with manganese). Dumont believed that molybdenum was precipitated with iron. It is even possible that a phosphate—molybdenum complex forms. Dumont showed experimentally that molybdenum would remain dissolved in aerobic solutions but that in anaerobic conditions it would precipitate. He suggested H_2S as a possible cause of the precipitation, but since he did not measure iron, manganese or H_2S and added a rather high concentration of molybdenum (5.6 mg l^{-1}), his suggestion cannot yet be accepted as applicable to natural waters. Dumont estimated that 0.4—2.5 μg l^{-1} of molybdenum occurred in phytoplankton, when there was 5 μg l^{-1} total molybdenum in the water. He found somewhat higher values for blue-green algae (up to 57 μg g^{-1}) and 6—22.5 μg g^{-1} of mixed phytoplankton and detritus.

Goldman discussed copper, vanadium, cobalt, manganese, zinc, boron, iron, sodium and calcium concentrations in lakes, but wrote little about effects on algal growth in nature. Addition of cobalt — an essential part of the vitamin B_{12} molecule — produced some response in lakes in South Island, New Zealand, and was found to be toxic in other lakes (Castle Lake). Vitamin B_{12} requirements have been reported for many marine phytoplankton species (Provasoli, 1960, 1963; Carlucci and Silbernagel, 1966a, b, 1967). Some freshwater algae seem to need either cobalt or vitamin B_{12}. Copper is more often reported as an algal toxin in polluted waters than as a limiter of algal growth (Steemann Nielsen and Wium-Andersen, 1971; TNO-Nieuws, 1972). Although it is well established that copper may be toxic to some animals, one cannot yet assess the importance of copper toxicity in algal communities. There are unsolved problems of methods and of algal adaptation.

Although the *solubility* of trace elements is low, large quantities may be transported adsorbed on silt. Erosion of rocks is one source of trace elements. The effluent from sewage works may be much more important in some cases; the discharged metals will then be adsorbed onto the silt if present.

Many heavy-metal concentrations have exceeded accepted toxic levels. Particularly serious cases are mercury, copper, cadmium and lead. The main problem when studying the metal content of silt is to know the level the silt would have in an unpolluted state. Comparison with data from suspected unpolluted or comparable old sediments may help, but the possibility of mobilisation from the consolidated sediments must be taken into account. Serruya (1969) measured — for geological evaluations — trace-metal concentrations in sediments from the Lake of Geneva (Table 11.2). Trace elements were associated with clay particles. At least two points important for limnology arise from her work. The first is that in measuring trace elements in phytoplankton one should be especially careful to avoid contamination by eroded mineral matter. Second, the *concentrations* found in water may be low when compared with the total *amounts* available in sediments. Other studies on trace elements in sediments are those of Gorham and Swaine (1965) and Groth (1971). Gorham and Swaine, comparing the elemental composition of oxidising and reducing muds, found that in the English Lake District the relatively organic lake muds exhibited the highest concentrations of copper, tin and zinc. Many of the oxidised crusts exhibited great enrichment in manganese, iron, barium, strontium, lead and zinc. Both manganese and molybdenum es-

TABLE 11.2

Trace elements of the clay fraction in sediments of Lake of Geneva (ppm) (Serruya, 1969)

Age of deposit	V	Mo	Pb	Zn	Cu	Cr	Ni	Co	Sr	No. of samples
Subatlantic Subboreal	169	2	10.8	115	45	217	102	23	120	11
Atlantic Boreal	184	1.2	5.3	99	24	188	78	18	141	6
Preboreal	192	3.0	10.7	112	42	215	107	23	113	10
Dryas	202	2.0	12.5	126	43	211	106	24	114	7

caped from the mud surface during the summer reduction of the profundal mud surface.

11.5. POLLUTION DUE TO HEAVY (TRACE) ELEMENTS

It should not be necessary to write about pollution by heavy trace elements, but in the industrial areas of the world the present disposal of heavy metals such as mercury, cadmium, lead, copper and arsenic evokes great concern because of increasing concentrations in many waters. One of the great dangers is the accumulation of these metals in food chains. The major part of these elements is adsorbed on silt (see Chapter 18) and accumulation in the food chain starts easily with the filter-feeding zooplankton. Smaller quantities may be present in a dissolved state being chelated by organic compounds. Due to the adsorption of for example mercury on silt, its concentration increased in Windermere sediments from 0.1—0.3 to 1.1 mg kg^{-1} and from 0.3 to 1.4 mg kg^{-1} in Lake Ontario (Förstner and Müller, 1974).

The worst examples of accumulation are reported from Japan, where hundreds of fishermen were killed by eating fish containing too much mercury (Minamata disease, Takeuchi, 1972) or cadmium (Itai-Itai disease, Kobayashi, 1971). These tragedies arose from the unfortunate coincidence of a mainly fish-eating population and fish which, though having high mercury concentrations, showed no symptoms themselves. Cases of such severe illness have not yet resulted from eating freshwater fish, but this is probably due to the warning from Japan. In Sweden several lakes were found with fish containing a few mg of mercury per kg body weight and many waters showed values of 1 mg kg^{-1} of fish (Swedish Royal Commission on Natural Resources, 1967). The main cause of the Swedish mercury problem was agriculture and the paper industry, where mercury (in inorganic and organic forms) was used as a fungicide (Lihnell, 1967). After strict prohibition of this practice, normal levels of mercury in these lakes are slowly returning (Jernelöv and Lann, 1973). A third case of severe mercury pollution is the River Rhine, in which 100 tonnes of mercury are transported per year per 70 km^3 (see subsection 18.2.4).

A special feature makes the mercury problem more serious. In anaerobic conditions — after sedimentation for example — mercury can be converted to the yet more toxic methyl-mercury compounds which, like the metal, can accumulate in organisms, probably adsorbed on -SH groups in enzymes, and also in food chains. The conversion to methyl mercury is a bacteriological conversion involving methane bacteria:

$$Hg^{2+} \rightarrow CH_3Hg^+ \rightarrow (CH_3)_2Hg$$

Due to this reaction mercury adsorbed onto sediments may be mobilised after their settlement (which will often lead to anaerobic conditions) and through food chains will accumulate in fishes. Accumulation of mercury in pike

(*Esox lucius*) and some other aquatic organisms has been described by Johnels et al. (1967). Hannerz (1968) reported on experimental investigations on mercury accumulations.

The main problems with mercury are not global but local, because industrial output is roughly equal to natural erosion, i.e. 5 000—10 000 tonnes per year. This natural erosion mobilises so much mercury that it is detectable even in Atlantic Ocean fishes (25—155 μg kg^{-1}). Concentrations in natural waters seem to be in the range between 10 and 50 ng l^{-1}, but near mercury-rich rocks they may be higher (Dall'Aglio, 1968; Klein, 1972). In the tributaries of the river Tiber, near a mercury mine, a river with 136 μg l^{-1} was found (Dall'Aglio, 1968). (In the Tiber mouth the concentration has already decreased to 40 ng l^{-1} (Van Urk, 1970).) Similar high concentrations were found in many waters near mercury ore (cinnabar) deposits even if not now worked. Dall'Aglio found no correlation between mercury content and river-water conductivity and pointed out that mercury is easily adsorbed on river alluvia — a phenomenon which enhances mercury mobility.

Accumulation can be demonstrated by the following examples:

Lake model			Minamata Bay			
water	0.00001	mg l^{-1}	water	0.0001—	0.001	mg l^{-1}
bream	0.1	mg kg^{-1}	diatoms 5	—10		mg kg^{-1}
fish eagle	10	mg kg^{-1}	fish		50	mg kg^{-1}

(Ui and Kitamura, 1969a, b)

Much mercury is still used in the production of caustic soda and caustic potash, as a catalyst in chemical industry, and as fungicide in the paper industry and in agriculture. Use in agriculture is now strictly limited in many countries, and direct disposal from industries is or ought to be completely forbidden.

Cadmium causes some problems similar to those caused by mercury, though there are few recorded cases apart from the Japanese one. The river Rhine transports roughly 200 tonnes per year, twice as much as of mercury. Cadmium is even more toxic than mercury; *Daphnia, Scenedesmus* and *E. coli* are very sensitive (Anderson, 1950; Bringmann and Kuhn, 1959) but cadmium in natural waters may be less toxic than in distilled water because of chelation or adsorption on silt. Bartlett and Rabe (1974) found growth inhibition of *Selanastrum capricornutum* at 50 μg l^{-1} of Cd, whereas complete inhibition occurred at 80 μg l^{-1}.

Another element which causes concern in industrially polluted areas is copper. About 1 g of copper causes acute illness in man. Copper is therefore less toxic than mercury and accidents like the Minamata disease are not likely to occur, nor it seems is poisoning of drinking water. The disposal of copper is so common and widespread, however, that copper concentrations in aquatic ecosystems may commonly be large enough to be harmful to organisms. In the U.S.A. a concentration of 1 mg l^{-1} in drinking water seems to be acceptable (see Table 19.5), but the European standard for running tap water specifies less than 50 $\mu g \, l^{-1}$. Some aquatic organisms may find tenths of micrograms per litre to be toxic. Levels of toxicity tolerance have decreased as detection methods have improved. Thus Bringmann and Kuhn (1959) reported that 0.15 mg l^{-1} is the threshold concentration which produces a noticeable effect on *Scenedesmus* . Liebmann (1951—1960) gave 80—800 $\mu g \, l^{-1}$ as the toxic level for freshwater fish. Steemann Nielsen and Wium-Andersen (1971) described an effect of 3 $\mu g \, l^{-1}$ of Cu on photosynthesis of cultures of *Nitzschia palea* and stated that even distilled water may contain toxic quantities, though they did not indicate how they removed the copper from the distilled water used. Nevertheless, most algal cultures are made in distilled water and adaptation to copper may protect many algae against effects such as those described by Steemann Nielsen.

Bartlett and Rabe (1974) found growth inhibition of *Selanastrum capricornutum* at 50 $\mu g \, l^{-1}$ of Cu, but they did not take into account the possible positive influence of the EDTA used for the medium. This may be the cause for the observed inhibition of copper toxicity by cadmium.

In cultures much more copper is often added, but the presence of Fe-EDTA may provide sufficient chelators to render copper harmless. Freshwater animals are probably more sensitive. Adema and De Groot-Van Zijl (1972) found that 50% of a population of the waterflea *Daphnia magna* died at concentrations ranging between 25 and 65 $\mu g \, l^{-1}$, the exact value depending on experimental exposure time and age of the animal. The influence of experimental time is probably a much more important factor than was originally realised. Experimental times of a few hours are too short. A suitable time would be a complete life cycle. Adema and De Groot-Van Zijl also found an influence of the age of the copper solution. If recently added to the solution copper was more harmful than after some ageing. It seems likely that ionic copper is the real toxic substance, while oxides or other colloidal particles or chelates are much less harmful.

The chemistry of copper in freshwater is rather complex. It may be chelated by several naturally occurring ligands, and several basic complexes such as $Cu(OH)^+$, $[Cu_2(OH)_2]^{2+}$, $Cu_3(OH)_2(CO_3)_2$ may form. Even in distilled water ionic copper, $[Cu(H_2O)_6]^{2+}$, will be transformed into one of these complexes; which one will depend in natural waters on alkalinity and pH. The process is not instantaneous but may take several weeks (Stiff, 1971).

Gächter et al. (1973) discussed the difficulties in measuring ionic copper in lakewater, in which it is easily masked by ligands. At present they believe it is not possible either to measure the concentration of free ions or to estimate it indirectly by calculation, as the nature of the ligands (and thus the formation constants) are not known. They suggested that differential pulse anodic stripping voltammetry is the best method available at present.

It is impossible therefore to generalise about natural concentrations of copper and at what concentration it is toxic. Man uses copper so commonly, however, that concentrations of hundreds of microgrammes per litre have been found locally. Although the major part will be adsorbed onto silt or will be chelated, more care in copper disposal is advisable. It is important to realise that levels of some metals that are acceptable for drinking water (see Chapter 19) are often higher than those that have no influence on aquatic organisms. For the protection of lake ecosystems different criteria — often much lower than those for drinking water — should be used or developed.

REFERENCES

Adema, D.M.M. en De Groot-Van Zijl, Th., 1972. De invloed van koper op de watervlo *Daphnia magna*. *TNO-Nieuws*, 27(9): 474—481 (English Summary).

Anderson, B.G., 1950. The apparent thresholds of toxicity to *Daphnia magna* for chlorides of various metals when added to Lake Erie water. *Trans. Am. Fish. Soc.*, 78: 96.

Arnon, D.I., 1956. Some functional aspects of inorganic micro-nutrients in the metabolism of green plants. In: A.A. Buzzati-Traverso (Editor), *Perspectives in Marine Biology*. Symp. at Scripps Institution of Oceanography, Univ. California, La Jolla, Calif., pp. 351—383.

Bartlett, L. and Rabe, F.W., 1974. Effects of copper, zinc and cadmium on *Selenastrum capricornutum*. *Water Res.*, 8: 179—185.

Bringmann, G. and Kuhn, R., 1959a. The toxic effects of waste water on aquatic bacteria, algae, and small crustaceans. *Gesundheitsingenieur*, 80: 115.

Bringmann, G. and Kuhn, R., 1959b. Water toxicology studies with protozoans as test organisms. *Gesundheitsingenieur*, 80: 239.

Carlucci, A.F. and Silbernagel, S.B., 1966a. Bioassay of seawater. Methods for the determination of concentrations of dissolved B_1 in seawater. *Can. J. Microbiol.*, 12: 1079.

Carlucci, A.F. and Silbernagel, S.B., 1966b. Bioassay of seawater. Distribution of vitamin B_{12} in the Northeast Pacific Ocean. *Limnol. Oceanogr.*, 11(4): 642—646.

Carlucci, A.F. and Silbernagel, S.B., 1967. Determinations of vitamins in seawater. *Proc. IBP-Symp. Amsterdam—Nieuwersluis, 1966*, pp. 239—244.

Christman, R.F., 1967. The chemistry of rivers and lakes: The nature and properties of natural product organics and their role in metal ion transport. *Environ. Sci. Technol.*, 1(4): 302—303.

Christman, R.F. and Minear, R.A., 1971. Organics in lakes. In: S.J. Faust and J.V. Hunter (Editors), *Organic Compounds in Aquatic Environments*. Marcel Dekker, New York, N.Y., pp. 119—143.

Dall'Aglio, M., 1968. The abundance of mercury in 300 natural water samples from Tuscany and Latium. In: L.H. Ahrens (Editor), *Origin and Distribution of Elements*. First Meet. Int. Assoc. Geochem. Cosmochem., Paris, 1967. Pergamon Press, Oxford, pp. 1065—1081.

Dumont, H.J., 1972. The biological cycle of molybdenum in relation to primary production and waterbloom formation in a eutrophic pond. *Verh. Int. Ver. Theor. Angew. Limnol.*, 18: 84—92.

Förstner, U. and Müller, G., 1974. *Schwermetalle in Flüssen und Seen als Ausdruck der Umweltverschmutzung*. Springer-Verlag, Berlin, 235 pp.

Gächter, R., Lum-Shue-Chan, K. and Chau, Y.K., 1973. Complexing capacity of the nutrient medium and its relation to inhibition of algal photosynthesis by copper. *Schweiz. Z. Hydrobiol.*, 35(2): 252—261.

Goldberg, E.D., 1952. Iron assimilation by marine diatoms. *Biol. Bull. Woods Hole*, 102: 243—248.

Goldman, Ch.R., 1964. Primary productivity and micro-nutrient limiting factors in some North American and New Zealand lakes. *Verh. Int. Ver. Theor. Angew. Limnol.*, 15: 365—374.

Goldman, Ch.R., 1965. Micronutrient limiting factors and their detection in natural phytoplankton populations. *Mem. Ist. Ital. Idrobiol.*, 18 (Suppl.): 121—135.

Golterman, H.L., 1960. Studies on the cycle of elements in fresh water. *Acta Bot. Neerl.*, 9: 1—58.

Gorham, E. and Swaine, D.J., 1965. The influence of oxidizing and reducing conditions upon the distribution of some elements in lake sediments. *Limnol. Oceanogr.*, 10: 268—279.

Groth, P., 1971. Untersuchungen über einige Spurenelemente in Seen. *Arch. Hydrobiol.*, 68: 305—375.

Hannerz, L., 1968. Experimental investigation on the accumulation of mercury in water organisms. *Rep. Freshwater Res. Drottingholm*, 48: 120—176.

Harvey, H.W., 1937. The supply of iron to diatoms. *J. Mar. Biol. Assoc. U.K.*, 22: 205—219.

Jernelöv, A. and Lann, H., 1973. Studies in Sweden on feasibility of some methods for restoration of mercury-contaminated bodies of water. *Environ. Sci. Technol.*, 7(8): 712—718.

Johnels, A.G., Westermark, T., Berg, W., Persson, P.I. and Sjöstrand, B., 1967. Pike (*Esox lucius* L.) and some other aquatic organisms in Sweden as indicators of mercury contamination in the environment. *Oikos*, 18: 323—333.

Klein, D.H., 1972. Some estimates of natural levels of mercury in the environment. In: R. Hartung and B.D. Dinman (Editors), *Environmental Mercury Contamination*. Science Publishers, Ann Arbor, Mich., pp. 25—29.

Kobayashi, J., 1971. Relation between the "Itai-Itai" disease and the pollution of river water by cadmium from a mine. *Adv. Water Pollut. Res. Proc. Int. Conf., 5th, San Francisco and Hawaii, 1970*, 1, I 25: 1—7.

Liebmann, H., 1951—1960. *Handbuch der Frischwasser- und Abwasserbiologie; Biologie des Trinkwassers, Badewassers, Frischwassers, Vorfluters und Abwassers*. Oldenbourg, Munich (2 Volumes).

Lihnell, D., 1967. The use of mercury in agriculture and in industry. *Oikos, Suppl.*, 9: 16—17.

Pringsheim, E.A., 1946. *Pure Cultures of Algae; Their Preparation and Maintenance*. Cambridge University Press, Cambridge, 119 pp.

Provasoli, L., 1960. Micronutrients and heterotrophy as possible factors in bloom production in natural waters. In: *Trans. Semin. Algae and Metropolitan Wastes*. U.S. Public Health Service, R.A. Taft Sanit. Eng. Center, Cincinnatti, Ohio, 9 pp.

Provasoli, L., 1963. Growing marine seaweeds. *Proc. Int. Seaweed Symp., 4th*. Pergamon Press, Oxford, pp. 9—17.

Rodhe, W., 1948. Environmental requirements of fresh-water plankton algae. Experimental studies in the ecology of phytoplankton. *Symb. Bot. Upps.*, 10: 1—149.

Scholte Ubing, D.W., 1970. *Milieuverontreiniging met kwik en kwikverbindingen*. TNO-Report A 60.

Serruya, C., 1969. Problems of sedimentation in the lake of Geneva. *Verh. Int. Ver. Theor. Angew. Limnol.*, 17: 209—218.

Shapiro, J., 1967a. Yellow organic acids of lake water: Differences in their composition and behavior. *Proc. IBP-Symp. Amsterdam—Nieuwersluis, 1966*, pp. 202—216.

Shapiro, J., 1967b. Iron available to algae: Preliminary report on a new approach to its estimation in lake water through use of the "ferrigram". *Proc. IBP-Symp. Amsterdam—Nieuwersluis, 1966*, pp. 219—228.

214

Siegel, A., 1971. Metal-organic interactions in the marine environment. In: S.J. Faust and J.V. Hunter (Editors), *Organic Compounds in Aquatic Environments*. Marcel Dekker. New York, N.Y., pp. 265—295.

Steemann Nielsen, E. and Wium-Andersen, S., 1971. The influence of Cu on photosynthesis and growth in diatoms. *Physiol. Plant.*, 24: 480—484.

Stiff, M.J., 1971. The chemical states of copper in polluted fresh water and a scheme of analysis to differentiate them. *Water Res.*, 5: 585—599.

Stumm, W. and Morgan, J.J., 1970. *Aquatic Chemistry; An Introduction Emphasizing Chemical Equilibria in Natural Waters*. Wiley-Interscience, New York, N.Y., 583 pp.

Swedish Royal Commission on Natural Resources (Editors), 1967. The mercury problem. In: *Symp. Mercury in the Environment, Stockholm, 1966. Oikos, Suppl.*, 9.

Takeuchi, T., 1972. Distribution of mercury in the environment of Minamata Bay and the inland Ariaka Sea. In: R. Hartung and B.D. DinMan (Editors), *Environmental Mercury Contamination*. Science Publishers, Ann Arbor, Mich., pp. 79—81.

TNO-Nieuws, 1972. Koper in het Nederlandse milieu. *TNO-Nieuws*, 27(9): 415—516.

Ui, J. and Kitamura, S., 1969a. *Sanitary Engineerings Approach to Minamata Disease*. Rep. Dep. Urban Eng., Univ. of Tokio and Rep. Dep. Public Health, Univ. of Kobe, Japan, 29 pp.

Ui, J. and Kitamura, S., 1969b. *Mercury Pollution of Sea and Fresh Water; Its Accumulation into Water Biomass*. Rep. Dep. Urban Eng., Univ. of Tokio, and Rep. Dep. Public Health, Univ. of Kobe, Japan. (Presented 4th Coll. Mar. Water Pollut., Napoli, 1969).

Van Urk, G., 1970. Literatuuroverzicht kwik. Opgesteld t.b.v. Werkgroep Kwik Commissie TNO voor onderzoek inzake nevenwerkingen van bestrijdingsmiddelen. Utrecht.

CHAPTER 12

ORGANIC MATTER

12.1. INTRODUCTION

All natural waters contain organic matter, both in dissolved and in particulate form; some have colloidal forms too. Most of the known biochemicals occur in water and most of them have been detected, at some time or other, although some are present only in minute quantities. Certain compounds are present in considerable quantities however and amongst these are products of hydrolysis of proteins and carbohydrates (Vallentyne, 1957). There are also many reports of the occurrence of purines, amino acids etc. These reports often give concentrations, but unless one knows the total concentration of organic carbon and the turnover times, it is difficult to evaluate the importance of the occurrence for example of 10 μg l^{-1} of acetic acid.

The concentration of organic carbon falls mostly within the range 0.1—10.0 mg l^{-1}, but of course exceptional cases occur. For example, in high mountain lakes or arctic lakes values may be lower; in lakes receiving much sewage or drainage water, values may be higher.

Although analytical methods are not discussed in this book, the number of methods commonly used for estimating "organic carbon" is so great and the results are so difficult to compare that a short compilation will be given here. Dissolved and particulate compounds can be separated by filtration and most methods can be used on both fractions. There is no sharp distinction between "dissolved" and "particulate"; the boundary depends on the filter used. Since international standardisation took place during the International Biological Programme there has been an increasing tendency to use 0.45 μm pore size filters. With such filters it is still practicable to filter sufficient water even in turbid lakes. During filtration the effective pore size will decrease owing to blocking of the pores. Even smaller particles will then be filtered off. For rather pure waters 0.2-μm filters are often used.

The oldest method for estimation of organic carbon is to measure the loss in weight on ignition. A sufficiently large quantity of water is evaporated to dryness (105°C) after which the residue is ignited at about 600°C (at higher temperatures $CaCO_3$ will decompose). The difference found represents the weight of the organic matter plus the loss of water from the clay.

The second method is based on the fact that most (perhaps nearly all) organic molecules can be oxidised by compounds such as $KMnO_4$, $K_2Cr_2O_7$ and by Ce^{4+}. (For analytical procedures see Golterman and Clymo, 1971.) Organic carbon and O_2 are indirectly related because the reaction is:

$$C_xH_{2y}O_z + \left[x + \frac{y-z}{2} \right] \cdot O_2 \rightarrow x\,CO_2 + y\,H_2O \; .$$

The amount of carbon oxidised could therefore be calculated from the amount of O_2 taken up if the ratio $(y - z)/x$ were known. This ratio is, however, unknown in most cases, and the results are therefore expressed as the amount of O_2 used per unit volume of water, the so-called Chemical Oxygen Demand or C.O.D. Some authors assume that 1 mole of O_2 gives rise to 1 mole of CO_2 and calculate the organic carbon content in this way. Carbon can also be determined directly by trapping and measuring the CO_2 produced after dry or wet oxidation and measuring it as CO_2 (e.g. conductrometrically, by gas chromatography or by infrared gas analysis). The C.O.D. method is, however, simpler than a carbon determination, and is directly comparable to the Biochemical Oxygen Demand (B.O.D.), which estimates the organic compounds available for bacterial oxidation. B.O.D. is often measured by incubating the lakewater in the dark at a given temperature for 2 or 5 days. C.O.D. has also a particular significance when photosynthesis is expressed in the same units (mg l^{-1} of O_2). The use of O_2 as a measure for primary production has the essential advantage that we can express the production chain in terms of numbers of electrons transferred. The determination of dissolved and particulate compounds is then only the determination of reducing capacity. Using O_2 — and therefore electrons — as a basis for studying the primary production processes, we can express measurements on organisms or dissolved organic substances easily and directly as proportional components of the energy-binding process.

KMnO$_4$ was introduced as an oxidising compound by Juday and Birge (1932). It has been widely used but many different methods are in use. The oxidation is carried out either at room temperature or at $100°C$, at an acid or an alkaline pH, and for a variety of times. The results of these different methods are not usually comparable. A second disadvantage is that KMnO$_4$ is subject to autooxidation:

$$2 \, MnO_4^- + 4 \, H^+ \rightarrow MnO_2 + 2 \, O_2 + 2 \, H_2O + Mn^{2+}$$

The reagent blanks therefore show a considerable and variable decrease in MnO_4^- concentration, especially after boiling, and it is uncertain whether or not this value should be subtracted from the final determination. The KMnO$_4$ method is being replaced by one using $K_2Cr_2O_7$, which is a more suitable oxidising agent. The reaction is:

$$Cr_2O_7^{2-} + 14 \, H^+ + 6 \, e^- \rightarrow 2 \, Cr^{3+} + 7 \, H_2O$$

so that 1.0 mole l^{-1} $Cr_2O_7^{2-} = 6.0N$.

The oxidising power of $K_2Cr_2O_7$ is slightly less than that of KMnO$_4$, so the results are sometimes lower than with KMnO$_4$, but they are more consistent. In subsection 19.5.2 an opposite case is described where permanganate oxidisability is only 40% of dichromate oxidisability. Rather higher values may be obtained by adding Ag^+ as a catalyst. This renders molecules such as acetic acid completely degradable.

The chemical determination of $Cr_2O_7^{2-}$ is often carried out by back titration. This leads to difficulties in the many cases when there is a low concentration of organic matter. It is now common practice therefore to use a $0.01N$ solution in stead of a $0.1N$, but this lowers the oxidation potential too much. It is better practice to use a $0.1N$ solution and to measure colorimetrically the concentration of the Cr^{3+} which is formed (see Golterman, 1971). The disadvantages of the dichromate method are the need for a high sulphuric-acid concentration (about $10N$) and for a high temperature ($140-150°C$).

Mackereth (1963) used the light extinction of lakewater at 360 nm as a direct measure of organic content. This gave a nearly linear regression with organic matter in one lake, but a different one in another lake. As a first approximation though it is a useful method.

Banoub (1973) used the light extinction at 260 nm to estimate organic matter. He found a linear regression between colour and organic carbon, but the variability was considerable. Ultraviolet absorption at 7 mg l^{-1} of C varied more than twofold in different water samples.

The organic compounds may be divided into two groups: the coloured and the colourless ones. Such a division is of course arbitrary, but it is in common use. Possession of colour is a subjective judgement, and most optically colourless compounds will absorb light in the near-ultraviolet.

12.2. THE GENERAL PROPERTIES OF ORGANIC MATTER

Organic compounds are constantly being produced by phytoplankton, either by excretion or by autolysis. Most compounds will rapidly be broken down by bacteria but refractory compounds do occur, both coloured and colourless ones. Higher plants and animals also produce organic compounds, but in the open lake their contribution is small.

Dissolved organic matter may have five different effects on organisms:

(1) It may supply energy or organic carbon, for example for bacteria, or blue-green algae (see Chapters 6 and 16).

(2) It may supply accessory growth factors, for example vitamins (see Chapters 14 and 16).

(3) It may have toxic effects (see Chapter 14).

(4) It may form organic complexes with trace elements, and these may have either a beneficial or a detrimental effect on organisms.

(5) It may absorb light and thus affect photosynthesis (see Chapter 4 and section 12.3).

Saunders (1957) listed 14 genera of algae (32 species) which are capable of using organic substances as energy sources and a list of 70 substances that can be used by algae for energy or growth. In view of the results of Hobbie and Wright (see section 16.6), it seems likely that these compounds, at the concentration at which they occur, will normally be used by bacteria. Very little information exists on turnover rates of these compounds.

An accessory growth factor may either be essential to an organism for its growth or it may stimulate growth when present in minute quantities. Yeast or soil extract is often supposed to have this function but it is not at present possible to say how much of the effect is owed to chelation of metals. (See subsection 17.3.3 for a discussion on growth effects of humic compounds.) Saunders gives a list of genera known to respond to growth factors and of these growth factors themselves. They include the vitamins, biotin, B_1, B_6, and B_{12} and in addition histidine, uracil, reduced sulphur compounds and some unkown factors. The ecological significance of these substances is not yet established. One might wonder how an alga with a demand for a special nutrient has managed to evolve in competition with other species with no special requirements. There seems to be an abundance of strains without special requirements.

Vitamins do occur in lakewater, although quantitative data are scarce. Carlucci and Bowes (1972), who reviewed older literature, found $5 - 38$ ng l^{-1} of thiamine and about 1 ng l^{-1} of biotin and of B_{12} at certain depths (but not all) of Lake Tahoe (California, U.S.A.).

The significance and possible ecological role of toxins has been discussed in Chapter 6, in the case of blue-green algae. Other known toxins are chlorellin produced by *Chlorella* and a toxin produced by *Prymnesium parvum* (a free-living brackish-water dinoflagellate), which kills fish and other gill-breathing organisms. The ecological implications or rather the lack of implications will be discussed further in Chapter 14.

Most organisms need many trace metals, such as iron, which are not normally very soluble in lakewater. Complexes may be formed with dissolved organic matter, which make the metal more soluble. Most of these compounds are coloured, though a few are colourless. The subject is considered in detail in section 11.2. Organic acids, amino acids, peptides and porphyrins may also act as chelating agents. Gerletti (1965) found a mean protein value equivalent to 0.24 mg l^{-1} of albumin from bovine serum, but he could not demonstrate a correlation with chelating capacity, probably owing to the low precision of the methods for measuring proteins and chelating capacity.

Wetzel (1972) studied the role of organic carbon in hard-water marl lakes in southern Michigan. The concentrations there are low to moderate (1—5 mg l^{-1} of C) and the compounds are mainly relatively refractory. In these lakes, e.g. Lawrence Lake, the organic matter is produced by submerged macrophytes as well as by phytoplankton. The former contribute about 48% of the total organic matter and this consists of both labile and refractory compounds. The labile ones are rapidly utilised by epiphytic and planktonic microflora (Wetzel and Allen, 1972). Some organic substances are removed by adsorption on and precipitation with $CaCO_3$ (Wetzel, 1970).

Concentration of iron in Lawrence Lake is generally within the range 1—10 μg l^{-1}, with a mean value of 5 μg l^{-1}. Iron salts which were added to the water precipitated immediately, and often coprecipitated the existing iron. Artificial chelators such as NTA and EDTA appeared to be able to keep the iron in solution if added in a certain ratio to the iron itself. If too little or too much chelator was added, iron could be filtered off. The fact that iron could no longer be detected if too much chelator was added may be owed to interference by EDTA and NTA with the analytic method for iron. The effect of calcium or of pH on the solubility of iron was not studied however (see section 11.2). Glycine was shown to be able to keep iron in solution, possibly by chelation. Applications of 10 μg l^{-1} and 1 000 μg l^{-1} of NTA did not increase $^{14}CO_2$ uptake by the natural population, the higher concentration probably because it chelates the iron strongly. A concentration of 100 μg l^{-1} of NTA enhanced photosynthesis. Natural amino acids such as glycine at concentrations of 100 μg l^{-1} were very effective in maintaining iron availability. Wetzel suggested, however, that labile organic compounds would be adsorbed onto the $CaCO_3$ particles, so that only refractory organic matter would be active in maintaining the iron in solution. Adsorption of artificially prepared yellow acids on $CaCO_3$ was demonstrated by Otsuki and Wetzel (1972).

Glucose and acetate could be used by heterotrophic bacteria living epiphytically on emergent and submerged aquatic macrophytes. It has been suggested therefore that this portion of the biotic community within a marl lake may be important in the cycling of dissolved organic matter (see Allen, 1969, cited in Wetzel, 1970). Rates of the turnover of glucose and acetate carbon are equivalent to 45—65% per day of the total dissolved organic carbon in the littoral zone. Wetzel and Manny (1972b) showed that secretion of dissolved organic carbon (DOC) by a submerged angiosperm *Najas flexilis* contributed an important part of the photosynthetically fixed carbon and that all DOC and dissolved organic nitrogen (DON) was easily oxidised during UV irradiation. Rapid oxidation during this treatment is a characteristic of simple, readily decomposable organic substrates. The lability towards UV irradiation depended on the medium used for the growth of *Najas*. Bacterial secretion products were highly UV-labile. Refractory DOC and DON seem to originate more from allochthonous terrestrial sources. Wetzel suggested that since bacteria are substrate-limited in hard-water lakes — just as they are in all other lakes — the production of DOC and DON by higher plants may accelerate bacterial growth and thus nutrient generation by mineralisation.

Wetzel et al. (1972) studied dissolved organic matter further in Lawrence Lake and showed that in these marl lake systems much more carbon may be dissolved than is present in the particulate (dead plus living) fraction. For the open lake water, the pelagial, this is probably not true, but quantitative data are scarce. The authors also discussed the occurrence of a detrital food chain using energy from dead organic matter. They found that the dissolved organic carbon pool was rather constant with a mean value of 5.6 mg l^{-1} of C (range 1.5—9.6) with maximal values in September or October, indicative of a progressive accumulation of largely refractory organic matter prior to the autumn overturn. These compounds were mainly from allochthonous sources and provided 21 g m^{-2} yr^{-1} of the 35.8 g m^{-2} yr^{-1} that left the lake. The rest came mainly from phytoplankton excretion and autolysis.

Wetzel et al. (1972) gave a carbon flow diagram for hard water with much macrophytic vegetation, but many of the estimates in this diagram are based on plausible hypotheses or on uncertain conversion factors.

Saunders (1972) examined extracellular release by phytoplankton in different lakes during a 24-h period from sunrise to sunrise. The relationship between photosynthesis and extracellular release of DOC is complicated by the effects of changing irradiance during the day. Using an argument based on kinetic considerations Saunders suggested that both simple organic molecules and large polymers are released; some compounds are released throughout the day, others are only produced during darkness. Saunders remarked that an exact interpretation of the kinetics of release awaits identification of the nature of the molecules released.

12.3. COLOURED ORGANIC MATTER

12.3.1. Occurrence and structure

The coloured organic matter is often referred to collectively by the term humic compounds or humic and fulvic compounds. These are polymeric products of carbohydrates, lignins, proteins and fats in various stages of decomposition. The exact nature, origin and properties are still obscure as are even the general properties. Because of the regularity and reproducibility with which they appear during isolation procedures, they might be considered to be one group. Very little is known about the origin of humic compounds. The old idea was that they were formed during a non-biological oxidation, but nowadays it is certain that a bacterial oxidation of lignin, cellulose, and protein is involved. The situation is nevertheless rather complex; for example cellulose that had passed through the digestive tract of *Chironomus* larvae appeared to be more available as a substrate for cellulytic bacteria than "raw" cellulose.

The old agricultural method of isolation used an alkaline extraction from the soil giving a dark brown solution. After acidification, a brownish precipitate formed, this containing the so-called humic acids. Those brown compounds which remained in solution were called fulvic acids. The brown compounds from water are often extracted after acidification with butanol, but not all the colour is thus extracted. De Haan (1972a) used slightly acid Al(OH)$_3$ as adsorbant to concentrate the brown compounds. The coloured compounds were redissolved with a phosphate buffer. As humic acids chelate metal ions (see below), it is possible that the strength of this complex binding is changed by the alkaline or acid conditions during the extraction, so that the brown compounds are no longer in their "native" state.

The quantities of humic compounds in lakewater may be quite large.

Ryhänen (1968) gave a mean concentration value of 13.2 mg l^{-1} of humic compounds for Finnish waters. As a result of run-off from the Finnish land-mass to the sea, an annual quantity of $1-1.25 \cdot 10^6$ tonnes of humic compounds is transported to the Baltic Sea.

The main organic substances to be found in natural waters are the fulvic acids; the term humic compounds will be used here however in a broad sense.

Two types of waters may be distinguished, those which are acid due to the presence of the fulvic acids (and carbon dioxide), and those like Tjeukemeer, which are alkaline due to the presence of $CaCO_3$ in the sediments and in the inflowing waters. The latter kind may also contain considerable quantities of dissolved organic carbon — mainly fulvic acid — in addition to the carbonate. Tjeukemeer has for example C.O.D. values of 10 mg l^{-1} of C, while additionally there are also large amounts of colloidal compounds, which are probably mainly humic acids. During rough weather even larger particles of humic compounds may be suspended in the water for several days.

The molecular structure of humic material is not yet known, but the elementary composition is roughly:

C, 50—60%; O, 30—40%; H, 5%; N, 3—5%; of the dry weight.

A controversy exists between organic chemists and ecologists studying humic compounds. Chemists use absence of nitrogen as a criterion for purity, but ecologists believe that nitrogen is necessarily present in quantities up to 5%. It is probable that the basic building unit contains no nitrogen, but the biologically active substance is probably a complex between this unit and adsorbed compounds containing nitrogen. The situation seems comparable with the chlorophyll—protein complex, where the chlorophyll molecule sensu stricto is the building unit, but this is not active unless it is bound to proteins.

In the humic-acid molecule several so-called "functional" groups may be recognised, i.e. groups that give the humic compounds their characteristic properties. Among these are:

carboxyl: —COOH   methoxyl: —OCH$_3$

phenol: ⬡—OH   ketone: ⟩C=O

quinone: ⬡=O   alkene: —C=C—

while aromatic rings, sugars and amino acids are also present. Owing to the presence of carboxyl and phenol groups, humic compounds will act as cation exchangers, and will form chelate complexes with metals:

$Fe(OH)_3$ + humic acid → Fe-humate
insoluble   soluble

Because of the different acidity of the various functional groups the number

TABLE 12.1

Iron holding capacity (meq. per 100 g of soil) and number of active groups per meq. (N) as function of pH of different soils (Kononova, 1961)

	pH = 4.5		pH = 6.4		pH = 8.1	
	meq.	N	meq.	N	meq.	N
Soil I	300	4	400	6	600	8
Soil II	240	3—4	400	5—6	550	7—8
Soil III	170	2—3	290	4	400	5

of active chelating groups depends both on the pH and on the type of soil or lakewater chemistry. Table 12.1 gives an indication of the binding capacity and number of active groups of different Russian soils at different pH values (Kononova, 1961).

It can be seen that with increasing pH the binding capacity and the number of active groups increase due to the dissociation of weak acids at higher pH values. Taking a mean value of 400 meq. per 100 g of humic acid, the iron binding capacity of Tjeukemeer water (containing 10 mg l^{-1} of humic compounds, pH = 7—8) should therefore be 40.0 $\times$ 56 μg = 2 mg l^{-1} of Fe, which is in good agreement with the measured values. Even phosphate, in combination with iron compounds may be bound to humic material, so that humic acids have a possible role in regulating the availability of phosphate for algal growth.

Schnitzer (1971) gave as a possible structure of the Fe^{3+} fulvic-acid complex the following formula:

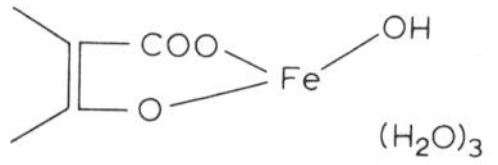

Many attempts have been made to construct a model of the humic-acid molecule. Some of the earlier ones are shown in Fig. 12.1A and B, while a more recent one for a lignin (Fig. 12.1C) is based on the derivative of phenylpropane as building unit:

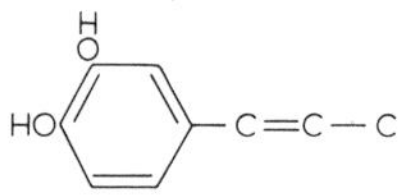

derivative of
phenylpropane

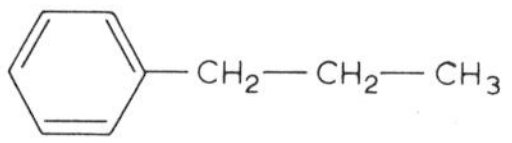

phenylpropane

Fig. 12.1.A. The structure of a humic-acid molecule, according to Dragunov (1958). (1) Aromatic ring of the di- and trihydroxyphenol type, part of which has the double linkage of a quinone grouping; (2) nitrogen in cyclic forms; (3) nitrogen of peripheral chains; (4) carbohydrate residues. B. The structure of the humic-acid molecule according to Fuchs (cited in Kononova, 1961). C. Structure of a lignin according to Freudenberg and Neish (1968).

A

B

C

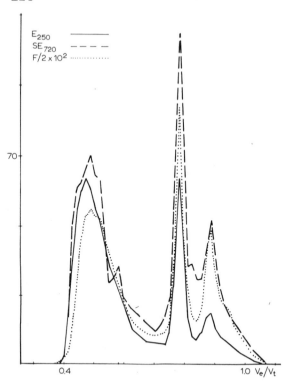

Fig. 12.2. Sephadex G-25 gel filtration of fulvic acids in phosphate buffer of pH = 7.0 and ionic strength 0.02. Humic compounds in the column effluent are estimated by the light extinction at 250, 470 and 720 nm (720 nm after reaction with a reagent for phenolic groups). (From De Haan, 1972a.)

12.3.2. Fractionation

Much attention is currently being given to devising methods of separation and characterisation of humic acids. Sephadex gel filtration is particularly useful (Povoledo and Gerletti, 1964; Gjessing, 1967; Shapiro, 1967; Ghassemi and Christman, 1968; De Haan, 1972a, b). Sephadex is a dextran gel in bead form. A tube is filled with Sephadex beads just as it might be with ion-exchange resin beads. The principle on which Sephadex acts is different however. The beads of Sephadex are porous on a molecular scale. Large molecules are excluded and are therefore eluted unretarded. Smaller molecules pass into and out of the beads to an extent depending on the molecule size and the pore size of the bead. Smaller molecules are therefore retarded more and elute later. Different pore sizes are obtainable (Sephadex G-10, G-15, G-25, G-50, G-75, G-100, G-150, G-200), so that for many compounds an approximate weight can be established if the columns are first standardised by passing through them compounds of known molecular weight and determining their elution rate. A typical elution pattern of soluble fulvic acids from Tjeukemeer is given in Fig. 12.2.

Gjessing (1967) found that the molecular weights of the humic compounds in Lake Aurevann (Norway) fell within a range of between 700 and 20 000, with the following distribution:

31%	$M > 20\,000$	4%	$700 < M < 2\,000$
33%	$5\,000 < M < 10\,000$	14%	$M < 700$
18%	$100 < M < 5\,000$		

Gjessing used distilled water as eluant; the elution pattern is dependent on pH and ionic strength. Theoretically no adsorption in the Sephadex should occur, but this is not always true in practice. Molecular-weight determinations made with Sephadex should therefore be regarded with the greatest caution (see Fig. 12.3). The influence of the pH may be due to polymer extension. The acidic phenol and carboxylic groups become more ionised and more H bridges can develop as the pH rises (Ghassemi and Christman, 1968).

Povoledo and Gerletti (1964) separated different fractions on Sephadex and monitored the eluate by making simultaneous measurements of colour,

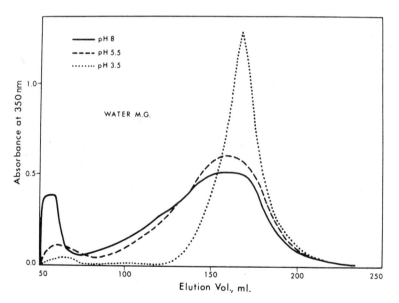

Fig. 12.3. Effect of pH on apparent molecular size on Sephadex G-75. Eluant: $0.01\,N$ NaCl, pH adjusted with HCl or NaOH. (From Christman and Minear, 1971.)

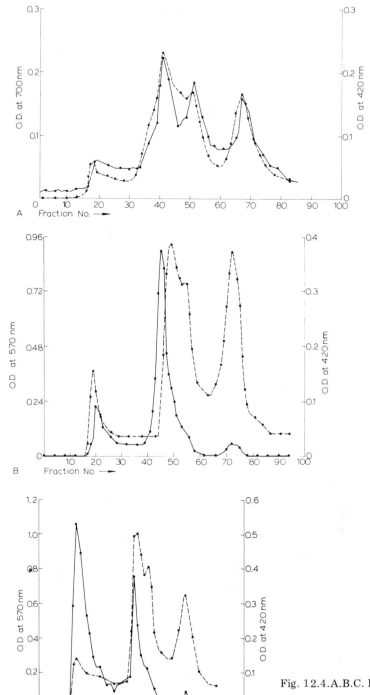

Fig. 12.4.A.B.C. For legend see p. 227.

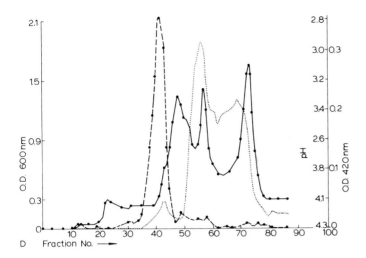

Fig. 12.4.A. Elution pattern in a gel filtration experiment made with humic matter on a Sephadex G-25 column. Legend: – – – –, absorbance at 420 nm (continuous monitoring); ————, absorbance at 700 nm after reaction of the fractions with the Folin-Ciocalteu phenol reagent. (From Povoledo and Gerletti, 1964.)
B. Elution pattern in a gel filtration experiment made with humic matter on a Sephadex G-25 column. Legend: – – – –, absorbance at 420 nm (continuous monitoring); ————, absorbance at 570 nm after development of 2-ml fractions with ninhydrin reagent, *before* hydrolysis.
C. As B, but ninhydrin reagent *after* hydrolysis.
D. Elution pattern in a gel filtration experiment made with humic matter on a Sephadex G-25 column. Legend: – – – –, absorbance at 420 nm (continuous monitoring); ————, absorbance at 600 nm after development of 2 ml fractions with a reagent for copper; ...., pH of each fraction.

fluorescence, UV absorption, concentrations of amino acids and sugars and copper-binding capacity (Fig. 12.4A, B, C and D). The results show that the elution pattern of the absorption at 420 nm is similar to the elution pattern of the phenolic compounds (Fig. 12.4A) and of amino acids before hydrolysis (Fig. 12.4B), while copper-binding capacity gives a different picture (Fig. 12.4D). Amino acids after hydrolysis (Fig. 12.4C) eluted at a different rate, thus appearing in other fractions than the phenolic compounds. The general conclusion from these elution patterns is that the amino acids are bound onto the larger molecular compounds from which they are separated to a fraction with a smaller molecular weight (fractions up to no. 45), but it is possible of course that separation between copper and humic acids occurred on the column.

Shapiro (1967) fractionated humic compounds from 22 Minnesota lakes and found 4 different elution patterns (see Fig. 12.5). Type I is a pattern given by samples consisting almost entirely of high molecular-weight components. In type II the high molecular-weight components are present, but their weight

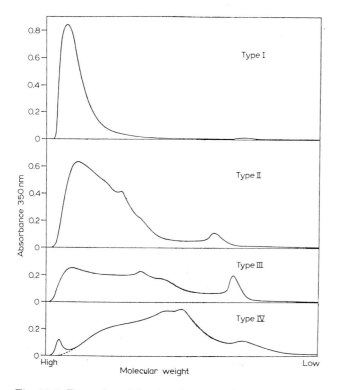

Fig. 12.5. Examples of the 4 main types of elution patterns found among 22 lakes in Minnesota. (Shapiro, 1967.)

range appears to be wider and two peaks of lower molecular-weight fractions appear. In type III some high molecular-weight components are still present, but the elution-pattern graph is quite flat, except for the peak of the lower molecular-weight fraction. In type IV there seems to be a predominance of medium molecular-weight substances. Shapiro also measured the iron binding capacity of the 22 lakewaters and noticed that type I waters have a higher capacity than those of type II and III, which themselves have a higher capacity than type IV. This work is especially relevant when considering the availability for algae of the different iron fractions in lakewater (see also section 11.3).

De Haan (1972b) studied the molecular size distribution of soluble fulvic acids from Frisian lakes (The Netherlands) by Sephadex gel filtration. On Sephadex G-25 all lakes showed the same three main fractions I, II and III. The ratio of the amount of larger molecules (fraction I) to that of the smaller ones (fraction II and III) differed from lake to lake. This difference in molecular size distribution was paralleled by the ratio of UV absorbance of the lakewater at 250 nm to that at 365 nm. Changes in the molecular size distribution of soluble fulvic acids in the lakes tested could be predicted by this ratio measurement.

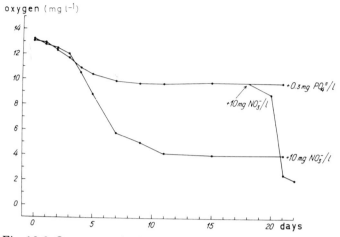

Fig. 12.6. Oxygen uptake from humic-rich lakewater with added phosphate and nitrate.
(From Ryhänen, 1968.)

12.3.3. Metabolism

Although the humic compounds are usually considered to be final products of metabolism, Ryhänen (1968) and De Haan (1972a) have shown that breakdown by bacteria can occur.

Ryhänen followed the O_2 uptake in water from Lake Hakojärvi (Finland) (Fig. 12.6) and found that following the addition of nitrate, oxygen uptake was greatly enhanced. He suggested that breakdown of organic matter (measured as oxygen uptake) was inhibited by nitrogen depletion. Although Ryhänen did not measure the amounts of humic compounds, he argued that the amount of autochthonous uncoloured material would be negligible because the water was collected from below snow-covered ice. He therefore concluded that the oxygen uptake was owed to oxidation of humic material.

Because of the complexity of the humus structure and the specialised metabolic functions of microorganisms, the complete degradation of humic material in freshwater will be the combined and perhaps stepwise work of several organisms. From what is known of the structure of humic matter, one can deduce that many of the bonds are not readily decomposable (for example aromatic nuclei and phenolic ethers). Thus, organisms restricted to humic compounds as a carbon and energy source will grow only slowly. The slow growth of an *Arthrobacter* species isolated from Tjeukemeer (De Haan, 1972a) when given humic compounds as the only organic carbon source, supports this view. During growth, changes occurred in optical absorbance of the filtered medium at 250, 365 and at 720 nm after reaction with the Folin-Ciocalteu phenol reagent. These changes indicate modifications in the structure of the humic molecules. It is likely that the *Arthrobacter* hydrolyses phenolic ether linkages, performing just one of the specific degradation steps.

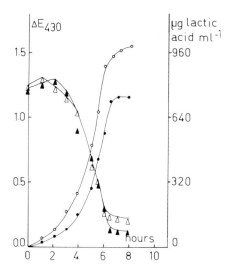

Fig. 12.7. Growth and lactate uptake of a *Pseudomonas* species from Tjeukemeer at 21°C. Black dots represent growth and black triangles lactate uptake in a medium containing lactate as sole source of organic carbon. White dots and white triangles the same in a medium containing lactate and an equal amount of fulvic acid. (From De Haan, 1974.)

Besides the direct degradation of humic compounds, an indirect oxidation by active metabolising organisms seems possible. There is experimental evidence (Horvath, 1972) for the so-called co-metabolism of very resistant compounds (for example herbicides, alkyl benzene sulphonates). The more rapid growth of a *Pseudomonas* species from Tjeukemeer after fulvic acid had been added to the medium has been explained by De Haan (1974) in terms of co-metabolism of fulvic acid (Fig. 12.7). The co-metabolism of fulvic acid must provide the organism with an extra carbon source or growth factor, because the *Pseudomonas* does not grow on fulvic acid alone and in yeast extract media the same effect of fulvic acid on growth is observed. So far as is known, co-metabolism and oxidation of two different compounds are performed by the same enzyme system of an organism. So the closer the structure of a refractory molecule is to that of the usual substrate for a given bacterium, the better the bacterium will be able to co-metabolise the refractory compounds. Because humic compounds contain much aromatic carbon De Haan (unpublished) studied oxidation of benzoate and co-metabolism of humic compounds, by organisms isolated from Tjeukemeer. One of the strains was able to grow slowly and with a low yield in media containing fulvic acid as the only organic-carbon source and on an enrichment medium containing benzoate. However, the addition of fulvic acid to a benzoate-containing medium increased the growth rate; the cell yield about doubled. In the case of lactate-grown *Pseudomonas* the cell-yield increase never exceeded 25%. So the benzoate-grown organism is able to co-metabolise fulvic acid to a greater extent than the lactate-grown *Pseudomonas*. In conclusion the work of De Haan shows the existence of co-metabolism of fulvic acid in freshwater. Under laboratory conditions this has been established for two different bacteria. There is no doubt that other genera will show co-metabolism of humic compounds also. The ecological significance is that humic compounds in freshwater may not be so refractory as is generally believed. The mineralisation of humic compounds in freshwater will depend on the number and activity of the heterotrophic bacteria present and thus on readily degradable organic matter. From the point of view of lake eutrophication, it might be interesting to know whether the release of easily decomposable organic matter in sewage water causes a decrease in the concentration of refractory compounds in the receiving lake. It seems more likely that this effect exists than that humic substances can increase the growth of algae (except through chelation effects such as those described in subsection 17.3.3) as is sometimes claimed.

Note: The chemistry of humic substances has been reviewed by Schnitzer and Khan (1972). Povoledo and Golterman (1975) edited a symposium volume including contributions concerned with both the structural and biological aspects of humic substances in aquatic as well as terrestrial systems.

REFERENCES

Allen, H.L., 1969. Chemo-organotrophic utilization of dissolved organic compounds by planktonic algae and bacteria in a pond. *Int. Rev. Ges. Hydrobiol. Hydrogr.*, 54: 1—33.

Banoub, M.W., 1973. Ultra violet absorption as a measure of organic matter in natural waters in the Bodensee. *Arch. Hydrobiol.*, 71(2): 159—165.

Carlucci, A.F. and Bowes, P.M., 1972. Determination of vitamin B_{12}, thiamine, and biotin in Lake Tahoe waters using modified marine bioassay techniques. *Limnol. Oceanogr.*, 17: 774—777.

Christman, R.F. and Minear, R.A., 1971. Organics in lakes. In: S.J. Faust and J.V. Hunter (Editors), *Organic Compounds in Aquatic Environments*. Marcel Dekker, New York, N.Y., pp. 119—143.

De Haan, H., 1972a. Some structural and ecological studies on soluble humic compounds from Tjeukemeer. *Verh. Int. Ver. Theor. Angew. Limnol.*, 18: 685—695.

De Haan, H., 1972b. Molecule-size distribution of soluble humic compounds from different natural waters. *Freshwater Biol.*, 2(3): 235—241.

De Haan, H., 1974. Effect of a fulvic acid fraction on the growth of a *Pseudomonas* from Tjeukemeer (The Netherlands). *Freshwater Biol.*, 4: 301—310.

Dragunov, S.S., 1948. A comparitive study of humic acids from soils and peats. *Pochvovedenie*, 7 (in Russian).

Freudenberg, K. and Neish, A.C., 1968. *Constitution and Biosynthesis of Lignin*. Springer Verlag, Berlin, 129 pp.

Gerletti, M., 1965. The soluble protein content of Lago Maggiore waters, with considerations on other organic substances. *Mem. Ist. Ital. Idrobiol.*, 18: 217—239.

Ghassemi, M. and Christman, R.F., 1968. Properties of the yellow organic acids of natural waters. *Limnol. Oceanogr.*, 13(4): 583—597.

Gjessing, E.T., 1967. Humic substances in natural water: Methods for separation and characterization. *Proc. IBP-Symp., Amsterdam—Nieuwersluis, 1966*, pp. 191—216.

Golterman, H.L. (with Clymo, R.S.) (Editors), 1971. *Methods for Chemical Analysis of Fresh Waters*. IBP Handbook no. 8, Blackwell, Edinburgh, 166 pp.

Hobbie, J.E. and Wright, R.T., 1965. Bioassay with bacterial uptake kinetics: Glucose in freshwater. *Limnol. Oceanogr.*, 10(3): 471—474.

Horvath, R.S., 1972. Microbial co-metabolism and the degradation of organic compounds in nature. *Bacteriol. Rev.*, 36(2): 146—155.

Juday, C. and Birge, E.A., 1932. Dissolved oxygen and oxygen consumed in the lake water of Northeastern Wisconsin. *Trans. Wis. Acad. Sci. Arts Lett.*, 27: 415—486.

Kononova, M.M., 1961. *Soil Organic Matter; Its Nature, Its Role in Soil Formation and in Soil Fertility*. Pergamon Press, Oxford, 450 pp. (transl. from the Russian).

Mackereth, F.J.H., 1963. *Some Methods of Water Analysis for Limnologists*. Sci. Publ.. Freshwater Biol. Assoc., Ambleside, 21, 70 pp.

Otsuki, A. and Wetzel, R.G., 1972. Coprecipitation of phosphate with carbonates in a marl lake. *Limnol. Oceanogr.*, 17: 763—767.

Povoledo, D. and Gerletti, M., 1964. Studies on the sedimentary, acid-soluble organic matter from Lake Maggiore (North Italy). I. Heterogeneity and chemical properties of a fraction precipitated by barium ions. *Mem. Ist. Ital. Idrobiol.*, 17: 115—150.

Povoledo, D. and Golterman, H.L. (Editors), 1975. *Humic Substances in Aquatic and Terrestrial Environments; Structure and Function. Proceedings of an IBP-Sponsored Meeting, Nieuwersluis, 1972*. Pudoc, Wageningen, 374 pp.

Ryhänen, R., 1968. Die Bedeutung der Humussubstanzen im Stoffhaushalt der Gewässer Finnlands. *Mitt. Int. Ver. Theor. Angew. Limnol.*, 14: 168—178.

Saunders, G.W., 1957. Interrelations of dissolved organic matter and phytoplankton. *Bot. Rev.*, 23: 398—409.

Saunders, G.W., 1972. The kinetics of extracellular release of soluble organic matter by plankton. *Verh. Int. Ver. Theor. Angew. Limnol.*, 18: 140—146.

Schnitzer, M., 1971. Metal-organic matter interactions in soils and waters. In: S.J. Faust and J.V. Hunter (Editors), *Organic Compounds in Aquatic Environments*. Marcel Dekker, New York, N.Y., pp. 297—315.

Schnitzer, M. and Khan, S.U., 1972. *Humic Substances in the Environment*. Marcel Dekker, New York, N.Y., 323 pp.

Shapiro, J., 1967. Yellow organic acids of lake water: Differences in their composition and behaviour. *Proc. IBP-Symp. Amsterdam—Nieuwersluis, 1966*, pp. 202—216.

Vallentyne, J.R., 1957. The molecular nature of organic matter in lakes and oceans, with lesser reference to sewage and terrestrial soils. *J. Fish. Res. Board Can.*, 14(1): 33—82.

Wetzel, R.G., 1970. Recent and postglacial production rates of a marl lake. *Limnol. Oceanogr.*, 15(4): 491—503.

Wetzel, R.G., 1972. The role of carbon in hard-water marl lakes. In: *Special Symp. Am. Soc. Limnol. Oceanogr.* Allen Press, Lawrence, Kansas, 1: 84—97.

Wetzel, R.G. and Allen, H.L., 1972. Function and interactions of dissolved organic matter and the littoral zone in lake metabolism and eutrophication. In: *Proc. IBP-UNESCO Symp. Productivity Problems of Freshwaters, Kazimierz Dolny, 1970*, pp. 333—347.

Wetzel, R.G. and Manny, B.A., 1972a. Decomposition of dissolved organic carbon and nitrogen compounds from leaves in an experimental hard-water stream. *Limnol. Oceanogr.*, 17: 927—937.

Wetzel, R.G. and Manny, B.A., 1972b. Secretion of dissolved organic carbon and nitrogen by aquatic macrophytes. *Verh. Int. Ver. Theor. Angew. Limnol.*, 18 (1): 140—146.

Wetzel, R.G., Rich, P.H., Miller, M.C. and Allen, H.L., 1972. Metabolism of dissolved and particulate detrital carbon. *Mem. Ist. Ital. Idrobiol.*, 29(Suppl.): 185—243.

CHAPTER 13

ALGAE AND THEIR PIGMENTS

13.1. STRUCTURE AND OCCURRENCE

All algae contain photosynthetic pigments. These are usually an integral part of the structure of the chloroplast lamellae, but sometimes, as in blue-green algae, they are homogeneously distributed throughout that part of the protoplasm called the "chromatoplasm".

Pigments are molecules which absorb light. The most efficient organic pigments have a molecular absorption which is one or two orders of magnitude greater than that of the inorganic pigments (such as cobalt blue, cinnabar, chrome yellow, etc.), which are used as paints.

Many pigments have a characteristic molecular structure of long carbon chains or closed rings linked by so-called "conjugated" double bonds. These bonds are particularly stable because they involve "resonance", a situation where two or more molecular configurations can exist simultaneously. Benzene for example "resonates" between the following classical structures:

Addition of oxygen or nitrogen to such a ring system increases the number of possible resonance positions. The simple heterocyclic pyrrole ring:

is one of the basic building units of many organic and biochemical pigments.

Chains or rings with few conjugated bonds have strong absorption at relatively short wavelengths (in the far UV region). Molecules with a larger number of conjugated bonds absorb strongly in the near-UV or in the visible region of the spectrum. Thus in benzene the first absorption band lies below 250 nm, while that of anthracene lies in the yellow region.

Strong light absorption by a molecule may be caused by its resonant structure. Owing to the occurrence of the different resonating configurations, several closely spaced energy levels are available. These lie between the ground state and the first excited state. The first absorption band then moves towards longer wavelengths (Fig. 13.1).

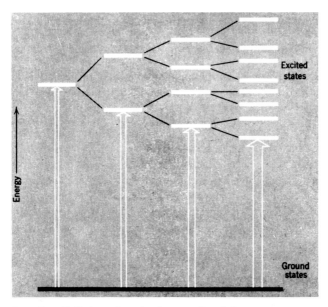

Fig. 13.1. Splitting of excited energy levels by resonance.

In photosynthetic cells the most commonly occurring pigment is chlorophyll a. The only exceptions are the photosynthetic bacteria which have bacteriochlorophyll (BChl). Chlorophyll a is present in brown, green and blue-green algae and in higher plants. Its yellow-green colour may be masked by colours of other pigments. It seems probable that chlorophyll complexes

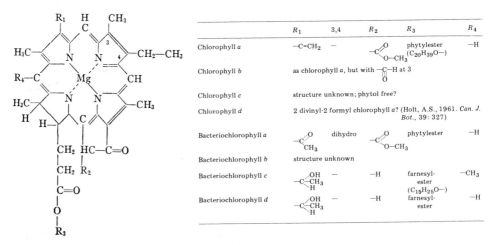

	R_1	3,4	R_2	R_3	R_4
Chlorophyll a	$-C=CH_2$	—	$-C\!\!<\!\!^O_{O-CH_3}$	phytylester $(C_{20}H_{39}O-)$	$-H$
Chlorophyll b	as chlorophyll a, but with $-C\!\!-\!\!H$ at 3				
Chlorophyll c	structure unknown; phytol free?				
Chlorophyll d	2 divinyl-2 formyl chlorophyll a? (Holt, A.S., 1961. *Can. J. Bot.*, 39: 327)				
Bacteriochlorophyll a	$-C\!\!<\!\!^O_{CH_3}$	dihydro	$-C\!\!<\!\!^O_{O-CH_3}$	phytylester	$-H$
Bacteriochlorophyll b	structure unknown				
Bacteriochlorophyll c	$-C\!\!<\!\!^{OH}_{CH_3}_{H}$	—	$-H$	farnesyl-ester $(C_{15}H_{25}O-)$	$-CH_3$
Bacteriochlorophyll d	$-C\!\!<\!\!^{OH}_{CH_3}_{H}$	—	$-H$	farnesyl-ester	$-H$

Fig. 13.2. Basic structure of chlorophyll and relationship between chlorophylls and bacteriochlorophylls (Pfennig, 1967).

235

function as the photoenzyme, although final proof is lacking. All other pigments (and a large part of BChl and of chlorophyll *a* itself) seem to serve solely as physical-energy suppliers. Pigments other than chlorophyll *a* are called accessory pigments. Green algae contain, in addition to chlorophyll *a*, the blue-green chlorophyll *b*. Diatoms and brown algae contain chlorophyll *c*, a partially oxidised derivate of chlorophyll *a*, but without the phytol chain. The structure of the chlorophyll molecule is shown in Fig. 13.2, and the occurrence and main absorption peaks of the different types are given in Table 13.1.

The chlorophylls *a* and *b* have a common basic structure consisting of four pyrrole rings, joined into a single master ring by CH bridges. The porphyrin structure (see Fig. 13.3) is related to that of the bilin pigments. Bacteriochlorophyll is related to tetrahydroporphyrin, having two fewer bonds and thus four more hydrogen atoms than porphyrin. The centre of the molecule is a magnesium atom, similar to the ferrous iron in the haem molecule. The function of magnesium is probably different: the ferrous haem molecule transports oxygen, whereas the chlorophyll molecule transports energy. It may be that the presence of a suitable central metal atom serves to stabilise

TABLE 13.1

Characteristic absorption peaks and occurrence of the chlorophylls (after Rabinowitch and Govindjee, 1969)

Type of chlorophyll	Characteristic absorption peaks		Occurrence
	In organic solvents (nm)	In cells (nm)	
chlorophyll *a*	420, 660	435, 670—680 (several forms)	all photosynthesising plants (except bacteria)
chlorophyll *b*	453, 643	480, 650	higher plants and green algae
chlorophyll *c*	445, 625	red band at 645	diatoms and brown algae
chlorophyll *d*	450, 690	red band at 740	reported in some red algae
chlorobium chlorophyll (also called "bacterioviridin")	two forms: (1) 425, 650 (2) 432, 660	red band at 750 (or 760)	green bacteria
bacteriochlorophyll *a* (BChl*a*)	365, 605, 770	red bands at 800, 850, and 890	purple and green bacteria
bacteriochlorophyll *b* (BChl*b*)	368, 582, 795	red band at 1 017	found in a strain of *Rhodopseudomonas*, (a purple bacterium)

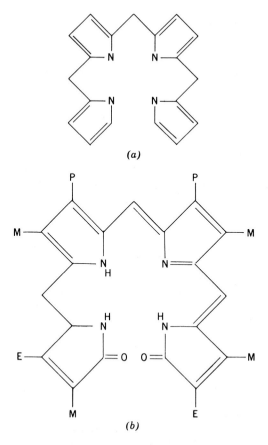

Fig. 13.3. Structure of bilin pigments: (a) bilan; (b) phycoerythrobilin. All unmarked corners are occupied by carbon atoms. *E*, ethyl; *M*, methyl; *P*, propionyl groups. (From Rabinowitch and Govindjee, 1969.)

the molecule. Following algal death, the magnesium is removed from the molecule probably during autolysis. In living cells the chlorophylls are bound to protein—lipid complexes just as most enzymes are. Extraction of the pigments by organic solvents such as acetone separates the chlorophylls from the proteins and has a considerable influence on their structure. Bacteriochlorophyll for example, which has three absorption bands at 800, 850 and 890 nm in vivo, shows only one band at 770 nm after it has been extracted by organic solvents. It may be that in vivo three separate complexes are formed with three proteins or that three different aggregation stages exist. Chlorophyll *a* in green algae shows a similar but less conspicuous polymorphism.

All photosynthetic cells contain, in addition to one or more chlorophyll pigments, an assortment of carotenoids. These are pigments related to that found in the root of the carrot plant. Carotenoids are either hydrocarbons

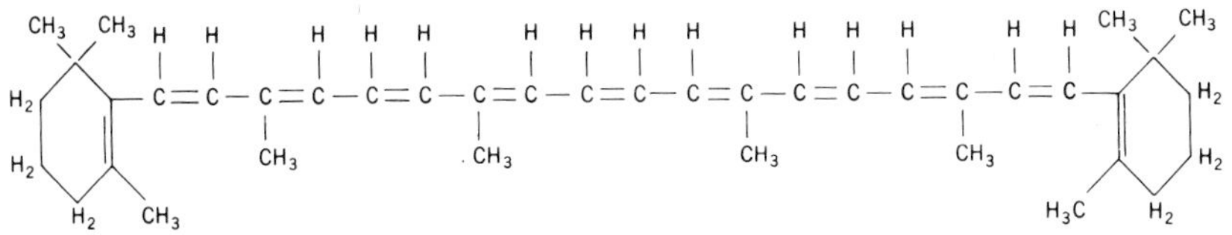

Fig. 13.4. Structure of β carotene (α and β forms are stereoisomers). All corners on the rings at the two ends are occupied by carbon atoms. (From Rabinowitch and Govindjee, 1969.)

(called α, β and γ carotenes; $C_{40}H_{56}$) or oxygen-containing compounds (called carotenols or xanthophylls). The structure of β carotene is shown in Fig. 13.4 (α and β carotene are "stereoisomers" distinguished only by their spatial configuration; γ carotene has one open ring structure at the end of the molecule).

The xanthophylls are composed mainly of varying amounts of five compounds: lutein (or luteol), violaxanthin, fucoxanthin, spirilloxanthin and neoxanthin. Of these there seems to be most lutein. It contains a —CHOH- group in a position occupied by a —CH$_2$ group in carotenoid. It is the major carotenoid of green algae.

Fucoxanthin occurs in diatoms (Bacillariophyceae) and brown algae (Phaeophyceae, of which the genus *Fucus*, from which the name derives, is found mainly on oceanic beaches). Fucoxanthin may serve to transfer light energy to chlorophyll. In addition diadinoxanthin also occurs in diatoms, but this pigment cannot transfer light energy. The possible function of diadinoxanthin has been mentioned in section 7.3. Zeaxanthin contains two, whilst violaxanthin and neoxanthin each have four oxygen atoms in the form of hydroxyl, carbonyl or carboxyl groups attached to the "ionone" ring. In a series of papers, Hertzberg and Liaaen-Jensen (cited in Hertzberg et al., 1971) reported a study of the carotenoid composition of blue-green algae, including *Oscillatoria* (several species) *Anabaena flos aquae*, *Aphanizomenon flos aquae* and *Phormidium* (several species). β carotene was the major carotenoid, and several less well-known xanthophylls occurred. The distribution patterns of the different pigments were compared and some correlations with the taxonomic status of the various algae were discussed. Thus zeaxanthin occurred in three species of *Phormidium* in amounts nearly equal to β carotene. In some species of *Anabaena* and *Nostoc* echinenone (= myxoxanthin) is more abundant than β carotene.

A third important group of algal pigments are the phycobilins (phycos = alga, bilin = bile pigment). The basic molecular structure is rather like that of chlorophyll. Phycoerythrin gives the red algae (Rhodophyceae) their typical brick red colour, while the bluish colour of blue-green algae (Cyanophyceae) is due to phycocyanin, which may be present in amounts equal to or greater than those of chlorophyll *a*.

The pigments are not distributed homogeneously over the photosystems

I and II. Chlorophyll *a* has been found in both systems, except in brown algae, where it is found only in system II. Fucoxanthin, phycobilin, and chlorophyll *b* and *c* are only found in system II, but β carotene and xanthophyll only in system I.

The chemical properties of the pigments determine the methods used to extract them. Phycobilins have no phytol chains and are covalently bound to water-soluble proteins. They are therefore easily extractable with pure water. Chlorophyll extractions require organic solvents such as methanol, ethanol or acetone. Most carotenoids are soluble in organic solvents such as petroleum, and they also dissolve in fats and oils.

13.2. ABSORPTION SPECTRA AND FUNCTIONS

Chlorophyll has one strong absorption band in the blue-violet — the so-called Soret band — and one in the red region of the spectrum. The Soret band is characteristic of the porphyrin derivatives. In ether the maximum absorbance is found at 430 nm but in living cells these bands lie at about 440 nm and 675 nm. At wavelengths between these two regions the absorption is weak, and this causes most land vegetation to appear green. Bacteriochlorophyll *a* also has two absorption bands, but they lie further apart (365 nm and 770 nm in methanol). Bacteriochlorophyll *b* has an infrared absorption at 1014 nm.

Chlorobium chlorophyll (also called bacterioviridin) absorbs at 750 or 760 nm. The auxiliary or accessory (non-chlorophyll) pigments absorb blue to green light. Red algae contain phycoerythrin with absorption bands in the middle of the visible spectrum (500, 545, 570 nm). Light energy absorbed by these pigments can be used for photosynthesis. Light of these wavelengths penetrates water further than does that of other useful wavelengths. It is often suggested that this enables red algae to live in deeper parts of the aquatic environment than can other light-dependent organisms. Marine red sea weeds are often found in lower zones of the intertidal shore region than are the green and brown ones. This ability to grow in deeper water is probably also attributable, at least partly, to their possession of the pigment phycoerythrin.

Phycocyanin has its main absorption band at 630 nm. It does not completely bridge the absorption gap in the green part of the chlorophyll spectrum, but it narrows it, as does chlorophyll *b*. The carotenoids absorb in the blue-violet part of the spectrum; β carotene for example has bands at 430, 450 and 480 nm. Fucoxanthin has a much broader absorption band than have other carotenoids. The absorption spectrum does not fall off sharply at about 500 nm but declines slowly, filling in the gap left by chlorophyll. Phycobilin shows the same pattern. Some of those pigments which absorb light at wavelengths which are not absorbed by chlorophyll seem to be active in light-energy transport. Thus, 80—90% of the quanta absorbed by phycoerythrin and fucoxanthin may be transferred to chlorophyll *a*. It has been shown that energy transfer from accessory pigments to chlorophyll is possible only for the pigments found in photosystem II. Much less efficient is the transfer from the carotenoids (ca. 20%). Transfer from β carotene and xanthophyll (both found in system I) seems to be impossible. The main function of these pigments remains obscure at present. It has been suggested that they have a protective role. In the presence of light and oxygen many photo-oxidations could occur and chlorophyll in particular can easily be destroyed. Experimental evidence for such a protective screening role of carotenoids is scarce. Sistrom et al. (1956) showed that a *Rhodopseudomonas spheroides* mutant which lacks the coloured carotenoids was rapidly killed in the air in the presence of light, while cells of the normal carotenoid-containing type thrived. A colourless mutant of *Halobacterium salinarium* grew far more slowly than did its parent strain so that in mixtures of mutant and wild-type parent the latter always became dominant in effect displacing its light-sensitive daughter population. Dundas and Larsen (1962), and Mathews and Krinsky (1965) also showed that carotenoids shield *Micrococcus* against injury by light. The carotenoids are not essential to these bacteria: the colourless mutant of *Rhodopseudomonas spheroides* functions normally as long as no oxygen is present.

It has been mentioned in Chapter 7 that diatoms, especially when silicastarved, are much more sensitive to light than are other algae. Diadinoxanthin synthesis in some diatom species decreases in darkness, but is maintained in the light, even during SiO_2 depletion. Moderate levels of light become then detrimental or lethal. The synthesis of the energy-transporting pigments chlorophyll and fucoxanthin ceased during this depletion.

Diatoms, which are amongst the most successful plants on earth (they occur in the upper layers of most oceans and may account for as much photosynthesis

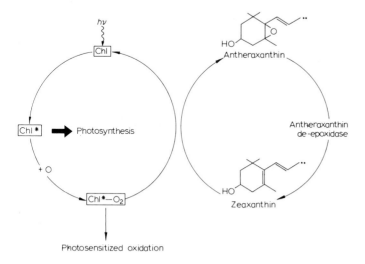

Fig. 13.5. Oxidation of zeaxanthin as a protective mechanism against photo-oxidation.

240

on earth as the green land plants), may have adapted themselves to lower irradiance during evolution. This is because they are able to grow only during the period that silica is present, i.e. the very early spring, during which light levels are low. The ability to use carotenoids to fill the gap in the absorption spectrum left by chlorophyll would then be of great value. A similar adaptation is found in green and purple sulphur bacteria (see subsection 16.4.1) and in blue-greens, which owing to their phycocyanin are able to grow under a layer of green algae.

It has been suggested that the protective mechanism acts via a reversible oxidation of carotenoids (Fig. 13.5). It has indeed been shown that at light levels high enough to inhibit photosynthesis, O_2 production was more inhibited than was CO_2 uptake.

13.3. RATIO BETWEEN CAROTENOIDS AND CHLOROPHYLL

If one extracts the pigments from algal populations of most temperate lakes, the extinction value in the 400-nm region of the spectrum is most likely to be 2—2.5 times greater than that at 665 nm. As the blue maximum of chlorophyll is roughly equal to that at 660 nm, the extra light absorption at 400—430 nm can be used to estimate the concentration of the carotenoids. Spectral diversity of the pigments (Fig. 13.6) is reflected in the ratio of the absorbances in the 400—430 nm region to that near 665 nm. The ratio between chlorophyll and carotenoids depends greatly on the physiological condition of the cells. Stress conditions also influence this ratio. Yentsch and Vaccaro (1958) compared chlorophyll and nitrogen concentrations in cultures of marine phytoplankton and found that when nitrogen was deficient so was chlorophyll. As soon as nitrogen was added, chlorophyll synthesis was resumed. During nitrogen starvation chlorophyll breakdown prevailed but carotenoid synthesis continued, even after the synthesis of chlorophyll ceased. The pigment ratio therefore changed. Only in conditions of extreme nitrogen deficiency did carotenoid synthesis decline. Yentsch and Vaccaro suggested therefore that this physiological measure could be used to estimate the rate of protein synthesis and cell division. Actual measurements based on the ratio of chlorophyll to carotenoids in natural populations gave disappointing results however. Theoretical ratios were calculated but these were not those obtained either in healthy or unhealthy cultures. The inconsistency was most probably attributable to the great diversity in natural populations.

Margalef (1965) used the ratios of pigments as an index to species diversity and tried to relate primary production to community structure. Diversity, D, was expressed by Margalef as bits per individual or:

$$D = - \Sigma p_i \log_2 p_i$$

where p_i denotes probability of occurrence of each species.

He found a strong positive correlation between A_{430}/A_{665} (absorption at

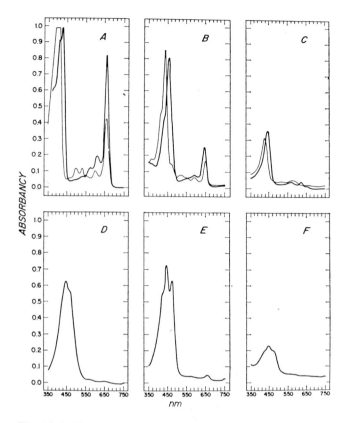

Fig. 13.6. Absorption curves for some chlorophylls and the corresponding phaeophytins and carotenoids in 85% acetone. A. Chlorophyll *a* found in all photosynthetic plants. B. Chlorophyll *b* found in higher green plants and green algae. C. Chlorophyll *c* found in higher brown algae, diatoms, dinoflagellates and coccolithophores. D. Fucoxanthin, the principal xanthophyll of brown algae and diatoms. E. Lutein, the principal xanthophyll of green plants. F. Carotene. (From Yentsch, 1967.)

wavelengths 430 and 665 nm, respectively) and diversity. Using data from man-made lakes and laboratory cultures, Margalef demonstrated that productivity per unit biomass was negatively correlated with the pigment ratio A_{430}/A_{665} ($r = -0.588$) and with diversity. Margalef found that production per unit biomass, P, could be expressed as:

$$\log P = 1.047 + 0.728 \log C - 0.615 \log (A_{430}/A_{665})$$

where C = concentration of chlorophyll.

Another form of this relation is:

$$P = 11.1 \, C^{0.728}/(A_{430}/A_{665})^{0.615}$$

and since C is approximately proportional to A_{665}, this equation may be written as:

$$P = 67.7 \, (A_{665})^{1.343}/(A_{430})^{0.615} \ .$$

From the data presented by Margalef one cannot tell whether this relationship is better than the classical hypothesis that production is related linearly to chlorophyll concentration or not. Margalef's fig. 2 shows two groups of data: one group has a A_{430}/A_{665} ratio of about 10 with a low productivity (10 mg C per g C per h) and the other has a A_{430}/A_{665} ratio of about 5 with high productivity values (> 20 mg C per g C per h).

In an earlier paper Margalef (1964) had computed the community diversity using the following definition:

$$d = \frac{1}{N} \, \log \frac{N!}{N_a! \, N_b! \, N_c! \, ... \, N_s!}$$

where $N_a, N_b, ..., N_s$ = numbers of cells of species a, b, ..., s, and N = total number of cells per sample. CO_2 uptake was found to be positively correlated with the amount of chlorophyll and negatively with pigment ratio.

Changes in the ratio A_{430}/A_{665} reflect changes not only in species composition of the community but also in the physiological state of single species populations. The sensitive response of the pigment ratio to nutrient depletion and supply makes it a useful parameter in the assessment of growth-limiting factors. High values of A_{430}/A_{665} may thus reflect high species diversity or stress due to nutrient depletion. In alcohol extracts of green algae values between 1.8 and 2.0 are commonly found. These values result from a mixture of chlorophyll a, which itself has a A_{430}/A_{665} ratio of 1.25 (Fig. 13.6), with yellow pigments. Relatively high values can also be found in unialgal populations in nature. We found a ratio of A_{430}/A_{665} of about 4 in a population of *Oscillatoria* sp. in crater lake in Uganda (Fig. 13.7). This blue-green alga gave the water a rather pink appearance and caused us to suppose that the lake contained purple bacteria. The presence of such an impressive quantity of carotenoids might be due to a combination of conditions; the lake received full sun, had a temperature of about 35°C, and contained 60 g l^{-1} of Cl$^-$ and the same amount of sulphate. Two or three subcultures from the lake population retained the pink-orange colour. The organism showed $^{14}CO_2$ uptake in the presence of H_2S (see section 16.4). The lakewater itself contained no detectable O_2.

Margalef (1965) found surprisingly high values of A_{430}/A_{665} : between 5 and 20 in his cultures and values of 4—7 in some Spanish reservoirs (Margalef, 1964). These values seem rather high but may be due to the extremely low chlorophyll concentrations in the reservoirs (ca. 1—2 mg m^{-3}) or to difficulties in the extraction.

During the evolutionary process diversity might be better achieved by proliferation of different species than by wide variation of pigments in one species.

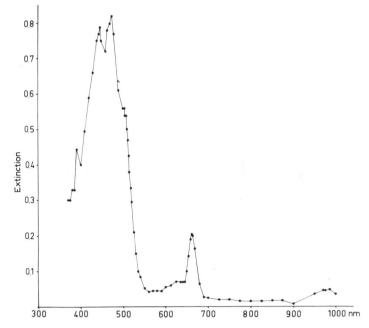

Fig. 13.7. Absorption spectrum of *Oscillatoria* sp. from a crater lake in Uganda.

In addition to the pigments described earlier, degradation products of chlorophyll are usually present. Yentsch and Ryther (1959) filtered large volumes of seawater from more than 100 m depth. In extracts of the filtered particles they found almost no chlorophyll but they did find decomposition products of chlorophyll. Two degradation pathways for chlorophyll breakdown are possible (Yentsch, 1967):

$$\text{chlorophyll} \xrightarrow{-\text{Mg}} \text{phaeophytin} \xrightarrow{-\text{phytol}} \text{phaeophorbide}$$
$$\searrow_{-\text{phytol}} \text{chlorophyllide} \xrightarrow{-\text{Mg}} \nearrow$$

The loss of phytol results in little or no change in absorption in the visible part of the spectrum, so that chlorophyllide will be indistinguishable from chlorophyll. In dead cells however, Mg will be quickly removed from the chlorophyll or chlorophyllide molecule, resulting in a shift of the blue band towards the ultraviolet while the absorption of the red light is shifted slightly towards longer wavelengths and decreases by almost 50%. Mg removal can be effected by mineral acids as well, so that solutions of chlorophyll may be distinguished from those of chlorophyllide by noting the changes in absorption spectrum following the addition of a (mineral) acid (F_o/F_a). Yentsch (1967) showed an increase in proportion of phaeophytin with increasing depth, both in the Indian and Atlantic Oceans; at depths greater than 100 m most of the pigment was phaeophytin. The conversion of chlorophyll to phaeopigment may take place

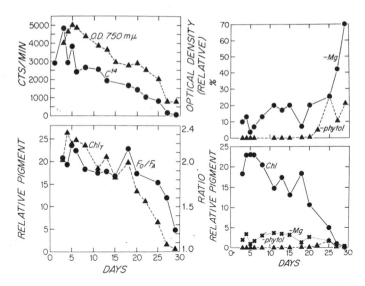

Fig. 13.8. Cell numbers, potential photosynthesis and pigments of the diatom *Skeletonema costatum* kept in darkness at 20°C. (Absorbance at 750 nm indicates cell number. Potential carbon fixation (^{14}C) is determined by removing a portion of the darkened culture and exposing it to light in the presence of radioactive carbonate.) (From Yentsch, 1967.)

as the result of grazing; the acidity of the gut of the animal seems to be sufficient to effect the change, but autolytic or bacterial processes may be just as important. Yentsch (1965, 1967) noted that prolonged periods of darkness (> 100 h) induced formation of phaeopigment in the diatom *Skeletonema costatum* (Fig. 13.8). During a 30-day period of continuous darkness, there was an initial increase of cell numbers and photosynthetic capacity, but later both decreased rapidly. Total chlorophyll and the ratio F_o/F_a declined at a slower rate during the first 15 days.

Yentsch found that both Mg and phytol were lost from the pigments in deeper water layers. He suggested that the loss of Mg is an "apparently reversible process" which can be explained as a rapid photo-oxidation of phaeopigment in light and a de novo synthesis of chlorophyll from some intact cells.

Yentsch also showed, by thin layer chromatography, that there are a variety of different pigments at different depths. Phaeophorbide appeared to be predominant in the deep waters. Although Yentsch studied these processes in deep oceans, the same processes may operate in deep, stratifying lakes. Diatoms, as soon as they are trapped in deep dark layers, will rapidly lose their photosynthetic pigments and thus their capacity for photosynthesis.

13.4. THE PHOTOSYNTHETIC UNIT

Not all chlorophyll molecules are active in light capture. Emerson and

Arnold (1932) made experiments in which the yield per light flash (using flashes lasting only a few micro seconds) appeared to be extremely low. Flash saturation was found to occur in normal cells when only one out of 2 500 chlorophyll molecules had received sufficient energy to reduce one molecule of CO_2 during the flash. This suggested the occurrence of a photosynthetic unit of 2 500 chlorophyll molecules, of which one is active as a photosynthetic rate-limiting enzyme (Gaffron and Wohl, 1936).

A structure which may correspond with such a unit, the quantosome, is distinguishable in electron micrographs of chloroplasts. Evidence indicating the existence of such a unit is the fact that a dense suspension of *Chlorella* will start to evolve oxygen immediately after light is switched on, although kinetic reasoning would lead one to suppose that each chlorophyll molecule in these conditions would have to wait an average of an hour to collect the quanta necessary for the reduction of one molecule of CO_2. The supposition of a transfer of energy from many pigment molecules to one active enzyme molecule provides an explanation for the absence of any lag in O_2 evolution. The transfer can be sufficiently rapid for this: one estimate is that 10 000 transfers can take place during 10^{-2} sec, which is the duration of the excited state of the chlorophyll molecule. The active centre where the absorbed energy is collected is probably concerned with the transfer of an electron and not directly with the reduction of the CO_2 molecule. Since the reduction of a single CO_2 molecule is estimated to require a minimum of 8 quanta it seems likely that the photosynthetic unit contains at least 300 (200—400) chlorophyll molecules. If isolated chloroplasts are broken down into progressively smaller units, these lose their characteristic photochemical activity when the fragments become smaller than about 200 chlorophyll molecules. Furthermore, it is found that one molecule of DCMU (see page 64) is sufficient to inactivate 200 chlorophyll molecules. Chlorophyll is not the only pigment which is able to transfer energy to the photosynthetic centre: fucoxanthin and phycocyanin may also do so. This has been demonstrated by measuring the quantum yield of photosynthesis at different wavelengths (Fig. 13.9; Tanada, 1951).

Quantum yield was found to be nearly constant between 520 and 680 nm, falling abruptly at longer wavelengths and showing a slight depression in the blue-green between 430 and 520 nm. It is clear that at 550 nm, where fucoxanthin is the pigment mainly responsible for the light absorption, the quantum yield is the same as at 600 nm, where absorption is entirely by chlorophyll. In a similar way the phycocyanin can be shown to be capable of transferring energy. The physical basis for the energy transfer is still unclear. Resonance may be important if absorption bands overlap, and the occurrence of crystal structures such as those in metals where electron orbits are shared is also possible.

These facts may have ecological significance. The high efficiency of the energy transfer from pigments absorbing light between the two chlorophyll peaks

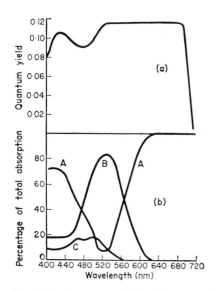

Fig. 13.9. Quantum yield of photosynthesis of the diatom *Navicula minima* as a function of wavelength (a), and the estimated distribution of light absorption among pigments in living cells of *Navicula minima* as a function of wavelength (b). *A*: chlorophylls *a* and *c*; *B*: fucoxanthin; *C*: other carotenoids. (From Fogg, 1968.)

raises the photosynthetic efficiency of diatoms (fucoxanthin) and blue-green algae (phycocyanin) above that of other competitors (such as green algae) lacking such pigments. One can also understand why change of photosynthesis with depth does not always parallel the most penetrating light component (see subsection 4.3.1)

When calculating primary production per unit of photosynthetic machinery (mg of O_2 per mg of chlorophyll) a correction for the presence of energy-transferring accessory pigments should be made, and when correlating primary production with diversity, a distinction must be made between energy-transferring and non-energy-transferring accessory pigments.

Photosynthesising bacteria also contain photosynthetic units, but the number of pigment molecules per unit is smaller than in green plants: about 50 instead of 300. This may be related to the smaller amount of energy needed to split H_2S (see subsection 16.4.2). The number of pigment molecules per photosynthetic unit is a species characteristic for normal, healthy cells.

It seems likely that the adaptation of algal cells to low irradiance by increasing the chlorophyll content (subsection 4.3.2) is related to the size of the photosynthetic unit, although the ecological significance of this has not yet been studied. More details about the physiology of photosynthesis are given by Forti (1965), Fogg (1968), and Rabinowitch and Govindjee (1969).

REFERENCES

Dundas, I.D. and Larsen, H., 1962. The physiological role of the carotenoid pigment of *Halobacterium salinarium. Arch. Mikrobiol.*, 44: 233—239.

Emerson, R. and Arnold, W., 1932. Photosynthesis in flashing light. *J. Gen. Physiol.*, 16: 191.

Fogg, G.E., 1968. *Photosynthesis.* English University Press, London, 116 pp.

Forti, G., 1965. Light energy utilization in photosynthesis. *Mem. Ist. Ital. Idrobiol.*, 18 (Suppl.): 17—35.

Gaffron, H. and Wohl, K., 1936. Zur Theorie der Assimilation. *Naturwissenschaften*, 24: 81—90; 103—107.

Hertzberg, S., Liaaen-Jensen, S. and Siegelmann, H.W., 1971. The carotenoids of blue-green algae. *Phytochemistry*, 10: 3121—3127.

Holt, A.S., 1961. Further evidence of the relation between 2-desvinyl-2-formyl-chlorophyll-*a* and chlorophyll-*d. Can. J. Bot.*, 39: 327—331.

Margalef, R., 1964. Correspondence between the classic types of lakes and the structural and dynamic properties of their populations. *Verh. Int. Ver. Theor. Angew. Limnol.*, 15: 169—175.

Margalef, R., 1965. Ecological correlations and the relationship between primary productivity and community structure. *Mem. Ist. Ital. Idrobiol.*, 18 (Suppl.): 355—364.

Mathews, M.M. and Krinsky, N.I., 1965. The relationships between carotenoid pigments and resistance to radiation in non-photosynthetic bacteria. *Photochem. Photobiol.*, 4: 813.

Pfennig, N., 1967. Photosynthetic bacteria. *Ann. Rev. Microbiol.*, 21: 285—324.

Rabinowitch, E. and Govindjee, 1969. *Photosynthesis.* John Wiley, New York, 273 pp.

Sistrom, W.R., Griffiths, M. and Stanier, R.Y., 1956. The biology of a photosynthetic bacterium which lacks colored carotenoids. *J. Cell Comp. Physiol.*, 48: 459—515.

Tanada, T., 1951. The photosynthetic efficiency of carotenoid pigments in *Navicula minima. Am. J. Bot.*, 38: 276—283.

Yentsch, Ch.S., 1965. The relationship between chlorophyll and photosynthetic carbon production with reference to the measurement of decomposition products of chloroplastic pigments. *Mem. Ist. Ital. Idrobiol.*, 18 (Suppl.): 323—346.

Yentsch, Ch.S., 1967. The measurements of chloroplastic pigments — thirty years of progress? In: *Proc. IBP-Symp. Amsterdam—Nieuwersluis, 1966*, pp. 255—270.

Yentsch, Ch.S. and Menzel, D.W., 1963. A method for the determination of phytoplankton chlorophyll and phaeophytin by fluorescence. *Deep-Sea Res.*, 10: 221—231.

Yentsch, Ch.S. and Ryther, J.H., 1959. Absorption curves of acetone extracts of deep water particulate matter. *Deep-Sea Res.*, 6: 72—74.

Yentsch, Ch.S. and Vaccaro, R.F., 1958. Phytoplankton nitrogen in the oceans. *Limnol. Oceanogr.*, 3: 445—448.

CHAPTER 14

SEASONAL PERIODICITY OF PHYTOPLANKTON

14.1. INTRODUCTION

One of the typical features of most phytoplankton populations is their relatively short duration. A population usually originates from a small number of cells, and numbers grow exponentially for a few weeks. It is common for a single species to dominate, although of course in natural situations many other species also occur in larger or smaller amounts. After it has reached a maximal density the population usually disappears rapidly, the duration of the whole cycle of growth and decline being 4—8 weeks. The maximal crop is sometimes determined by the initial amount of growth yield-limiting nutrients. The best known example is the spring maximum of diatoms in temperate waters. This will be discussed in detail in section 14.2. After this spring maximum successive populations of other species may grow and then decline at various periods through the year. The term "phytoplankton succession" is commonly used for this phenomenon, because successions of particular groups of species occur during particular seasons, the species composition and order in the sequence often remaining the same for many years. This constant reappearance year after year of similar algal blooms in many lakes is a very impressive phenomenon.

Limnological use of the term succession is of course different from that in terrestrial ecology; it would probably be better to refer only to seasonal periodicity although this may be confused with the changes in growth rates as the seasons progress. For example, there is a well-known spring maximum of photosynthetic rate followed by lower rates during summer (owing to depletion of nutrients) and finally a second maximum during autumn. A good example of these changes was recorded by Findenegg (1964) in the Lunzer See (Austria). Although population growth rate and photosynthesis are related, it is useful to consider them separately since in some situations photosynthesis may continue, and yet population numbers may change very little.

When considering algal populations in lakes two main problems must be solved:

(1) What is the origin of the "inoculum" of a population of a species which suddenly becomes abundant?

(2) What factor or factors allow such a species to grow and how do they control the density of the population of that species?

Concerning the first question, two quite distinct origins are possible. First the phytoplankton species may be present in the lake throughout the year, but usually only in very small amounts, or secondly the cells may be washed anew into the lake with the inflow water or from sheltered bays, etc., or perhaps they may be resuspended from the bottom deposits or grow from resting (overwintering) stages. Situations suggestive of both origins are to be found. Lund (1949) found no evidence that *Asterionella formosa*, the diatom that causes the spring maximum in Lake Windermere, was actually introduced into that lake at the beginning of each seasonal bloom, and he never observed resting stages. By bringing lakewater from Windermere into the laboratory, he could always get growing cultures of this diatom: in winter by increasing light and temperature; in summer by increasing the silicate content. In the same lake, however, the occurrence of *Melosira italica* provides an example of re-inoculation since resuspension by wind causes the return of individual cells to the open water thus renewing the population there. Both examples will be discussed in more detail below.

We turn now to the second question — what controls the population growth? The factors that control spring and other maxima can be put into three groups:

Physical	(Bio)chemical	Biological
light	inorganic nutrients	parasitism
temperature	organic nutrients	predation
turbulence	other organic compounds	competition
through flow	(such as growth-promoting and inhibiting substances, vitamins, chelating substances, antibiotics)	

Every algal bloom is the result of a favourable combination of several factors and it is rare for one factor alone to be recognised as the main cause of a bloom. Increasing irradiance and temperature are often the dominant factors controlling the onset of the spring outburst of diatom growth, but the absence of predation cannot be neglected. Since it is not yet possible to give an overall description of the interaction of all the factors, a study of some particular case histories may provide an understanding of the way in which certain factors can operate.

In section 14.2 it will be explained how increasing irradiance and temperature give rise to the spring bloom of diatoms, but consideration of the individual effects of temperature and light is not at present possible. In the older literature optimal temperatures for growth of many algae are recorded, but the effects observed may have been due to particular conditions of light or nutrient supply or both. A few attempts have been made to separate the effects of light and temperature on growth. For example, Rodhe (1948) found that *Melosira islandica* grew fastest at 20°C, but the culture died rapidly. The culture at 10°C developed faster than that at 5°C, but declined somewhat. For continued growth 5°C was the best of these three temperatures. Other approximate optimal temperatures found were:

Melosira islandica	5°C	*Ankistrodesmus falcatus*	25°C
Synura uvella	5°C	*Scenedesmus, Chlorella*	20—25°C
Asterionella formosa	10—20°C (ca. 15°C)	*Coelastrum, Pediastrum*	
Fragilaria crotonensis	15°C		

The correspondence between optimal temperature found in culture and actual temperature occurring in a lake during growth of these species is striking. But temperature is less likely to be important in a lake than it is in nutrient-rich cultures. In his observations on *Asterionella* Lund (see below) found one division per day at 10°C and two at 20°C in cultures, but in Lake Windermere the growth rate was only one-sixth of these values owing to a nutrient limitation. It has been suggested that comparison of the values for temperature and light during the spring and autumn maxima, when the same temperature does not occur at the same irradiance, would enable the individual effects of light and temperature on growth rates to be separated. Unfortunately, however, nutrient supply is also quite different, and because of this it is impossible to separate the individual effects of light and temperature, nor could interactive effects of light and temperature be discovered in this way. The optimal temperatures given above should therefore be considered as valid only for culture conditions. In lakes there are large differences in temperature optima for populations of the different (sub)-species. *Melosira italica* for example has a maximal growth rate in Lake Windermere between 2°C and 5°C (Lund, see below). The maximal rate of *M.granulata* in the Nile (Talling, see below) falls between 20°C and 30°C, while West (1909, cited in Pearsall, 1923) found a maximal rate between 10° and 23°C for *M.italica* though Lund suggests that West's alga was really *M.granulata*.

14.2. THE DIATOM SPRING MAXIMUM

14.2.1. Physical and chemical factors

The most detailed records of an algal maximum are for *Asterionella formosa* in Windermere (Lund, 1949, 1950). As soon as irradiance and day length increase significantly (early March), *Asterionella* begins to grow exponentially and this process continues until the maximum has been reached, usually at the end of May, or early June (Fig. 14.1). The maximum population density attained varies in different years averaging about 10^7 cells per litre. This normally means an increase of 1000-fold or about 10 divisions. The maximum is related to the SiO_2 concentration in winter because SiO_2 is the growth yield-limiting factor, as was demonstrated by Lund. Thus:

$$\frac{dN}{dt} = \beta \frac{K - N}{K} \cdot N \qquad (14.1 = 10.5)$$

where K = maximum number reached; N = number of *Asterionella formosa*; and β = growth constant.

Lund found the SiO_2 content of *Asterionella formosa* to be 0.14 mg per 10^6 cells, so K can be calculated from the winter SiO_2 content. At the end of the logarithmic growth period the decline in population is even more rapid than the preceding rise. The decline starts as soon as the silicon concentration

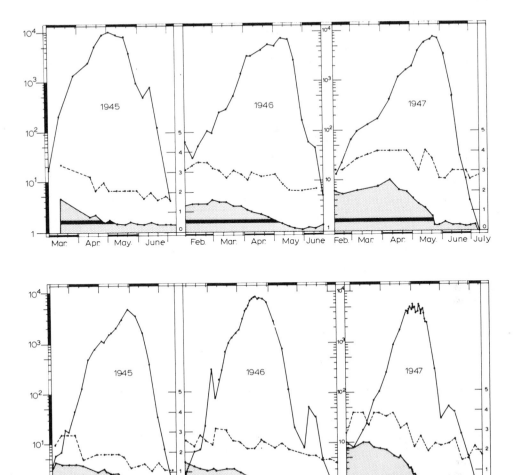

Fig. 14.1. Population density (live cells per ml) of *Asterionella* in Windermere, North Basin (upper figures) and South Basin (lower figures), in 1945, 1946, 1947 (solid line). Nitrate nitrogen ($\times$ 10; interrupted line) and SiO_2 (mg l^{-1} shaded, with 0.5 mg l^{-1} solid black). (After Lund, 1949, 1950.)

falls below 0.4—0.5 mg l^{-1} of SiO_2. Lund argued that at this point the silica concentration is insufficient to permit one more cell division because with 10^7 cells per litre this would require 1.4 mg l^{-1} of SiO_2. He suggested that subsequent cell divisions after this density has been reached result in the formation of naked cells which must die because the necessary silicification cannot take place (see Chapter 7).

One may wonder why the cell divisions do not just cease and the natural population remains constant — except for the effects due to sinking — since

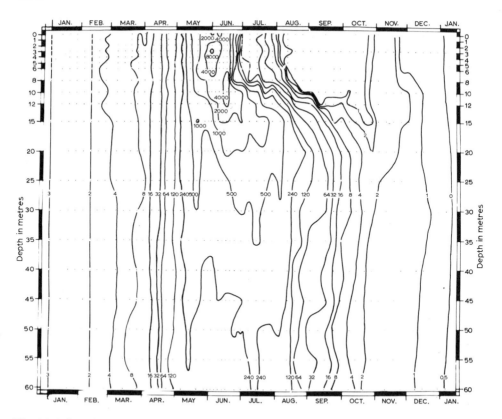

Fig. 14.2. Population density (live cells per ml) of *Asterionella* in Windermere North Basin, during 1947. In Fig. 9.3 is shown the depth—time distribution of the greatest vertical density gradient, associated with the thermocline. (From Lund et al., 1963.)

this often happens in *cultures* if for example they are placed in dim light. Most diatom cultures maintain constant cell density if placed in dim light after nutrient depletion, although numbers decline if the culture is kept at high irradiance. This behaviour is quite different from that of many laboratory cultures of Chlorococcales, which will often maintain their numbers under conditions of high irradiance together with nutrient stress.

It has been suggested (e.g. Golterman, 1960, following the more general reasoning of Findenegg, 1943, 1947) that it is the combination of high irradiance with relatively high temperature which is lethal but only because the cells are silica-depleted and therefore not at their physiological optimum. Addition of more silica to the medium at the moment of maximum cell density would certainly cause the diatom numbers to increase. Other possible causes of the sudden decline were excluded by Lund, who found no significant grazing or parasitism by fungi, nor outflow or sedimentation. Loss by outflow is probably the main factor responsible for reducing the winter population. This population grows slowly because the photosynthetic gains are little greater than respiratory losses during the winter.

In a detailed study of the vertical distribution of *Asterionella formosa* (Figs. 14.2 and 14.3), it has been shown (Lund et al., 1963) that sedimentation of cells from the epilimnion increases after nutrient depletion.

After the spring maximum higher cell densities are found ($0.5-1 \cdot 10^6$ cells per litre) at greater depths ($10-20$ m) than in the upper layers of the epilimnion ($10-30 \cdot 10^3$ cells per litre). This might be explained either by sedimentation or by continued cell division in the deeper layers, where the light level may be more favourable than in the $0-10$-m water layer. Sedimentation of live cells does not seem to increase suddenly after the spring maximum, but sedimentation of dead cells does increase and may sometimes reach nearly 25% of the total number of cells present.

A rough estimate of the influence of sedimentation is given by Lund in the same paper. The losses from the epilimnion were estimated by calculating the

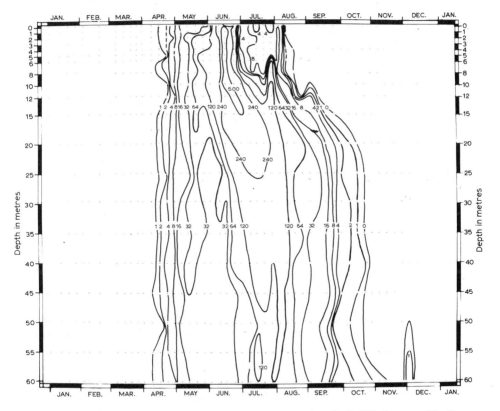

Fig. 14.3. Population density (dead cells per ml) of *Asterionella* in Windermere North Basin, during 1947. In Fig. 9.3 is shown the depth—time distribution of the greatest vertical density gradient associated with the thermocline. (From Lund et al., 1963.)

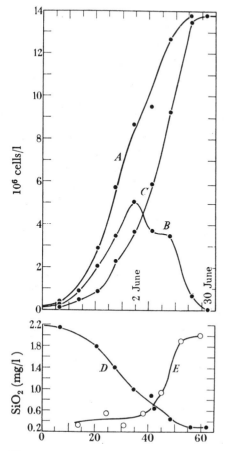

Fig. 14.4. *Asterionella* production and loss in Windermere, 28 April to 30 June 1947. The epilimnion is assumed to occupy the top 8 m. The curves represent: *A*, the cumulative total of cell production computed from silicate uptake; *B*, the mean concentration of cells in the epilimnion (standing crop); *C*, the cumulative loss of cells from the epilimnion (*A* minus *B*); *D*, the epilimnetic silicate concentration; *E*, the relative rates of loss of cells from unit epilimnetic population (arbitrary scale). (From Lund et al., 1963.)

cumulative cell total from the decrease in SiO_2 concentration and comparing this with the standing crop in the epilimnion (Fig. 14.4). It can be seen that the loss of cells sinking from the epilimnion increases greatly after the maximum population density is reached but is low during the growth period, indicating that live cells sediment less rapidly than dead cells. A computer programme may perhaps provide a good picture of gains and losses which occur after the spring maximum.

Hughes and Lund (1962) found that the addition of small amounts of phosphate to Windermere water stimulated *Asterionella formosa* to take up all the silicate and so it seems likely that the final level of silicate is controlled by phosphate concentration. It is difficult in this case to decide which nutrient is the growth *yield*-limiting factor. Both phosphate and silicate concentrations are finely balanced between consumption and supply. The fact that a nutrient is present in low concentration does not necessarily mean that it is in short supply. Although the influence of phosphate concentration on silicate uptake cannot be neglected, the initial silicate concentration certainly seems to limit the maximum numbers of *Asterionella formosa* and the uptake of an extra $0.4-0.5$ mg l^{-1} of silicate has little effect on the maximum numbers. Certainly phosphate is important as a growth rate-limiting factor. The growth rate of *Asterionella formosa* in Windermere (see Fig. 14.1) is remarkably constant, with doubling times in lake conditions of $5-7$ days. This is much longer than the time taken in laboratory conditions (9.6 h at $20°$C: Lund, 1949). This difference is attributable both to different light conditions in the lake and to low phosphate concentrations. It seems likely that the growth constant in eq. 14.1 could be replaced by:

$$\beta = \beta_{\max} \frac{[PO_4]}{C_1 + [PO_4]} \qquad (14.2 = 10.3)$$

It would be interesting to see whether growth rates in different lakes (Windermere, Esthwaite, Blelham Tarn) studied by Lund (1972) follow the same relationship. Circumstantial evidence for this is presented in Chapter 17, so that 14.1 becomes:

$$\frac{dN}{dt} = \beta_{\max} \cdot \frac{[PO_4]}{C_1 + [PO_4]} \cdot \frac{N_{Si} - N}{N_{Si}} \cdot N \qquad (14.3)$$

where N = number of *Asterionella* cells, and N_{Si} = maximal number of *Asterionella* cells calculated from winter SiO_2 concentration.

Qualitatively it has been shown (Lund, see Chapter 17) that the growth rate in Blelham did increase: the maximum was reached earlier as phosphate concentration increased over the course of several years.

Lund (1950) obtained growth rates in 12-l glass tanks (in both December and February) which were similar to those normally occurring in Lake Windermere. Each week 10 l of the water in the tank were replaced by filtered lakewater, while the *Asterionella* cells were filtered off and washed back. Although the same illumination was used in both cases, growth in December—February (curve A in Fig. 14.5) appeared to be faster than in February—April (curve B). It seems likely that this might be explained by the occurrence of a higher phosphate concentration in December. Although there were differences in illumination between the lake and tank, growth rates were similar in the February—April experiment (curves B and b), which indicates that it is not the light which controls the growth rate.

Talling (1955) showed that photosynthesis is light-saturated in the upper 5 m of Windermere, so phosphate may be the main factor controlling growth

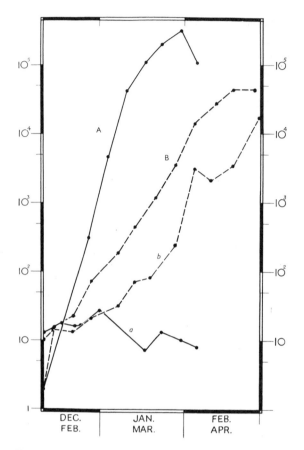

Fig. 14.5. Growth of *Asterionella* in Windermere North Basin water in the laboratory and in the lake itself. Lines *A* and *B*: growth in glass vessels containing 12 l of lakewater, 10 of which are renewed weekly. Lines *a* and *b*: density of the population in the 0—5-m water column at the Windermere North Basin buoy during the period covered by the experiments *A* and *B*. Vertical axis: cells per 10 ml. Horizontal axis: time scales for first experiment (*A, a*) above; second (*B, b*) below. (From Lund, 1949.)

rate there. Talling obtained growth rates of up to 7 divisions per week by suspending bottle cultures of *Asterionella* in Chu's no. 10 medium at 1 m depth in Lake Windermere. High nutrient concentration might explain the increased growth rate here, although the low values in January—March are probably due to shortage of light. In these experiments both day length and temperature should be regarded as important factors with day length probably being the more important. An increase of day length both increases photosynthesis and decreases the respiration losses overnight, so that productivity increases.

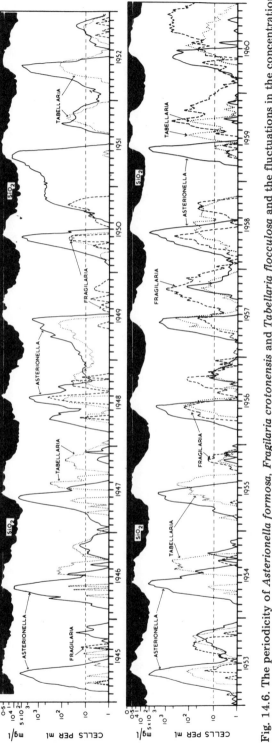

Fig. 14.6. The periodicity of *Asterionella formosa*, *Fragilaria crotonensis* and *Tabellaria flocculosa* and the fluctuations in the concentration of dissolved silicate in the 0—5-m water column of the North Basin of Windermere from 1945 to 1961. (From Lund, 1950.)

258

14.2.2. Competition

After the *Asterionella* in Lake Windermere had flourished and then declined, other species of diatoms remained. Maxima of either *Fragilaria* sp. or *Tabellaria* sp. then occur, these species being present in smaller amounts during the preceding *Asterionella* maximum (Fig. 14.6). Lund (1964) considered the problem of why *Asterionella* itself predominates during the spring period, rather than either of the other species. He suspended cultures of all three diatom species at 1 and 6 m below the lake surface. From data shown in Figs. 14.7 and 14.8, it appeared that *Fragilaria* has about the same, and *Tabellaria* has a lower growth rate than *Asterionella*. The growth rates were shown by Lund to be related to cell size, *Asterionella* and *Fragilaria* being equal in size

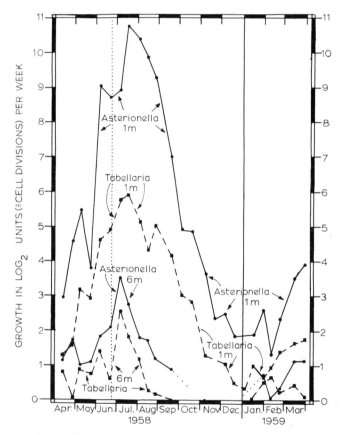

Fig. 14.7. The growth of *Asterionella formosa* and *Tabellaria flocculosa* in cultures suspended at 1 and 6 m below the surface of Windermere. Each exposure lasted one week or, occasionally, 4 or 8 days. Growth expressed as cell divisions (i.e. on a $\log_2$ basis) per week. (From Lund, 1950.)

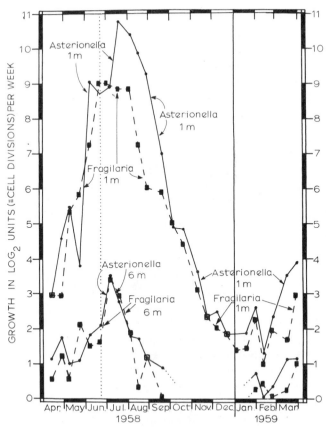

Fig. 14.8. The growth of *Asterionella formosa* and *Fragilaria crotonensis* in cultures suspended at 1 and 6 m below the surface of Windermere. Each exposure lasted one week or, occasionally 4 or 8 days. Growth expressed as cell divisions (i.e. on a $\log_2$ basis) per week. (From Lund, 1950.)

and *Tabellaria* being twice as large. Thus *Asterionella* will grow faster than *Tabellaria* if both diatoms have all the nutrients they need. On the other hand, *Fragilaria* starts at a disadvantage, having a low population density because relatively more of the cells have sunk during the winter. A puzzling feature concerns the source of silicate during the summer predominance of *Fragilaria* and *Tabellaria* which both give rise to small summer crops. Lund suggested that perhaps they are able to utilise low concentrations of nutrients more efficiently than *Asterionella*, and it seems likely that both *Fragilaria* and *Tabellaria* may use nutrients released by the decaying population of *Asterionella*. Silica release from dead *Asterionella* cells has been demonstrated by Moed (1973). The summer crops usually occur at the same time as rainy periods, when there is an increased inflow, which will supply extra phosphate and silicate. Although in general the silicate concentration limits the spring maximum of *Asterionella*, there are exceptions to this, e.g. in 1948 (see Fig. 14.6). Lund suggested that

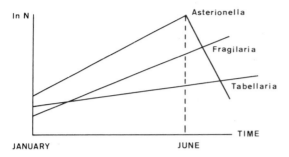

Fig. 14.9. Outcome of competition, in cell numbers (ln N) per unit volume, of populations of *Asterionella*, *Fragilaria* and *Tabellaria* due to differences in cell number of the inoculum and in growth rates. If the inoculum of *Fragilaria* (= winter concentration) is much lower than that of *Tabellaria*, a situation may develop where *Tabellaria* overtakes *Asterionella* because then the *Tabellaria* curve may cross the *Asterionella* one earlier than the *Fragilaria* curve.

some other nutrient limited the growth in this year and related this limitation to a period of drought, because growth resumed after the rain had increased the flow of rivers again. Populations of other algae generally increase either slowly or not at all during the *Asterionella* maximum (whether this is a normal maximum in relation to silicate or not), so Lund suggested that another nutrient limited the crop of phytoplankton as a whole. The large number of *Asterionella* cells places them in a favourable competitive position in relation to this supposedly growth-limiting substance. The competition between *Asterionella*, *Fragilaria* and *Tabellaria* is represented schematically in Fig. 14.9, in which the different growth rates and sizes of inocula are depicted.

The influence of one phytoplankton population on another one is often attributed to inhibition by toxic substances. There are no well-authenticated cases of this in nature and it is probable that competition for nutrients is usually the main operative factor in natural conditions. Nutrient limitation may also occur in cultures although this may be easily overlooked. Toxic substances have been shown to occur in culture experiments but such results should not be regarded as necessarily applicable to natural conditions in lakes.

Mutual inhibition between algae is too often invoked as an explanation of otherwise insoluble difficulties as Lund has emphasised.

In numerous experiments with *Oscillatoria agardhii* in Esthwaite Water, he showed that if this species has any influence at all on the growth of *Asterionella*, it would be expected to be stimulatory, although in the lake itself growth of *Asterionella* is suppressed during growth of the blue-green alga. Thus Lund aptly stated: "In explaining the rise and fall of populations in nature, inhibition is as easy to suggest as it is difficult to prove".

A detailed study of such a growth interaction — that between *Chlamydomonas globosum* and *Chlorococcum ellipsoideum* — was made by Kroes (1971, 1972). In the first paper Kroes described a newly developed filter culture in which two separate cultures of different species are connected via a filtering system through which medium is exchanged while the cells themselves are kept separate. In this device *Chlorococcum* inhibited the growth of *Chlamydomonas*. In mixed cultures this inhibition was also found but only in an unbuffered medium. Kroes, who reviewed and criticised older work, found that pH was an important factor in this inhibition. High pH values may occur during photosynthesis of algae under natural conditions, so his work seems relevant to ecological situations, although no combination of both algae in nature has been described. In his second paper he found some influence of extracellular compounds from *Chlorococcum* on *Chlamydomonas*. Steam volatile substances (probably short-chain fatty acids such as acetate) showed a stimulatory effect, which seems understandable. The lipophilic group inhibited growth initially but had no lasting effect. An isolated fraction containing yellow pigments showed an understandable chelating effect. Kroes correctly pointed out that the high concentrations of organic matter he used are not likely to occur in natural conditions and stated that factors such as nutrient depletion and pH are more important than extracellular compounds.

14.3. COMPETITIVE EXCLUSION

If an algal species is a better competitor for a given compound or light, it might easily be thought that this species will prevent growth of all other ones. This problem will be discussed based partly again on observations on Windermere. Growth rates play an important role in this "model", and as growth rates are partly dependent on algal size this latter characteristic enables small algae to grow faster. The chance of being consumed by zooplankton increases equally however, so that faster growth may not automatically induce a survival in competition. The smallest nanoplankton species of algae — the μ algae — are another group which is abundant during the spring in Lake Windermere. Their periodicity is caused by grazing (Edmondson, 1965; see also subsection 15.3.3) by their capacity for heterotrophic growth (but see Lund's warning concerning mutual inhibition!), and by the upper limit of photosynthesis, i.e. by competition for nutrients. When considering the fluctuations in numbers of the main species of algae present in the spring maxima in Esthwaite, Lund found that although the numbers differ from year to year the amount of carbon incorporated into the total crop was similar, and this can only be explained by assuming the existence of an upper limit of production and standing crop. The same is true for the total chlorophyll concentration, which remains much more constant than do the actual numbers of algae, e.g. Tjeukemeer (Fig. 14.10).

From Lund's estimate of how many algal species contribute significantly to primary production, it appears that in winter and spring this figure is lower than in summer and autumn with or without the inclusion of μ algae. This problem has caused Hutchinson (1961) to wonder how it is possible that during the summer a number of species may co-exist in a relatively isotropic

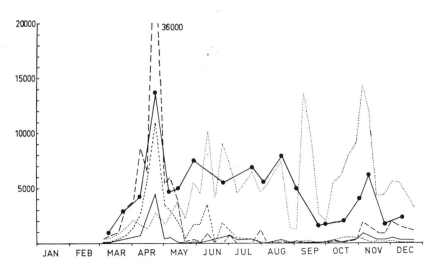

Fig. 14.10. Seasonal periodicity of dominant algae in Tjeukemeer in 1970. Legend: ———, pennate diatoms (number of cells per ml); ———, centric diatoms excluding *Melosira* (number of cells per ml); - - - - *Melosira* (number of cells per ml); . . . . . . , *Scenedesmus* (number of colonies per ml). Black solid line: chlorophyll.

or unstructured environment, all competing for the same kind of material, of which the deficiency is so great that competition is likely to be severe. This is not a real problem, however, because neither the environment nor the phytoplankton are in fact unstructured. The environment is certainly not uniform. The distributions of phytoplankton and especially their predators are never homogeneous and neither are light or nutrient supply. Nearly every species of alga has at least one parasite species which can afflict it and most are grazed — the nanoplankton largely by Crustacea and Rotifera and the net-plankton by Protozoa. The former animals engulf their prey, the latter enter their prey and ingest the cells (e.g. colonial algae) or attach themselves to the algae and extract their contents (e.g. filamentous algae). Growth will further-more be optimal only at a certain depth and may be inhibited by too much light above that zone or too little light below. Richerson et al. (1970), working on Lake Tahoe (California), demonstrated the occurrence of a distinct patchiness in the spatial distribution of many phytoplankton species, indicating that the rate of mixing is slow enough in relation to algal growth rate to enable micro-niches to occur. They used a contemporaneous disequilibrium model to explain the diversity. In any particular micro-niche (patch) one species is at a competitive advantage relative to others.

Incomplete mixing is not necessarily the cause of this patchiness because in shallow lakes where mixing is certainly more complete than in deeper ones, the same phenomenon is found. In deeper lakes the phytoplankton is not uniformly distributed either. Some algae are apparently better competitors than others and may reach light-saturated photosynthesis at lower irradiance than others, probably by making more efficient use of the nutrients available, or by differences in I_k value (subsection 4.2.2). The slower-growing algae (for example, many desmids) are unable to compete with the species which form spring blooms (e.g. diatoms) most of which grow faster than the desmids at lower irradiance and temperatures. By summer the light and temperature conditions are suitable for a wider range of algal species than those occurring in spring. It has often been suggested that competition between two species with similar ecology will result in exclusion of one of them, e.g. Hardin (1960) who challengingly reformulated the principle as "complete competitors exist". Lund's example shows that complete competition can and does occur, but only for a limited period. Thereafter one or more factors may change (e.g. silicate content) and the competition may have a different outcome. Slobodkin (1961) also demonstrated that grazing may alter the course or outcome of competition. The "phytoplankton paradox" is based on the supposition of equilibrium situations, whereas the natural situation probably consists of rapid changes from one state to a new one. The limitation of growth *rate* (of *Asterionella*) by phosphate, and of *yield* by silicate (p. 255) emphasises that the exact nature of the competition must be specified and that this can be done only after careful experimentation. Vagueness of statement results in endless fruitless discussions. The fact that recycling occurs and that nutrients released from a decaying population (mineralisation) can be used by a newly growing population demonstrates the weakness of the different points in the competition-exclusion principle. Since both these populations are not in equilibrium, they do not compete, although comparison of cell numbers may give the impression that both populations are growing simultaneously.

Although the pattern of rise and fall of *Asterionella* in Windermere is well defined, its behaviour in other lakes may depend upon different factors. Thus for drinking water reservoirs of the Metropolitan Water Board in London, Gardiner (1941) found that *Asterionella* disappeared within the same period as in Windermere, although the silicate concentration was still at $3-5$ mg l^{-1}. In 1938 Gardiner found that after the decline of the *Asterionella* population *Stephanodiscus astraea* appeared. These diatoms depleted the remaining silica and then the population declined. In 1939 *Asterionella* followed a bloom of *Fragilaria crotonensis* and declined after silica depletion. From Gardiner's figures it appears that silica mineralisation takes place. This is reused in spring but not in summer. Gardiner concluded that some factor other than the concentrations of silica or phosphate must operate to cause the decline of the diatom population, but he did not suggest any likely factors. As he did not measure all the relevant parameters it is not possible to do much retrospective analysis of his results, but it seems likely that irradiance may have been high enough to inhibit diatom growth in summer.

It was mentioned earlier that the larger diatom *Tabellaria* has a lower growth rate than either *Asterionella* or *Fragilaria* (Fig. 14.11A). Although no

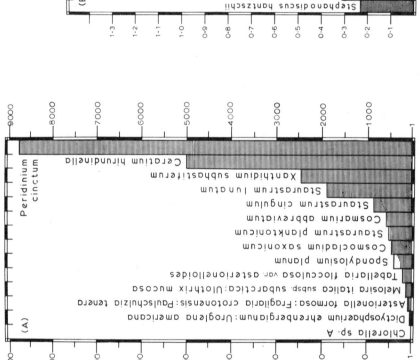

Fig. 14.11. A. Average dry weights of certain plankton algae of Windermere based on estimates from natural or cultivated populations. B. Approximate dry weight of 1 mm³ of the live cells of certain freshwater and marine plankton algae from natural or cultivated populations. (From Lund, 1964).

detailed figures are available, this might be explained by assuming diffusion of nutrients to cells was growth-limiting. The smaller the algae the greater their surface/volume ratio, and therefore small cells would be expected to be more efficient in nutrient uptake than larger ones.

It can be seen from Fig. 14.11A that the average dry weight of algal cells falls in the range between 1 and 9 000 "*Chlorella* units". The dry weight per unit volume also varies considerably (Fig. 14.11B). Although quantitative data are scarce, it is obvious that both these features must play an important role in competition.

Manny (1972) separated net- from nanoplankton by filtering the water through a 10-μm cloth filter before trying to establish their respective contributions to plankton growth. He found the total particulate organic nitrogen to range between 40 and 120 μg l^{-1}, of which the nanoplankton component contained between 50 and 75%. The same ratio was found for the chlorophyll content. Nanoplankton cells were 10—50% of the total cell number with occasional samples reaching 100%. Furthermore Manny found the nitrogen content per 10^8 μm^3 of cell volume to be 10—100 μg of N for the nanoplankton and 1—10 μg for the netplankton. He attributed the difference to a higher metabolic activity of the nanoplankton owing to a higher surface area/volume ratio.

Kalff (1972) studying a small naturally eutrophic lake (Lac Hertel, Canada) found that 75—79% of the primary production was synthesised by plankton smaller than 64 μm, and about 50% by organisms smaller than 20 μm, while the netplankton (larger than 64 μm) contributed more than half of the dry weight in the epilimnion showing the relatively large growth rates of the smaller phytoplankton. Kalff pointed out that oligotrophic lakes generally have a planktonic flora containing a high proportion of nanoplankton, while eutrophic lakes have a greater proportion of netplankton, although in those lakes nanoplankton may be responsible for a larger part of the photosynthesis (see, however, Loch Leven, section 15.5). Findenegg (1965) compared populations from different lakes and showed that nanoplankton is more active in assimilation than are the larger species. He also demonstrated that increasing population density diminished relative assimilation rate. In a second paper (1971) he compared photosynthesis in situ of many natural populations, selecting those which contained mainly (70—99%) a single species. Findenegg calculated an activity coefficient, i.e. the carbon assimilated per carbon fresh weight — the latter being taken as 12% of the total fresh weight—, for both the plankton in the layer of optimal photosynthesis and for the plankton in all layers where it was found (Table 14.1). The variation in the activity coefficient per species (not shown in Table 14.1) is large, as might be expected for populations in different lakes, and this might be due partly to variation in the conversion factors (volume/dry weight; dry weight/carbon). It seems not to be due to variation in the irradiance, as the activity coefficient for one species is not closely correlated with the light values. It should be noted however that by

TABLE 14.1

Activity coefficients of phytoplankton species in different layers and irradiance in those layers (Findenegg, 1965)

Species	Optimal layer		Mean layer	
	Activity coefficient	Irradiance (mcal. cm^{-2} 3 h^{-1})	Activity coefficient	Irradiance (mcal. cm^{-2} 3 h^{-1})
Ankistrodesmus acicularis	3.70	563	2.47	509
Oocystis lacustris	1.00	380	0.80	582
Uroglena americana	0.33	517	0.77	442
Gloeococcus schroeteri	5.15	700	1.75	361
Fragilaria crotonensis	0.86	333	0.74	328
Cryptomonas erosa	6.57	444	3.63	326
Cyclotella comensis	1.22	352	0.78	430
Dinobryon sociale and divergens	1.76	322	1.46	409
Ceratium hirundinella	1.48	346	1.18	396
Anabaena flos aquae	1.70	271	1.45	372
Anabaena planctonica	0.95	419	0.53	291
Stephanodiscus hantzschii	0.79	230	0.55	271
Oscillatoria rubescens	0.58	121	0.27	132

comparing assimilation rates of phytoplankton populations in different lakes or at different moments, a second variable besides the species present is introduced, i.e. the nutrient supply, for which Findenegg gives little information. This extra parameter can be ruled out by comparing different filtrates from the same lake, as was done by Manny and Kalff.

Situations in other lakes may be quite different from that in Windermere. Berardi and Tonolli (1953) showed that in Lago Maggiore chlorophyll content is greatest in October—November, when diatoms and blue-green algae (*Fragilaria crotonensis* and *Gomphosphaerica lacustris*) have their maxima. Lago Maggiore (Italy) shows an overturn only once every five years or so, so that nutrient depletion of the hypolimnion occurs over a more protracted period than in Windermere. If nutrients are washed into the lake during the autumn, an algal peak may occur as the light conditions are more favourable than those in Windermere at that time of the year.

14.4. OTHER FACTORS CONTROLLING ALGAL SUCCESSION

Of the many factors (mentioned in section 14.1), which control the spring bloom of diatoms in temperate lakes, some of the physical and chemical ones are the best understood at the present time. Much less is known of the biological and biochemical ones. Some of the other factors will now be discussed.

14.4.1. Turbulence

Lund (1954, 1955, 1966) has shown that the seasonal cycle of the plank-tonic diatom *Melosira italica* is correlated with the turbulence of the water. Based on the same kind of research as that carried out for *Asterionella*, he suggested that *Melosira* can only produce a large population in the euphotic zone if there is sufficient vertical mixing. Mixing is particularly strong in iso-thermal conditions with wind-generated turbulence. Diatoms sink relatively rapidly, so that during summer stratification cells will sink to the bottom and develop into "resting stages", some of which can survive for many years in the dark and under anaerobic conditions.

Interactions between nutrient demands and light, similar to those described for *Asterionella*, are found. *Melosira* has been observed to multiply in waters in the English Lake District only when there is more than 300 μg l^{-1} of SiO$_2$-Si, but this is not an absolute limiting concentration. In shallow lakes (see below) *Melosira* depletes SiO$_2$-Si to 50 μg l^{-1} of SiO$_2$-Si; phosphate con-centrations are normally much higher there. A very clear demonstration of the influence of turbulence occurred in Blelham Tarn, where a late summer in-crease of cell numbers was caused by artificial destratification. The cell popula-tion density was one of the highest ever found in that lake (Lund, 1971a).

Melosira is a genus with a widespread geographical distribution. It occurs in Lake Baikal, the deepest lake in the world, and in shallow lakes like Tjeuke-meer. Lund (1966) suggested that sedimentation in Lake Baikal is prevented by turbulence which may occur even under the ice, due to the hydrology of the lake. Its occurrence in shallow lakes like Tjeukemeer seems to be unrelated to the turbulence or at least its disappearance is not caused by a lack of turbu-lence. *Melosira* species are responsible for the spring diatom bloom in Tjeuke-meer, but they disappear as soon as silica is depleted (Fig. 14.10). It seems likely that the population dies off just as *Asterionella* does, because summer or autumn peaks are rarely found even after rough weather in Tjeukemeer.

Moed et al. (1975) recorded a single sudden increase up to about 30% of the original maximum of *Melosira* spp. after a severe storm in April 1973. *Diatoma elongatum* also reappeared, although 40—50% of the cells were thought to be dead. The *Diatoma* population almost doubled in April—May, showing a second maximum, while the *Melosira* population declined slowly. It seems likely that *Diatoma* is a better competitor for the small amounts of silicate that entered the lake. Apparently turbulence, although being a factor involved in the *Melosira* bloom in Tjeukemeer, is certainly not the major one.

It has been suggested that interactions between light and nutrients are more important than sedimentation in shallow lakes. Reynolds (1973) examined the seasonal periodicity of four species of diatoms in a small shallow eutrophic lake (Crose Mere, Cheshire Meres, Great Britain), which does not stratify. He found that the occurrence of appropriate light values determined both the onset of the spring bloom and the maximum size of the population, whilst turbulence regulated the vertical distribution. Thus physical rather than chemical control regulated the populations. In the presence of excess nutrients relative specific differences in growth requirements determined which species dominated, the dominance being modified by fungal parasites. Reynolds showed that the presence of the parasite was tolerated by the host population as long as the latter was growing. The parasite assumed epidemic proportions if growth of the host was restricted by other agents. These events may however be explained also by assuming that a fast-growing host may outgrow the parasite.

Changes in numbers of *Asterionella formosa* and *Fragilaria crotonensis* showed a striking similarity clearly because they did not compete for silicate, in contrast to Windermere. It is remarkable that their seasonal pattern was not disturbed in 1968, the only year that silicate became depleted. Both diatoms showed overwintering populations in 1969–1970 and 1970–1971 but not in the earlier years of the observation period. In 1970–1971 silicate concentration remained lower than in those earlier years, which may be an indication of silicate remineralisation during winter. Reynolds (1972) showed that a large population of *Anabaena circinalis* occurred in May–June, when diatom numbers were low. It is clear that this replacement cannot be caused by a lack of light, so that the influence of light on these populations remains obscure. The seasonal succession of phytoplankton in similar meres was described by Reynolds in 1973. Diatoms appear to be dominant generally in early spring and are normally succeeded by large populations of *Eudorina* and *Pandorina* in late spring and by blue-green algae in summer. *Ceratium* is present in late summer, when diatoms may produce a second maximum. Cryptomonads were abundant at most times of the year. Reynolds suggested that this sequence may be taken to represent a regional type of succession. It is probably a succession typical for temperate eutrophic waters (see below).

14.4.2. Other chemical factors such as phosphate or nitrogen supply

Shortage of phosphate or nitrogen or both are often mentioned as factors controlling seasonal periodicity by their influence on the growth rate. It seems unlikely, however, that a low growth rate causes seasonal succession, although the density of the population may well be limited by the phosphate concentration.

The sequence of algal species in Tjeukemeer (Fig. 14.10) is fairly typical for eutrophic waters. In early spring, diatoms such as *Melosira* spp. and *Diatoma elongatum* dominate, in some years simultaneously, in some years one following the other. If rainfall is high during the early autumn (October), extra silicate may be supplied during a period with sufficiently high irradiance to allow an autumn peak to occur. This does not happen if the silicate is provided in November, presumably because light values are too low by then. When silicate is depleted during the summer, the green algae *Scenedesmus* or *Pediastrum* or both exhibit short (sharp) blooms. Although phosphate and nitrogen concentrations are rather low, they do not appear to control these peaks in cell densities, because the nutrient concentrations are low throughout the whole summer period of algal population flux. In some years blue-green algae appear, but in the five years during which records have been kept no two years have been similar. Originally it was thought that N_2 fixation by blue-green algae could be a regulatory mechanism, but it was found that phosphate and nitrogen concentrations were much the same, whether green or blue-green algae were present, and in some years blue-green algae populations occurred that had no capacity for nitrogen fixation. It seems likely that predation by grazing controls the numbers of algae like *Scenedesmus* but not those of *Pediastrum* or the blue-green algae. The decline of the first populations of *Scenedesmus* may be caused by zooplankton but the November maximum cannot be ended in this way because there are few planktonic animals then. The decline is too steep to be caused by decreasing irradiance or an outwash from the lake. The combination of these effects may be suggested, but there is no evidence to support the suggestion.

An unusual succession for temperate shallow waters was described by Gibson et al. (1971) from Lough Neagh (Ireland). Here *Oscillatoria redekei* dominated in April—May 1969 and 1970, while *Oscillatoria agardhii* was found in the autumn. Diatom crops were small and were demonstrated by *Melosira italica* in May 1969, but by *Stephanodiscus astraea* in February 1970. From the silicate concentrations and the silicate content of the diatoms, the authors calculated that the gross production of diatoms was perhaps three times greater than the net production indicated by the counting technique used. The rather unusual peak of *Oscillatoria redekei* seems to have been caused by an overwintering population.

Bush and Welch (1972) compared different algal species assemblages with corresponding chemical composition of the waters from various localities in Moses Lake (Washington) over a period of 26 months. Three very common algal populations were identified. Population I was composed of diatom genera and was often dominated by species of *Navicula*, *Fragilaria* and *Cyclotella*. This population persisted throughout the study period in an upper basin near the river inlet and was sometimes found during March—April at other places also. Although no silicate quantities are reported, it seems likely that silicate concentrations control this population. The concentration of total phosphates was high, compared with the concentration of chlorophyll *a* (7 : 1). In the major portion of the lake during the summer months population II, of blue-green algae, developed. It had a phosphate to chlorophyll ratio of 10 : 1. The main species were *Aphanizomenon flos aquae* and *Microcystis aeruginosa*, with smaller numbers of *Anabaena circinalis* and *Coelosphaerium naeglianum*. Green algae such as *Scenedesmus quadricauda* and *Oocystis* sp. occurred only in the northern portion of one of the bays of the lake. Population III usually had a normal phosphate to chlorophyll ratio of about 2 : 1, and contained mostly green algae such as *Scenedesmus quadricauda* and *Oocystis* sp.

From a consideration of the concentrations of inorganic and total phosphate and nitrate, it seems possible in this lake that nitrate is the limiting factor for algal growth, but the negative correlation between inorganic phosphate and biomass suggests that phosphate is likely important. Unfortunately no data are given either for total organic nitrogen or for ammonia. Cluster analysis of plankton data proved to be an effective method in describing the composition of the algal populations. Bush and Welch pointed out correctly that the observed negative correlation between blue-green algal biomass and low concentrations of phosphate and nitrate is the effect of the blue-green algal population and not the cause (see section 6.5.3). Blue-green algal chlorophyll concentration however was related to the initial (winter) phosphate concentration.

The populations of diatoms and of green algae occurred mainly in parts of the lake of which the waters showed high values for alkalinity, conductivity, phosphate and nitrate (inter alia). It is not clear however, whether these higher nutrient values occurred because these algal populations were limited by some other factor (for example light penetration) and could not therefore utilise them fully, or because the blue-green algae can deplete their surrounding nutrients to lower values. Nutrient release from the sediments plays a role in the nutrient supply, but the upper layers may be unaffected because of thermal stratification. Unfortunately Bush and Welch did not record in which parts of the lake stratification occurred. Other factors which may favour the blue-green algae are the ability to migrate vertically using gas vacuoles and the ability to fix nitrogen. This study once again strongly suggests that the blue-green algae do not occur at places where other populations persist. The wax and wane of blue-green algal populations must be explained not only by their physiological characteristics but also by those of other algae. Such studies of micropopulations in heterogeneous lakes may help to elucidate the factors which give rise to successions of algal populations.

14.4.3. Predation, grazing, parasitism

Phytoplankton serve as food for zooplankton; heavy grazing may cause a considerable reduction of algal numbers. In a simple two-organism system a decrease of phytoplankton will be followed by a decrease of the predator, but eventually an equilibrium may establish and the phytoplankton will not be eliminated. In natural conditions, however, the decrease of one phytoplankton organism may be followed by an increase of another species. If this new species can also support zooplankton growth, the zooplankton may continue to reduce the numbers of the first species too. (In Chapter 15 food chains and predation by grazing will be discussed in detail.) The most important animal groups which graze on phytoplankton are the Crustacea (mainly Copepoda and Cladocera) and the Rotatoria. The latter graze mainly on μ algae: population fluctuations, however, have been little investigated.

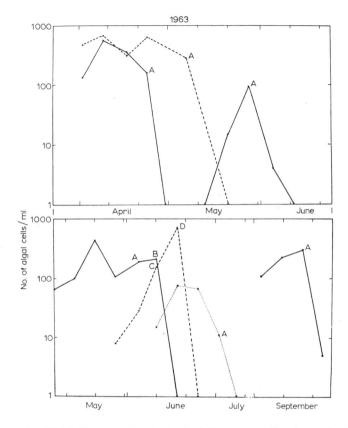

Fig. 14.12. Decreases in algal populations in relation to grazing by Protozoa during 1963. Upper figure, continuous left-hand line: *Eudorina elegans*; continuous right-hand line: *Gemellicystis imperfecta*; broken line: *Synura* sp., all in Esthwaite Water. Lower figure, continuous left-hand line: *Gemellicystis imperfecta*; continuous right-hand line: *Dictyosphaerium pulchellum*; broken line: *Paulschulzia tenera* and *P. pseudovolvox*, all Windermere, South Basin; dotted line, as broken line, but in Windermere, North Basin. *A*, date when protozoans were first seen; *B*, 98%; *C*, 1%; and *D*, 88% of the algal colonies contained Protozoa. (From Canter and Lund, 1968.)

An example of parasitism is given by Canter and Lund (1968), who found that the protozoan *Pseudospora* infects colonies of the green alga *Gemellicystis*. (Fig. 14.12). They found that of 14 infections in May—June in 9 cases the algae decreased by 99%, in 4 cases the algae decreased between 96 and 99%, and in 1 case the algae decreased by 93%.

It seems typical of parasitism that the host populations are greatly reduced in numbers (> 90%). In this respect parasitism is quite different from grazing, which will usually lead to the establishment of an equilibrium. Parasitism is also an important factor governing the periodicity of some desmids in Windermere (Lund, 1971b).

The desmids *Cosmarium contractum*, *C. abbreviatum* and *Staurastrum lunuatum* increase in number in May and decrease in September or October, population densities being more associated with the temperature-curve changes than with changes in irradiance. The populations occur mainly in the epilimnion when the lake is mixed. All three species have parasites which have a marked effect on the relative abundance of the algae but not on the overall pattern of seasonal periodicity. By exposing cultures in situ Lund showed that *Cosmarium contractum* grows more slowly and reaches its maximum later (September—October) than the other two species (July—August). A species which has a maximum at a later date can spread through a deeper epilimnion. In Lake Windermere this is about 20 m deep in September. *C. contractum* reached a much higher relative abundance possibly because the greater dilution in the deeper epilimnion allowed a larger growth rate. Another advantage of a late maximum is that the activity of parasites is restricted as a result of lower temperatures.

The relative influence of sinking, dilution and grazing is difficult to assess. Uhlmann (1971) used a "model" experiment with sewage ponds and aquarium vessels. His equation is:

$$\frac{dN}{dt} = D \cdot N_0 + \beta \cdot N - D \cdot N - G \cdot N - B \cdot N$$

where N = cell numbers; N_0 = cell number of incoming water; β = growth constant; and D, G and B are dilution, grazing and sedimentation constants, respectively (per day). This equation assumes that all processes proceed at a rate which is proportional to cell numbers. In batch experiments simulating sewage ponds — where mineralisation of phytoplankton by zooplankton adds no extra nutrients to those which are already present in excess — Uhlmann found a value of G for *Chlorella* of up to 1.36 d^{-1}. Sinking rate B for *Scenedesmus* was only 0.06—0.6 d^{-1}.

In a few lakes periodicity of algae may be controlled by that in the river flowing into the lake. There are very few studies of periodicity of algae in rivers, presumably because, in addition to the factors already mentioned, two other factors are also relevant: distance from source or mouth and river discharge. A few detailed studies (Prowse and Talling, 1958; and Talling and Rzóska, 1967) have been made on the algae in reaches of the White Nile and the Blue Nile. Both rivers are affected by reservoirs where climatic factors (such as light and temperature) can have had little effect on periodicity as seasonal changes of light are small. Two dense populations, dominated by the diatom *Melosira granulata* and the blue-green algae *Anabaena flos aquae* developed, species composition being governed by the abundance of these algae in the riverwater entering the reservoir basin. *Melosira* usually reached its maximum density rapidly during and immediately following the filling of the reservoir basin. Some part of the increase may have been due to resuspension of live cells from the mud. The decline — in which two morphological forms behaved differently — was caused by depletion of inorganic nitrogen. A second maximum, often inversely related to that of *Melosira*, was caused by *Anabaena* frequently accompanied by populations of similar cell density of *Lyngbya limnetica*.

As the *Lyngbya* cells are much smaller than those of *Anabaena*, they contributed less to the biomass. *Anabaenopsis* was usually prominent during the final stages of the decline of *Anabaena*. The blue-green algae obtain their nitrogen from sources other than dissolved inorganic nitrogen. The release of storage water in March—April sometimes prevented later growth of dense phytoplankton. However, population decline had frequently begun before this release of water and this must have been caused by other factors.

REFERENCES

Berardi, G. and Tonolli, V., 1953. Clorofilla, fitoplancton e vicende meteorologische (Lago Maggiore). *Mem. Ist. Ital. Idrobiol.*, 7: 165—187.

Bush, R.M. and Welch, E.B., 1972. Plankton association and related factors in a hypertrophic lake. *Water Air Soil Poll.*, 1: 257—274.

Canter, H.M. and Lund, J.W.G., 1968. The importance of Protozoa in controlling the abundance of planktonic algae in lakes. *Proc. Linn. Soc., Lond.*, 179: 203—219.

Edmondson, W.T., 1965. Reproductive rate of planktonic rotifers as related to food and temperature in nature. *Ecol. Monogr.*, 35(1): 61—111.

Findenegg, I., 1943. Untersuchungen über die Ökologie und die Produktions-Verhältnisse des Planktons im Kärntner Seengebiet. *Int. Rev. Hydrobiol.*, 43: 366—429.

Findenegg, I., 1947. Über die Lichtansprüche planktischer Süsswasseralgen. *Sitzungsber. Akad. Wiss. Wien, Math.-Naturwiss. Kl. (Abt.1)*, 155: 159—171.

Findenegg, I., 1964. Produktionsbiologische Planktonuntersuchungen an Ostalpenseen. *Int. Rev. Ges. Hydrobiol. Hydrogr.*, 49(3): 381—416.

Findenegg, I., 1965. Limnologische Unterschiede zwischen den österreichischen und ostschweizerischen Alpenseen und ihre Auswirkung auf das Phytoplankton. *Vierteljahresschr. Naturforsch. Ges. Zürich*, 110(2): 289—300.

Findenegg, I., 1971. Die Produktionsleistungen einiger planktischer Algenarten in ihrem natürlichen Milieu. *Arch. Hydrobiol.*, 69(3): 273—293.

Findenegg, I., 1972. Das Phytoplankton des Reither Sees (Tirol, Österreich) im Jahre 1971. *Ber. Naturwiss.-Med. Ver. Innsbruck*, 59: 15—24.

Gardiner, A.C., 1941. Silicon and phosphorus as factors limiting development of diatoms. *J. Soc. Chem. Ind. Lond.*, 60: 73—78.

Gibson, C.E., Wood, R.B., Dickson, E.L. and Jewson, D.H., 1971. The succession of phytoplankton in L. Neagh 1968—70. *Mitt. Int. Ver. Theor. Angew. Limnol.*, 19: 146—160.

Goldman, Ch.R., 1972. The role of minor nutrients in limiting the productivity of aquatic ecosystems. *Limnol. Oceanogr., Spec. Symp. Vol.*, 1: 21—38.

Golterman, H.L., 1960. Studies on the cycle of elements in fresh water. *Acta Bot. Neerl.*, 9: 1—58.

Hardin, G., 1960. The competitive exclusion principle, an idea that took a century to be born has implications in ecology, economics, and genetics. *Science*, 131: 1292—1297.

Hughes, J.C. and Lund, J.W.G., 1962. The rate of growth of *Asterionella formosa* Hass. in relation to its ecology. *Arch. Mikrobiol.*, 42: 117—129.

Hutchinson, G.E., 1961. The paradox of the plankton. *Am. Nat.*, 95: 137—145.

Kalff, J., 1972. Net plankton and nanoplankton production and biomass in a north temperate zone lake. *Limnol. Oceanogr.*, 17(5): 712—720.

Kroes, H.W., 1971. Growth interactions between *Chlamydomonas globosa* Snow and *Chlorococcum ellipsoideum* Deason and Bold under different experimental conditions, with special attention to the role of pH. *Limnol. Oceanogr.*, 16: 869—879.

274

Kroes, H.W., 1972. Growth interactions between *Chlamydomonas globosa* Snow and *Chlorococcum ellipsoideum* Deason and Bold: The role of extracellular products. *Limnol. Oceanogr.*, 17: 423—432.

Lund, J.W.G., 1949. Studies on *Asterionella*. I. The origin and nature of the cells producing seasonal maxima. *J. Ecol.*, 37(2): 389—419.

Lund, J.W.G., 1950. Studies on *Asterionella formosa* Hass. II. Nutrient depletion and the spring maximum. *J. Ecol.*, 38: 1—14.

Lund, J.W.G., 1954. The seasonal cycle of the plankton diatom *Melosira italica* (Ehr.) Kütz. subsp. *subarctica* O. Müll. *J. Ecol.*, 42: 151—179.

Lund, J.W.G., 1955. Further observations on the seasonal cycle of *Melosira italica* (Ehr.) Kütz. subsp. *subarctica* O. Müll. *J. Ecol.*, 43: 90—102.

Lund, J.W.G., 1964. Primary production and periodicity of phytoplankton. *Verh. Int. Ver. Theor. Angew. Limnol.*, 15: 37—56.

Lund, J.W.G., 1966. The importance of turbulence in the periodicity of certain freshwater species of the genus *Melosira. Transl. Bot. Zhurnal S.S.S.R.*, 51: 176—187.

Lund, J.W.G., 1971a. An artificial alteration of the seasonal cycle of the plankton diatom *Melosira italica* subsp. *subarctica* in an English lake. *J. Ecol.*, 59: 521—533.

Lund, J.W.G., 1971b. The seasonal periodicity of three planktonic desmids in Windermere. *Mitt. Int. Ver. Theor. Angew. Limnol.*, 19: 3—25.

Lund, J.W.G., 1972. Eutrophication. *Proc. R. Soc. Lond, Ser. B*, 180: 371—382.

Lund, J.W.G., Mackereth, F.J.H. and Mortimer, C.H., 1963. Changes in depth and time of certain chemical and physical conditions and of the standing crop of *Asterionella formosa* Hass. in the North Basin of Windermere in 1947. *Philos. Trans. R. Soc., B*, 246: 255—290.

Manny, B.A., 1972. Seasonal changes in organic nitrogen content of net- and nanophytoplankton in two hardwater lakes. *Arch. Hydrobiol.*, 71(1): 103—123.

Moed, J.R., 1973. Effect of combined action of light and silicon depletion on *Asterionella formosa* Hass. *Verh. Int. Ver. Theor. Angew. Limnol.*, 18(prt. 3): 1367—1374.

Moed, J.R., Hoogveld, H. and Apeldoorn, W., 1975. Dominant diatoms in Tjeukemeer. II. Silica depletion. *Freshwater Biol.* (in prep.).

Pearsall, W.H., 1923. A theory of diatom periodicity. *J. Ecol.*, 9(2): 165—182.

Prowse, G.A. and Talling, J.F., 1958. The seasonal growth and succession of plankton algae in the White Nile. *Limnol. Oceanogr.*, 3(2): 222—237.

Reynolds, C.S., 1972. Growth, gas-vacuolation and buoyancy in a natural population of a blue-green alga. *Freshwater Biol.*, 2: 87—106.

Reynolds, C.S., 1973. Phytoplankton periodicity of some north Shropshire meres. *Br. Phycol. J.*, 8: 301—320.

Richerson, P., Armstrong, R. and Goldman, Ch.R., 1970. Contemporaneous disequilibrium, a new hypothesis to explain the "paradox of the plankton". *Proc. Natl. Acad. Sci. U.S.A.*, 67: 1710—1714.

Rodhe, W., 1948. Environmental requirements of freshwater plankton algae. Experimental studies in the ecology of phytoplankton. *Symp. Bot. Upps.*, 10(1): 149 pp.

Slobodkin, L.B., 1961. *Growth and Regulation of Animal Populations.* Holt, Rinehart and Winston, New York, N.Y., 184 pp.

Talling, J.F., 1955. The light relations of phytoplankton populations. *Verh. Int. Ver. Theor. Angew. Limnol.*, 12: 141.

Talling, J.F. and Rzóska, J., 1967. The development of plankton in relation to hydrological regime in the Blue Nile. *J. Ecol.*, 55: 637—662.

Uhlmann, D., 1971. Influence of dilution, sinking and grazing rate on phytoplankton populations of hyperfertilized ponds and micro-ecosystems. *Mitt. Int. Ver. Theor. Angew. Limnol.*, 19: 100—124.

West, G.S., 1909—1911. The algae of the Yan Yean Reservoir, Victoria. *J. Linn. Soc. (Bot.)*, 39.

CHAPTER 15

ENERGY AND MASS TRANSPORT THROUGH FOOD CHAINS

15.1. INTRODUCTION

A small fraction (less than 0.01) of the radiant energy incident on a lake is converted by the phytoplankton into chemical energy in the form of organic matter. Some of this energy will then be transferred to zooplankton, the so-called primary or herbivorous consumers. From here either directly or indirectly via secondary carnivorous consumers it may flow to small fish such as pope, smelt and others and then finally to the larger predators. Both primary producers (algae) and primary consumers may be utilised directly by plankton-feeding fish. Fig. 15.1 shows a widely occurring food web. Many examples are known which deviate from this scheme and, in some of these, allochthonous flora, fauna or detritus may be of some importance. In lake waters it is quite common for several food chains to co-exist. For example, some of the algae — living or dead — will be eaten by such benthic animals as *Chironomus* spp., oligochaete worms, and *Gammarus* spp. (the freshwater shrimp). At the other end of the web certain predatory fish will consume a variety of smaller fish species; pike, for example eat smelt, perch pope and char, whilst trout eat perch (see Fig. 15.2), minnows, eel, bullheads and may even be cannibal. Cannibalism is also common among perch.

It is unfortunate that there is no single unit in universal use to express growth rates and efficiency of energy transfer in food chains.

Solar energy is expressed in calories or Joules per unit area per unit time, or as milliWatts per unit area, while algal growth is expressed as the quantity of oxygen or electrons produced or of carbon (^{14}C method) consumed per unit area per unit time (see Chapter 4). Zooplankton growth is expressed mainly as fresh or dry weight per volume or per unit area per unit time (for example as gram per m^2 per week, month, or season), while fish production is expressed mostly as kg ha^{-1} yr^{-1} on fresh weight basis. Fortunately it is usual nowadays to express the results of fish analysis as carbohydrates, lipids and proteins and these can be used to calculate calories since 1 g of carbo-

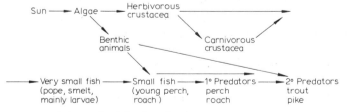

Fig. 15.1. Schematised food web.

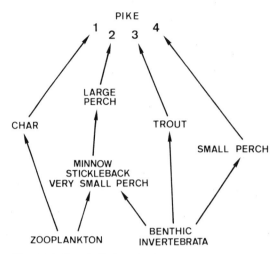

Fig. 15.2. Food chains of pike (Worthington, 1949).

hydrates and 1 g of proteins each yield 4 kcal., while 1 g of lipids yields 9 kcal. As a mean value for "live" material a figure of 5 kcal. per gram dry weight is often used, and since carbon is often 50% of the dry weight, 1 g of carbon is assumed to be equivalent to 10 kcal.

For the zooplankton the same units as those used for fish are applicable or alternatively the biomass may be estimated by a wet oxidation process which gives C.O.D. (chemical oxygen demand) — the amount of O_2 necessary for a complete oxidation (see Chapter 12). If the quantities are expressed as O_2, then comparison with the photosynthetic unit is easy (the principle in both cases is to measure the number of electrons required for the reduction or oxidation), while if photosynthesis is measured as carbon, the C.O.D. value may be converted to carbon by assuming 1 mole of O_2 to be equivalent to 1 mole of CO_2.

Thermodynamically speaking a food chain is a series of chemical processes in which, though energy is never lost, the conversion of energy in one compound to energy in another is never completely efficient. A food chain can therefore be described as a series of energy dilutors, the energy which fails to reach the next level being dissipated as heat or movement. Most of the energy contained in the food of animal consumers is used by them during the processes of food collection, digestion and other essential metabolic activity. The overall efficiency of a food chain is therefore rather low and depends mainly on the number of different participants and thus on the length of the food chain. Generally speaking the greater the complexity of the food web the greater is the efficiency of the system as a whole.

The greatest energy loss occurs during the first step: the conversion of solar energy into chemical energy by algae. Biologically available energy is the energy absorbable by the chlorophyll and other plant pigments (the range of

350—700 nm, which contains 50% of the total radiant energy). Rabinowitch (1951) has estimated that only 2% of this is used by plants, and therefore only 1% of the total energy is used (see Chapter 4).

Riley (1944) estimated 0.2% to be the efficiency of energy conversion in marine algae. In freshwater, efficiencies ranging between 0.1 and 1% are often reported. In Tjeukemeer for example the primary production is 2—3 g m^{-2} d^{-1} of C, or 400—600 g m^{-2} of C per growing season (or per year), which is equivalent to 4 000—6 000 kcal. m^{-2}. The total incident solar energy at this latitude is roughly 600 000 kcal. m^{-2} per season, so the efficiency is 0.6—1%. In artificially fertilised water bodies, higher efficiencies may be achieved. Values of 2% for *Chlorella* in ponds or fish ponds, and a few exceptionally high values of up to 4 or 5% have been reported. This efficiency is low when compared with the yield of sugar cane plantations (11 g m^{-2} d^{-1}), but production in aquatic environments usually has a relatively high protein content: about 50% of algal weight is high-quality protein (high quality here means a high concentration of essential amino acids). Even in eutrophic lakes only a minor fraction of the primary production may be converted into fish, because of low efficiencies in the energy conversion and long food chains. Data from Tjeukemeer illustrate this point.

In Tjeukemeer harvestable fish production is about 4 g m^{-2} yr^{-1} (or 40 kg ha^{-1}, fresh weight) which is equivalent to 2 kcal. m^{-2} or 0.05% of the gross algal energy production or 4% of the zooplankton production. The latter is estimated to be above 1 000 kg ha^{-1} or 100 g m^{-2} of fresh weight (= 5 g m^{-2} of C = 1% of gross primary production). The low fish production in Tjeukemeer is probably due to this low efficiency of zooplankton growth and to the rather long food chain. In other freshwater lakes yields of 200 kg ha^{-1} are often obtained under natural conditions. Fertilisation with inorganic nutrients to increase the primary production can increase fish yield to 2 000 kg ha^{-1}, while with cheap organic fertilisers (e.g. soya meal) yields of 5 000 kg ha^{-1} are reported in Israel. Another means of increasing the yield is by the introduction of herbivorous fish, such as *Tilapia nilotica*, a species which has been used extensively in Africa. *Tilapia* can digest larger-sized plants than most other fish are able to and can also consume organisms such as blue-green algae which are not easily digested (see section 15.5).

The main factors which determine the efficiency of food transfer are temperature, energy, and protein supply. Proteins form the major part of digestible algal and of animal cells, and animals depend on receiving an adequate supply with their food since they are unable to synthesise some (the so-called essential) amino acids. Energy was for a long time regarded as the most important factor when considering food chains. In the following paragraphs it will be shown that the supply of proteins is at least as important, and so also is animal behaviour. If an animal has to hunt for its prey, more energy is used than when the animal lies passively awaiting it. The migration of many fish and invertebrates between open water and reed beds also requires energy, which is thus lost in the food chain.

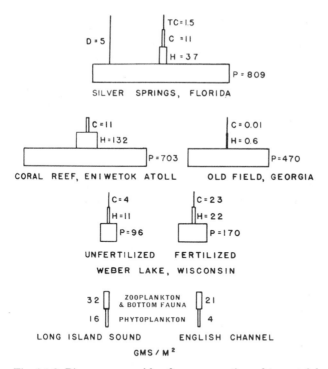

Fig. 15.3. Biomass pyramids of some aquatic and terrestrial ecosystems. Pyramids are drawn approximately to same scale, and figures are grams of dry biomass per square metre. P, producers; H, herbivores; C, carnivores; TC, top carnivores; and D, decomposers. (Data from Odum and Odum, 1960.)

Food chains used often to be depicted as pyramidal arrangements either of numbers or of standing crop (Fig. 15.3). In cases where both metabolism and size of the organisms are similar at all levels, the same overall relationship will be found. But when the producers are small in size and the top organisms (final consumers) are large, the pyramid for biomass may be inverted. The mass of bacteria in a fish pond, for example, will normally be much smaller than that of the fish, whereas the amount of bacterial metabolic turnover may equal that of the fish. It is important to note that the smaller the organism the larger is its metabolic rate (per unit mass). This might perhaps be due to the larger surface area/volume ratio of the smaller organisms, resulting in greater diffusion capacity and thus to a greater nutrient uptake.

Following ideas developed by Hutchinson, Lindeman (1942) emphasised the trophic dynamic viewpoint which describes an ecosystem on the basis of energy transport or available-energy relationships within the community unit. Although he was not the first person to suggest these ideas, Lindeman defined the trophic level (Λ) in a food chain as the number of steps between any organism on that level and the solar energy, which itself is called level Λ_0.

Algae are therefore at level Λ_1 and primary consumers at level Λ_2, etc. The system does not exclude the possibility that a given species may be at two levels, e.g. a young herbivorous stage and an adult carnivorous stage. Omnivorous animals are less easily placed, but these practical difficulties do not invalidate the concept.

The rate of change in the energy content, Λ_n, may therefore be divided into a positive and a negative component:

$$\frac{d\Lambda_n}{dt} = \lambda_n + \lambda_n' \qquad (15.1)$$

where λ_n is by definition positive and represents the rate of contribution of energy from Λ_{n-1} to Λ_n, while λ_n' is negative and is minus the sum of the rate of energy dissipation from Λ_n and of the rate of energy transmission to the next level Λ_{n+1}.

Lindeman pointed out that the amount of energy lost by respiration at different levels varies considerably according to differences in food level, temperature, and to the particular stage reached in the life cycle of the various organisms. He estimated 33% as being probably the best available respiratory loss coefficient:

$$\frac{\text{respiration}}{\text{growth}} \times 100\%$$

for lacustrine producers, a figure which nowadays seems rather low. For primary consumers, such as *Tubifex*, Lindeman cited a value of 60% for the respiratory loss coefficient, and for the aquatic predators (predatory yearling fishes) he gave values of 120—140%. More recent data are given in Table 15.2 and in subsection 15.3.4. Lindeman related the higher coefficient at the higher food-chain levels to the activity of organisms whilst obtaining their food and discussed the principle of predation and decomposition correction. These two processes, both of which lead to unavoidable energy losses, may enter the food cycle at lower levels. He also defined the progressive efficiency of a trophic level as:

$$\frac{\lambda_n}{\lambda_{n-1}} \cdot 100\%$$

From the two examples which Lindeman gave, he concluded that the progressive efficiency increases with the number of the trophic level. Although this would seem to contradict the previous generalisation (that of increasing respiratory loss), it need not necessarily do so. The amount of energy which an animal needs to catch its prey may increase, but once the prey has been caught the efficiency of its utilisation may be high. Energy taken in will in general be used for assimilation plus production of progeny, while a part of it will not be used (faeces). Thus:

intake = growth + respiration + reproduction + egestion

15.2. THE ORGANISMS IN THE FOOD CHAIN BETWEEN FISH AND PHYTOPLANK-
TON

15.2.1. Introduction

The details of animals given here are included principally for students of biochemistry, interested in food chains. For detailed taxonomic information the following books are recommended: Hutchinson (1967), Dussart (1966, 1967, 1969). In general the animals may be either benthic organisms, i.e. those living in sediments and on plants (e.g. *Gammarus* and *Chironomus*), or organisms of the open water, the zooplankton.

The freshwater shrimp *Gammarus* (Crustacea) is often found living amongst littoral vegetation such as reeds, sedges, *Potamogeton* spp., etc. It feeds on dead leaves, epiphytic diatoms and probably on dead animals as well. It can be up to 20 mm long, and may produce several generations per year. Chirono-mid larvae occur in great quantities in sediments of the open water and in the reed-bed mud. Population densities of 10 000 per m^2 are sometimes found. Both of these organisms are important in the diet of many kinds of fish, the sheltered reed-bed populations often being especially important for the young fish.

In the open-water zooplankton, many members of the crustacean sub-orders Copepoda and Cladocera are found. Classification of these groups of animals is shown in Table 15.1.

15.2.2. Copepoda

Two important genera of Copepoda are commonly found both in tropical and in temperate freshwater lakes — *Diaptomus* (Calanoida) and *Cyclops* (Cyclopidae). *Diaptomus* is a typical so-called filter feeder. A current of water is pumped towards the mouth by movements of the antennae; particles are then filtered out from the water as it passes through the body at considerable speed. *Cyclops* on the other hand uses jerky movements of its antennae to collect its food actively. Females of both genera carry their eggs in one or more egg sacs, attached to the posterior part of their bodies. Two kinds of eggs are laid by some species; a parthenogenetic one which contains a female but no male gamete, and whose development is completed in 10—14 days; and a gamogenetic one, which contains the usual female gamete ferti-lised by a male gamete and is a resting type capable of surviving an unfavour-able period. The latter kind are called winter or "dauer" eggs. Male copepods do sometimes occur. These are smaller than the females and seem to be un-necessary during the growing season. They are necessary though to fertilise the gamogenetic eggs. The egg hatches into a nauplius larva. Up to six separate nauplius stages may occur followed by one or perhaps several copepodite stages before the adult form is finally achieved.

TABLE 15.1

Classification of some freshwater animals

Phylum	Class	Subclass	Order	Suborder or super-family	Family	Genus
Rotatoria						*Keratella, Asplanchna,* etc.
Arthropoda	Crustacea	Malacostraca	Peracarida	Amphipoda	Gammaridae	*Gammarus*
		Branchiopoda	Diplostraca	Cladocera	(*Daphnia*)	*Daphnia,* etc.
				Copepoda	Calanoida, etc.	*Diaptomus, Cyclops,* etc.
	Insecta		Diptera	Nematocera Calitoidea	Chironomidae	*Chironomus, Chaoborus*

15.2.3. Cladocera (often called waterfleas)

Commonly occurring genera of Cladocera in freshwater lakes are: *Daphnia* (*D. magna, D. pulex,* and *D. obtusa*), *Ceriodaphnia, Bosmina, Polyphemus* and *Leptodora.* The species *Sida crystallina* and *Diaphanosoma brachyurum* also occur in quantity. Just as in the Copepoda, some Cladocera species are filter-feeding and some hunt actively. Their movement through the water consists of a series of jerks or hops, a phenomenon which gives rise to their popular name of waterfleas. They swim mainly by moving their antennae. If these movements cease even briefly, the animals will sink. Like the Copepoda, the Cladocera produce eggs throughout both summer and winter if conditions are favourable, these eggs being parthenogenetic (requiring no male gamete for fertilisation). When unfavourable (winter) conditions occur some of the eggs hatch into males and these males subsequently fertilise small numbers of specially produced eggs. These eggs hatch into winter eggs. Only the winter eggs of *Leptodora* give rise to nauplius larvae. Cladocera vary quite considerably in size: *Bosmina* belongs to the smaller zooplankton organisms (ca. 1 mm) but *Leptodora* may reach a length of 12 mm. Large differences in size are also found in the genus *Daphnia. D. magna* can reach 5.5 mm, while *D. hyalina* normally does not exceed 2 mm.

15.2.4. Rotatoria

A very large number of species of the rotifers occur in freshwater. Some of the main genera found are *Asplanchna, Synchaeta, Keratella, Kellicottia* and *Brachionus.* During most of the breeding period, only females occur, producing parthenogenetic unfertilised eggs. In the autumn smaller eggs are produced from which males hatch. These males fertilise the winter eggs; a parallel with the Cladocera. The males are little more than swimming sacs of sperms; in some species no males are found at all. It is therefore not surprising that in productivity studies males are very seldom mentioned; they appear to have a much lower rate of metabolism than the females. The study of the feeding of rotatoria has been very much neglected. In subsection 15.3.3 a study of their feeding is discussed.

15.3. STUDY OF ZOOPLANKTON IN THE FOOD CHAIN

A study of the growth of zooplankton populations (which includes food interrelationships, life cycles, behaviour and population dynamics) presents many more difficulties than those which occur with phytoplankton work, where all species effect much the same process, i.e. the reduction of CO_2 to carbohydrates, proteins and lipids. Furthermore, the amount of sampling required to establish the population density is tedious since animals are often not distributed randomly but tend to cluster together. The occurrence of groups of animals in some samples from a given place may require special statistical treatment. Studies of zooplankton are therefore still at the pioneer stage; the few papers to be discussed in this chapter are selected mainly to indicate the lines along which current research is developing. Feeding and population dynamics will be treated separately and data given here are not presented chronologically.

15.3.1. Feeding of zooplankton

One of the most detailed studies of feeding is described in a series of papers dealing with a marine copepod, *Calanus* (Corner, 1961; Cowey and Corner, 1963; Corner and Cowey, 1964; Corner et al., 1965; Butler et al., 1970). Much of the success of this work is due to the fact that the animal was studied in a semi-continuous culture, thereby avoiding many of the disadvantages such as accumulation of waste products and changing food concentrations which are inherent in a flask culture. The animals used in these experiments were freshly collected and were cultured in Plymouth seawater. The experiments were performed in two flasks, one with 96 and one with 48 animals, with an average volume of 12 ml of seawater for each animal. Cultures were kept at 10°C in darkness, and 100 ml of seawater per animal was flushed through the flask during the experimental period of four days; use of longer periods would have caused problems due to bacterial growth, while shorter periods would have given less reliable results. The difference between the flasks containing 96 and those containing 48 animals is presumably entirely due to the effect of the additional 48 animals, since the flasks received identical treatment. Faeces were collected separately and were measured when necessary. Food was collected on membrane filters from at least 20 l and analysed. In the first series of experiments the amount of particulate matter before and after passage through the culture was measured and the inorganic and the organic portions were estimated separately. The results were (in mg):

	Inorganic	Organic	Total
before	224	281	505
after	215	246	461
	9	35	44

Of the organic portion (mostly diatoms) 13% was ingested, but only 4.5% of the inorganic part was ingested. This suggests that the organic part is preferred and that feeding may be more selective than is commonly supposed. A filtering rate of 10—36 ml per animal per day can be calculated from these results which is an impressive achievement for an animal with a dry weight of only about 0.1 mg. It is not surprising that the filtering rate varies over quite a large range.

Females eat more than males, and different filtering rates are found for different kinds of food. Thus, *Calanus* filters off the diatom *Ditylum bright-wellii* at a greater rate than the flagellate *Chlamydomonas*. In these experiments the rate of assimilation of the food particles was measured, just as was the constitution of the food (carbohydrates 70%, lipids 20—25% and proteins 8—12%), of which 80% was assimilated. The necessary O_2 consumption could therefore be calculated:

	Uptake per day per animal
carbohydrates	18 μg = 13 μl O_2
lipids	6.5 μg = 13 μl O_2
proteins	2.7 μg

(proteins were not considered to be a fuel and therefore were not included in the O_2 balance).

The oxygen uptake per animal shows seasonal variation. For animals collected in April a value of around 90 μl of O_2 per mg dry weight per day was found, but this fell to about 25 μl for animals collected in January. As the measurements were all made at 8°C, these results may indicate an adaptation to low winter temperatures. Most respiration values fell within the range of 3—13 μl per animal per day. The value 13 μl is low compared with 26 μl calculated from the assimilated food, but this discrepancy was partly explained by the authors in terms of growth or of storage in the animal. In any case this discrepancy indicated that the particulate food intake provided more than sufficient substrate to account for the O_2 uptake of these animals. This strongly suggests that Pütters hypothesis — that the zooplankton use dissolved organic matter — does not hold for *Calanus*. Corner et al. (1965) also suggested that Pütter used erroneous figures for his calculations.

The amino-acid composition of *Calanus* in the seawater near Plymouth was fairly constant during both summer and winter. This was not the case for the amino-acid composition of the particulate matter which fluctuated considerably closely following that of the diatom *Skeletonema*. Although *Calanus* is thought not to use proteins as fuel, not all the ingested protein will be used for production of new protein. Some will be used for natural replacement processes which occur in the metabolism of any animal. *Calanus* excretes the nitrogen from the metabolised proteins as ammonia. If any animal metabolises proteins, nitrogen will be excreted. Most mammals excrete urea whilst certain fishes form substances such as methylamines. *Calanus* excretes considerable quantities of ammonia, the amount varying according to the different forms and stages in larval development, and also to other factors. Thus the copepodite stages II, III and IV produced 21 μg of NH_3-N, while stage V and adult females produced 7 and 10 μg of NH_3-N (all per mg dry weight per day). The constitution of the food is also an important factor. Females grazing on cultures of *Brachiomonas*, *Cricosphaera* and *Skeletonema* (with nearly the same particulate —N content) produced respectively 12.4, 9.0 and 6.4 μg of NH_3-N per mg per day, but these values increased to 14.6, 11.4 and 19.7 if the particulate —N content of the food was increased (by increasing the cell density) to ten times the original value. Apparently the proteins were not equally digestible; a feature which may be connected with the variable individual amino-acid content of the different algae.

A study of winter conditions using starvation experiments was made by Corner et al. (1965). Starving winter and starving summer animals lost amino acids to the extent of 1.8 and 2.2% of their body weight per day. A filtering rate of 30 ml (in winter) and 50 ml (in summer) per animal per day must be sustained in order to replenish this daily loss of the essential amino acids. Ammonia-nitrogen excretion doubled between experimental temperatures of 5° and 15°C, but rose only slightly further at 25°C. From the mean value of 13.5 μl of O_2 consumed and the amount of N excreted, the authors calculated a gross efficiency (amount of food ingested and converted into animal weight) of 36% and a net efficiency (assimilated food as % of ingested food, i.e. ingested minus faeces) of 60%. A plentiful food supply increased the excretion rate both for freshly caught well-fed animals and for starving animals. Winter and summer excretion differed not only in nitrogen content but also in phosphate content.

	Winter	Summer
% body N excreted daily	2.6 (s.d. = 0.9)[1]	4.6 (s.d. = 1.5)
% body P excreted daily	6.5 (s.d. = 2.3)	10.3 (s.d. = 4.1)[2]
N/P excreted	6.6	5.0

After normally maintained animals had been fed, the N-excretion rate decreased rapidly within 20 h for instance from nearly 20 to 5 μg of N per mg dry weight per day. When the animals had been fed previously on an inadequate (semi-starvation) diet, the initial N excretion was only 8 μg, which decreased to 4 μg, within the 20-h experimental period.

Considerable amounts of assimilable nitrogen and phosphate are necessary each day to ensure growth or even survival of the animals. The efficiency of assimilation may be as high as 62% for N or even 77% for P. The filtering rates needed (if daily requirements of N were to be provided) were 29 ml per day in spring and 47 ml per female and 49 ml per copepodite stage V in winter, the winter values being higher than the summer values because of lower phytoplankton density. For P the calculated values were 58 ml per female and 37 ml for copepodite stage V (summer and winter values). These values are comparable with those already given for the essential amino acids.

The distribution of ingested nitrogen and phosphate between the fractions that were excreted or were removed with faecal pellets or were incorporated in body constituents (growth) in the experiments with *Calanus* was:

	Excreted in soluble form	Faecal pellet	Growth
N	35.7%	37.5%	26.8%
P	60 %	23 %	17 %

[1] s.d. = standard deviation; [2] may be as high as 28%.

Thus the amount of nitrogen and phosphate excretion is quantitatively significant and is of considerable importance as a source of nutrient supply for the phytoplankton. Ketchum (1962) using data from Harris (1959) calculated that in three cases 77%, 66% and 43% of the N needed for algal growth was provided by the zooplankton. Phytoplankton growth is therefore both limited by grazing by the zooplankton, and yet at the same time stimulated by the release of nitrogen and phosphate compounds which themselves may be present in growth-limiting amounts.

Cowey and Corner (1963) recognised that in long-term starvation experiments protein catabolism may be accelerated because the carbon skeletons of the amino acids are used as sources of energy. Normally energy would be supplied by fats and carbohydrates in the food. Nevertheless as shown above increasing food concentrations increased the rate of N excretion. Harris (1959) also found rather high rates of N excretion occurring even in normal conditions. Thus it seems likely that there is a very rapid turnover of proteins. Therefore not only energy supply but also amino-acid supply should be considered as a regulating factor in food chains.

The resemblance between N excretion by zooplankton and bacterial mineralisation is striking. In the case of bacteria a larger part of the algal matter is used for their growth; due to the high rate of bacterial metabolism, a large part of the nitrogen is released. Thus although their mechanisms are somewhat similar, the two mineralisation processes are obviously competitive. Nothing is yet known concerning those factors which control this competition, but a possible situation could be that bacteria mineralise that part of the phytoplankton which is not eaten by the zooplankton, either because the algae are produced at the wrong moment, or are too large or have the wrong taste (blue-green algae). This recycling of nutrients is as important for understanding ecosystem mechanisms as is the energy flow. In addition to the study of natural animal populations, feeding experiments using one species with a well-defined food supply are desirable. Such experiments would probably yield food-balance data which could give greater insight into the processes involved.

The experiments of Ryther (1954) will therefore now be discussed. He used *Daphnia* as a zooplankton organism feeding on cultures of *Chorella* and of other algae. Twenty-five *Daphnia* were placed in 100 ml of algal suspension for one hour. Algae were counted at the beginning and end of the experiment. Assuming that the animals filter at a constant rate, the reduction in number of algae can be described using the equation:

$$N_t = N_0 e^{-kt}$$

where N_t = cell population density at time t; N_0 = cell population density at the beginning; and k = "decay" constant. This is the growth equation (see Chapter 10) again, but applied here to a decreasing population. One could use the growth equation of Chapter 10 and give the growth constant β a negative value, but it is preferable to show that a decreasing population is involved by putting the negative sign in the equation, and by replacing β by k. The numeric value of k is then positive.

If v is the volume of water per individual *Daphnia* (4 ml in this case) the filtering rate F ($= \text{ml animal}^{-1} \, t^{-1}$) is:

$$F = vk = v \, \frac{\log N_0 - \log N_t}{0.434 \, t} \tag{15.2}$$

No correction was made for growth during the experimental time. Feeding rate is the number of cells taken up and may be calculated thus:

$$f = F \cdot \frac{N_t + N_0}{2} \text{ or } F \cdot \sqrt{N_0 N_t}$$

Using a value for N_0 of $0.2 \cdot 10^6$ cells per ml, Ryther found the food consumption per mg dry weight of animal decreased linearly with the size of the *Daphnia* (expressed as dry weight). If food consumption were expressed per animal, it was shown that a maximum amount was used in animals with a dry weight of 0.11—0.12 mg (2.7—2.8 mm in length). In Ryther's filter experiments, for which only the size class 2.5—2.9 mm was used, the results were expressed as food uptake per animal. Food concentrations also influenced the feeding rate. With increasing *Chlorella* concentrations a decrease in the filtration rate was found, both with starved animals and with animals adequately fed beforehand. From Ryther's table 1 it can be seen, however, that although the filtration rate decreased, the feeding rate increased with degree of prior nutrition although not linearly. Ryther assumed that even with increasing food concentrations the filtering rate remained constant, and therefore suggested that an inhibitory substance, chlorellin, might be responsible. It seems much more likely however that factors such as pH and O_2 concentration, which were not measured, and which will change drastically due to the photosynthesis of the algal cells, caused the inhibitory effects. Oxygen is a well-known inhibitory substance and might reach an oversaturation of up to 300% in this kind of experiment. O'Brien and De Noyelles (1972) described mortality in nutrient-enriched ponds which was caused by photosynthetically elevated pH values. They suggested that such factors are more often the cause of animal mortality than toxins. This may also be the cause of low feeding rates on senescent cells. Similarly it would explain why in experiments with *Scenedesmus* and *Navicula* as food source a decrease was found with increasing algal cell concentrations. *Daphnia* fed most rapidly on *Chlorella* (cell size 3.5 μm^3), less rapidly on *Navicula* (9.2 μm^3) and still more slowly on *Scenedesmus* (25.5 μm^3). Reduction in the number of cells ingested as food may have been compensated by the larger size of the cells.

During the same period as that during which the work of Corner and others was being carried out at Plymouth, Rigler and coworkers studied the mechanisms of food uptake in *Daphnia* (Rigler, 1961; McMahon and Rigler, 1963, 1965; McMahon, 1965; and Burns and Rigler, 1967).

Rigler fed the yeast *Saccharomyces cerevisiae* labelled with ^{32}P to *Daphnia magna* and found that below 10^5 cells per ml the feeding rate of *D. magna* was proportional to the concentration of food. Rigler defined "feeding rate" as the number of cells consumed by a zooplankton animal. Thus feeding rate (cells per animal per hour) equals filtering rate (ml per hour) times concentration of food (cells per ml). If feeding rate is proportional to the concentration of food, then this is the same as saying that the filtering rate is constant and independent of the food concentration.

Rigler pointed out that Ryther's interpretation of his results is not the only possible one and suggested that when filter-feeding crustaceans encounter low concentrations of food the feeding rate is limited by the filtration rate, but that above a critical concentration of food the feeding rate becomes constant and is determined by the ability of the animal to ingest or digest the food cells. Rigler discussed critically the possible errors inherent in this kind of work (e.g. feeding time, leaching or excretion of isotopes). He found, as Conover (1962) did with *Calanus*, that feeding rate measured by providing radioactive food gives a value which is only a small fraction of the rate determined by measuring decrease of algal cells. Rigler suggested that a decreased feeding rate could perhaps be due to a rejection of the collected food or to a reduced filtration rate and concluded tentatively that the latter hypothesis was more likely to be correct.

The great mobility of ^{32}P was not perhaps sufficiently appreciated at that time, but it is now known that ^{32}P may exchange quite rapidly and spontaneously by physical rather than biological processes.

McMahon and Rigler (1963) showed that when an animal was transferred from a *Chlorella* suspension below the critical food concentration level to one above it, the animal at first ingested food at an abnormally high rate but after a while the collection rate decreased and rejection began. Behaviour of the animals was studied microscopically in an observation chamber in which *Daphnia* was fastened with stopcock grease. The same authors (1965) measured the feeding rate of *Daphnia magna* by determining the radioactivity of animals fed in labelled pure cultures of *Escherichia coli, Saccharomyces cerevisiae, Chlorella vulgaris* or *Tetrahymena pyriformis*, and confirmed the existence of a critical food level. They also showed that the maximum volume of the various foods eaten per unit time is not the same and is probably determined more by digestibility than by the size of the food cells. Filtering efficiency of *Daphnia magna* was independent of the size of the food between 0.8 μm^3 and $1.8 \cdot 10^4$ μm^3. *Chlorella* cells were not observed to inhibit feeding when in log-phase growth but senescent cells caused *Daphnia magna* to decrease its filtering rate.

Burns and Rigler (1967) measured the filtering rate of *D. rosea* in natural lakewater and in a pure culture of a yeast by adding ^{32}P-labelled yeast cells to both media. Filtering rate increased with increasing body length of the *Daphnia* and with increasing temperatures up to 20°C, above which it declined again. Furthermore they found that in natural lakewater (Hear Lake, Canada) the filtering rate was less than maximal between June and November and the values found were well below the maximum possible for their body length and the water temperature.

D. *rosea* behaved differently from D. *magna* in that its filtering rate was constant only below a concentration of $0.25 \cdot 10^5$ cells per ml; above this level the filtering rate decreased gradually. In D. *magna* the filtering rate was constant at all food levels.

Richman (1958) measured an energy balance in cultures of *Daphnia pulex* feeding on the green alga *Chlamydomonas reinhardtii*. He used the relationship:

intake = $\underbrace{\text{growth + respiration + reproduction}}_{\text{assimilation}}$ + egestion

The intake was measured by counting the number of algae taken up and converting it into calories using an average calorific value of 5 249 cal. g^{-1}, which was equivalent to $1.3 \cdot 10^{-6}$ cal. per cell. The calorific value per gram was rather close to the mean value derived from various sources in the current literature, this being 5 340 cal. g^{-1}. The filtering rate at 20°C of 0.7-, 1.3- and 1.8-mm *Daphnia* at 4 food concentrations (25 000—100 000 cells ml^{-1}) was found to be independent of the algal concentrations. Filtering rate increased with *Daphnia* size, from 0.90 to 5.15 ml per animal per day. (Richman, unlike Rigler, regarded filtering rate as being equal to feeding rate.) Respiration was measured as O_2 consumption and growth was estimated from the body weight—body length regression curve.

Respiration could be expressed as:

$$\log O_2 = \log 0.0014 + 0.881 \log W \qquad (15.4)$$

where W = body weight (mg per animal), and O_2 = O_2 uptake in μl of O_2 per animal per hour. The equivalent calorific value of 1 μl was taken to be 0.005 cal. No attempts were made to correct for protein respiration, and in fact the RQ found (1.03) suggested that carbohydrate was the main respiration substrate. Growth was measured as the calorific value of the newly generated (or released) animals.

Egestion was measured as energy consumed minus that used in growth and respiration. Energy consumed increased with increasing food levels (constant filtering rate) and age of animals, while egestion increased with increasing uptake (Table 15.2). It can be seen from this table that due to an increase in the egestion component, assimilation and growth showed a relative decrease as measured on a percentage basis with increasing food levels. After 6 days 4—13% and after 34 days only 0.4—1.25% of the food energy supplied was used for growth.

As the production of young progeny increased, the percentage used for growth decreased. The percentage of energy assimilated that was used for

TABLE 15.2

Food uptake, assimilation, growth, and production of young in percentage of food uptake of *Daphnia pulex* (from Richman, 1958)

	Food level: Cells per ml			
	25 000	50 000	75 000	100 000
0—6 days old				
Uptake (energy consumed in cal.)	0.47	0.58	1.39	1.91
Assimilation (% energy consumed)	23.9	15.8	8.4	6.6
Growth (% energy assimilated)	13.2	9.1	4.8	3.9
6—34 days old				
Uptake (energy consumed in cal.)	5.67	13.0	19.3	27.3
Assimilation (% energy consumed)	31.7	20.2	16.8	14.2
Growth (% energy consumed)	1.25	0.78	0.57	0.43
Growth (% energy assimilated)	3.95	3.85	3.41	2.99
Young (% energy consumed)	16.5	12.5	10.9	10.0
Young (% energy assimilated)	52.0	61.8	66.8	70.5
Average number of young produced	*76.3*	*132*	*177*	*223*
0—40 days old				
Uptake (energy consumed in cal.)	6.1	13.6	20.7	29.2
Assimilation (% energy consumed)	31	20	16	14
Growth + young (% energy consumed)	17.4	13.1	11.4	10.0
Growth + young (% energy assimilated)	56	65	70	73
Respiration (% of growth + young)	78.6	53	43	37

growth was about 56% after 6 days, and 3—4% after 34 days, while the amount of energy assimilated that was used for production of young progeny was 52—70% after 34 days. Since growth after 40 days remained the same as that after 34 days, it can be seen that the major part of the assimilated energy during that period was incorporated in the progeny and that the production of growth plus young expressed as percentage of growth decreased with increasing food levels.

Richman (1964) used lakewater from which netplankton had been removed and found that *Diaptomus* showed much lower filtering rates with this both in situ (0.067 ml d^{-1}) and in the laboratory (0.12 ml d^{-1}). He did not discuss possible reasons for the differences found between these and previous measurements. It may be that the rather high standing crop of phytoplankton was a contributory cause, but there is no direct evidence to support this hypothesis. In later work Richman (1966) fed *Diaptomus oregonensis* with two species of green alga labelled with ^{14}C and measured both filtering and feeding rates. The results (see Figs. 15.4 and 15.5) show that maximum filtering rate was maintained while food levels increased up to a certain cell concentration ($40 \cdot 10^3$ cells per ml) and that feeding rate increased to the

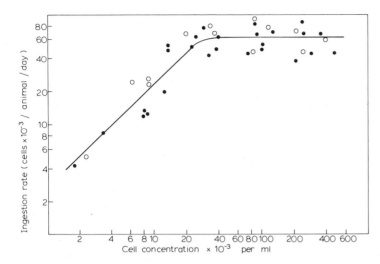

Fig. 15.4. The ingestion rate of *Diaptomus oregonensis* in relation to cell concentration. Both axes logarithmic; •, *Chlamydomonas*; ○, *Chlorella*. (From Richman, 1966.)

same extent up to the same food concentration. Filtering rates were 1.4—2.5 ml per animal per day with *Chlamydomonas* and 0.3—1.4 ml per animal per day with *Chlorella*, and these are in good agreement with other reported values.

The use of the ^{14}C isotope as a tool in the study of nutrition of aquatic animals is elaborately described in a manual by Sorokin (1968).

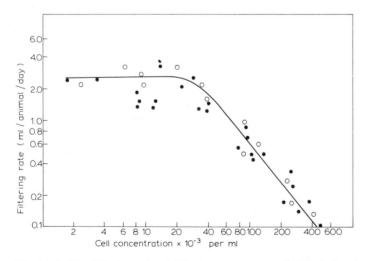

Fig. 15.5. The filtering rate of *Diaptomus oregonensis* in relation to cell concentration. Both axes logarithmic; •, *Chlamydomonas*; ○, *Chlorella*. (From Richman, 1966.)

Arnold (1971) compared the food value of several species of blue-green algae (*Anacystis nidulans, Synechococcus elongata, S. cedrorum, Merismopedia* sp., *Anabaena flos aquae, Synechocystis* sp. and *Flococapsa alpicola*) with that of green algae when used as a diet for *Daphnia pulex*. The toxicity of several genera of blue-green algae to higher animals is well established, but few reports of their toxic effects on aquatic invertebrates exist. It has been frequently reported that zooplankton stop grazing if fed with blue-green algae such as *Microcystis aeruginosa* and *Anabaena*. In all the cases which Arnold studied blue-green algae provided less food for *Daphnia* than green algae (such as *Chlorella, Ankistrodesmus* and *Chlamydomonas*). Reproduction rate of *Daphnia* fed on blue-green algae was always lower than of *Daphnia* fed on green algae. There were large differences between individual species of blue-green algae in their effects on ingestion, assimilation and reproduction rate of *Daphnia*. No agreement was found between ingestion plus assimilation values and reproduction. This might be explained by the toxicity of the blue-green algae. Some of these did in fact show some apparently toxic effects on *Daphnia* (*Anacystis nidulans, Merismopedia* and *Synechocystis*). *Synechococcus cedrorum* caused the highest survival rate of *Daphnia*, while *Synechococcus elongata* caused no reproduction at all, mainly because it was not ingested. Ingestion was high, but assimilation low with *Anacystis nidulans*.

Some of the effects found by Arnold (1971) may however be attributed to the high densities of the blue-green algal cultures he used, but it may be tentatively concluded that in natural conditions blue-green algae will probably be less grazed than green algae. This may be one of the reasons why blue-green algae form blooms. Once the bloom develops, the high concentration of organisms may further reduce the removal rate of blue-green algae by zooplankton. No evidence suggesting the occurrence of selective filter-feeding from mixed populations of green and blue-green algae seems to be available at present.

15.3.2. Field studies (population dynamic approach)

The growth of a population depends not only on its food supply but also on daily birth and death rates (rate of natality and mortality, both natural and by predation). The total process is described by the unsuitable word "production". Because of both vertical and horizontal migration the numbers of animals in natural populations in the field are difficult to estimate and only a few fairly complete studies have been made.

The following definitions are used in this discussion:

B = biomass of population = weight per animal times number of animals per m^2 or m^3. When given as weight per m^2 it is understood that a vertical column through the whole lake depth is intended.

P = net production of biomass per unit time per m^2 or m^3.

From these parameters the following can be derived:

P/B = production per biomass unit in a given period, T. It is a measure of the intensity of production. The period chosen may be day, month or growing season.

t = turnover time, the average time to replace the whole population, by elimination or growth or both. This is the average life span of individuals or the average development time for juveniles under natural conditions. Because of predation t is shorter than the physiological life.

If P/B of a given animal is estimated — either in the laboratory or in some lakes — a first approach towards measuring the productivity of many other lakes may be made by measuring the average biomass $\overline{B}$ and calculating P as:

$$P = P/B \cdot \overline{B}$$

Furthermore:

$$t = \frac{T}{P/B} = \frac{T \cdot B}{P} \tag{15.5}$$

From this equation we can see that the turnover time is inversely proportional to P and increases linearly with B.

Estimation of zooplankton production can be made using two basic criteria: (1) measurement of t and B (15.3.3); and (2) measurement of daily increase in number and weight of individuals and juvenile stages (15.3.4).

15.3.3. Measurement of t and B

Stross et al. (1961) studied the turnover time and production of planktonic Crustacea at two sites in a bog lake (Wisconsin, U.S.A.). To one of these lime had been added, the other was untreated.

They used what was essentially eq. 15.5, but with different definitions and measured turnover time of individuals $\overline{T}_n$. They then calculated the production (within time span T) in weight units:

$$P_{(T)} = \frac{\overline{B}_{(T)} \cdot T}{T_n} \tag{15.6}$$

Population changes were calculated using the equation:

$$\Delta N_i = \beta_{i-1} N_{i-1} - \alpha N_i + \beta_i N_i \tag{15.7}$$

where N_i = the number in the ith age category at the beginning of the experimental time Δt; α = a coefficient of mortality for that category, considered to be characteristic of Δt; and β_{i-1} = a coefficient of recruitment (or growth) for the $i-1$st category, also considered to be characteristic of Δt. Thus each age category recruits individuals as they grow into the ith from the preceding (smaller) category and loses them either through death or growth into the succeeding (larger) category.

They found a turnover time of 2.1 weeks for *Daphnia longispina*, which inhabited the lime-treated lake site and 4.6 weeks for *Daphnia pulex* in the control area. Slightly more crustacean biomass was produced in the treated site (8.7 g m^{-2}) than in the untreated (7.2 g m^{-2}) presumably, at least partly, because of the more rapid turnover. Biomass in the treated site was roughly half that of the control. After the lime treatment was stopped, there was a change in the age distribution of the *Daphnia* population and a brief reappearance of the crustacean *Holopedium gibberum*, intolerant of lime. Whether or not the concentrations of calcium, magnesium or bicarbonate changed is not known.

This work also illustrates well the heterogeneous distribution of certain species of Crustacea and their mutual influence. The population of *Cyclops* was distributed near the surface in May. A short time later it was dispersed throughout the aerobic layer, but when *Daphnia* and *Diaptomus* became abundant, almost the entire population of *Cyclops* was found to be limited to a 2 m deep zone in the thermocline, the other two genera being dispersed above it. Stross et al. suggested competitive exclusion as a possible mechanism which could produce these effects. Different species became progressively dominant. After the ice had melted (late April) copepods were dominant and occupied the warmer water near the surface. In June *Holopedium* became numerically dominant in the untreated portion until July when the numbers of *Daphnia pulex* increased and those of *Holopedium* decreased. In the treated portion *Diaptomus* became more numerous than *Cyclops* and persisted until July when it was replaced by *Daphnia longispina*. *Diaptomus* did not decline when *Daphnia* became dominant and nor did *Holopedium* in the treated part of the lake. Both species of *Daphnia* continued to dominate during the summer and early autumn.

Hall (1964) measured a turnover time of four days during July and early August for *Daphnia galeata mendotae* in Base Line Lake (Michigan, U.S.A.). The slightly higher average temperature in this lake than in the situations just described would be expected to cause a shorter population turnover time than that measured by Stross et al. In laboratory conditions, *Daphnia galeata mendotae* has a maximum rate of increase of $r = 0.51$ at 25°C ($r =$ daily growth rate, see subsection 15.4.2). Assuming that predation does not change the age distribution, Hall calculated a theoretical turnover time for *Daphnia* of slightly less than 1.5 days. The observed turnover rate of once every 4 days indicates that practically all the biomass produced by this population is passed on rapidly to the next trophic level: the animals are rapidly eaten.

The numbers of animals per litre are given in Fig. 15.6 (lower half) which has a typical bimodal form. The percentage occurrence of the different stages is also indicated. The bimodal population distribution is generally interpreted as reflecting a food-limiting situation. Phytoplankton becomes abundant in spring and the zooplankton responds to this increase in food

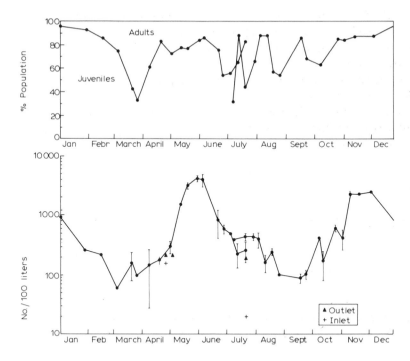

Fig. 15.6. Upper figure: percent juveniles and adults in the population. Lower figure: population size (*Daphnia galeata mendotae*/100 l) in 1960—1961. Vertical bars show standard deviations. (From Hall, 1964.)

level. After the food supply has declined (owing to predation or other causes), the zooplankton population decreases. In summer the phytoplankton may become abundant again, but because it consists then mainly of blue-green and filamentous algae, which are less available as a food source, the zooplankton density remains low. During autumn mixing (see Chapter 9) more nutrients for unicellular phytoplankton growth become available and the zooplankton density may subsequently reflect this and rise again. Hall found evidence to suggest that the zooplankton-population size is probably more regulated by predation pressure than by scarcity of food. The predators can therefore be considered as removing the central portion from a basically unimodal population curve (see also Dodson's work in subsection 15.3.4).

Because the effects of food and temperature are separable, Hall was able to provide evidence which showed that temperature exerts a greater influence than do food levels in determining the nature of zooplankton growth. He found that frequency of moulting and reproduction, duration of egg development, and physiological life span are all influenced mainly by temperature. Growth per instar, maximum carapace length and brood size however are influenced mainly by food.

The same method was applied by Edmondson (1960, 1965, 1968b) to some rotifers. He used the following definitions and calculations:

E = fertility = eggs per female from lake samples.

N = population number.

D = duration of egg development (measured in laboratory studies as a function of temperature).

B = daily population reproductive rate = E/D being a function mainly of food supply and temperature.

As one female produces E/D juveniles per day the increase of population per day is:

$$1 + \frac{E}{D} = 1 + B$$

Assuming an exponential increase:

$$N_t = N_0 e^{bt}$$

where b is the birth rate, we find:

$$b = \ln(B + 1)$$

as in 1 day the population would grow by $\ln(B + 1) - \ln 1 = \ln(B + 1)$.

Not all juveniles will reach full development due to the death rate, d, so that:

$$N_t = N_0 e^{(b - d)t} \tag{15.8}$$

As b is known from B and as the total change under natural conditions can be measured, N_t and N_0 are known so that d can be calculated.

Using these calculations and a **regression** analysis of the birth rate with the presence of food, Edmondson found that small flagellates served as food especially for *Keratella cochlearis* and *Kellicottia longispina*, while bacteria were apparently of no great nutritional importance. *Polyarthra vulgaris* depended on the presence of the green alga *Cryptomonas* and to a lesser extent on smaller organisms. A negative correlation between the occurrence of Rotatoria and *Chlorella* was found; with high population density of *Chlorella* low densities of Rotatoria were found. There is no obvious explanation for this. It may be that *Chlorella* is not a suitable food for Rotatoria. The phenomenon may not necessarily be connected directly with *Chlorella* at all but with another common factor which affects both organisms.

In the calculations of the birth and death rate the assumption is made that the proportions of the young and old eggs remain a constant percentage of the total number. If this is the case, the number of eggs surviving from a given batch will decrease linearly with time; if the age distribution of egg numbers is variable however, there will be certain periods with high hatching rates and others with lower rates. Edmondson (1972) discussed this problem further.

15.3.4. Measurement of daily increases

The death rate, d, may also be estimated by using eq. 15.8. The birth rate, b, may also be calculated using either an exponential growth rate model or a non-continuous (discrete) model. If r is the observed instantaneous growth rate, then:

$$d = r - b$$

By comparing (correlation analysis) d with the predator density, the population dynamics of the prey may be evaluated. In this way Hall (1964) showed that the decrease of *Daphnia* which he found in Base Line Lake correlated with increased production by *Leptodora* and not with resource limitation, since the observed summer maximum was accompanied by a large actual birth rate. Furthermore he showed that death rate was correlated with predator density.

Wright (1965) computed birth rate, population rate of change and mortality rate for a population of *Daphnia schodheri* in an artificial impoundment, the Canyon Ferry Reservoir (Montana, U.S.A.). Birth rate was significantly correlated with chlorophyll content, while the rate of population change corresponded with the birth rate, except when *Leptodora* was present. Mortality rate was strongly and positively correlated with the density of *Leptodora*. The gross primary production was 1.1 g m^{-2} d^{-1} of C, whilst the net production of *Daphnia* was 0.15 g m^{-2} d^{-1} of C. Thus the loss due to *Leptodora* predation was 0.05 g m^{-2} d^{-1} of C which is 10% of the *Daphnia* assimilation.

Wright described the population density in terms of the equation:

$$N_t = N_0 e^{(b-d-p)t} \tag{15.9}$$

where $b = 0.025$ µg of chlorophyll per litre, and the predation rate, p, related to the number (L) of *Leptodora* per litre; $p = 0.0095\,L$. On the basis of the average mortality rates observed when *Leptodora* was not present a constant death rate, $d = 0.088$, was assumed. Eq. 15.9 then becomes (disregarding dimensions):

$$N_t = N_0 e^{(0.025\,\text{Chl.} - 0.088 - 0.0095\,L)t}$$

The calculated and observed curves are shown in Fig. 15.7. The discrepancies were attributed by Wright to the wrong assumption of a constant death rate. While high mortality was observed when phytoplankton density was low, the natural death rate was low during predation by *Leptodora*.

Dodson (1972) found that for the last four generations of six *Daphnia* populations *Chaoborus* (Insecta) predation accounted for about 90% of the daphniid mortality. Dodson used an accurate working model and compared the *Daphnia* death rate with the number of *Chaoborus* larvae present. Mortality was estimated by Edmondson's technique (eq. 15.8), but he estimated

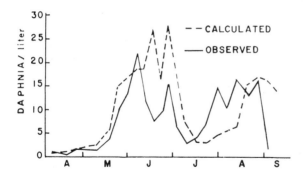

Fig. 15.7. Calculated population density of *Daphnia schodleri* compared with the observed population density. (From Wright, 1965.)

egg development time, D, not as a function of temperature (because diurnal changes occurred) but from field data of the growth of cohorts. He also measured the predator—prey encounters and the *Chaoborus*-feeding rate. He discussed the importance of size-selective predation because this will determine which of a group of crustacean herbivores will be present in a given body of water.

Herbivorous zooplankton populations are therefore regulated by three factors: food level, predation pattern and temperature. The large influence predation may have explains why the effect of increasing the food levels to zooplankton is often so small.

In 1963 Nauwerck made an extensive study on population dynamics and feeding of *Eudiaptomus graciloides* in Lake Erken (Sweden). Number of eggs, nauplii, copepodites and adults were estimated during a year, while develop-

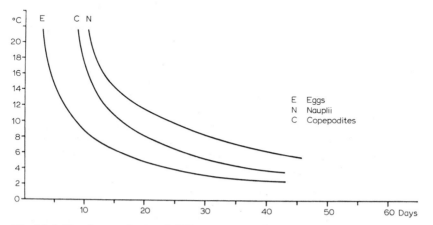

Fig. 15.8. Development rate of different stages of *Eudiaptomus graciloides* in relation to temperature calculated for Erken, 1957. (From Nauwerck, 1963.)

ment times were measured in the laboratory. To facilitate the procedure a month was taken as time span for his calculations.

Nauwerck used the renewal coefficient E_k (Erneuerungs coefficient) instead of turnover time:

$$E_k = \frac{\text{month (in days)}}{\text{development time span (in days)}}$$

Development time varied around 1.5 days for nauplii (stage $x \to x + 1$). From 1st nauplius stage to 1st copepodite stage took about 11 days (at room temperature, see Fig. 15.8).

Temperature appeared to be the most important factor governing the time taken for nauplius development, while for copepodites both temperature and food supply affected E_k.

The mean results obtained over 12 months can be summarised:

	E_k	Development time (in days)
eggs	5.6	
nauplii	1.6	19
copepodites	2.1	14
adults		11
		44

From these data and the number per litre in the lake the total yearly production per litre was calculated: eggs 202 (+ 20 "dauer" eggs + 75 loss); nauplii 280; copepodites 202; and adults 150.

Using the average weight of the animals, Nauwerck calculated a total production of zooplankton of 43 mg cm^{-2} yr^{-1} (fresh weight) of which 60% was in *Eudiaptomus*. The food pyramid consisted of the following compartments:

phytoplankton[1]	160 mg cm^{-2} yr^{-1} (fresh weight)
herbivorous *Diaptomus*	43 mg cm^{-2} yr^{-1} (fresh weight)
secondary consumers	7 mg cm^{-2} yr^{-1} (fresh weight)

The conversion of filtration rate to estimated feeding rate produced results which led Nauwerck to suggest that the animals did not ingest sufficient algae to account for their own total body weight production. He supposed that bacteria and detritus might be supplementary food sources. The energy from these two sources comes ultimately of course from decaying algae, unless allochthonous material is also involved. It seems likely that the necessary experimental pretreatment of the animals may have influenced his results. Thus *Eudiaptomus* was separated from the other zooplankton by filtration which might also have removed the large phytoplankton organisms and thereby have caused the filtration rate to increase.

---

[1] Of which 61 mg is nanoplankton; value not corrected for respiration and mineralisation.

Patalas (1972), whose work is discussed in relation to eutrophication (see subsection 17.4.1 and Fig. 17.11), found that temperature had no effect on development time of Crustacea at low food concentrations (chlorophyll $a <$ 2—5 mg m^{-3}), but had a large influence at high food concentrations (> 5 mg m^{-3}).

A study of food-transport efficiency in the Atlantic Ocean was made by McAllister (1969). He measured a mean annual primary production (based on alternate 6 weekly periods over 6 years) of 48 g m^{-2} yr^{-1} of C, a figure which he increased to 60 g to compensate for estimated grazing losses. Secondary production appeared to be sensitive to zooplankton respiration and assimilation. A value of 13 g m^{-2} yr^{-1} of C was accepted as being the most likely estimate of mean annual secondary production, so that the transfer efficiency was about 20%. By using combinations of factors such as plant growth constant, zooplankton respiration and grazing (continuous or nocturnal), a range of gross growth efficiency between 23 and 36% was found, the most likely value being about 25%.

Energy transformation and zooplankton production was studied in detail by Comita (1972) in Severson Lake (Minnesota, U.S.A.) during the ice-free period (184 days) of 1955. He found the following partition of energy ("energy pyramid"):

I Solar input	89 586 cal. cm^{-2} (184 days)	
II Gross phytoplankton production	582 cal. cm^{-2} = 0.65% of I	
III Phytoplankton respiration	309 cal. cm^{-2}	
IV Net phytoplankton production	272 cal. cm^{-2} = 0.25% of I	
Respiratory coefficient	52%	
V Zooplankton net production		
Diaptomus	1.39%	
Daphnia	0.85%	
Bosmina	0.75%	
Mesocyclops	0.37%	
Diaphanosoma	0.35%	
Brachionus	0.35%	
Keratella quadrata	0.14%	
Polyarthra sp.	0.07%	
Keratella cochlearis	0.03%	
Filinia	0.02%	
Asplanchna, Synchaeta, Chaoborus	0.42%	
	4.74 = 1.6% of IV	
Animal respiration	23.91	
Animal assimilation	28.65 = 0.03% of I	

Lindeman's respiratory loss coefficient for this situation was $(23.9/4.7) \times 100\% = 500\%$. The efficiency of the conversion of solar energy by the primary consumers was low (0.03%), for which Comita (1972) gave several reasons. Two of these are: the energy of the eggs produced was not included, except for those of copepods; some population densities may have been underestimated and average life span overestimated. But no combination of these errors would increase the efficiency estimate by a factor of two, so that the overall result still remains inexplicably low. The low efficiency is due almost entirely to the low net zooplankton production (relative to that of net phytoplankton production).

The secondary consumers converted 10% of the net production available to them. To account for the 1.2 cal. cm^{-2} found as gross production, the secondary consumers must therefore have consumed at least 2.4 cal. cm^{-2} (if one assumes 50% assimilation).

Besides his productivity studies Comita correlated 23 zooplankton variables and 3 phytoplankton variables as pairs in all possible combinations. The correlation coefficients obtained were then used to indicate the interactions which occur between the populations. Thus the two *Keratella* (rotifer) populations (*K. cochlearis* and *K. quadrata*) coincide very closely in time and are both negatively correlated with the occurrence of *Diaphanosoma*. While Comita often attributed negative correlations to competition for food (e.g. *Bosmina*

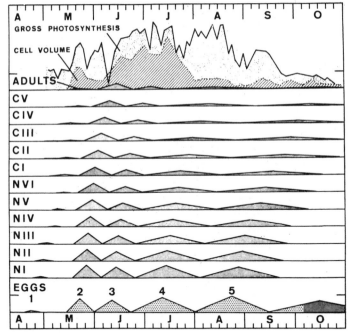

Fig. 15.9. Idealised development of *Diaptomus siciloides* generations expressed as numbers. Idealised instar development of the *Diaptomus siciloides* population expressed as relative numbers m^{-3} derived from survival curve equation. The triangle shown for the egg population of generation 2 represents 155 910 eggs m^{-3}, that for the adults that survive in this generation represents 62 787 m^{-3}. In the upper graph are shown gross photosynthesis (solid line) at the 0.5-m level as mg O$_2$ l^{-1} and cell volume (dotted line and diagonally hatched area) as mm^3/100 ml lakewater at the surface. (From Comita, 1972.)

with *Keratella quadrata* and *Filinia* with *Keratella quadrata*), he saw no obvious reason for this particular negative correlation since both *Keratella* spp. and *Diaphanosoma* are very different in size. The mean egg size and mean number of eggs per attached sac in *Diaptomus siciloides* are negatively correlated. As *Anabaena* is the dominant phytoplankton genus when egg size is largest, this suggests that *Anabaena* in the lake must be assimilable at least to *Diaptomus*. The mean number of eggs was significantly negatively correlated with numbers of *Daphnia* present. The total number of eggs produced per day by *Mesocyclops edax* correlated very well with the daily values of gross and net production and also with the cell volume of phytoplankton in the surface water layers. The total number of *M. edax* copepodites and especially the nauplii were strongly correlated with integral gross photosynthesis (ΣA per m^2) and phytoplankton cell volume in the surface water (but less so with integral net photosynthesis, but this itself is a dubious figure so that the low correlation found can probably be neglected). This is additional evidence to suggest that these larval stages are herbivorous.

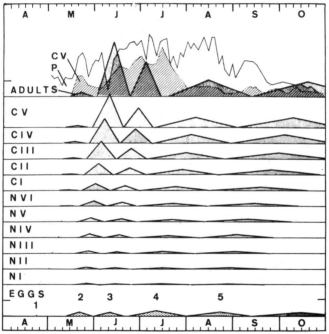

Fig. 15.10. Idealised development of *Diaptomus siciloides* expressed as calories. Idealised instar development of the *Diaptomus siciloides* population expressed as relative amounts of calories. The number m^{-3} is derived from survival curve equations and converted into calories. The triangle shown for the egg population of generation 2 represents 104.7 cal. m^{-3}, that for adults that survive in this generation represents 1 467.6 cal. m^{-3}. In the upper graph are shown gross photosynthesis (solid line) at the 0.5-m level as mg l^{-1} of O$_2$ and cell volume (dotted line and diagonally hatched area) as mm^3/100 ml lakewater at the surface. (From Comita, 1972.)

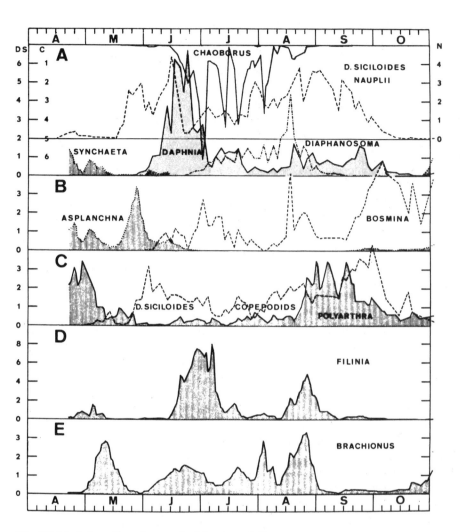

Fig. 15.11. Interactions in zooplankton populations.

A: *Chaoborus* population as an inverted curve, C, inside left ordinate as hundreds m^{-3}. The *Daphnia, Diaphanosoma* and *Synchaeta* populations, DS, outside left ordinate × 21 172 = numbers m^{-3}; and the *Diaptomus siciloides* nauplius populations, total of all instars, N, outside right ordinate × 21 172 = numbers m^{-3}. (From Comita, 1972.)

B: *Asplanchna* and *Bosmina* populations, both on left ordinate × 21 172 = number m^{-3}.

C: *Diaptomus siciloides* copepodite population, all immature instars; the shaded curve is the *Polyarthra* population; both populations on outside left ordinate × 21 172 = numbers m^{-3}.

D: *Filinia* population, left ordinate × 21 172 = numbers m^{-3}.

E: *Brachionus* population, left ordinate × 211 715 = numbers m^{-3}.

Although Comita used the measured temperature of the water in his calculations of the life span of the animals, no estimate of the effect of variations in temperature on the observed correlations was made. Hall's (1964) results, showing the temperature to be the most prominent factor affecting the growth rate, indicate that inclusion of the effects of temperature in Comita's data could increase the statistical significance of his correlation calculations.

Comita's schemes of idealised development of successive *Diaptomus siciloides* generations are shown in Fig. 15.9, while that development expressed as calories is shown in Fig. 15.10. Interaction is shown in Fig. 15.11.

In some current research, more attention is being given to differential grazing, i.e. the fact that filtering Crustacea prefer to ingest certain types of phytoplankton cells rather than others. Porter (1973) measured grazing inside three polythene bags (0.5 m^3) filled with lakewater. From the first bag the larger grazing animals were removed, the second had the fauna enriched with grazers obtained from vertical net hauls, whilst the third was unaltered and used as a control. The major effect of grazing was shown to be the suppression of small algae, primarily the flagellates and nanoplankton, and of large diatoms. Desmids, dinoflagellates and chrysophytes were largely unaffected. Numbers of large green algae increased, this being probably due to a shift in their favour of various competitive influences. Cells from the groups whose numbers were unaffected were hardly ever found in the gut of the zooplankton animals. On the other hand, cells from groups which were suppressed constituted the main bulk of those species actually found in the gut of animals. *Anabaena affinis* and *A. flos aquae* were among those species which were not ingested. The group which increased most during the presence of grazers in these experiments was dominated by species of green algae, which are encased in thick gelatinous sheaths, such as *Sphaerocystis schroeteri* and *Elakatothrix gelatinosa*. They passed through the gut of animals, but remained viable.

Haney (1973) examined the grazing activity of natural limnetic zooplankton in Heart Lake, a small eutrophic lake (Ontario, Canada) using ^{32}P-labelled yeast, *Rhodotorula* sp., a bacterium, *Pseudomonas fluorescens*, and the green alga *Chlamydomonas reinhardtii*. He compared his results with those obtained from a *Sphagnum*-rich bog (brown water, pH = 4.5) and from a deep oligotrophic lake, the grazing profiles of which are given in Fig. 15.12 and 15.13. It seems likely that the greatest intensity of grazing occurs at the same depth as the photosynthetic maximum. The occurrence of swarming, however, imposes a limit on the degree of accuracy which can be attained in work of this kind. Haney recorded for instance that one sample taken from the bog lake contained 20 000 *Bosmina* per litre and had a grazing intensity of 785% per day. Diurnal vertical migration is also a factor which may decrease sampling accuracy. Haney calculated a grazing rate in terms of mg C per day per mg C biomass and expressed this value as a percentage. This may be compared with the phytoplankton renewal coefficient. (If mg C assimilated per day per mg of C algal biomass = 1, then the renewal coefficient = 100%.) Monthly mean grazing rates

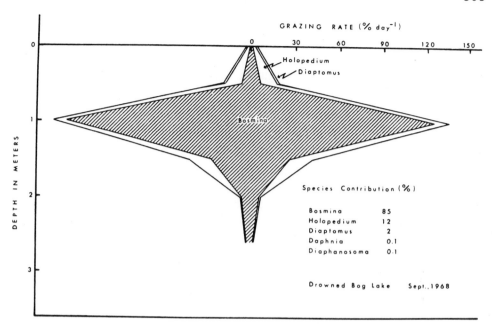

Fig. 15.12. Grazing profiles in Drowned Bog Lake, showing species contribution at each depth, September 1968. (From Haney, 1973.)

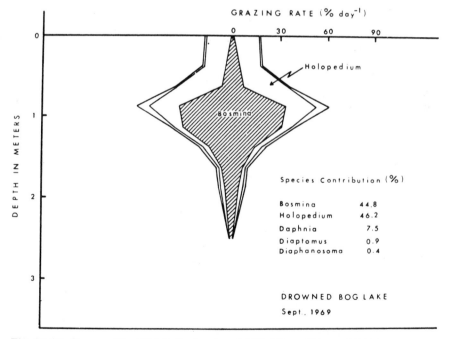

Fig. 15.13. Same as Fig. 15.12, September, 1969. (From Haney, 1973.)

in Heart Lake exceeded 100% d^{-1} only during July, and other average values were 80% for June—September and 19% for January—May. Haney suggested that phytoplankton renewal rates in oligotrophic lakes are generally far in excess of zooplankton grazing rate, whereas in eutrophic lakes the rates are about the same. There seems to be very little evidence to support this generalisation however.

The problem of whether the species and types of organisms used were suitable for establishing filtering rates needs to be considered, and so also does the validity of using ^{32}P as an isotope marker in this kind of study.

Grazing was measured by Haney in the natural populations before, during, and after stratification; grazing activity exhibited great vertical heterogeneity during stratification with maximum rates occurring at or near the depth where oxygen values were at 1 ppm. It was at this depth that Sorokin (1965) showed that in Lake Belovod (Russia) purple bacteria provided more food for grazing animals than did algae. Nevertheless, there is no evidence to suggest that this aggregation of grazers is the result of active migration of the zooplankton.

15.4. THE LARGER ANIMALS

There is very little detailed information available on the energy transport through larger organisms such as fish and their prey in lakes. There are perhaps two main reasons for this. First, most fish population studies have dealt predominantly with the population dynamic aspects, i.e. the study of the factors controlling numbers or densities. Secondly, it is difficult to catch young fish (larvae and those up to 1 year old) in sufficient quantity to allow a true estimate of their role in energy and mass transport. The studies of fishery and limnology tend to have polarised into two separate disciplines and mostly take place in different institutes often with different financial and administrative allegiances and direction*.

One special technique which is useful for the student of general limnology is the cohort approach. Of every year class of fish (or life class of other shorter-living organisms such as *Gammarus*), the mean individual weight is plotted against the number per unit area (Fig. 15.14) (Allen, 1951; Neess and Dugdale, 1959). The latter authors presented a theoretical model for insects with a single pulsed input of young per year or at least an input separable in time, this being termed the cohort.

As the cohort ages, the animals grow and the numbers decrease because some animals die naturally or are eaten. Production (P_T) is the product of growth increment (ΔW) per unit time and the mean numerical density ($\overline{N}_T$); or:

$$P_T = \overline{N}_t \Delta W$$

---

* The International Biological Programme has proposed plans for their re-integration; the IBP results will be summarised by LeCren (in preparation).

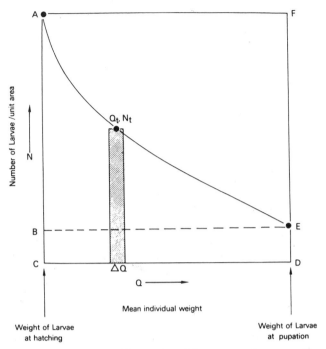

Fig. 15.14. Age-specific survivorship curve and age-specific growth function. (From Edmondson, 1971.)

Integration of the area under the curve gives the production for that generation. The area $ACDE$ in Fig. 15.14 was called by Nees and Dugdale the actual net production, while the area $BCDE$ equals the pupal crop. The area ABE was called the directly recycled production, i.e. that part which remains and is mineralised in the aquatic ecosystem. Excluding mortality, the area $ACDF$ is thus the potential production and the area AEF is therefore the lost potential net production. In addition to knowing the age distribution of the population, it is also necessary to know the absolute distribution densities. The most common method is to mark a few thousand fish with a metal, plastic or "coloured spot" tag and to measure the ratio of marked to unmarked fish in sample hauls which are caught after the marked individuals have been released. This ratio times the number of marked fish actually released gives a measure of the absolute density. It is, of course, necessary to obtain some estimate of the loss of tags and of mortality amongst the marked fish. In some, especially shallow, lakes it is sometimes possible to get a check by use of a large enclosure net. It is also necessary to know the length and body weight of the fish. These variables often show a linear correlation. It is also necessary to know the age of individuals. Age of fish can be measured by counting ring formation on the scales and also in some special cases from the otoliths, percula or other bones of the fish.

After this consideration of some of the difficulties encountered in correlating primary and secondary production, it should be quite clear that work on the study of food chains has not yet reached an adequately quantitative stage. The outcome of a few IBP sponsored studies will therefore be given (in section 15.5) as an illustration of the progress of some current or recent research.

15.5. SOME CASE HISTORIES

15.5.1. *Lake George*

Lake George (East Africa), a shallow lake on the equator, was chosen as one of the sites in the British IBP Programme. The results have been summarised in a symposium (Greenwood and Lund, 1973). The following account is based on that symposium.

Viner and Smith (1973) described the geographical, historical, and physical aspects. The lake covers 250 km^2 with a mean depth of 2.4 m and receives solar energy of 1970 J cm^{-2} ± 13% per day. Surface water temperature is about 30°C. A continuous flow of water from a mountainous catchment causes a flushing rate of 2.8 times the mean lake volume per year. Nocturnal turbulence and mixing is caused by local winds but in day time stratification occurs. Wind-initiated rotary currents may be the cause of the occasional concentric distribution of plankton. Organic enrichment has occurred throughout the lake's history (3 600 ± 90 years), but during the last 700 years near-equilibrium has existed. Nutrients were transported into the lake with the water inflow from the surrounding mountains. The concentrations of these in the inflow streams are given in Table 15.3. Particulate matter was not recorded.

The product of total rainfall and mean concentration in 19 samples indicates that about 25% of nitrogen and about 66% of phosphate is supplied in the rain. (The proportion for phosphate seems unbelievably high.) The phosphate which comes in as dust is probably mostly as particles which are not immediately soluble. Annual output of the lake is 73 000 tonnes dry matter, 25 000 tonnes carbon, 3 400 tonnes nitrogen, 220 tonnes phosphate phosphorus, and 300 tonnes chlorophyll *a*. On a daily basis the amounts fixed must average approximately 200 tonnes dry weight, 70 tonnes carbon, 9.2 tonnes nitrogen, 0.6 tonne phosphorus, and 0.8 tonne chlorophyll *a*.

TABLE 15.3

Input of nutrients to Lake George (μg l^{-1})

	Range	Mean
NH$_3$-N	3 — 135	24.4
NO$_3$-N	0.0—2 390	533
PO$_4$-P	2.5— 215	78.8

The net amount of carbon which is fixed photosynthetically is equivalent to 0.28 g m^{-2} d^{-1} (but see p. 310 for gross primary production).

In discussing the ecological stability of the lake, Ganf and Viner (1973) pointed out that the concentration of phytoplankton remained fairly constant around 30 g of C per m^2. The biomass of the dominant zooplankton organism *Thermocyclops hyalinus* also remained reasonably stable with fluctuations only ranging within a twofold limit. Ganf and Viner assumed, rather surprisingly, that this low variation reflects an ecologically stable situation; the term is more often used to describe the consistency of seasonal successions during prolonged periods.

They described phytoplankton changes of biomass using Uhlmann's (1971) equation:

$$\frac{dx}{dt} = \mu x + Ix - (Gx + Sx + Ex)$$

where x = biomass; t = time; μ = growth constant; I = input; G, S and E are grazing, sedimentation, and export constants, respectively.

In the case of Lake George, the biomass input is negligible, so:

$$\mu' = \mu - (G + S + E)$$

where μ' is used for dx/dt expressed as a fraction of x. The authors took no account of natural physiological death rate (i.e. without grazing). They calculated that μ' varied between +0.05 and −0.03 d^{-1}. They pointed out that as the population varies only between 20 and 40 g of C per m^2, a μ' value of 0.05 d^{-1} could be sustained for only 7 days. After comparing photosynthesis and respiration rates, Ganf and Viner concluded that small positive and negative values for μ' may occur during short periods depending upon daily radiation and light penetration characteristics.

No experimental data on either process are included in their paper but respiration rate can be calculated from the weighted mean O$_2$ uptake (1.1 mg h^{-1}) per mg chlorophyll. Ganf (1974) gave about the same value for community respiration (this included zooplankton respiration which was 5% of the community respiration). Since the mean chlorophyll concentration is 500 mg m^{-2}, the O$_2$ uptake is therefore 13.2 g m^{-2} d^{-1}. Photosynthetic rates are not presented but can be estimated from the mean maximum rate (20 ± 4 mg O$_2$ per mg chlorophyll per h; Ganf (1974) gave 25 mg O$_2$ per mg chlorophyll per h) and the chlorophyll concentration. The estimate is:

$$A_{max} = 20 \cdot 200 = 4 \text{ g m}^{-3} \text{ h}^{-1} \text{ of O}_2$$

As $\Sigma A = A_{max} Z'$ (eq. 4.9), Z' must be calculated from eq. 4.10:

$$\tfrac{1}{2} I_k = I_0 e^{-\epsilon Z'}$$

(For symbols see section 4.3.) I_0 = 1 000 J cm^{-2} d^{-1} minus 10% reflection = 900 J cm^{-2} d^{-1} or 90 J cm^{-2} h^{-1}. As $1/2 I_k$ = 7.5 J cm^{-2} h^{-1}:

310

$$e^{-\epsilon Z'} = \frac{7.5}{90} = 0.0083$$

The mean chlorophyll value is 200 mg m^{-3} so that ϵ can be derived from Ganf and Viner's data to be 5 m^{-1} and thus $Z' = 0.5$ m. Assuming that a full rate of photosynthesis can take place throughout a 10-h period per day:

$$\Sigma A = A_{max} \cdot Z' \cdot 10 = 4 \cdot 0.5 \cdot 10 = 20 \text{ g m}^{-2} \text{ d}^{-1} \text{ of } O_2$$

(a value which lies near a preliminary value of 12 g m^{-2} d^{-1}: Ganf, 1972). This is equivalent to 7.5 g m^{-2} d^{-1} of C. As 1 g of C represents $4.2 \cdot 10$ J, the energy fixed during photosynthetis is $7.5 \cdot 42$ J = 300 J m^{-2} d^{-1}, which amounts to 3% of the available light and thus 1.5% of the total solar irradiance.

The difference between the O_2 production and uptake in the water column may be attributed to O_2 uptake by both sediments and animals, as the system is in equilibrium and no net increase of phytoplankton takes place in the long term. Ganf and Viner followed Talling's (1957, 1971) calculation for estimating the length of the water column at which respiration, R, would equal photosynthesis, A, in terms of O_2 balance. Talling suggested that there is a critical value which defines the column compensation point when the ratio R/A is unity (no net production) and derived the following expression for determining this critical value:

$$q_c = \frac{[l.d.h.]^{day}}{\epsilon_{min} Z_m / \ln 2} = 32 \, r$$

where q_c is the critical value of the ratio when the column compensation point is reached; $[l.d.h.]^{day}$ = light division hours, a logarithmic expression of daily incident radiation defining the total daily incident radiation available for A (see section 4.7.2); $\epsilon_{min} Z_m / \ln 2$ = optical depth of the mixed layer (see Chapter 4); and $r = R/A_{max}$.

If $q_c > 32 \, r$, the column compensation point has not been reached and the phytoplankton will have a positive production, while if $q_c < 32 \, r$, the production is negative. With $r = 0.005$ in Lake George, Ganf and Viner demonstrated that the measured chlorophyll concentration is predictable given the lake depth, the optical depth, and the solar energy. It is obviously also possible to predict the depth where zero net production occurs from the chlorophyll concentration.

The authors demonstrated that both nitrogen and phosphate are in short supply and serve to limit growth of the phytoplankton. Viner (1973) demonstrated markedly different responses to nitrogen enrichment by different algal-size groups and showed that more photosynthetic products could be formed in the lake than could be immediately incorporated into protein. The blue-green alga *Microcystis*, in particular, did not show the same rapid uptake of nitrogen as the smaller algae did. Horne and Viner (1971) have suggested that N_2 fixation accounted for 30% of the particulate nitrogen annually flushed out from the lake. It is obvious that the high rate of gross production must be sustained by other nutrient sources because nutrient concentrations in the water are negligible. Ganf and Viner estimated mineralisation by the zooplankton biomass to be 36 mg m^{-2} d^{-1} of NH$_3$-N and 7 mg m^{-2} d^{-1} of PO$_4$-P. The O$_2$ uptake of 13.2 g m^{-2} d^{-1} is not due solely to respiration by the algae, but also to bacterial oxidation of decaying algal cells and dissolved organic matter (Talling's r value also includes bacterial respiration). Golterman (1971, see Chapter 4) showed that bacterial respiration of dissolved organic matter accounted for a large fraction (0.5) of the oxygen uptake, whilst algal respiration accounted for only a minor part. Ganf (1974) estimated that the bacterial respiration may contribute up to 60% of the total oxygen uptake.

Assuming that the mineralisation can be described as:

$$5 \, O_2 + C_5H_7NO_2 \rightarrow 5 \, CO_2 + NH_3 + 2 \, H_2O$$

(where C$_5$H$_7$NO$_2$ stands for "mean algal composition", see section 6.2.2), it can be calculated that the oxygen uptake should be equivalent to $0.5 \cdot 13.2 \cdot 14/160 = 590$ mg m^{-2} d^{-1} of NH$_3$-N, which is more than an order of magnitude higher than the part which is actually mineralised by the zooplankton. Golterman (1971, unpublished) measured a decrease of 600 mg m^{-2} d^{-1} in the particulate nitrogen, which was equivalent to a phytoplankton turnover rate of 10% per day.

Ganf (1974) found nitrogen recycling values of the same magnitude in some of his experiments, but not in all. His experiment 9 showed a recycling only if nitrogen content of the phytoplankton was increased artificially. It may be that if the phytoplankton is starved of nitrogen, the inhibitor used, dinitrophenol, does not completely inhibit nitrogen uptake by the living algae. The high turnover rate of the phytoplankton in Lake George cannot be caused by seasonally changing physical or chemical factors. It may be suggested that photo-oxidative death in the upper water layers is the cause of the algal death. According to Abeliovich and Shilo (1972a, b), this will take place if blue-green algae are exposed to light in the presence of oxygen in a CO$_2$-free medium.

The mean standing crop of the main organisms in the open water and sediments is given by Burgis et al. (1973) in the same volume (Table 15.4).

The zooplankton although having a low biomass comprises large numbers (up to $2.5 \cdot 10^6$ m^{-2}) of small-sized individuals ($< 650 \, \mu$m). Food supply does not appear to be limiting for adults, which were seen to ingest and assimilate *Microcystis*, but food may be limiting for the young nauplius stages. Oxygen uptake by the zooplankton can be calculated from its dry weight, numbers per litre, and Richman's curve for O$_2$ uptake as function of dry weight of *Daphnia* as: $2.4 \cdot 10^5$ animals m$^{-2} \cdot 0.72 \, \mu$l d$^{-1} = 0.2$ l m^{-2} d^{-1} of O$_2$. Assuming a $Q_{10} = 2$ for converting from Richman's temperature (20°C) into the temperature of Lake George (30°C) and as 1 l of O$_2$ = 1.5 g, we can estimate an O$_2$

TABLE 15.4

Standing crop of the main organisms in Lake George (excluding hippopotamus)

Organisms	$g\,m^{-2}$ of C
Phytoplankton	46.8*[1]
Zooplankton (Crustacea)	0.8*[2]
(*Thermocyclops hyalinus*)	0.6
Chaoborus	0.2
Other benthic fauna	0.3
Herbivorous fish	0.9
Carnivorous fish	0.5

*[1] 30% *Microcystis*, 20% *Anabaenopsis* spp., 10% *Anabaena flos aquae* and *Lyngbya* spp., and 10% *Synedra berolinensis*.
*[2] 80% *Thermocyclops hyalinus* (herbivorous cyclopid copepod), 10% *Mesocyclops leuckarti* (carnivorous cyclopid copepod).

uptake for the zooplankton of 0.6 g m^{-2} d^{-1}, which is small compared with the O$_2$ uptake of the whole community (13.2 g m^{-2} d^{-1} of O$_2$). Ganf and Blazka (1974) estimated that the zooplankton consumed ca. 570 mg m^{-2} d^{-1} of O$_2$.

Population dynamics of the zooplankton were studied by Burgis (1973, 1974). The population is stable because of the near-equality of birth and death rates, b and d, which are 0.28 and 0.29, respectively. Although the population turnover time was short, the net production, as C, calculated from standing crop biomass and turnover time was only 19.6 mg m^{-2} d^{-1}, which is low (0.5%) compared with the primary production (7.5 g m^{-2} d^{-1}). The net production of *Thermocyclops hyalinus* was also low when calculated by a method involving recruitment time (calculated from clutch size E, see section 15.4) and new data on dry weight (Burgis, 1973). With this method a production of 59 mg m^{-2} d^{-1} of C was found. Burgis (1974) believed that the first estimate (20 mg m^{-2} d^{-1}) is more reliable, as the egg ratio method was originally developed for Rotatoria, which have a less complex life cycle. It is, however, not clear whether her second method measured gross or net production. Burgis and Walker (1972) and Burgis (1974) compared the population dynamics of the zooplankton population with those of the IBP site at Loch Leven (Scotland), a shallow eutrophic temperate lake. They used Hrbáček's hypothesis (which proposes that predators on zooplankton select the larger forms, leaving a residue of smaller forms to proliferate and dominate) to explain the absence of larger zooplankton organisms in Lake George. This can only be true if, after equilibrium has been established, the predator can then subsequently live on the smaller zooplankton organisms. Otherwise the predator would starve and the larger zooplankton would become re-established. There is no indication, however, that the death rate of the zooplankton in Lake George is abnormal.

The zooplankton in Lake George is eaten by the larvae of *Chaoborus* and by many young fish, mainly *Haplochromis* sp. and *Protopterus*, which together constitute 40% of the total mass of fish.

Sixty percent of the fish population is herbivorous, mainly *Haplochromis nigripinnis* and *Tilapia nilotica*. In fact, *Tilapia nilotica* forms 80% of the commercial fish yield (maximally 5 000 tonnes per year). No production figures for fish and benthic organisms can yet be given, but it is clear that the short food chain to *Tilapia* is responsible for the large fish productivity.

Feeding and grazing by various fish and zooplankton has been described by Moriarty et al. (1973), who showed that enzymic digestion of blue-green algae occurred in the intestine but only after acid hydrolysis had previously taken place in the stomach. This is contrary to previous reports in the literature. *Tilapia nilotica* and *Haplochromis nigripinnis* could assimilate 70—80% of ingested carbon from *Microcystis*, while *Tilapia nilotica* could only assimilate up to 50% of the carbon from *Chlorella*. Values obtained from mixed lake phytoplankton were somewhat lower than the values of laboratory cultures of blue-green algae. As the diatoms from the phytoplankton (20%) are easily digested, it seems that the blue-green algae from the lake are less available than those from the cultures.

Copepodites and adults of *Thermocyclops hyalinus* assimilated 35% of the ingested carbon from *Microcystis*, while the nauplii digested 58% (Tevlin, unpublished). There is however some contradiction with Tevlin's statement (in Burgis et al.), that the nauplii refused *Microcystis*. It is possible that the laboratory culture has smaller cells or is less toxic than the naturally occurring population in the lake.

The amounts eaten by *Tilapia nilotica* and *Haplochromis* were estimated; daily ingestion followed a diurnal pattern. For *T. nilotica* the amount of food and the amount ingested could be described by:

$$y = 271 + 13\,x$$

where y = dry weight ingested (mg d^{-1}), and x = wet weight of fish. Applying this to the total population, and estimating the ingestion of phytoplankton into the zooplankton, Moriarty et al. (1973) derived a daily ingestion value of 500 mg m^{-2} d^{-1} of C into the zooplankton and 35 mg m^{-2} d^{-1} into the herbivorous fish. As the zooplankton population remains stable, this carbon uptake should be equivalent to $1/3 \cdot 500 \cdot 32/12 = \sim 450$ mg m^{-2} d^{-1} of O$_2$ assuming the assimilation to be 1/3 of the amount ingested. This is in good agreement with the O$_2$ uptake. The estimated carbon uptake of 500 mg m^{-2} d^{-1} is extremely low when compared with the primary production. Benthic organisms were estimated to use 1 g m^{-2} d^{-1} of C, a value which seems rather high when compared with their standing crop (Table 15.3). Of this quantity 300 mg m^{-2} d^{-1} is probably faeces of planktonic herbivores, so that 700 mg m^{-2} d^{-1} probably comes directly from primary production. Therefore, a maximum of 1.2 g m^{-2} d^{-1} of the primary production (7.5 g of C m^{-2} d^{-1}) is used for secondary production, which is low if compared with standing crop. Although it has been established in experiments that blue-green algae can be digested and assimilated in Lake George, it is still possible that these algae do not provide an optimal diet for herbivores at all stages in their life cycles. Secondary production is low compared with primary production and this can only be explained by the low food value of the blue-green algae. The problem of low animal production when feeding on blue-green algae remains unsolved.

15.5.2. Loch Leven

Loch Leven is a shallow eutrophic temperate lake. It was also one of the IBP sites. A summary of the work has been published in the Proceedings of the Royal Society of Edinburgh (1974). The following account is based on that symposium.

Loch Leven (Scotland) lies in an agricultural plain of Old Red Sandstone overlain by glacial drift. The lake basin was formed at the end of the Pleistocene Epoch from kettle holes formed in the glacial drift after the retreat of the ice. In contrast to the flora and fauna of Lake George, the Loch Leven system is extraordinarily unstable. It varies quantitatively and qualitatively both from year to year and in the long term. Populations of macrophyte species have varied and so also have those of some of the fauna. *Daphnia hyalina* disappeared and then reappeared, whilst *Endochironomus*, formerly an abundant species, has disappeared. Nutrient chemistry has been described by Holden and Caines (1974). The main source providing phosphate is a woollen mill, but there is also considerable input from sewage and rivers. The latter also supply the major part of nitrogen, mostly in the form of nitrate, probably derived from agricultural fertiliser application in the catchment. Nitrogen is not usually the limiting factor for algal growth. Nitrate declines from a late winter maximum to a spring minimum; the algal cellular nitrogen accounts for only a part of the nitrate which disappears (see subsections 6.2.1 and 6.2.3).

Bailey-Watts (1974) described the algal successions. Diatoms have contributed more to the high biomass and production than any other group and predominate in the benthos. Blue-green algae (*Synechococcus* n. sp. and *Oscillatoria* spp.) dominated the plankton for various periods in each of the four years 1968—1971.

Two green algae (*Steiniella* sp. and *Dictyosphaerium pulchellum*) dominated a few of the dense crops which occurred from time to time. The algal growth is not related in any obvious way to the supply of nitrogen and phosphate, although concentrations of these nutrients are often low enough to be growth-limiting.

Diatom growth is controlled by the silica concentration. Supply of silica from the inflow and from solution within the interstitial waters of the mud is probably sufficient to account for this phenomenon. An interesting feature of Loch Leven phytoplankton is that they are mostly very small in size, despite the eutrophic situation; many rotifers are found in the zooplankton.

Grazing seems to be an important factor controlling the details of algal succession. For example, during the period when *Daphnia hyalina* was abundant (60 per litre), their guts were packed with algae and algal density and Secchi-disc readings were respectively the lowest (3 μg l^{-1}) and highest (4.75 m) recorded. Bailey-Watts reported that utilisation of living algal cells by Fungi, Protozoa and Crustacea seemed to be insignificant. Most of the phytoplankton must therefore autolyse or be broken down by bacteria. The density of the herbivorous *Chironomus* feeding on (dead or living) algae is not known.

Recently Bindloss (1974) investigated the primary production to which phytoplankton makes a considerable and perhaps dominant contribution. She pointed out that macrophytes were sparse. Although considerable quantities of benthic algae are also found, it seems unlikely that they or the macrophytes contribute to a very significant extent to the primary production. The productivity of neither has been measured though.

Photosynthesis of phytoplankton usually followed Talling's model (section 4.3) with ϵ_{min} varying between 0.5 and 3.1 ln units m^{-1}. Depth of the euphotic zone, Z_{eu}, was inversely proportional to ϵ_{min}:

$$Z_{eu} \approx 3.7/\epsilon_{min}$$

and varied between 1.2 and 7.4 m. The Secchi-disc depth was normally one-third of Z_{eu}. The value of ϵ_{min} varied with the chlorophyll concentration, the increment of ϵ_{min} per mg chlorophyll being 0.0086 ± 0.0011. Gross photosynthesis varied according to the season and values were in the range 0.4—21.0 g m^{-2} d^{-1} of O_2, with values of 10—15 g m^{-2} d^{-1} being fairly common.

Values less than 5.0 g m^{-2} d^{-1} were restricted to autumn and winter months and were strongly correlated with the temperature. Mean daily photosynthesis (over the 4 years 1968—1971) was 5.8 g m^{-2} d^{-1} of O_2 or 2.2 g m^{-2} d^{-1} of C. Self-shading — as in Lake George — is an important factor controlling the gross photosynthetic rate. Chlorophyll *a* ranges from 15 to 456 mg m^{-2}; owing to self-shading the theoretical upper limit for Loch Leven is 430 mg m^{-2}. Photosynthesis causes the pH to vary between 7.5 and 10.0 with an alkalinity value of 1.0—1.6 meq. l^{-1} (the theoretical pH in equilibrium with air is 8.1—8.3). Bindloss (1974) concluded that pH values above 9.0 inhibited the photosynthetic rate. Annual gross productivity converted $1 \pm 0.1\%$ of total solar radiation into algal biomass (radiation: $372—397 \cdot 10^7$ J m^{-2} yr^{-1}; primary productivity: $25—40 \cdot 10^6$ J m^{-2} yr^{-1}).

Respiration rate fell within the range 0.1—3.9 mg O_2 per mg chlorophyll per hour, values greater than 2 mg being uncommon. Talling's *r* value (see subsection 15.5.1) varied between 1:2 and 1:14. Negative values for primary production were found, even during periods when algal populations were increasing, but animal respiration seems unlikely to have caused this discrepancy.

316

Zooplankton was studied by Johnson and Walkers (1974). In 1969 the crustacean zooplankton was nearly a mono-culture of the carnivorous cyclopid
copepod *Cyclops strenuus abyssorum* and filter feeders were nearly absent.
Diaptomus gracilis (Sars) and two predatory cladocerans formed less than 5%
of the zooplankton. In 1970 and 1971 increasing amounts of the filter-feeder
Daphnia hyalina were found, which in 1972 co-dominated with *Cyclops*.

Cannibalism was probably an important mode of feeding amongst the
Cyclops, with larger individuals preying on nauplii and early copepodite stages,
although the rotifer *Keratella*, diatoms, and green algae, with a few small
chironomid larvae, were also seen in the gut contents of *Cyclops*. *Daphnia*
reached maximum numbers of over 200 per litre in June 1972 and declined
later that month, this decline being correlated with a sharp increase in density
of the blue-green alga *Anabaena* sp. Secondary productivity has not yet been
measured, but it is probably small compared with primary productivity. A
tentative estimate of crustacean production in 1969 gave 18 g m^2 (Johnson
and Walkers, 1974). Maitland and Hudspith (1974) found that the macro-
invertebrate fauna in the sandy littoral area (42% of the Loch bed) is dominated by larval Chironomidae. *Glyptotendipes* gave an annual production of
40.5 and 5.0 g m^{-2} in 1970 and 1971 respectively, while *Stictochironomus*
produced 1.2 g m^{-2} and 10.2 g m^{-2} for 1970 and 1971, respectively. A
speculative estimate of the entire zoobenthos production in 1970 was 46.5
g m^{-2}.

Endochironomus disappeared from the Loch in 1969; the causes are unknown. The change from *Glyptotendipes* towards *Stictochironomus* — the
former eats mainly planktonic and the latter benthic algae — occurred in the
same year that *Daphnia* appeared. Changes such as these may occur spontaneously and frequently in nature, but their occurrence remains undetected because observations are rarely continued for long enough. It should be noted

TABLE 15.5

Production and energy content estimates for 1970 and 1971 based on Allen curve data for
Limnochironomus, *Glyptotendipes* and *Stictochironomus* in Loch Leven

	Limnochironomus		Glyptotendipes		Stictochironomus	
	mg m^{-2}	J m^{-2}	mg m^{-2}	J m^{-2}	mg m^{-2}	J m^{-2}
1970 A	137	3 067	9 643	212 917	313	6 326
1970 B	457	10 232	30 876	681 742	893	18 048
Total	594	13 299	40 519	894 659	1 206	24 374
1971 A	163	3 650	2 557	56 459	483	9 761
1971 B	221	4 948	2 463	54 383	9 737	196 785
Total	384	8 598	5 020	110 842	10 220	206 546

A: 1st generation; B: 2nd generation in each year.

TABLE 15.6

Estimated production of trout and perch in Loch Leven (from Thorpe, 1974)

A. Preliminary estimates of adult perch production (wet weight)

Date (May)	Age group	N	Sex ratio (male/female)	$\overline{W}$ (g)	G	B (kg)	$\overline{B}$ (kg)	P (kg)
1970	2	327 387	1/1.14	48.2	0.933	33 769	22 930	21 394
1971		43 396		130.2		12 091		
1970	3	73 104	1/1.03	100.7	0.640	14 944	14 506	9 284
1971		36 300		190.9		14 067		
1970	4	10 319	1/1.03	153.1	0.423	3 207	2 937	1 243
1971		5 621		233.8		2 668		
1970	5—9	37 326	1/1.65	245.7	0.233	24 303	21 971	5 119
1971		23 875		310.4		19 639		
Total	2—9							37 040

B. Estimated production by 0—2 year old perch (wet weight)

	Group	Date (May)	N	$\overline{W}$ (g)	G	B (kg)	$\overline{B}$ (kg)	P (kg)
Maximum	0	year 0	$265 \cdot 10^9$	0.001	8.517	2 650	78 175	665 816
		year 1	$30\,740 \cdot 10^3$	5.0		153 700		
	1	year 1	$30\,740 \cdot 10^3$	5.0	2.266	153 700	85 470	193 675
		year 2	356 941	48.3		17 240		
							Total	859 491
Minimum	0	year 0	356 941	0.001	8.517	357	1 071	9 122
		year 1	356 941	5.0		1 785		
	1	year 1	356 941	5.0	2.266	1 785	9 513	21 556
		year 2	356 941	48.3		17 240		
							Total	30 678

TABLE 15.6 (*Continued*)

C. Production of trout, 1968—1972 (kg wet weight)

Year	April—October		October—April	April—April
	Somatic	Gonadal	Somatic	Total
1968—9	6 458	5 281	−695	11 044
1969—70	−1 243	3 624	3 312	5 693
1970—1	21 242	4 501	−2 536	23 207
1971—2	10 605	5 497	−1 751	14 351

that the total numbers of chironomids varies between 4 000 and 25 000 per m^2. Table 15.5 shows a summary of the production estimates based on Allen curve data. The production of the benthic animals is 2.5—4.0% of the primary production. The annual variation of the production of *Glyptotendipes* and *Stictochironomus* is large. The migrations of the larval chironomids are discussed by Davies (1974), but are not reviewed here.

Thorpe (1974) studied several species of the fish fauna which consists mainly of brown trout (*Salmo trutta* L.) and perch (*Perca fluviatilis* L.). Pike (*Esox lucius* L.), minnow (*Phoxinus phoxinus* L.) and stickleback (*Gasterosteus aculeatus* L.) are present in small numbers. Other species occur occasionally. Juvenile trout (5—20 cm) enter the Loch from nursery streams during the autumn and winter months at the end of their first and second growing seasons. Computation of the production of trout was based on the Allen technique. Population density was measured using the mark-recapture method, while the age of the fish was determined by scale growth. Mortality was determined experimentally. Gonad quantity increased from a basic 1.0-g wet weight for gonads of each sex, up to maximal values (when spawners were leaving the lake) that could be expressed as:

male: $y = 0.03316x - 0.65$
female: $y = 0.1522x + 0.78$ (1967—1969)
  $y = 0.2378x - 42.54$ (1970—1971)

where y = fresh weight of gonads, and x = fresh weight of fish.

TABLE 15.7

Food consumption by trout and perch in Loch Leven, June—September, 1971 (from Thorpe, 1974)

Food	Total consumed by trout		Total consumed by perch	
	0/000*	% of total	0/000*	% of total
Perch fry	9 500	29.9	8 850	13.5
Asellus	7 670	24.2	22 200	33.8
Chironomids	4 800	15.1	8 650	13.1
Molluscs	3 800	11.9	—	—
Daphnia	2 900	9.1	9 600	14.6
Trichoptera	1 400	4.5	—	—
Bythotrephes	700	2.2	1 800	2.7
Leeches	500	1.6	8 500	12.9
Gammarus	70	0.2	4 000	6.1
Miscellaneous	370	1.2	2 100	3.3
Total	31 700	100.0	65 800	100.0

* Expressed as parts per 10 000 (0/000) of fresh weight of fish.

TABLE 15.8

Food conversion by trout and perch in Loch Leven, June—September, 1971 (from Thorpe, 1974)

Period	Mean intake (% body weight per day)		Mean growth rate (% body weight per day)		Conversion efficiency (%)		Overall conversion efficiency June—Sept. (%)	
	trout	perch	trout	perch	trout	perch	trout	perch
June—July	3.20	6.60	0.53	0.19	16.6	2.88		
July—Aug.	2.46	5.93	0.14	0.19	5.7	3.20	10.3	2.89
Aug.—Sept.	2.01	4.18	0.16	0.10	8.0	2.39		

A summary of trout production is given in Table 15.6. Negative production implies the use of body reserves for maintenance. The productivity of trout varies between 5 000 and 23 000 kg per year (wet weight) for the whole lake (1968—1972). The productivity of perch older than 2 years was 37 000 kg per year (1970—1971), while that of 0—2 years old perch is estimated between 30 700 and 860 000 kg per year. This larger production of perch is notable, because they use mainly the same food (Table 15.7) and trout has a higher overall conversion efficiency (Table 15.8). The total food consumption of both trout and perch has been estimated from field experiments (Table 15.7). Stomach-filling followed a diurnal pattern; composition (quantitative and qualitative) of the stomach contents was determined, and the rate of evacuation was estimated. Perch completes its life cycle wholly within the Loch. Spawning occurs between late April and early June within that area of the Loch which has a depth of less than 5 m. Mark-recapture methods were used to assess total population density. Tag loss was determined by double marking and appeared to be only 1.5%. Perch production is summarised in Table 15.6.

Food conversion efficiency is given in Table 15.8. From Tables 15.7 and 15.8 it can be seen that perch and trout eat mainly the same prey, with some variation in relative quantities. A considerable quantity of perch fry is eaten by trout (and perch itself). The only other organism which forms more than 30% of the total food consumed is *Asellus*.

15.6. CONCLUDING REMARKS

From the description of these two lakes, it will be clear that simple food chains do not exist. A quantitative approach to a food web is much more difficult. A study of population dynamics, e.g. of *Chironomus* in Loch Leven which was thought to be a simple problem, was complicated by the difficulties of measuring growth and food uptake. Further difficulties occurred because the chironomids were eaten by at least two different major fish populations and because their numbers fluctuated markedly from year to year. Achievement of the ultimate goal of productivity studies — the measurement and understanding of the efficiency of food chains — still lies far in the future. Short cuts by the use of biological models, valuable as they may be for the more pragmatic approach of the practical water manager, contribute relatively little towards scientific understanding of the aquatic ecosystem. It can not be overemphasised that a biological model can only be a "short cut" when all the parameters and food chain efficiencies are measured.

There has been a considerable amount of work done on productivity in lakes in Russia, and this is summarised by Winberg (1972). The Russian approach relies on the use of estimated conversion factors such as dry weight, P/B values, and correlations between such processes as O_2 uptake and body weight. Because the work has recently been reviewed by Winberg (1972), Alimov and Winberg (1972), Moskalenko (1972), Sokolov (1972) and Romanenko (1973) and also because much of the original work is in Russian, it is not considered here.

REFERENCES

Abeliovich, A. and Shilo, M., 1972a. Photooxidative reactions of c-phycocyanin. *Biochim. Biophys. Acta*, 283: 483—491.

Abeliovich, A. and Shilo, M., 1972b. Photooxidative death in blue-green algae. *J. Bacteriol.*, 111(3): 682—689.

Allen, K.R., 1951. The Korokiwi stream. A study of trout population. *Fish. Bull. N.Z. Mar. Dep.*, 10: 231 pp.

Alimov, A.F. and Winberg, G.G., 1972. Biological productivity of two northern lakes. *Verh. Int. Ver. Theor. Angew. Limnol.*, 18(1): 65—70.

Arnold, D.E., 1971. Ingestion, assimilation, survival, and reproduction by *Daphnia pulex* fed seven species of blue-green algae. *Limnol. Oceanogr.*, 16(6): 906—920.

Bailey-Watts, A.E., 1974. The algal plankton of Loch Leven, Kinross. *Proc. R. Soc. Edinb., B*, 74: 135—156.

Bindloss, M.E., 1974. Primary productivity of phytoplankton in Loch Leven, Kinross. *Proc. R. Soc. Edinb., B*, 74: 157—181.

Burgis, M.J., 1973. The ecology and production of copepods, particularly *Thermocyclops hyalinus*, in the tropical Lake George, Uganda. *Freshwater Biol.*, 1: 169—192.

Burgis, M.J., 1974. Revised estimate for the biomass and production of zooplankton in Lake George, Uganda. *Freshwater Biol.*, 4(6): 535—541.

Burgis, M.J. and Walker, A.F., 1972. A preliminary comparison of the zooplankton in a tropical and a temperate lake (Lake George, Uganda and Loch Leven, Scotland). *Verh. Int. Ver. Theor. Angew. Limnol.*, 18(2): 647—655.

Burgis, M.J. et al., 1973. The biomass and distribution of organisms in Lake George, Uganda. *Proc. R. Soc. Lond., Ser. B*, 184: 271—298.

Burns, C.W. and Rigler, F.H., 1967. Comparison of filtering rates of *Daphnia rosea* in lake water and in suspensions of yeast. *Limnol. Oceanogr.*, 12: 492—502.

Butler, E.I., Corner, E.D.S. and Marshall, S.M., 1970. On the nutrition and metabolism of zooplankton. VII. Seasonal survey of nitrogen and phosphorus excretion by *Calanus* in the Clyde sea-area. *J. Mar. Biol. Assoc. U.K.*, 50: 525—560.

Comita, G.W., 1972. Seasonal zooplankton cycles, production and transformation of energy in Severson Lake, Minnesota. *Arch. Hydrobiol.*, 70: 14—66.

Conover, R.J., 1962. Metabolism and growth in *Calanus hyperboreus* in relation to its life cycle. *Rapp. P.-V. Réun. Cons. Perm. Int. Explor. Mer*, 153: 190—197.

Corner, E.D.S., 1961. On the nutrition and metabolism of zooplankton. I. Preliminary observations on the feeding of the marine copepod *Calanus helgolandicus* (Claus). *J. Mar. Biol. Assoc. U.K.*, 41: 5—16.

Corner, E.D.S. and Cowey, C.B., 1964. Some nitrogenous constituents of the plankton. *Ann. Rev. Oceanogr. Mar. Biol.*, 2: 147—167.

Corner, E.D.S., Cowey, C.B. and Marshall, S.M., 1965. On the nutrition and metabolism of zooplankton. III. Nitrogen excretion by *Calanus. J. Mar. Biol. Assoc. U.K.*, 45: 429—442.

Cowey, C.B. and Corner, E.D.S., 1963. On the nutrition and metabolism of zooplankton. II. The relationship between the marine copepod *Calanus helgolandicus* and particulate material in Plymouth sea water, in terms of aminoacid composition. *J. Mar. Biol. Assoc. U.K.*, 43: 495—511.

Davies, B.R., 1974. The planktonic activity of larval Chironomidae in Loch Leven, Kinross. *Proc. R. Soc. Edinb., B*, 74: 275—283.

Dodson, St.I., 1972. Mortality in a population of *Daphnia rosea. Ecology*, 53(6): 1011—1023.

Dussart, B., 1966. *Limnologie; l'étude des eaux continentales.* Gauthier-Villars, Paris, 618 pp.

Dussart, B., 1967. *Les Copépodes des eaux continentales d'Europe occidentale, 1. Calanoides et Harpacticoïdes.* Boubée, Paris.

Dussart, B., 1969. *Les Copépodes des eaux continentales d'Europe occidentale, 2. Cyclopoïdes et biologie.* Boubée, Paris.

Edmondson, W.T., 1960. Reproductive rates of rotifers in natural populations. *Mem. Ist. Ital. Idrobiol.*, 12: 21—77.

Edmondson, W.T., 1965. Reproductive rate of planktonic rotifers as related to food and temperature in nature. *Ecol. Monogr.*, 35: 61—111.

Edmondson, W.T., 1968a. Lake eutrophication and water quality management: The Lake Washington case. In: *Water Quality Control.* University of Washington Press, Seattle, pp. 139—178.

Edmondson, W.T., 1968b. A graphical model for evaluating the use of the egg ratio for measuring birth and death rates. *Oecologia*, 1: 1—37.

Edmondson, W.T. (Editor), 1971. *A Manual on Methods for the Assessment of Secondary Productivity in Fresh Waters.* IBP Handbook no. 17. Blackwell, Oxford, 358 pp. (in collaboration with G.G. Winberg).

Edmondson, W.T., 1972. Instantaneous birth rates of zooplankton. *Limnol. Oceanogr.*, 17: 792—795.

Ganf, G.G., 1972. The regulation of net primary production in lake George, Uganda, East Africa. In: Z. Kajak and A. Hillbrecht-Ilkowska (Editors), *Productivity Problems of Freshwaters.* Polish Scientific Publications, Warszawa—Kraków, pp. 693—708.

Ganf, G.G., 1974. Community respiration of equatorial plankton. *Oecologia*, 15: 17—32.

Ganf, G.G. and Blažka, P., 1974. Oxygen uptake, ammonia and phosphate excretion by the zooplankton of a shallow equatorial lake (Lake George, Uganda). *Limnol. Oceanogr.*, 19: 313—325.

Ganf, G.G. and Viner, A.B., 1973. Ecological stability in a shallow equatorial lake (Lake George, Uganda). *Proc. R. Soc. Lond., Ser. B*, 184: 321—346.

Golterman, H.L., 1971. The determination of mineralization losses in correlation with the estimation of net primary production with the oxygen method and chemical inhibitors. *Freshwater Biol.*, 1: 249—256.

Greenwood, P.H. and Lund, J.W.G., 1973. A discussion on the biology of an equatorial lake: Lake George, Uganda (organized by P.H. Greenwood and J.W.G. Lund, held on 14 February 1973). *Proc. R. Soc. Lond., Ser. B*, 184: 227—346.

Hall, D.J., 1964. An experimental approach to the dynamics of a natural population of *Daphnia galeata mendotae*. *Ecology*, 45(1): 94—110.

Haney, J.F., 1973. An in situ examination of the grazing activities of natural zooplankton communities. *Arch. Hydrobiol.*, 72: 87—132.

Harris, E., 1959. The nitrogen cycle in Long Island Sound. *Bull. Bingham Oceanogr. Coll.*, 17: 31—65.

Holden, A.V. and Caines, L.A., 1974. Nutrient chemistry of Loch Leven, Kinross. *Proc. R. Soc. Edinb., B*, 74: 101—121.

Horne, A.J. and Viner, A.B., 1971. Nitrogen fixation and its significance in tropical Lake George, Uganda. *Nature*, 232: 417—418.

Hutchinson, G.E., 1967. *A Treatise on Limnology*, 2. *Introduction to Lake Biology and Limnoplankton*. John Wiley, New York, N.Y., 1115 pp.

Johnson, D. and Walkers, A.F., 1974. The zooplankton of Loch Leven, Kinross. *Proc. R. Soc. Edinb., B*, 74: 285—294.

Ketchum, B.H., 1962. Regeneration of nutrients by zooplankton. *Rapp. P.-V. Réun. Cons. Perm. Int. Explor. Mer*, 153: 142—147.

Lindeman, R.L., 1942. The trophic-dynamic aspect of ecology. *Ecology*, 23(4): 399—418.

Maitland, P.S. and Hudspith, P.M.G., 1974. The zoobenthos of Loch Leven, Kinross, and estimates of its production in the sandy littoral area during 1970 and 1971. *Proc. R. Soc. Edinb., B*, 74: 219—239.

McAllister, C.D., 1969. Aspects of estimating zooplankton production from phytoplankton production. *J. Fish. Res. Board Can.*, 26(2): 199—220.

McMahon, J.W., 1965. Some physical factors influencing the feeding behaviour of *Daphnia magna* Straus. *Can. J. Zool.*, 43: 603—611.

McMahon, J.W. and Rigler, F.H., 1963. Mechanisms regulating the feeding rate of *Daphnia magna* Straus. *Can. J. Zool.*, 41: 321—332.

McMahon, J.W. and Rigler, F.H., 1965. Feeding rate of *Daphnia magna* Straus in different foods labeled with radioactive phosphorus. *Limnol. Oceanogr.*, 10: 105—113.

Moriarty, D.J.W. et al., 1973. Feeding and grazing in Lake George, Uganda. *Proc. R. Soc. Lond., Ser. B*, 184: 299—319.

Moskalenko, B.K., 1972. Biological productive system of Lake Baikal. *Verh. Int. Ver. Theor. Angew. Limnol.*, 18(2): 568—573.

Nauwerck, A., 1963. Die Beziehungen zwischen Zooplankton und Phytoplankton im See Erken. *Symb. Bot. Ups.*, 17(5): 1—163.

Neess, J.C. and Dugdale, R.C., 1959. Computation of production for populations of aquatic midge larvae. *Ecology*, 40(3): 425—430.

O'Brien, W.J. and De Noyelles, F., 1972. Photosynthetically elevated pH as a factor in zooplankton mortality in nutrient enriched ponds. *Ecology*, 53: 605.

Odum, E.P. and Odum, H.T., 1960. *Fundamentals of Ecology*. Saunders, Philadelphia, Pa., 2nd ed., 546 pp.

Patalas, K., 1972. Crustacean plankton and the eutrophication of St. Lawrence Great Lakes. *J. Fish. Res. Board Can.*, 29(10): 1451—1462.

Porter, K.G., 1973. Selective grazing and differential digestion of algae by zooplankton. *Nature*, 244: 179—180.

Rabinowitch, E.I., 1951. *Photosynthesis and Related Processes*, 2. (Part 1). *Spectroscopy and Fluorescence of Photosynthetic Pigments, Kinetics of Photosynthesis*. John Wiley, New York, N.Y., 605 pp.

Richman, S., 1958. The transformation of energy by *Daphnia pulex*. *Ecol. Monogr.*, 28: 273—291.

Richman, S., 1964. Energy transformation studies on *Diaptomus oregonensis*. *Verh. Int. Ver. Theor. Angew. Limnol.*, 15: 654—659.

Richman, S., 1966. The effect of phytoplankton concentration on the feeding rate of *Diaptomus oregonensis*. *Verh. Int. Ver. Theor. Angew. Limnol.*, 16: 392—398.

Rigler, F.H., 1961. The relation between concentration of food and feeding rate of *Daphnia magna* Straus. *Can. J. Zool.*, 39: 857—868.

Riley, G.A., 1944. Carbon metabolism and photosynthetic efficiency. *Am. Sci.*, 32: 132—134.

Romanenko, V.I., 1973. Abundance and production of bacteria in Latvian lakes. *Verh. Int. Ver. Theor. Angew. Limnol.*, 18 (part 3): 1306—1310.

Ryther, J.H., 1954. Inhibitory effects of phytoplankton upon the feeding of *Daphnia magna* with reference to growth, reproduction, and survival. *Ecology*, 35(4): 523—533.

Sokolov, A.A., 1972. Hydrological investigations of lakes and reservoirs in the USSR. *Verh. Int. Ver. Theor. Angew. Limnol.*, 18 (part 2): 779—786.

Sorokin, Y.I., 1965. On the trophic role of chemosynthesis and bacterial biosynthesis in water bodies. *Mem. Ist. Ital. Idrobiol.*, 18 (Suppl.): 187—205.

Sorokin, Y.I., 1968. The use of ^{14}C in the study of nutrition of aquatic animals. *Mitt. Int. Ver. Theor. Angew. Limnol.*, 16: 1—41.

Stross, R.G., Neess, J.C. and Hasler, A.D., 1961. Turnover time and production of planktonic crustacea in limed and reference portion of a bog lake. *Ecology*, 42(2): 237—245.

Talling, J.F., 1957. The phytoplankton population as a compound photosynthetic system. *New Phytol.*, 56: 133—149.

Talling, J.F., 1971. The underwater light climate as a controlling factor in the production ecology of freshwater phytoplankton. *Mitt. Int. Ver. Theor. Angew. Limnol.*, no. 19: 214—243.

Thorpe, J.E., 1974. Trout and perch populations at Loch Leven, Kinross. *Proc. R. Soc. Edinb., B*, 1974: 295—313.

Uhlmann, D., 1971. Influence of dilution, sinking and grazing rate on phytoplankton populations of hyperfertilized ponds and micro-ecosystems. *Mitt. Int. Ver. Theor. Angew. Limnol.*, no. 19: 100—124.

Viner, A.B., 1973. Responses of a mixed phytoplankton population to nutrient enrichments of ammonia and phosphate, and some ecological implications. *Proc. R. Soc. Lond., Ser. B*, 183: 351—370.

Viner, A.B. and Smith, I.R., 1973. Geographical, historical and physical aspects of Lake George. *Proc. R. Soc. Lond., Ser. B*, 184: 235—270.

Winberg, G.G., 1972. Etudes sur le bilan biologique, énergetique et la productivité des lacs en Union Soviètique (Edgardo Baldi Memorial Lecture.) *Verh. Int. Ver. Theor. Angew. Limnol.*, 18 (part 1): 39—64.

Worthington, E.B., 1949. An experiment with populations of fish in Windermere, 1939—1948. *Proc. Zool. Soc. Lond.*, 120: 113—149.

Wright, J.C., 1965. The population dynamics and production of *Daphnia* in Canyon Ferry Reservoir, Montana. *Limnol. Oceanogr.*, 10: 538—590.

CHAPTER 16

BACTERIAL LIMNOLOGY

16.1. INTRODUCTION

In previous chapters we have referred to processes operated by bacteria. So far little attention has been given to the bacteria themselves. Like the algae they depend entirely on their environment; the processes they govern are at the same time the *raison d'être* of these bacteria.

The ecology of these organisms has not been studied in much detail. Bacteria may obtain their energy, like algae, from solar radiation or they may obtain it from dissolved inorganic or organic molecules. The photosynthetic bacteria differ from the algae in that they use other inorganic or organic molecules and not H_2O as reducing agent. Such molecules are, in general, easily oxidisable, so their concentration in water will always be low, except under anaerobic conditions. Thus most bacteria are more restricted than algae, both in population numbers and in the places where they can occur.

The bacteria can conveniently be grouped by which energy source and electron donor they use:

(I) *Chemotrophic bacteria* (using chemical energy): (a) chemolithotrophic — energy from oxidation of inorganic compounds such as NH_3, H_2S and Fe^{2+}; (b) chemoorganotrophic — energy from oxidation or fermentation of organic compounds.

(II) *Phototrophic bacteria* (using solar energy): (a) photolithotrophic — growth depends on an inorganic electron donor, e.g. H_2S; (b) photoorganotrophic — growth depends on an organic electron donor.

Group Ib used to be called heterotrophic bacteria, while groups Ia and IIa were referred to as autotrophic bacteria. These terms should be avoided, but in the following discussion they are sometimes used when the original authors did so.

A bacterium may be obligately autotrophic, which means that it cannot grow heterotrophically, or it may be facultative, which means that it may grow either autotrophically or heterotrophically. A rather peculiar metabolism is the so-called "mixotrophic" type, in which certain organic compounds can be used but only in combination with (weak) light. Examples are blue-green algae growing in dim light on glucose (subsection 6.5.1), and the purple non-sulphur bacteria (subsection 16.4.3). Mixotrophic chemolithotrophs are mentioned in subsection 16.2.1.

In most textbooks the ecology of bacteria is hardly discussed. Sometimes a chapter is dedicated to "the Environment", and here such subjects as the influence of sugar on growth are discussed, stating that 0.5—2% gives maximal growth, while 20—40% is inhibitory! A review of Russian work on the ecology of aquatic bacteria is published by Kuznetsov (1968) and a review of biochemical ecology of microorganisms by Alexander (1971).

The lithotrophic bacteria were reviewed by Kelly (1971), who regarded the presence of a functional Calvin-Benson reductive pentose-phosphate cycle (see section 4.4) as a characteristic feature of all lithotrophs. The enzymes ribulose diphosphate carboxylase (RDP-ase) and phosphoribulokinase occur in all lithotrophs but not in the non-photosynthetic heterotrophs nor in the sulphate-reducing lithotrophs. They do occur in the photo-organotrophic Athiorodaceae after photosynthetic growth and are inducible in *Micrococcus denitrificans* if growing as a hydrogen bacterium. In *Chlorobium* the major mechanism for CO_2 fixation is a reversed tricarboxylic acid (TCA) cycle with ferredoxin-dependent carboxylations of acetyl CoA to pyruvate and of succinyl CoA to α-ketoglutarate — without NADPH — and carboxylations of α-ketoglutarate and pyruvate (Wood and Workman reaction, section 4.4). The product of the reversed cycle is oxaloacetate. The mechanism for regulation of this TCA cycle is still unknown.

One of the regulatory mechanisms of the Calvin cycle is the suppression of RDP-ase synthesis by organic nutrients during heterotrophic growth, and this suppression is complete in some species of *Thiobacillus*. But the suppressions are always measured at substrate concentrations much higher than the naturally occurring ones. In many heterotrophic bacteria and in some facultative autotrophic ones the TCA and glyoxylate cycle provide energy, but the TCA cycle in lithotrophs serves a biosynthetic function. In the obligate chemolithotrophs and in the photolithotrophs, such as the green bacteria and *Chromatium*, an incomplete TCA cycle is found, lacking α-ketoglutarate dehydrogenase so that acetate can only be converted into α-ketoglutarate, which gives rise to glutamate, proline and arginine. Whether or not lack of one of the TCA-cycle enzymes is the cause of obligate lithotrophy is still an open question. The same explanation has been proposed for the failure to oxidise NADH but NADPH seems to be oxidised though at a rather low efficiency. Kelly suggested that the apparent failure or this low efficiency might be the result of procedures used in the preparation of cell-free extracts, while the low efficiency of ATP formation may occur only in the living cell. He also suggested that poor permeability into resting cells rather than a failure of the respiratory apparatus may be the cause of lithotrophy.

Many studies deal with the question: "What is the biochemical basis of obligate chemolithotrophy?" Smith et al. (1967) reviewing this subject demonstrated that in blue-green algae and *Thiobacillus* the tricarboxylic-acid cycle is blocked at the level of α-ketoglutarate oxidation. The bacteria lacked not only α-ketoglutarate dehydrogenase but also NADH oxidase, so that no ATP can be generated from organic substrates, and showed low levels of malic and succinic dehydrogenases (see Appendix III). Thus these bacteria cannot grow heterotrophically, but may make use of some simple organic acids, e.g. acetic acid. An older hypothesis about the biochemical basis of chemolithotrophy, i.e. general impermeability (see subsection 6.5.1) is therefore inaccurate. Acetate, although it is more rapidly assimilated than other organic compounds tested, cannot contribute more than 10% of the newly synthesised carbon of the cell, so that e.g. the blue-green algae remain strongly dependent on the assimilation of CO_2. Naturally occurring mixtures of compounds have not yet been tested, although they may be significant from an ecological point of view.

In photolithotrophic organisms, NADH and ATP can be generated by cyclic and non-cyclic photophosphorylation, while for chemolithotrophs Smith et al. suggested that NADH may be generated by the ATP-mediated reversed electron transport. Hooper (1969) showed that *Nitrosomonas* cannot grow heterotrophically, and is thus obligately lithotrophic. It also lacked α-ketoglutaric dehydrogenase and a KCN-sensitive NADH oxidase, but it did possess a NADH oxidase of the peroxidase type which was not coupled with ATP synthesis. Succinic dehydrogenase activity was low as well.

The reduction of NAD by an inorganic nitrogen compound requires the concomitant hydrolysis of ATP. Borris and Ohlmann (1972) showed that citrate synthetase — another enzyme of the TCA cycle — is the key enzyme in the regulation of heterotrophic and autotrophic metabolism of *Rhodopseudomonas*. The enzyme is strongly inhibited by light-generated NADH, which may cause a repression of the TCA cycle by the reductive pentose-phosphate cycle (Calvin cycle, see Appendix III).

It should be remarked that some bacteria originally thought to be "obligate" lithotrophs, for example *Nitrobacter agilis* and some strains of *Thiobacillus ferrooxidans*, will in fact grow heterotrophically.

Some (or even many) bacteria depend on organic material, though this does not mean that these compounds are used as such. Although the chemoorganotrophic bacteria may be considered to be carnivores, they differ in that they will normally decompose the organic compounds which they take up because they use these compounds as sources of C, H, N, P, S, O and of energy and do not often use the carbohydrates, fats and amino acids directly. The energy incorporated in the metabolised molecules will generally be converted into ATP before it can be used again, although small organic molecules such as acetic acid may be used as such. There is one exception to this general principle; if bacteria need vitamins and purines, they will incorporate these compounds directly without breaking them down first. The need for vitamins in the lithotrophic groups of bacteria does not therefore render them organotrophic.

Finally most if not all bacteria need Mg^{2+}, Mn^{2+}, Fe^{2+}, Co, and MoO_4^-, which all serve in an active prosthetic group of an enzyme (see Table 11.1). The quantities required are therefore minute.

In the next paragraphs the occurrence of different groups in the aquatic habitat is discussed. For detailed information on their biochemistry the reader is referred to the appropriate textbooks.

16.2. CHEMOLITHOTROPHIC BACTERIA

16.2.1. Introduction

Chemolithotrophic bacteria derive their energy from the oxidation of inorganic compounds, such as H_2, NH_3, NO_2^-, S and Fe^{2+}, and use CO_2 as source of carbon. In a lake there may be much potential energy in this form during certain periods and at certain places; sometimes production of these bacteria may even be greater than the primary production.

Several Russian microbiologists tried to estimate the extent of chemosynthesis, while at the same time they also tried to distinguish between chemolitho- and chemoorganotrophic (which they called heterotrophic) growth.

Sorokin (1965) measured the total dark $^{14}CO_2$ uptake in the normal way as described in Chapter 4. The concomitant heterotrophic assimilation of CO_2 was estimated from the rate of production of bacterial biomass, estimated from direct microscopic counts, and from K, the relative average amount of CO_2-carbon, which was used for biosynthesis by the usual heterotrophic microflora. The possible assimilation of CO_2-carbon by heterotrophic microflora, H, was then calculated:

$$H = \frac{24 \cdot K \cdot 0.07 \cdot V \cdot N}{G} \text{ mg l}^{-1} \text{ d}^{-1} \text{ of C}$$

where V = average volume of cells; N = initial number of bacteria counted; G = generation time in hours; and K = average value of the relative heterotrophic assimilation of CO_2. K was estimated using the radiocarbon method (based on specific activities) and the organic content of bacterial cells. These were grown in pure cultures in a complex medium with a large amount of HCO_3^-. K was found to be 0.02—0.03. Using microscopic measurements of the biomass, a value of about 0.04 was found. Romanenko (1964, cited in Sorokin) used a value of 0.06. He found that the relation between the O_2-consumption, heterotrophic assimilation of CO_2 and production of bacteria in terms of carbon was roughly 1 000 : 6 : 100. The generation time, G, was estimated from the increase of bacterial cells in bottles suspended in situ, but enclosing a water sample in a bottle renders most bacteria more active and thus G is shorter. Moreover, since experimental error is great when calculating a value for V, it is obvious that the method is a fairly crude one, giving results of only limited validity. Sorokin detected chemosynthetic CO_2 uptake in the range of 5—30 mg m^{-3} d^{-1} of C in the boundary layers of aerobic zones of some Russian reservoirs. In Cheremshan bay of the Kuybyshev reservoirs (Fig. 16.1) he found a maximum of 20 mg m^{-3} d^{-1} at 9 m depth, just below the metalimnion. The main sources of energy for chemosynthesis in the reservoirs were the gases H_2, H_2S and CH_4. Addition of these gases stimulated chemosynthesis strongly, while their removal caused a great decrease of the rate of $^{14}CO_2$ assimilation. Romanenko (1966) quoted a pro-

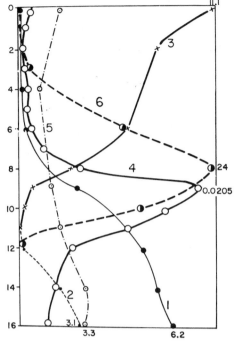

Fig. 16.1. Chemosynthesis and other factors connected with it in Cheremshan bay of Kuybyshev reservoir in August 1958. (From Sorokin, 1965.)

Legend

1: $CH_4 (mg\ l^{-1})$
2: $H_2S\ (mg\ l^{-1})$
3: $O_2\ (mg\ l^{-1})$
4: Chemosynthesis $(mg\ l^{-1} d^{-1}$ of C)
5: Total number of bacteria $(10^9\ l^{-1})$
6: $Daphnia$ (individuals l^{-1})

duction of 166 g m^{-2} during the growth period in the Rybinsk reservoir. At least 5 times as much organic matter must have been destroyed to permit this production, and as primary production amounted to 216 g m^{-2} of dry organic matter in the same period, one must conclude that in the reservoir considerably more allochthonous organic matter was oxidised than autochthonous, the primary production.

Romanenko attempted to compare the values of bacterial assimilation of CO_2 with the number of H_2- and CH_4-oxidising bacteria, but found no obvious correlation, although considerable numbers of these bacteria were present.

Although in nature chemolithotrophic growth cannot be distinguished from chemoorganotrophic growth, chemolithotrophic growth is certainly important in anaerobic hypolimnia and in silt. It should be emphasised, however, that there is no *additional* production of organic matter because the energy to produce the reduced compounds H_2, NH_3, NO_2 and Fe^{2+} is derived from primary production. This may be from algal material in the lake or from external allochthonous sources.

Table 16.1 summarises the types of chemolithotrophic bacteria which may be found in the aquatic habitat, together with their energy sources and metabolic end-products.

TABLE 16.1

Chemolithotrophic bacteria with energy sources and products of oxidation

Bacterium	Energy source		End-product of oxidation from	
	electron donor	electron acceptor	electron donor	electron acceptor
Beggiatoa spp.	H_2S	O_2	S	H_2O
Desulphovibrio desulphuricans	H_2	SO_4^{2-}	H_2O	H_2S
Micrococcus denitrificans	H_2	NO_3^-	H_2O	N_2
Hydrogenomonas	H_2	O_2	H_2O	H_2O
Nitrobacter spp.	NO_2^-	O_2	NO_3^-	H_2O
Nitrosomonas spp.	NH_3	O_2	NO_2^-	H_2O
Thiobacillus thiooxidans	S^*	O_2	SO_4^{2-}	H_2O
Thiobacillus denitrificans	S^*	NO_3^-	SO_4^{2-}	N_2
Thiobacillus ferrooxidans	Fe^{2+} +	O_2	Fe^{2+}	H_2O
	thiosulphate	O_2	SO_4^{2-}	H_2O
Ferrobacillus ferrooxidans	Fe^{2+}	O_2	Fe^{3+}	H_2O
Methanomonas	CH_4	O_2	CO_2	H_2O
Alcaligenes ⎫ *Achromobacter* ⎬	CH_4	NO_3^-	CO_2	N_2

* Or thiosulphate and other inorganic S compounds.

The chemolithotrophic bacteria are characterised physiologically by their ability to use CO_2 as a sole carbon source using the energy for the reduction from one of the oxido-reduction reactions given in Table 16.1. In many cases they are completely unable to oxidise organic compounds and then they are obligately chemolithotrophic. Some exceptions exist.

Taylor et al. (1971) showed that *Thiobacillus denitrificans* which oxidises thiosulphate as an obligate chemolithotroph used ^{14}C-labelled acetate for its cell carbon, while NO_3^- was reduced to N_2. About 10% of newly synthesised cellular carbon was provided by the acetate, which was incorporated mainly into lipids and into the amino acids: glutamate, proline, arginine and leucine. From Table 16.1 it can be seen that most reduced compounds can be oxidised either with O_2 or with NO_3^- or SO_4^{2-}. Thus *Hydrogenomonas* uses O_2, while *Desulphovibrio* uses SO_4^{2-} to oxidise H_2. *Beggiatoa* and most *Thiobacillus* species use O_2 to oxidise sulphur, whereas *T. denitrificans* uses NO_3^- for this purpose. This means that in the hypolimnion O_2 can be used so long as it is present and that after its depletion SO_4^{2-} or NO_3^- will be reduced (see Chapter 9).

For a long time chemolithotrophic bacteria were considered to be obligate lithotrophs because many organic compounds strongly inhibit their growth. This widely held belief was strongly criticised by Rittenberg (1969), who discussed alternative hypotheses which explain the inhibitory effect of compounds, such as glucose, peptone and asparagine. Basing his argument on the effects of certain compounds on growth of *Thiobacillus*, Rittenberg believed that the inhibition by some amino acids (e.g., valine and phenylalanine, $10^{-3} M$) could be caused by amino-acid imbalance. Rittenberg suggested that other compounds, such as acids from the TCA cycle, could inhibit by repressing energy generation or the ribulose-diphosphate cycle or both (see Appendix III). However, many other organic compounds are readily assimilated. Ecological studies are needed using mixtures of compounds at much lower concentrations. It is not impossible that the amino-acid imbalance would then completely disappear. A few cases of slow heterotrophic growth are summarised by Rittenberg, who believed that true heterotrophic growth is restricted to a few cases only and that growth in these cases is rather slow.

Rittenberg classified the chemolithotrophs in three groups:

Group I: obligate; i.e. using only inorganic energy and an inorganic carbon source.

Group II: mixotrophic; i.e. concurrent use of inorganic and organic energy or the concurrent assimilation of CO_2 via the ribulose-diphosphate cycle and organic carbon.

Group III: chemolithotrophic heterotrophs, such as *Desulphovibrio*, which get their carbon from organic sources but have a chemolithotrophic potential.

16.2.2. Sulphur bacteria

The sulphur-oxidising chemolithotrophic bacteria belong to two different groups.

The first comprises the colourless filamentous *Beggiatoa* (which much resembles certain blue-green algae, especially *Oscillatoria*) and *Thiothrix*, which is less common and resembles *Beggiatoa*, but occurs, characteristically, in a rosette form. *Beggiatoa* is common near sulphur springs and on the surface of black mud but only in mud which is in contact with the atmosphere. It oxidises H_2S to elemental sulphur, which is stored in the cells as droplets or granules and these give *Beggiatoa* cells their characteristic appearance. When the H_2S is depleted, the sulphur will be oxidised to SO_4^{2-}.

The second group contains *Thiobacillus*, which is a small, gram-negative, polarly flagellated, eubacterium and reproduces by binary fission. Most species can oxidise sulphur and thiosulphate, the latter compound being often used in cultures because it is easier to handle. It could be regarded as a source of very finely suspended sulphur. Because these bacteria depend for their growth on the oxidation of H_2S (either by O_2 or NO_3^-), conditions for their existence are rare; H_2S and O_2 only occur together in unstable conditions, e.g. in a hypolimnion containing O_2 in the upper but not in the lower layers. Some species in this group use O_2 and others use NO_3^- as electron acceptor, so that the chemical state of the hypolimnion will determine which species of *Thiobacillus* will actually occur there.

T. denitrificans cannot use nitrate as a nitrogen source, since it is not able to reduce it to ammonia.

16.2.3. Hydrogen bacteria

The enzyme hydrogenase, which catalyses the reaction:

$$H_2 \rightleftharpoons 2 H^+$$

occurs in a wide variety of bacteria including purple and green bacteria, sulphate reducers, *Pseudomonas, Clostridia, Micrococci,* coliform bacteria and *Azotobacter.* If these bacteria also contain enzymes which reduce CO_2, they can grow chemolithotrophically with H_2 as energy source. This ability is widespread among the polarly flagellated, gram-negative rods of the *Pseudomonas* group.

These and other chemolithotrophic H_2 oxidisers can also grow organotrophically, i.e. at the expense of organic compounds. When doing so they often grow better than as chemolithotrophs. Those pseudomonads which are potentially able to oxidise H_2 are placed in a special genus, *Hydrogenomonas,* but they are indistinguishable from other organotrophic pseudomonads. True obligate H_2-oxidising chemolithotrophs do not seem to occur at all. Rittenberg and Goodman (1969) established mixotrophic growth conditions for *Hydrogenomonas eutropha* by adding lactate to cultures in a gaseous environment of H_2, O_2 and CO_2. Growth rate and yield were greater than in heterotrophic cultures and it was shown that the complete autotrophic and heterotrophic modes of nutrition functioned simultaneously under mixotrophic conditions. There is a great need for research on the ecology of these bacteria, which may be expected in the top layers of the hypolimnion.

Micrococcus denitrificans can use both O_2 and NO_3^- as electron acceptor, although O_2 is preferred if both are present.

16.2.4. Iron bacteria

The iron bacteria (including the genus *Ferrobacillus*) oxidise ferrous to ferric iron:

$$4 Fe^{2+} + 4 H^+ + O_2 \rightarrow 4 Fe^{3+} + 2 H_2O$$

Other iron bacteria are *Gallionella* and *Leptothrix.* They occur in oxygenated water layers and cannot oxidise Fe^{2+} in a deoxygenated hypolimnion. They may occur in the top layers of a hypolimnion, although they have rarely been looked for. Sokolova (1961, cited in Kuznetsov, 1968) studied the iron cycle in Lake Glubok, where iron bacteria formed about 10% of the total bacterial population. Eleven species were identified. *Spirothrix pseudovacuolata* was present in all plankton samples and was the most numerous species. It also occurred near the bottom in the oxygen-free zone. Sokolova has shown that iron bacteria play a most important part in the iron cycle of mesotrophic lakes by mediating oxidation and mineralisation of iron humates with subsequent sedimentation of $Fe(OH)_3$. Other organisms found were *Gallionella* and *Metallogenium personatum*. The first did not occur in the oxygen-free zone. *Metallogenium* did not occur in the epilimnion, but in the deep bottom layer $340 \cdot 10^6$ cells per litre were found.

Siderocapsa, which is able to mineralise humates, appeared during rainy periods when many iron humates were brought into the lake. Gorlenko and Kuznetsov (1972) found *Metallogenium personatum* at a depth of 10 m in the micro-aerobic zone of Lake Kononjer. The occurrence of the iron-oxidising bacteria in oxygen-free waters is difficult to explain, unless it is caused by sedimentation of cells which have become heavier owing to accumulation of $Fe(OH)_3$.

Shafia et al. (1972) were able to convert (irreversibly) the chemolithotroph *Ferrobacillus ferrooxidans* into an obligate organotrophic strain by replacing Fe^{2+} with 0.5% glucose. In this respect the species differs from *Thiobacillus* and *Nitrobacter*, which can revert to an autotrophic mode of life after continued cultivation on organic matter. This reversibility may be ecologically important during long periods without ferrous iron present.

16.2.5. Methane-oxidising bacteria

Although methane is an organic compound, *Methanomonas* is customarily regarded as chemolithotrophic since its metabolism in some ways resembles that of the chemolithotrophs. It is almost as specialised physiologically as the nitrifying bacteria, being unable to grow at the expense of any other oxidisable compound. In this it contrasts with the organotrophic *Pseudomonas* spp., which have a very wide range of different organic substrates (see section 16.3). The association of *Methanomonas* with *Hydrogenomonas* in hypolimnia encourages one to consider them both as chemolithotrophic.

Davies (1973) isolated various species of *Alcaligenes* and of *Achromobacter* that are able to oxidise CH_4 with nitrate. This is interesting: all the oxidations of Table 16.1 *can* be performed by O_2 or NO_3^-. Davies discussed the importance of these findings for practical purposes, i.e. denitrification of treated sewage.

Occurrence of these bacteria in hypolimnia may be restricted because CH_4 production during stratification will not start before nitrate is depleted.

16.2.6. Nitrifying bacteria

Nitrifying bacteria were originally isolated in 1890 by Winogradsky, who showed that there are two kinds: *Nitrosomonas*, which oxidises NH_3 to NO_2^-, and *Nitrobacter*, which converts NO_2^- into NO_3^-. Both nitrifying bacteria were considered to be obligate autotrophs for a considerable time because many organic compounds inhibit nitrification. This phenomenon is however less common than was once thought.

Rittenberg (1969) classified *Nitrosomonas* as obligately chemolithotrophic but *Nitrobacter* as mixotrophic. *Nitrosomonas* has very small gram-negative polarly flagellated cells, which reproduce by binary fission. *Nitrobacter* is rod-shaped and multiplies by a special process of budding. The bud grows directly out of one pole of the mother cell and divisional stages consequently have a spear-shaped form.

The biochemistry of nitrifying bacteria was reviewed by Wallace and Nicholas (1969). One of their most important conclusions from an ecological point of view is that, although *Nitrobacter* can grow chemolithotrophically, it can also metabolise acetate via the Krebs tricarboxylic-acid cycle. During this facultative chemoorganotrophic growth, acetate may suppress the CO_2-fixing enzymes of the pentose-phosphate pathway, although generation times remain as long as they are when nitrite is used. If this happened at natural concentrations of organic compounds it would certainly help them to survive during NO_2^- depletion. Long lag phases found after adding nitrite to lakewater may be caused by cells having to adapt and need not necessarily be due to growth of a whole new population.

Another biochemical phenomenon which may have ecological importance is the low yield of *Nitrobacter* and of most other chemolithotrophic organisms, together with the associated long generation time. Wallace and Nicholas cited as one possible cause the low activity of the Krebs tricarboxylic-acid cycle in most chemolithotrophic bacteria. In contrast to *Escherichia coli* for example, the TCA activity does not increase after acetate uptake. This may have the effect of retarding the supply of carbon skeletons for amino acids, etc. Another factor which Wallace and Nicholas considered is the low yield of ATP per mol. O_2 taken up (0.15 mol. ATP per mol. O_2 or P/O = 0.3). This contrasts with the chemoorganotrophic bacteria, but is in agreement with the low energy production of the two oxidations. If for example glucose is oxidised, 56 kcal. are produced per O atom, while for NH_3 or NO_2 oxidation only 20—25 kcal. are produced.

It is important to note that generation of NADH for the reduction of CO_2 requires ATP, and this feature makes the chemolithotrophic system uneconomical. This too may contribute to their slow growth.

Details of the electron transfer chain, which includes flavines and quinones, are not yet well established, although it seems likely that cytochromes *a* and *o* are terminal oxidisers. Nevertheless, the chain must have some remarkable features as it is able to oxidise toxic substances such as NH_2OH and HNO_2 more effectively than it can NADH or succinate.

Daubner and Ritter (1973) related the activity of nitrifying bacteria to environmental parameters — mainly O_2 concentration — in two gravel pits. The nitrite-oxidising bacteria decreased less with lack of oxygen than the ammonia-oxidising ones (which need more O_2).

16.2.7. Desulphovibrio

Desulphovibrio desulphuricans is strictly speaking not a chemolithotrophic bacterium as it cannot derive its cell carbon from CO_2 alone, although it derives its energy, like the chemolithotrophic species, from two inorganic sources:

$$H_2SO_4 + 4 H_2 \rightarrow H_2S + 4 H_2O$$

Organic compounds may also be used as H donor, e.g.:

$$\underset{\text{lactic acid}}{2 \; CH_3-CHOH-COOH} + SO_4^{2-} \rightarrow \underset{\text{acetic acid}}{2 \; CH_3COOH} + 2 \; CO_2 + H_2S + 2 \; OH^-$$

Like other chemolithotrophic bacteria this species does not have a complete tricarboxylic-acid cycle. Therefore the cells cannot oxidise actetate and so acetic acid is excreted.

There is a difference of opinion whether *Desulphovibrio* can reduce CO_2 in small quantities or not at all. Sorokin (1966) demonstrated $^{14}CO_2$ uptake but this may have been due to an isotopic exchange process. Rittenberg quotes specific activity of cell carbon to be usually 5—12% (exceptionally 29%) of that of the $H^{14}CO_3^-$ initially added, with or without yeast extract.

Van Gemerden (1967) found some growth in the presence of CO_2, with formate as electron donor, but he considered that this was due to impurities present in the added components. Formate, and ethyl alcohol are used as electron donors but not as a source of carbon. Van Gemerden showed that 13% of the added lactic acid was used as carbon source. As the reducing capacity of the lactic acid was fully used, it seems likely that this carbon is taken up via acetate. One might hypothetically compare this acetate usage with that of blue-green algae (see subsection 6.5.1) and that of *Thiobacillus* (subsection 16.2.2). Yeast extract itself can be used as carbon source; a fact not always recognised.

Van Gemerden made two important observations of ecological significance. In media with yeast extract as carbon source he showed that sulphide is produced at a constant rate after depletion of the carbon sources and that sulphate and sulphur are reduced simultaneously. This second observation may imply that H_2S production in the mud may continue, even after sulphate is depleted. For this depletion to occur considerable amounts of organic matter are necessary. In Lake Vechten 15 g m^{-3} of SO_4^{2-} is reduced in about three months, for which an amount of $15/96 \cdot 2 \cdot 90 = 30$ g m^{-3} of lactic acid equivalent must be used or about 0.3 g m^{-3}d^{-1}, or — assuming a layer of 1 m from the bottom to be depleted — 0.3 g m^{-2}d^{-1}, which is roughly equivalent to 0.1 g m^{-2} d^{-1} of C. This is quite a considerable proportion of the primary production (annual mean 0.4 g m^{-2} d^{-1}), most of which is in any case mineralised in the epilimnion. When comparing these quantities, it should be remembered that primary production occurs during more than seven months and that the H_2S produced by *Desulphovibrio* may give rise to further organic production by photo- or chemolithotrophic bacteria. Both facts increase the quantity of organic matter available for sulphate reduction. Fischer (1972) followed seasonal changes of the production of chemolithotrophic bacteria both in bottom sediments of a natural pond (112 m^2 × 2 m deep) in a forest area and in a newly formed small reservoir (4 m^2 × 120 cm). Uptake of $^{14}CO_2$ was monitored both in dark and light, while the bacterial cells were counted directly on membrane filters. The coefficient of production was calculated using the equation:

$$\gamma = \frac{\Delta B}{N}$$

where γ = coefficient of production; ΔB = 24 h increment of bacterial biomass in unit volume; and N = initial number of bacterial cells. The coefficient designates the yield efficiency of 10^6 bacterial cells in producing chemolithotrophically a new biomass of bacteria in 24 h. Fischer did not count bacterial cell numbers in bottles after the incubation periods as he believed (correctly) that for several reasons this is misleading in mixed natural populations. Marked seasonal differences were noted both for γ and for N, although in different ways. When populations were small, a high production of biomass was recorded and the low numbers were probably the result of grazing. The opposite situation was found in winter when production is low owing to low tempera-

TABLE 16.2

Enrichment conditions for some chemolithotrophic bacteria[*]

Common features	Additions to medium		Special environmental features		Organism enriched
			atmosphere	pH	
	NH$_4$Cl	1.5			
	CaCO$_3$	5.0	air	8.5	*Nitrosomonas*
Basal medium					
MgSO$_4$·7 H$_2$O 0.2	NaNO$_2$	3.0	air	8.5	*Nitrobacter*
K$_2$HPO$_4$ 1.0					
FeSO$_4$·7 H$_2$O 0.05	NH$_4$Cl	1.0	85% H$_2$		Hydrogen
CaCl$_2$ 0.02			10% O$_2$	7.0	bacteria
MnCl$_2$·4 H$_2$O 0.002			5% CO$_2$		
NaMoO$_4$·2 H$_2$O 0.001					
	NH$_4$Cl	1.0	air	7.0	*Thiobacillus*
	Na$_2$S$_2$O$_3$·7 H$_2$O				
		7.0			
Environment					
In the dark,	NH$_4$NO$_3$	3.0	none	7.0	*Thiobacillus*
temperature,	Na$_2$S$_2$O$_3$·7 H$_2$O		(stoppered		*denitrificans*
25°—30° C		7.0	bottle)		
	NaHCO$_3$	5.0			

[*] The components of the medium are given in grams per liter.

TABLE 16.3

Primary environmental factors that determine the outcome of enrichment procedures for photosynthetic microorganisms

Light as source of energy	absence of organic compounds	absence of sulphide	N_2 as sole nitrogen source	blue-green algae
			presence of combined nitrogen	green algae
		presence of sulphide anaerobic conditions	high sulphide concentration	green sulphur bacteria
			low sulphide concentration	purple sulphur bacteria
	presence of organic compounds	anaerobic conditions		non-sulphur purple bacteria

ture, although grazing pressure is also low. The high summer values coincided with high values of algal photosynthesis. No control was performed to indicate whether $^{14}CO_2$ uptake resulted only from chemosynthesis or partly from chemoorganotrophic uptake (see page 328).

Sulphate, nitrate or carbon dioxide-reducing bacteria can be obtained easily in a so-called "enrichment" culture, i.e. a culture in which they dominate. *Desulphovibrio* can, for example, be obtained by adding sulphate with formate to lakewater (sugars should not be used as they stimulate heterotrophic growth). Ammonia should be given as nitrogen source since NO_3^- will stimulate growth of nitrate reducers. Using an isolation, e.g. on agar plates, pure cultures may then be obtained. Enrichment conditions for various bacteria are given in Tables 16.2 and 16.3 (Stanier et al., 1963). Enrichment cultures are reviewed by Schlegel and Jannasch (1967).

16.3. CHEMOORGANOTROPHIC BACTERIA

This group — originally called heterotrophic bacteria — contains a wide variety of different bacteria with quite different nutrient requirements. With respect to the element carbon, for example, a wide spectrum of nutritional types exists in nature. At one end of this spectrum are those organisms which can utilise relatively simple substances like formic and acetic acids, at the other end the obligately parasitic forms need many complex organic compounds. Between these two extremes are organisms which need from one to a dozen or more organic substrates of various degrees of complexity.

Many of these bacteria have a common structure: they are polarly flagellated, gram-negative rods and are thus collectively referred to as *Pseudomonas* (for example *Nitrosomonas, Thiobacillus, Hydrogenomonas* and *Methanomonas*, mentioned in 16.2). Amongst the chemoorganotrophic bacteria is one outstanding group: the fluorescent pseudomonads. These are all aerobes, but some can also grow anaerobically, using denitrification as energy supply. They have simple nutrient requirements and do not require growth factors. Apart from their fluorescence, their most remarkable property is the very wide range of different organic compounds which they can use as sources of carbon and energy. This substrate versatility is greater than that of any other biological group. A single strain of *Pseudomonas fluorescens* can oxidise well over 100 different organic substances, amongst which are alcohols, acids, amino acids, carbohydrates, and several ring compounds. This versatility results from the ability to form the necessary enzymes in response to the presence of a particular substrate. This process is called enzyme induction. Not all compounds are oxidised completely; when growing on sugars for instance these bacteria may produce large quantities of keto acids. In nature these keto acids will probably induce other strains to grow.

Some of these pseudomonads produce other pigments — besides the fluorescent ones — of a different chemical nature, belonging to the group of organic dyes known as phenazines. For example *Pseudomonas aeruginosa* produces a deep blue-green colour in its culture medium.

A second group, members of which are often found in water, are the coliform bacteria. This group contains two genera of gram-negative, peritrichously flagellated or immotile, rod-shaped bacteria, *Escherichia* and *Aerobacter*. They can grow aerobically and in these conditions they can metabolise sugars, organic acids, amino acids and other simple substrates. Anaerobically they can ferment sugars and some of them can use organic acids. During fermentation of sugars the coliform bacteria yield a variety of end-products such as H_2, formic, acetic, lactic and succinic acids and ethyl alcohol.

Several members of the *Escherichia coli* group are carriers of water-borne diseases such as typhoid, cholera and dysentery. The presence of a "high temperature" strain of *E. coli* is widely used as an indication of sewage pollution (see Chapter 19).

Owing to the wide variety of substrates and conditions, a whole spectrum of different bacterial species and bacterial processes occur simultaneously in the water. The commonly reported counts of the total number of heterotroph ic bacteria present are therefore of little value. Sometimes a correlation with phytoplankton biomass is found (Overbeck, 1972). This is to be expected because the phytoplankton is the main producer of organic matter. Sometimes no correlation is found. This probably means that the primary system has been disturbed by other processes, such as dilution, grazing or addition of allochthonous organic matter.

Attempts to correlate algal with bacterial populations are reviewed by Jones (1971, 1972), who listed a number of papers, in some of which correla-

TABLE 16.4

Mean percentages of different types of bacteria in water and mud

Type of bacterium according to exoenzymes produced:	Windermere North Basin			Esthwaite Water		
	amylase	protease	lipase	amylase	protease	lipase
Water	13.7	9.0	3.6	6.6	7.0	3.4
Mud	9.7	7.5	2.2	24.0	6.0	3.5

tions were found and gave a number of critical suggestions why direct relationships do not always exist. Gunkel (1968) for example pointed out that maxima of phytoplankton appeared before those of bacteria, and Drabkova (1965) suggested that peaks of bacterial numbers were in response to decaying but not to viable algae. The depth and time at which samples are taken will of course affect the overall picture which the results convey. Jones himself examined two lakes, one of which (Esthwaite Water) is much richer in nutrients than the other (Windermere North Basin). In Esthwaite Water major factors found to correlate with bacterial populations were temperature, O_2 concentration and pH. In Windermere a positive correlation was noted, not only with temperature and pH but also with particulate matter and rainfall. No correlation was found between viable bacteria and chlorophyll a concentration, but enzyme activity occasionally increased with, or soon after, phytoplankton density. High pH is caused by phytoplankton photosynthesis, however, so that numbers of bacteria seem to show a better correlation with primary production than with algal biomass. Populations of amylase-producing bacteria showed two peaks: one after the spring diatom maximum and one during the August—September maximum of Ceratium and blue-green algae. Protease-producers showed a slight peak in both lakes after the spring maximum but not after that in August—September on Esthwaite Water. In autumn physicochemical factors probably control the bacterial population. The mean percentages of three different types of bacteria are shown in Table 16.4. No quantitative information on the metabolic activity is available.

Dominant genera of exoenzyme-producing bacteria were:

Protease

Achromobacter sp.
Acinetobacter (4 spp.)
Bacillus (spp.)
Corynebacterium (3 spp.)
Enterobacter sp.
Flavobacterium (13 spp.)
Micrococcus sp.
Pseudomonas (8 spp.)
Vibrio (2 spp.)
Xanthomonas (7 spp.)
yeasts (3 spp.)

Amylase

Acinetobacter sp.
Serratia sp.
Xanthomonas sp.
yeast

Lipase

Bacillus (2 spp.)
Pseudomonas (2 spp.)
Xanthomonas (2 spp.)
yeast (4 spp.)

Later Jones (1972) demonstrated a higher correlation between "viable" bacteria and phytoplankton than between total bacterial numbers and phytoplankton. Some algal groups correlated more strongly than others (blue-green algae > filamentous green and blue-green algae > diatoms > dinoflagellates). A positive correlation was found between bacterial populations ("viable" and direct counts) and water temperature when all samples were included in the analysis. Jones also measured the alkaline phosphatase activity of paper-filtered samples and demonstrated a positive correlation between concentrations of this enzyme with both algal and bacterial numbers. The amount of enzyme activity of membrane-filtered samples was significantly correlated only with numbers of phosphatase-producing bacteria, and not with the total bacterial population. Much more information of this kind is needed to provide a better understanding of the processes which take place in lakes and of how they vary with season.

Oxidation of dissolved organic compounds can be studied by use of radioactive compounds. In this way it is possible to count different bacterial groups according to their specific metabolites. Thus proteolytic bacteria were found by Jones (and ourselves) to form only 1—10% of the total number of the heterotrophic bacteria, although they are essential to start the mineralisation process. Apparently heterotrophic bacteria derive a large part of their food from dissolved organic compounds excreted either by other bacteria or by algae or produced during autolysis. Overbeck (1972) used the method described by Romanenko (see section 16.2) for measuring "heterotrophic production". He found bacterial dark CO_2 fixation to average about 7% of total dark CO_2 fixation in the Pluszsee (Germany). He stated that exact physiological investigations especially into the influence of dissolved organic substances on the rate of CO_2 fixation are still needed. Overbeck gives no information about his method for separating dark CO_2 fixation in bacteria from that in algae (e.g. blue-greens) or from CO_2 uptake by chemolithotrophic bacteria. Overbeck also found heterotrophic bacterial production of a magnitude similar to that of primary production. Such bacterial production could be achieved only if there were input of allochthonous organic matter in amounts of about 10 times primary production.

16.4. PHOTOTROPHIC BACTERIA

16.4.1. Introduction

Photosynthetic bacteria are able to perform photosynthesis with H_2, reduced sulphur or organic compounds as electron donor, but not with H_2O. All are able to reduce CO_2 and to use it as a sole source of carbon. Because they depend on reducing compounds which are easily oxidised by O_2 and because they do not themselves produce O_2, they are all strictly anaerobic. Both cylic and non-cyclic photophosphorylations have been demonstrated in a wide variety of purple bacteria (see section 4.4). In green plants, from algae upwards, cyclic photosynthesis (photosystem I) produces ATP, but no NADPH or O_2. In non-cyclic photosynthesis H_2O is used to produce O_2 and NADPH with a concomitant production of ATP. In the photosynthetic bacteria, however, compounds other than H_2O, such as H_2S, H_2, S, and succinic acid, can be used as electron donors. Some of these oxidisable substrates do not have sufficient energy to serve directly in the enzymatic reduction of NADP. Other compounds do have sufficient energy and can be used directly for the generation of NADPH. In such cases the function of the light may be solely to provide ATP through cyclic photophosphorylation and not to provide reducing power. Like the chemotrophic bacteria, the photosynthetic bacteria can be divided into two groups: (a) those which use inorganic electron acceptors (photolithotrophic); and (b) those which use organic electron acceptors (photoorganotrophic). These groups will be discussed separately, although they have many common features.

16.4.2. Photolithotrophic bacteria

Important groups occurring in lakewaters are the purple sulphur bacteria (Thiorhodaceae, with *Chromatium* as an important genus) and the green bacteria (e.g. *Chlorobium*). Although the two groups show a variety of cell shapes and modes of cell division, they share two common structural properties: both are gram-negative and, if motile, both are polarly flagellated. They can be distinguished from one another by their photosynthetic pigments. In purple

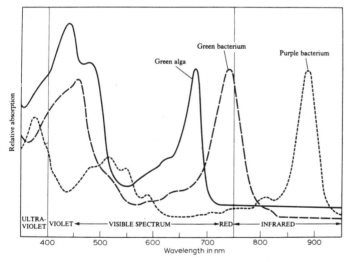

Fig. 16.2. Light absorption by a green alga, a green bacterium and a purple bacterium. Note the complementarity of light absorption in the red and infrared regions of the spectrum by these three photosynthetic organisms. (From Stanier et al., 1963.)

bacteria a special kind of chlorophyll, bacteriochlorophyll, is found. It has maximal absorption around 875 nm. Purple bacteria also have aliphatic (open-chain) carotenoid pigments. The green bacteria contain either one or two chlorophylls with a maximal absorption around 740 nm accompanied by alicyclic (closed-ring) carotenoids (see Chapter 13). The different absorption spectra of pigment extracts of these bacteria are given in Fig. 16.2. Thus the different photosynthetic bacteria may use light of different wavelengths for their photosynthesis.

The photolithotrophic bacteria use H_2S and not H_2O as electron donor for this photosynthesis:

$$CO_2 + 2\ H_2S \xrightarrow{\text{light}} (CH_2O) + H_2O + 2\ S$$

$$3\ CO_2 + 2\ S + 5\ H_2O \xrightarrow{\text{light}} 3\ (CH_2O) + 4\ H^+ + 2\ SO_4^{2-}$$

The similarity between the first reaction above and the gross reaction of plant photosynthesis is striking. In fact the study of the purple bacteria contributed greatly to our understanding of plant photosynthesis (see section 4.4).

If a water sample contains purple bacteria, with sulphur particles, and no H_2S, then the sulphur disappears, demonstrating that the oxidation proceeds beyond the level of sulphur.

The reducing power in the steps both of H_2S to S and from S to H_2SO_4 can be used to reduce CO_2 for the synthesis of cell material (Van Gemerden, 1967, 1968). He showed that, contrary to the existing ideas, the second step begins as soon as the first sulphur has been formed and not only after all H_2S has been oxidised to sulphur. Van Gemerden suggested that about one quarter of all the cell material is synthesised using energy from the H_2S step and three quarters is formed as a result of using sulphur as electron donor. The basis for this ratio is the number of electrons transferred during the two oxidation stages (H_2S to S to SO_4^{2-}).

As early as 1937 the occurrence of accumulations of *Lamprocystis* and *Chromatium* in the hypolimnia of Krottensee (Upper Austria) was reported by Ruttner (1937). Kuznetsov (1968) found a sharply developed *Chromatium* "plate" at a depth of 15 m in Lake Belovod (Fig. 16.3). In both these lakes the bacteria develop in zones with light values well below the compensation point for algae. Nevertheless, Kuznetsov's data show that photosynthesis of *Chromatium* per unit volume of water reached a much higher value than that of phytoplankton, but the integral photosynthesis per unit area seemed to be about equal. The rather low algal photosynthesis value does not agree with the apparent O_2 maximum at 5 m; the O_2 concentration indicates a much higher rate of photosynthesis. Total chemosynthesis showed a remarkably high value. When comparing the depth where these bacteria occur with the compensation depth for algal photosynthesis, it should be remembered that the bacterial chlorophyll of the purple sulphur bacteria has a maximal light absorption at 900 nm, which is far into the infrared region of the spectrum. It is to be expected that algae occurring in the upper layers above those where the purple bacteria occur will absorb proportionately less light of this wavelength than they do in the visible region. No measurements are available of depth penetration and absorption of light of different wavelengths in these lakes. From an ecological and evolutionary point of view it is interesting to note that the infrared light has less energy per quantum. Less energy is required to split the H_2S molecule (heat of formation 9 kcal.) than to split the H_2O molecule (heat of formation 68 kcal.). The ecological significance of energy-transferring carotenoids needs to be studied. It might be that the shift into the infrared of the bacterial chlorophyll absorption peak is related to the red-wards shift of the absorption peak of the bacterial carotenoids (450—550 nm).

Growth rates of both green and purple sulphur bacteria were studied by Takahashi and Ichimura (1970), who found that the main factors determining the growth rates were H_2S concentration in the upper parts of their range and light in the deeper parts. The rates of growth observed in lakes conformed to those predicted by calculation using laboratory data.

In both the Krottensee and Lake Belovod the densest populations of these bacteria were just below the layers where oxygen could be detected. The thick plate formed in this position reflects the delicate balance of conditions required for growth: absence of O_2, presence of light from above and of H_2S from below (where it is produced near the bottom by *Desulphovibrio*). It is striking that both in the Krottensee and in Lake Belovod the maxima of the

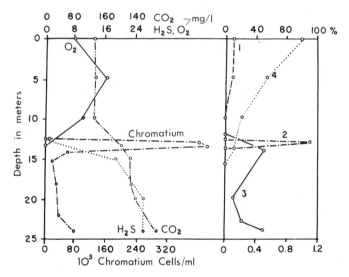

Fig. 16.3. Hydrochemical analysis of Lake Belovod and the stratification of photosynthesis and chemosynthesis (July 1954). 1: Photosynthesis of phytoplankton in mg CO_2 l^{-1} d^{-1}. 2: Photosynthesis of *Chromatium* spp. in mg CO_2 l^{-1} d^{-1}. 3: Total chemosynthesis in mg CO_2 l^{-1} d^{-1}. 4: Percent penetration of light. (From Kuznetsov, 1968.)

Thiorhodaceae occurred at the same depth as mass accumulations of the co-lourless sulphur bacterium *Achromatium mobile.*

Gorlenko and Kuznetsov (1972) described the occurrence of a "plate" of the blue-green alga *Oscillatoria prolifica* at 10.5—12 m depth, just below the thermocline of Lake Kononjer (Fig. 6.4) together with photosynthetic bacteria of ten different species. The bacteria were either purple, greenish or brownish. Above this layer a population of a reddish-coloured *Oscillatoria prolifica* developed and still higher — in the micro-aerobic zone — the bacteri-um *Metallogenium personatum* (see subsection 16.2.4). Gorlenko and Kuz-netsov suggested that the numbers and kinds of the different species were controlled by the H_2S content and pH value of the water and by light attenu-ation. The occurrence of a reddish-coloured *Oscillatoria* species was also noted by Golterman (unpublished, see section 13.3) in a crater lake in Ugan-da. These organisms could photosynthesise using H_2S. It has been suggested that some blue-green algae resemble photosynthetic sulphur bacteria in their photosynthetic abilities; but their photoheterotrophic growth on organic compounds resembles that of the purple non-sulphur bacteria.

Knobloch (1969) suggested that H_2S could be used as electron donor for photosynthesis, not only in bacterial metabolism but also in photosynthetic reactions of green plants, for example *Chlorella* and *Scenedesmus.* The figures which he gave could equally well be explained, however, by inorganic oxidation of H_2S by oxygen either from the air or produced during algal photosynthesis.

The second type of photosynthetic bacterium which must be mentioned is *Chlorobium* (Chlorobacteriaceae). It can oxidise H_2S only as far as sulphur, which it deposits extracellularly. It is not clear however whether or not its remarkable carbon-reduction pathway — the reverse Krebs tricarboxylic-acid cycle (section 16.1) — is related specifically to the production of S instead of sulphate.

Species of *Chlorobium* occur in shallow ponds polluted with organic mat-ter, where H_2S is produced by anaerobic decomposition of proteins, or by sulphate reduction. *Chlorobium* is often limited to shallow water layers. This can be explained by the strong absorption of light in this kind of water in the near infrared region. The peculiar CO_2-reduction cycle may be related to the organic-rich environment. However, *Chlorobium* also occurs in deep lakes, where populations may be so dense as to colour the water greenish. In these conditions they seem to contain more carotenoids (Pfennig, 1967) and are then often brownish. As light in the wavelength range between 450 and 550 nm is the most penetrating type of light, the carotenoid pigments may en-hance photosynthesis; in cultures the carotenoids are photosynthetically active, although the efficiency varies between 30% and 98% (Goedheer, 1959).

The presence of both H_2S-generating and -consuming organisms makes the hypolimnion a "sulphuretum", a term applied by Baas Becking (1925) to associations of organisms affecting the oxidation state of sulphur.

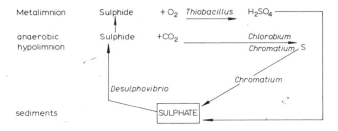

Fig. 16.4. Schematised sulphur cycle in a hypolimnion.

Desulphovibrio occurring in or near the sediments produces H_2S, which diffuses upwards. If sufficient light is present in the anaerobic layers, the H_2S is oxidised by phototrophic bacteria; otherwise it rises higher and meets a layer with O_2, where it is oxidised by chemolithotrophic bacteria. Sorokin (1965) noted that oxidation of H_2S in Lake Belovod occurred as a result of the combined activity of *Thiobacilli* and *Chromatium*. In the intermediate layer, large diurnal changes were reported both for H_2S and O_2 concentrations as well as in the redox potential. The purple bacterial plate sank from 10 to 12 m between 5.00 and 13.00 h. The zones containing both O_2 and H_2S meet at about 9 m depth at 5.00 h; but at 13.00 h O_2 and H_2S are hardly detectable. The situation is summarised schematically in Fig. 16.4.

The sulphuretum was partially simulated by Van Gemerden (1967) with mixed cultures of *Desulphovibrio* and *Chromatium*, the former producing H_2S, which was oxidised to sulphate by the latter. In light the rate of the sulphur cycle reactions was not influenced by a low total sulphur content. In an inorganic medium electron donor molecules for *Desulphovibrio* were not provided by *Chromatium*, but dead *Chromatium* cells did provide carbon sources. *Chromatium* produced an electron donor, probably H_2, for *Desulphovibrio* from yeast extract components. In the light, the sulphur cycle was not influenced by ferric ions, though it was affected by ferrous ions, demonstrating that H_2S concentrations were sufficient to precipitate Fe^{2+}, but not to reduce Fe^{3+}. Following dark periods with ferric ions, the growth of *Chromatium* was inhibited in the subsequent light period due to the production of ferrous ions in the dark, which demonstrated that in darkness sufficient H_2S was produced for the reduction of Fe^{3+}. Dead cells of *Scenedesmus* could also be used as carbon source by *Desulphovibrio*. It seems likely that with these kinds of experiments with mixed cultures of *Desulphovibrio*, *Chromatium* and *Chlorobium* it may be possible to discover what causes one of the latter two bacteria to develop in lakes.

16.4.3. Photoorganotrophic bacteria (purple, non-sulphur bacteria)

This second group of purple bacteria, somewhat different in their physiolo-

346

gy from the first group, can reduce CO_2 but uses organic compounds not only as electron donor but also as a carbon source. Acetic acid is a substrate which is often used in cultures. This is reduced and converted into intracellular reserve material, poly-β-hydroxybutyric acid under anaerobic conditions in the light. The reduction requires an input of two electrons (H → H$^+$ + e):

$$2n\ C_2H_4O_2 + 2n\ [H] \rightarrow (C_4H_6O_2)_n + 2n\ H_2O$$

The electrons are derived from the anaerobic oxidation, through reactions of the tricarboxylic-acid cycle, of other molecules of acetic acid:

$$C_2H_4O_2 + 2\ H_2O \rightarrow 8\ [H] + 2\ CO_2$$

Thus for each molecule of acetic acid oxidised, eight other molecules can be converted into cell material. The reduction reaction requires not only reducing power but also ATP. The ATP is generated through photophosphorylations. The overall equation for the formation of the polymer from acetic acid can be written as:

$$9\ CH_3COOH \xrightarrow{\text{light}} 2\ CO_2 + 4\ (C_4H_6O_2) + 6\ H_2O$$

Thus it is possible for nearly 90% of the carbon from acetic acid to be converted into cell material as can be shown by using ^{14}C-labelled substrate. This high efficiency can never be achieved in respiratory or fermentation metabolism and is owed here to the generation of ATP in the light. Because ATP can be produced by use of light energy in non-cyclic phosphorylation, the electron donor can be used exclusively for the reduction and conversion of external CO_2 into cell material. The electron donor is not required for the generation of ATP. With CO_2 as the carbon source, the photosynthesis of these bacteria can be represented as:

$$CO_2 + 2\ H_2 \xrightarrow{\text{light}} (CH_2O) + H_2O$$

Alternatively if acetic acid is used, the reaction is:

$$2\ C_2H_4O_2 + H_2 \xrightarrow{\text{light}} (C_4H_6O_2) + 2\ H_2O$$

With molecular hydrogen providing the reducing power, light energy is used

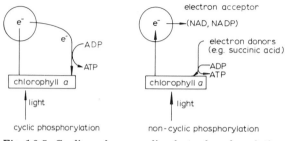

Fig. 16.5. Cyclic and non-cyclic photophosphorylation.

solely via cyclic photophosphorylation in the generation of ATP. With reducing compounds such as thiosulphate or succinic acid this is not the case:
light is used in non-cyclic photophosphorylation as well. The two processes
are shown schematically in Fig. 16.5.

Pfennig (1967) reviewed the literature dealing with photosynthetic bacteria and discussed in detail aspects of their ecology, enrichment and isolation,
morphology and growth habitats, physiology and pigments.

16.5. THE PROBLEM OF ADAPTATION

Many organisms show a remarkable ability to adapt themselves to changing
conditions. We have seen how algae can adapt their chlorophyll concentration
to changes in the irradiance, while pseudomonads make use of induced enzymes to use a variety of different organic substrates for their growth.

This adaptability proves to be a considerable problem when one is studying
the ecology of microorganisms, since the organisms studied may have properties quite different from those occurring under natural conditions. The work
of Ul'yanova illustrates this point (1961a, b, c). She compared the activity
of the bacterium $Nitrosomonas$ (which oxidises NH_3 to NO_2^-) in "enrichment" and "pure" cultures both of which were repeatedly subcultured. From
Table 16.5 it can be seen that the activity of the enriched culture is several
times higher than that of the pure culture, although the activity of both cultures remained constant during subculturing.

The different strains came from different soils, activated sludge, etc., the
latter showing the highest activity, whilst sandy soils yielded low activity
strains.

The influence of the origin persisted in all strains but the activity was considerably less in the pure cultures. Ul'yanova also noted that a positive reaction on nitrite and the time at which the maximal nitrite concentration was
found in pure cultures was always much later than that found with the en-

TABLE 16.5

Concentration of NO_2^- (in mg l^{-1}) after 10 days in 5 subcultures of "enrichment" and
"pure" cultures of $Nitrosomonas$

Strain	Enrichment cultures					Pure cultures				
	I	II	III	IV	V	I	II	III	IV	V
24a	2 000	2 000	2 000	2 000	2 000	18	19	19	19	30
73b	1 000	1 000	1 000	1 100	1 100	19	19	19	19	29
15S	800	800	800	800	800	18	18	17	18	18
6C	600	700	700	700	700	16	16	16	15	17
102a	550	570	570	570	570	13	13	14	13	15
18S	250	270	270	270	270	7	8	8	8	9

richment culture. Furthermore the pure cultures often became inactive and died, whilst the enrichment cultures remained active for more than 2.5 years. Ul'yanova suggested that the other species of bacteria in the enrichment cultures have a positive effect by reducing the redox potential or by producing growth-stimulating substances. Concerning the latter possibility, since Winogradsky's experiments were published, *Nitrosomonas* has always been grown in pure cultures, without organic substances. Nobody has yet compared the activity of these with that in an enrichment culture.

Ul'yanova (1961b) also measured the pH optimum for the different strains and noted that, although no strict correlation was found, the optimal pH varied with that of the original substrate (soil). From the more acid soils strains were isolated which had a more alkaline pH optimum. Unfortunately, she gave no details of the pH of her enrichment medium.

The different strains also showed differences in their optimum ammonium-ion concentrations ranging between 0.01 and 0.4 M l^{-1}. This ammonium optimum was also found to be influenced by phosphate, sulphate and malate ions. Strains from acidic sandy soils showed higher optimal NH_4^+ concentrations at lower concentrations of organic matter. Soil extracts stimulated the growth but only if extracted from the soil from which the strain was isolated. Extracts from other soils sometimes had no apparent effect at all.

The origin of the strain has also an influence on the concentration of those organic compounds which inhibit the ammonia oxidation. Since Winogradsky's time it has been known that glucose and peptone have an inhibitory action at concentrations of 0.2% (but see subsection 16.2.6.). Strains from manure heaps however were inhibited by glucose only at concentrations as high as 4—10%. It is well known that these organisms have a high temperature optimum and this may reflect their origin.

16.6. MEASURING ACTIVITIES IN NATURAL CONDITIONS

The estimation of the growth rate of bacteria and their rate of metabolic activity remains one of the greatest problems in bacterial limnology. The method of keeping the lakewater samples in bottles and calculating the generation time from the observed increase in numbers is open to severe criticism because enclosure in the bottle causes the bacteria to increase their growth rate, perhaps owing to the relatively large surface. Moreover, only the difference between growth rate and rate of mortality is measured. Much more promising is the approach of Jannasch (see subsection 10.5.2), who measured growth rate in continuous cultures with extremely low substrate concentrations. Although the concentration of asparagine nitrogen used may in fact be as low as the total organic-nitrogen concentration, the main difficulty remains that the origin of these compounds is unknown and many nitrogen compounds may not be used at all. One must hope that eventually lakewater with a known composition will be used in continuous culture experiments and that the growth constants will be estimated for many isolated bacterial strains from lakes. Even approximate measurements would serve as a basis for comparison with results obtained using other methods.

An important step in this direction has been made by Jannasch (1967) who metered natural water, without filtration or storage, directly into a chemostat. He measured the growth rate of a test organism in the presence of the natural populations by comparing dilution rate and washout rate. Specific growth rates were found to be as low as 0.005 h^{-1} (generation times of 20—200 h). It has been shown in subsection 10.5.2 that definite minimum growth rates must exist in a chemostat. In many cases the test strains ceased to grow when the dilution rate exceeded their maximal growth rate. (Washout rate then equals dilution rate.) In other cases the test strains continued to grow. Jannasch calculated the actual growth rate in those cases where washout rate reached a constant value. In his experiment two washout rates were found. He attributed this to a population effect similar to that observed in continuous culture experiments where the concentration of the growth-limiting substrate was lowered stepwise at an unchanged dilution rate. When a certain minimum population density was reached, growth ceased and complete washout occurred. Jannasch (1968) explained this population-density effect by postulating the production of growth-stimulating substances by the bacterial cells.

In growth experiments with marine bacteria, threshold concentrations of some limiting carbon sources were in fact found (Jannasch, 1970). The chemostat should be regarded as a useful means of measuring growth rates of microorganisms, especially when growth is slow. It should not be expected to reflect natural conditions since it takes no account of grazing, sedimentation and formation of resting stages which always occur in nature. The chemostat removes from the system all species, except those which are successful in their competition for the limiting substrate.

Metabolic activity involving one single compound can be measured by the method of Hobbie and Wright (Hobbie and Wright, 1965a, b; and Wright and Hobbie, 1965). In this method small quantities of ^{14}C-labelled and unlabelled substrates are added to the lakewater sample and incubated for a fixed period at a constant temperature. After incubation the samples are filtered through membrane filters (0.5 μm pore size) and the activity on the filters is measured. The general equation for the uptake of any solute is:

$$v = (S_n + A)c/C \, \mu t \qquad (16.1)$$

where v = velocity of uptake (mg l^{-1} h^{-1}); $c/C \, \mu t$ = percent of the added isotope taken up per hour; S_n = natural substrate concentration (mg l^{-1}); and A = added substrate concentration (mg l^{-1}).

The uptake of glucose was measured in cultures of *Chlamydomonas* sp. and a strain of a planktonic bacterium, EY-19 (motile rods) (Fig. 16.6). It can be seen that the rate of uptake by the algae was directly proportional to the sub-

350

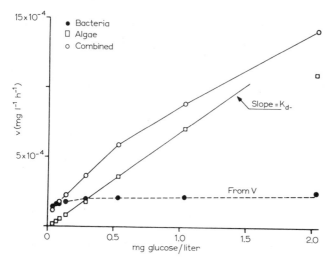

Fig. 16.6. The velocity of glucose uptake for bacterium EY-19 and *Chlamydomonas* sp., and the two combined. (From Hobbie and Wright, 1965b.)

strate concentration and could be described by:

$$v_d = k_d (S_n + A)$$

(16.2)

where $v_d = v$ for glucose; k_d = diffusion constant (h^{-1}) = the slope of the line; and S_n = the natural substrate concentration (which is zero in this case). The bacterial uptake v_t increased with increasing concentration of A, up to a concentration at which it was maximal. This type of uptake is probably due to the operation of a transport system which follows the Michaelis-Menten kinetics:

$$v_t = V \frac{S_n + A}{K_t + S_n + A}$$

(16.3)

where K_t = transport constant similar to the Michaelis-Menten constant, and V = maximum uptake velocity. Again S_n is zero.

Eq. 16.3 may be rewritten as:

$$\frac{S_n + A}{v_t} = \frac{K_t + S_n}{V} + \frac{A}{V}$$

(16.4)

Combination of eq. 16.1 with 16.4 gives:

$$\frac{C\mu t}{c} = \frac{K_t + S_n}{v} + \frac{A}{V}$$

Therefore when $C\mu t/c$ is plotted against A, the slope of the line is $1/v$, while the intercept on the ordinate is $(k_t + S_n)/V$ and on the abscissa $- (K_t + S_n)$. The data from eq. 16.6 are plotted in this way in Fig. 16.7 where bacterial uptake now shows a straight line, while the combined uptake gives a satura-

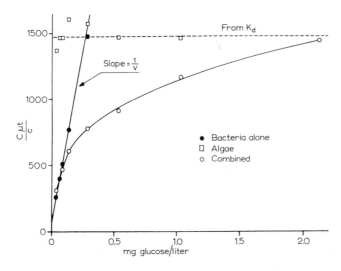

Fig. 16.7. Uptake of glucose for bacterium EY-19 and *Chlamydomonas* sp., and the two combined. Plotted according to eq. 16.4. (From Hobbie and Wright, 1965b.)

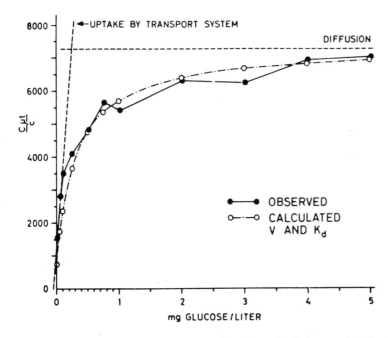

Fig. 16.8. Uptake of glucose in Lake Erken (1 m, 14 February 1964), plotted according to eq. 16.4. Theoretical curves calculated from V and K. (From Hobbie and Wright, 1965b.)

tion-like curve. Thus, uptake due to the different systems can be separated.

Hobbie and Wright found the same type of curve for substrate uptake in water of Lake Erken (Sweden) (Fig. 16.8). They pointed out that it is not necessary to measure uptake velocities over such a wide range of substrate concentrations. When concentrations within the range 0.5—2.0 mg l^{-1} are used, the rate of uptake is mainly controlled by diffusion to the algae and K_d is a measure of this. With low substrate concentrations (60 μg l^{-1}), the uptake will be controlled mainly by the bacteria. Fig. 16.9 illustrates results obtained when the same technique was used to study acetate uptake. The value of $(K_t + S_n)$ is 9 μg l^{-1} in Fig. 16.9A and 20 μg l^{-1} in Fig. 16.9B, which sets an upper limit for S_n, while values of V of 13 and $2 \cdot 10^{-5}$ mg l^{-1} h^{-1}, respectively, set the upper limit for uptake velocity. Using cultures with a known K_t, a value of 60 μg l^{-1} of glucose was found in polluted waters,

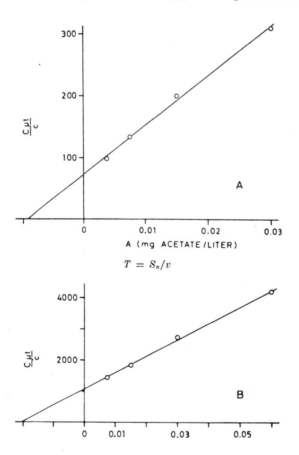

Fig. 16.9. Uptake of acetate at low substrate levels in Lake Erken, plotted according to eq. 16.4. Samples from 1 m on 14 October and 1 m on 7 January, 1965. (From Hobbie and Wright, 1965b.).

while in unpolluted waters the concentrations were usually less than 10 μg l^{-1}.

Only acetate was measured in Lake Erken and it was found to be always less than 20 μg l^{-1}. At these low concentrations bacterial transport seems always to be more effective than the diffusion mechanism of algae.

Knowing the values for S_n and v, the turnover time, T, can be calculated, assuming a constant supply:

$$T = S_n/v$$

Assuming a natural substrate level of 10 μg l^{-1} of glucose, the bacterial population in the laboratory experiment described above would have a turnover time, T_t, of 130 h and the algae a turnover time, T_d of 1 470 h, despite their 250 times greater biomass.

Turnover times in nature can be measured easily, even though S_n and v are unknown. For algae $T_d = 1/k_d$ (from eq. 16.2). For the bacteria, the ordinate is the T_t in hours, when the data are plotted using eq. 16.4. This intercept is $(K_t + S_n)/V$ and from eq. 16.3 this equals S_n/V_t, when $A = 0$. Thus in Fig. 16.6, T_t was 70 h in October and 1 000 h under ice in January. Hobbie and Wright studied the specificity of these systems for glucose and found that, of a large range of amino acids, fatty acids and sugars, only mannose interfered to some extent. It is suggested that two improvements could be made. The uptake by bacteria, which Hobbie and Wright consider to be the main one at the very low concentrations occurring in natural waters, is measured essentially only at three low substrate concentrations, very near to one another. Moreover, no correction is made for loss of $^{14}CO_2$ by respiration. Realising that much of both the acetate and the glucose will certainly be used for respiration, one wonders whether a correction could be made by measuring the $^{14}CO_2$ produced in the medium.

Hobbie (1967), using the same technique, measured seasonal concentrations of glucose and acetate and their turnover times in Lake Erken. The $(K + S_n)$ value showed no seasonal cycle, although 40-fold seasonal changes were found in the uptake velocity, with two peaks for glucose in June and September (highest value $45 \cdot 10^{-5}$ mg l^{-1} h^{-1} for glucose). Acetate uptake followed the same general pattern with slight differences (a slight increase in December and a large increase in April) with a maximum of about $55 \cdot 10^{-5}$ mg l^{-1} h^{-1}. Concentrations of glucose and acetate were less than 6 μg l^{-1} and 10 μg l^{-1}, respectively, and turnover times varied between 10 and 1 000 h. Hobbie calculated that with a turnover rate of 10 h 2.2 mg of glucose would have been taken up in six months, assuming a mean concentration of 5 μg l^{-1} of glucose. Although this represents a significant proportion of the total dissolved organic carbon (10 mg l^{-1} of C), it is still a minute fraction of the carbon turned over in photosynthesis.

Allen (1971) studied the uptake of glucose and acetate ^{14}C by freshwater planktonic algae and bacteria in size-fractionated samples. The reduction in algal biomass by filtration through a 20-μm and a 58-μm net was reflected in

354

a decrease in the diffusion constant K_d, which describes the algal uptake. Uptake due to active transport — bacterial uptake — was only slightly affected by the 58-μm filtration but altered considerably as a result of the 20-μm filtration. Further size-fractionated filtration revealed that organisms between 3 and 8 μm were responsible for the majority of the active uptake of both glucose and acetate, while organisms smaller than 1.2 μm played a minor role. The organisms between 4 and 9 μm were predominantly small microflagellates. Allen suggested that nanoplanktonic algae with surface/volume ratios of the same magnitude as those of bacteria may compete with bacteria for their substrate. The significance of these results will, however, be seriously affected by lack of data on respiration, as the green algae may lose less of the consumed material via respiration than do bacteria since they probably use products of their photosynthesis as well. Furthermore, no counts were made in order to establish whether the bacteria were indeed separated from the algae; all kinds of association often occur naturally.

Pearl and Goldman (1972) studied water movement in Lake Tahoe (California, U.S.A.). By measuring assimilation rates, they found that the inflowing riverwater could be located because higher uptake occurred there. They showed by autoradiographical studies that bacteria are largely responsible for ^{14}C-acetate uptake during a 1 hr incubation period. Bacteria attached to detritus particles also became labelled. This demonstrates the dangers of a filtration step to separate bacteria from algae.

REFERENCES

Alexander, M., 1971. Biochemical ecology of microorganisms. *Ann. Rev. Microbiol.*, 25: 361—392.
Allen, H.L., 1971. Dissolved organic carbon utilization in size-fractionated algal and bacterial communities. *Int. Rev. Ges. Hydrobiol.*, 56(5): 731—749.
Baas Becking, L.G.M., 1925. Studies on the sulphur bacteria. *Ann. Bot. Soc. Lond.*, 39: 613—650.
Borris, R. and Ohlmann, E., 1972. Citrate synthase as a key enzyme in the regulation of heterotrophic and autotrophic metabolism of *Rhodopseudomonas. Biochem. Physiol. Pflanz.*, 163: 328—333.
Daubner, I. and Ritter, R., 1973. Bakteriengehalt und Stoffumsatzaktivität einiger physiologischer Bakteriengruppen in zwei künstlichen Grundwasserseen (Baggerseen). *Arch. Hydrobiol.*, 72(4): 440—459.
Davies, T.R., 1973. Isolation of bacteria capable of utilizing methane as a hydrogen donor in the process of denitrification. *Water Res.*, 7: 575—579.
Drabkova, V.G., 1965. Dynamics of the bacterial number, generation time, and production of bacteria in the water of a red lake. *Mikrobiologiya*, 34: 933—938 (translation).
Fischer, E., 1972. A yearly cycle of changes in dynamics of production of the chemoautotrophic bacteria in bottom sediments of a water body. *Polskie Arch. Hydrobiol.*, 19(4): 343—359.
Goedheer, J.C., 1959. Energy transfer between carotenoids and bacteriochlorophyll in chromatophores of purple bacteria. *Biochim. Biophys. Acta*, 35: 1—8.
Gorlenko, W.M. and Kuznetsov, S.I., 1972. Über die photosynthetisierenden Bakterien des Kononjer-Sees. *Arch. Hydrobiol.*, 70(1): 1—13.

Gunkel, W., 1968. Die Bakterien und ihre Beziehungen zum Plankton in den Tumpeln der Helgoländer Dume nach der schweren Sturmflut im Februar 1962. *Mitt. Int. Ver. Theor. Angew. Limnol.*, no. 14: 31—42.

Hobbie, J.E., 1967. Glucose and acetate in freshwater: Concentrations and turnover rates. *IBP Symp., Amsterdam—Nieuwersluis, 1966*, pp. 245—251.

Hobbie, J.E. and Wright, R.T., 1965a. Bioassay with bacterial uptake kinetics: Glucose in freshwater. *Limnol. Oceanogr.*, 10: 471—474.

Hobbie, J.E. and Wright, R.T., 1965b. Competition between planktonic bacteria and algae for organic solutes. *Mem. Ist. Ital. Idrobiol.*, 18(Suppl.): 175—185.

Hooper, A.B., 1969. Biochemical basis of obligate autotrophy in *Nitrosomonas europaea*. *J. Bacteriol.*, 97: 776—779.

Jannasch, H.W., 1967. Enrichments of aquatic bacteria in continuous culture. *Arch. Mikrobiol.*, 59: 165—173.

Jannasch, H.W., 1968. Growth characteristics of heterotrophic bacteria in seawater. *J. Bacteriol.*, 95: 722—723.

Jannasch, H.W., 1970. Threshold concentrations of carbon sources limiting bacterial growth in seawater. In: D.W. Hood (Editor), *Organic Matter in Natural Waters*. Symposium held at the University of Alaska, 1968, pp. 321—328.

Jones, J.G., 1971. Studies on freshwater bacteria: factors which influence the population and its activity. *J. Ecol.*, 59: 593—613.

Jones, J.G., 1972. Studies on freshwater bacteria: Association with algae and alkaline phosphatase activity. *J. Ecol.*, 60: 59—77.

Kelly, D.P., 1971. Autotrophy: Concepts of lithotrophic bacteria and their organic metabolism. *Ann. Rev. Microbiol.*, 25: 177—210.

Knobloch, K., 1969. Sulphide oxidation via photosynthesis in green algae. *Progr. Photosynth. Res.*, 2: 1032—1034.

Kuznetsov, S.I., 1968. Recent studies on the role of microorganisms in the cycling of substances in lakes. *Limnol. Oceanogr.*, 13(2): 211—224.

Overbeck, J., 1972. Distribution pattern of phytoplankton and bacteria, microbial decomposition of organic matter and bacterial production in a eutrophic, stratified lake. In: Z. Kajak and A. Hillbricht-Ilkowska (Editors), *Productivity Problems of Freshwaters. Proc. IBP—UNESCO Symp., Kazimierz Dolny (Poland), 1970*, pp. 227—237.

Pearl, H.W. and Goldman, Ch.R., 1972. Heterotrophic assays in the detection of water masses at Lake Tahoe, California. *Limnol. Oceanogr.*, 17(1): 145—148.

Pfennig, N., 1967. Photosynthetic bacteria. *Ann. Rev. Microbiol.*, 21: 285—324.

Rittenberg, S.C., 1969. The role of exogenous organic matter in the physiology of chemolithotrophic bacteria. *Adv. Microbiol. Physiol.*, 3: 159—196.

Rittenberg, S.C. and Goodman, N.S., 1969. Mixotrophic growth of *Hydrogenomonas eutropha*. *J. Bacteriol.*, 98(2): 617—622.

Romanenko, V.I., 1964. Heterotrophic assimilation of CO_2 by bacterial flora of water. *Microbiol. U.S.S.R.*, 33(4): 610—614 (translation).

Romanenko, V.I., 1966. Microbiological processes in the formation and breakdown of organic matter in the Rybinsk Reservoir. In: B.K. Shtegman (Editor), *Production and Circulation of Organic Matter in Inland Waters*. Israel Program for Scientific Translations, Jerusalem, pp. 137—158.

Ruttner, F., 1937. Limnologische Studien an einigen Seen der Ostalpen. *Arch. Hydrobiol.*, 32: 167—319.

Schlegel, H.G. and Jannasch, H.W., 1967. Enrichment cultures. *Ann. Rev. Microbiol.*, 21: 49—70.

Shafia, F., Brinson, K.R., Heinzman, M.W. and Brady, J.M., 1972. Transition of chemolithotroph *Ferrobacillus ferrooxidans* to obligate organotrophy and metabolic capabilities of glucose-grown cells. *J. Bacteriol.*, 111(1): 56—65.

Smith, A.J., London, J. and Stanier, R.Y., 1967. Biochemical basis of obligate autrotrophy in blue-green algae and *Thiobacilli. J. Bacteriol.*, 94(4): 972—983.

Sokolova, G.A., 1961. Seasonal changes in the structure and numbers of iron bacteria and the cycling of iron in Lake Glubok. *Gen. Hydrobiol.*, 2: 5—11.

Sorokin, Y.I., 1965. On the trophic role of chemosynthesis and bacterial biosynthesis in water bodies. *Mem. Ist. Ital. Idrobiol.*, 18(Suppl.): 187—205.

Sorokin, Yu.I., 1966. Role of carbon dioxide and acetate in biosynthesis by sulphate-reducing bacteria. *Nature*, 210: 551—552.

Stanier, R.Y., Douderoff, M. and Adelberg, E.A., 1963. *General Microbiology*. MacMillan, London, 753 pp.

Takahashi, M. and Ichimura, Sh., 1970. Photosynthetic properties and growth of photosynthetic sulfur bacteria in lakes. *Limnol. Oceanogr.*, 15(6): 929—944.

Taylor, B.F., Hoare, D.S. and Hoare, S.L., 1971. *Thiobacillus denitrificans* as an obligate chemolithotroph. *Arch. Mikrobiol.*, 78: 193—204.

Ul'yanova, O.M., 1961a. Nitrifying activity of pure and enriched cultures of *Nitrosomonas* isolated from different natural substrates. *Microbiol. U.S.S.R.*, 30(1): 41—46 (translation).

Ul'yanova, O.M., 1961b. Adaptation of *Nitrosomonas* to existence conditions on various natural substrates. *Microbiol. U.S.S.R.*, 30(2): 236—242 (translation).

Ul'yanova, O.M., 1961c. Ecology of *Nitrosomonas. Microbiol. U.S.S.R.*, 30(3): 550—564 (translation).

Van Gemerden, H., 1967. *On the Bacterial Sulfur Cycle of Inland Waters*. J.H. Pasmans, Den Haag, 110 pp.

Van Gemerden, H., 1968. Utilization of reducing power in growing cultures of *Chromatium. Arch. Mikrobiol.*, 64: 111—117.

Wallace, W. and Nicholas, D.J.D., 1969. The biochemistry of nitrifying microorganisms. *Biol. Rev. Cambridge*, 44: 359—391.

Wright, R.T. and Hobbie, J.E., 1965. The uptake of organic solutes in lake water. *Limnol. Oceanogr.*, 10: 22—28.

CHAPTER 17

NUTRIENT BUDGETS AND EUTROPHICATION

17.1. INTRODUCTION

One of the most menacing threats to water quality is man's impact on nutrient budgets in those lakes which are situated in the more densely populated areas of the world. Sewage effluent containing organic matter provides all the nutrients essential for algal growth. It may enter the lakes directly, or indirectly via inflowing rivers, and will eventually become mineralised to leave a residue of inorganic nitrogen- and phosphorus-containing salts. The extra input of these nutrients will usually enhance algal growth which is normally limited by much lower natural levels. It does not help the situation if the organic matter is oxidised during the sewage treatment process (see Chapter 19); the effluent will still contain most of the growth-enhancing nutrients in an inorganic form. The algae will produce an amount of organic matter roughly equivalent to that which was originally broken down either in the river or in the lake or in the sewage treatment. It is often difficult to differentiate between the human input and the natural input; even in undisturbed oligotrophic lakes, input is never negligible, because of erosion from the watershed. Nevertheless the relative importance of the human input must often be known in order to justify the cost of measures to remove it.

Input of nutrients nearly always exceeds output. Accumulation in sediments will account for the difference. This accumulation should never be regarded as a loss because there are interactions between sediments and overlying waters. The only considerable losses are outflow, evaporation, or harvest. Of these, outflow is usually the largest. In a nutrient budget the following items may contribute:

Input	Output
river water, erosion in watershed	river outlet
direct run-off, drainage water	evaporation (NH_3?)
direct and indirect waste water sources	harvest
precipitation and dustfall	seepage
agricultural activities	(denitrification)
(nitrogen-fixation)	

difference: input − output = sedimentation

Direct measurement of the sedimentation rate is not easy; in most cases it will be estimated as the difference between input and output. This is an unsatisfactory situation. Compounds such as calcium carbonate and silicates (clay) can be measured by using traps but substances such as those containing phosphate, nitrogen, iron and organic carbon are more difficult to estimate because the amounts may be influenced by uptake or release by organisms growing on the organic matter which collects in the trap. For silicate no chemical distinction can be made between that from erosion products and that from diatom cell walls. In traps near the bottom resuspension of bottom material may induce serious errors.

If the nutrient budget shows a positive balance the concentration of solutes in the lake will rise, and if an algal growth rate-limiting factor is concerned algal growth will increase. This will lead to increased sedimentation, which together with increased outflow will establish a new equilibrium.

17.2. NUTRIENT BUDGET CALCULATIONS

17.2.1. Theoretical models

Several attempts have been made to calculate the amount of a substance present in a lake by measuring input and water flow (Biffi, 1963; Piontelli and Tonolli, 1964; Vollenweider, 1964; all three being summarised in Vollenweider, 1969).

Biffi calculated the outwash rate of a substance, not involved in biological uptake or precipitation, on the assumption that:

$$-\frac{dx}{dt} = (a/b)x - c \tag{17.1}$$

where a = mean water outflow volume ($m^3\ d^{-1}$); b = lake volume (m^3); c = supply ($kg\ d^{-1}$); and x = total amount present in lake (kg).

Integration of eq. 17.1 gives:

$$\ln \frac{ax_0 - bc}{ax - bc} = \frac{a}{b}t \tag{17.2}$$

where x_0 = amount present at time t = 0. Using eq. 17.2, a half-life time ($T_{1/2}$) can be calculated, i.e. the time needed to remove 50% of a given concentration (thus $x = \frac{1}{2}x_0$):

$$T_{1/2} = \frac{b}{a} \ln \frac{ax_0 - bc}{\frac{1}{2}ax_0 - bc} \tag{17.3}$$

Using this equation, Golterman (1973) calculated the half-life time for phosphate in the Lake of Zürich and found it to agree with that measured by Thomas (1968) (see Fig. 17.5), indicating that during the period of measurement (March—June) sedimentation contributed much less to the disappearance of the phosphate than did through-flow. It is important to realise that if water flows only through the epilimnion (as in the case of the Lake of Zürich), b must be taken as the volume of the epilimnion and not that of the whole lake. If the latter volume were used, the half-life time for the Lake of Zürich would be found to be much too long.

Biffi's equation can be used to calculate the water balance of a lake, for example by using chloride concentrations if those in the lake are different from those in the riverwater.

Piontelli developed an equation which takes sedimentation into account. He assumed that a constant fraction of the supply precipitates, the amount being independent of the lake volume:

$$\frac{dM_w}{dt} = (1-r)J - Q_d[m_w] \tag{17.4}$$

where M_w = amount present in the lake (kg); $[m_w]$ = concentration in the lake (kg m^{-3}); Q_d = water outflow (m^3 d^{-1}); J = daily supply (kg d^{-1}); and r = portion which sediments.

The assumption that a constant fraction of the supply precipitates is not usually valid, however. In calcium-rich lakes soluble-phosphate concentrations may reach an upper limit set by the solubility product. After this has been reached, any additional phosphates will precipitate. The two processes, precipitation and algal uptake, compete for phosphate (see Fig. 5.1) and the outcome varies according to seasonal and other factors.

Calcium carbonate also precipitates in some lakes, the amount being variable rather than a constant proportion of the supply. Control by the solubility product operates here, and factors such as pH and algal growth play an important part in the system. Calcium-carbonate precipitation may be greater as the result of an increase of pH following photosynthesis of a growing algal population. Both Biffi's and Piontelli's equations describe the concentration changes after a sudden change of C or J and can be used to calculate the final equilibrium concentration. If C or J do not change, Biffi's equation gives a final concentration:

$$X_\infty = \frac{bc}{a} \tag{17.5}$$

where X_∞ = concentration in equilibrium state, while Piontelli's equation (17.4) becomes:

$$[m_w] = (1-r)J/Q_d \tag{17.6}$$

The relationship between the amount in the lake and time can be calculated from 17.4:

$$M_w = (1-r)J\frac{V}{Q_d}(1 - e^{(Q_d/V)t}) \tag{17.7}$$

where V = lake volume.

Vollenweider assumed that the amount that precipitates follows the mechanics of a monomolecular reaction:

$$\frac{d[m_w]}{dt} = (1-s)\frac{J}{V} - \gamma[m_w]$$ (17.8)

where s = fraction in outflow, and γ = precipitation-reaction constant.

Using this equation Vollenweider (1969) calculated the annual loading ($g\ m^{-2}\ yr^{-1}$) of several Swiss lakes and tried to calculate the relationship between loading and total phosphate concentration and to establish which factors control the sedimentation rate. The s value must be measured; it depends theoretically on Q_d, V and $\gamma[m_w]$, but cannot be calculated from eq. 17.8. The equation 17.8 used by Vollenweider seems, on closer inspection, to enshrine the same assumptions as that of Piontelli (eqs. 17.4 or 17.7). The assumption that sedimentation is a monomolecular reaction is similar to the assumption that a constant fraction of the supply sediments, because the concentration in the lake is a function of the net supply. The two assumptions are identical under equilibrium conditions, i.e. when $[m_w]$ is constant or $d[m_w]/dt = 0$, then:

$$0 = J - rJ - Q_d[m_w]$$ (17.4 for equilibrium)

and:

$$0 = J - sJ - \gamma[m_w]V$$ (17.8 for equilibrium)

Thus:

$$Q_d[m_w] = (1-r)J$$

and:

$$\gamma[m_w]\cdot V = (1-s)J$$ (17.9a)

or:

$$\frac{[m_w]}{J} = \frac{1-r}{Q_d} = \frac{1-s}{\gamma V}$$

or:

$$\frac{Q_d}{V} = \frac{(1-r)\gamma}{1-s}$$

and:

$$\frac{1}{T_r} = \frac{s\gamma}{r}$$ (17.9b)

because at equilibrium, $s = 1 - r$.

Eq. 17.9b gives the theoretical relation between the sedimentation and outflow constants and water retention time (T_r), *for equilibrium conditions*. The values γ, r, T_r and s do not depend upon whether or not **equilibrium** exists. The identity of eqs. 17.4 and 17.8 for equilibrium conditions can be demonstrated by substituting eq. 17.9a in 17.4:

$$\frac{dM_w}{dt} = J - rJ - Q_d[m_w]$$
$$= J - (1-s)J - (1-r)J$$
$$= J - \gamma[m_w]V - sJ$$

and as $M_w = V[m_w]$, this becomes:

$$\frac{d[m_w]}{dt} = (1-s)\frac{J}{V} - \gamma[m_w] \qquad\qquad (= 17.8)$$

Eq. 17.8 used by Vollenweider is therefore no different from that given by Piontelli under equilibrium and near-equilibrium conditions, i.e. when $dM_w/dt \ll J$. Both formulations are useful, however, because they allow different measurements to be made.

Assuming time-independent (t → ∞) mean chemical conditions, Vollenweider reformulated eq. 17.8, after integration over time:

$$[\overline{m_w}] = L\frac{1}{\sigma\overline{z} + q_s} \qquad\qquad (17.10)$$

where L = annual supply (g m^{-2} yr^{-1}; $J = L \cdot V/1\,000\,\overline{z}$; σ = monomolecular sedimentation rate (t^{-1}); $\overline{z}$ = mean depth; and q_s = flow through rate, expressed in height of water column (m yr^{-1}). Since $[m_w]$, L, q_s, and $\overline{z}$ can be measured, eq. 17.10 can be used to calculate σ. For the Swiss lakes Vollenweider found that the few data available did not accord with results expected from eq. 17.10 and sedimentation was found to be greater than the predicted amount. Better results were obtained if an equation based on the assumption that sedimentation depends on both concentration and annual (net) loading was used.

Assuming firstly the sedimentation to be a function of the total quantity of a substance in the lake:

$$\frac{dM_s}{dt} = \sigma M_w \qquad\qquad (17.11a)$$

Vollenweider arrived at:

$$\frac{dM_w}{dt} = J - \sigma M_w - Q[m_w] \qquad\qquad (17.11b)$$

where M_s = quantity of substance M in sediments. As $M_w = [m_w] \cdot V$:

$$\frac{dM_w}{dt} = J - (\sigma + \rho)M_w$$

or:

$$\frac{d[m_w]}{dt} = \frac{L}{z} - (\sigma + \rho)[m_w] \qquad\qquad (J = L \times \text{surface} = \frac{L \cdot V}{z})$$

Relating sedimentation to concentration *and* loading, eq. 17.1 can be written as:

$$\frac{dM_s^\star}{dt} = \sigma[m_w] \cdot \bar{z} + \tau' \cdot L \tag{17.12}$$

where $M_s^\star = M_s$ per m^2 ; and τ' = fraction of L sedimenting. Vollenweider calculated τ' (from experimental data?) for the individual lakes. It fell in the range between 0.075 and 0.345 (theoretical range 0—1). In the Aegerisee τ' was about 0. No indication is given by Vollenweider how τ' varies over different years for any one lake. To obtain a better value for τ', the loading was corrected for outflow:

$$L' = L(1 - q_s/\bar{z}) = L(1 - \rho) \tag{17.13}$$

Eq. 17.12 becomes then:

$$\frac{dM_s^\star}{dt} = \sigma \cdot [m_w] \cdot \bar{z} + \tau' L(1 - q_s/\bar{z}) \tag{17.14a}$$

Eq. 17.14a was rewritten by Vollenweider as:

$$\frac{dM_s^\star}{dt} = L\{1 + \tau'(1 - q_s/\bar{z})\} - q_s[m_w] \tag{17.14b}$$

which is only true if $L - q_s[m_w] = \sigma[m_w]\bar{z}$ (i.e. if the change of the concentration in the lake is negligible). As has been shown above, this was the condition that Piontelli's equation (17.4) and Vollenweider's (old) equations are identical.

Vollenweider found a satisfactory agreement between eq. 17.14 and the experimental data in some Swiss lakes. The agreement depends however on the deviation of the individual values of τ' from the mean value and is found as a result of sedimentation and outflow quantities. The observed agreement could thus be called a *petitio principii*. In future equations τ' should be related to properties of the lake. One may assume for example that τ' for phosphate depends on the calcium concentration and on the fraction of cellular phosphate that is not mineralised per season in the epilimnion. If these values of τ' were known — which at present they are not — they could be applied to other lakes.

17.2.2. Individual lake models (field data)

O'Melia (1972) calculated the carbon flow for the Lake of Lucerne. He estimated from the decrease of alkalinity in the epilimnion in June that 13 moles of $CaCO_3$ (1300 g) per m^2 per year precipitated towards the hypolimnion. It is unlikely, however, that the rate of downwards transport is the same throughout the whole 12 months as in June. Of the 13 moles $CaCO_3$, 12 moles reached the sediments during the summer season of each year, but

9 moles of this redissolved during winter circulation. It is not clear how the alkalinity in winter can increase by 0.15 meq. l^{-1} through the whole column (60 m), whereas the decrease in the summer is only 0.15 meq. l^{-1} in the epilimnion (15 m).

The average production in the epilimnion itself is 4 mol. m^{-2} yr^{-1} and 4 mol. m^{-2} yr^{-1} of allochthonous organic carbon is added. Of this total of 8 mol. m^{-2} yr^{-1}, 7 mol. m^{-2} yr^{-1} sinks into the hypolimnion, from where it returns to the epilimnion as $CO_2 + HCO_3^-$ while 1 mol. m^{-2} yr^{-1} becomes sediment. Although the calculations are based on many assumptions (for example the total input of organic matter is calculated from oxygen disappearance in the hypolimnion), the model may serve to stimulate further work and thought and seems to show that both Vollenweider's and Piontelli's equations are too simple.

O'Melia's model for phosphate flux in the epilimnion of a stratifying lake is:

$$V_e \frac{d[P_t]_e}{dt} = W - Q[P_t]_e + k_z A_e \frac{d[PO_4\text{-}P]}{dz} - \sigma V_e [PP]_e \qquad (17.15)$$

where W = input, V_e = volume of epilimnion (l), $Q[P_t]_e$ = rate of removal in lake discharge (from epilimnion), $\sigma V_e[PP]_e$ = rate of sedimentation of part-P to the hypolimnion, σ = sedimentation coefficient (yr^{-1}), $[PP]_e$ = concentration of part-P in epilimnion (g l^{-1}), k_z = vertical mixing coefficient $(m^2\ yr^{-1})$, A_e = horizontal area of thermocline (m^2), $d[PO_4\text{-}P]/dz$ = gradient of phosphate across the thermocline. The term $k_z A_e (d[PO_4\text{-}P])/dz$ describes the input of phosphate to the epilimnion by diffusion from the hypolimnion. This vertical flux can exceed input of phosphate from land run-off during the summer period of stagnation. O'Melia calculated that if $k_z = 0.05$ cm^2 sec^{-1} in a thermocline 5 m thick with a gradient of 20 μg l^{-1} per 5 m, then the vertical flux is:

$$0.05\ cm^2\ sec^{-1} \cdot 20 \cdot 10^{-6} g\ l^{-1} \cdot 0.2\ m \cdot \frac{10^3\ l}{m^3} \cdot \frac{m^2}{10^4\ cm^2} \cdot 3.16 \cdot 10^7\ sec\ yr^{-1} =$$
$$0.6\ g\ m^{-2}\ yr^{-1}\ of\ PO_4\text{-}P$$

This is equal to the estimated rate of addition of phosphate from the land to the Lake of Lucerne. It seems likely that k_z is really much greater; eddy diffusion has been ignored (see Chapter 9), while gradients may be much higher than 20 μg l^{-1} per 5 m. Direct solution of eq. 17.15 is not possible as $[P_t]_e$, $[PP]_e$ and $d[PO_4\text{-}P]/dz$ are interrelated, vary with time, and are not measurable directly.

Assuming steady-state conditions, $V_e(d[P_t])/dt = 0$, and that input, W, equals lake discharge, $Q[P_t]$, eq. 17.15 becomes:

$$k_z A_e \frac{d[PO_4\text{-}P]}{dz} = \sigma V_e[PP]_e \quad or \quad \frac{[P_t]_e}{[P_t]_h} = \frac{K_z}{Z_e Z_t} \qquad (17.16)$$

as $[PO_4\text{-}P]_h = [P_t]_h$ and $[PP]_z = [P_t]_e$. With $\sigma = 0.02$ d^{-1} and using measurements obtained by Gächter (1968) O'Melia found that $[P_t]_e/[P_t]_h = 0.3$, which seems a reasonable value and suggests that eq. 17.15 is a fairly good description of the real situation.

If total phosphate behaves conservatively, then:

$$\bar{t}_{H_2O} = \bar{t}_{P_t}$$

where $\bar{t}_{H_2O} = V/Q$,

$$\bar{t}_{P_t} = \frac{P_t}{dP_t/dt}$$

If phosphate sediments, $\bar{t}_{P_t}$ will be less than $\bar{t}_{H_2O}$. If the phosphate concentration increases, the reverse will be true. When $\bar{t}_{P_t}/\bar{t}_{H_2O}$ is greater or less than 1, $[P_t]_e/[P_t]_h$ should be greater or less than 1, which is consistent with eq. 17.16. O'Melia suggested that $[P_t]_e/[P_t]_h$ and therefore $\bar{t}_{P_t}/\bar{t}_{H_2O}$ are both functions of k_z/σ.

Slow sedimentation (small σ) and rapid vertical mixing (large k_z) will result in accumulation of phosphate in the water and the reverse situation will result in loss of phosphate from the water towards the sediments.

Sedimentation rate depends on water current, calcium concentration, and mineralisation rate in the epilimnion. Vertical mixing depends on eddy diffusion and other local currents, so that the results of calculations for the Swiss lakes should be applied only very cautiously, if at all, to other situations.

The model does not allow for the possibility that water from the hypolimnion may escape upwards round the edges of an oscillating thermocline (see section 9.3). A more pragmatic way of calculating a nutrient budget is to make daily or weekly measurements of the inputs and outputs. Clearly this is only possible in cases where there are few inputs and outputs.

Hetling and Sykes (1973) constructed a phosphate and nitrogen budget for Canadarago Lake (New York, U.S.A.), a eutrophic calcium-rich lake. The lake has a surface area of 7.6 km^2, a mean depth of 6.7 m and a maximum depth of 12.8 m. The magnesium, potassium and chloride content, total and soluble and particulate phosphate could be described by:

$$L = AQ^{B+1} \tag{17.17}$$

where L = load via stream (kg d^{-1}); Q = stream flow (ft.^{3} sec^{-1}); and A and B = regression coefficients.

Particulate phosphate concentrations [part-P] (μg l^{-1}) were related to flow:

$$[\text{part-P}] = 27.22 \, Q^{-0.023}$$

while the variation of nitrate and nitrite was given by:

$$L = (A + BQ)Q \tag{17.18}$$

TABLE 17.1

Estimated total phosphorus inputs in Lake Canadarago

Source	Annual values[1]		Growing season value[2]	
	input $(kg\ yr^{-1})$	input (%)	input (kg)	input (%)
Village of Richfield Springs	2 660	44.1	890	66.4
Lake shore cottages	140	2.3	110	8.4
Subtotal	2 800	46.4	1 000	74.8
Gauged tributaries	2 550	42.4	260	19.2
Ungauged tributaries	570	9.4	60	4.3
Subtotal	3 120	51.8	320	23.5
Rainfall	100	1.7	25	1.9
Total input	6 020	100.0	1 340	100.0
Oaks Creek output	4 660	77.5	550	40.9
Net accumulation	1 360	22.5	790	59.1

[1] April 15, 1969, through April 14, 1970; [2] June 1, 1969, through September 30, 1969.

The estimated phosphate input is given in Table 17.1. Most attention was given to figures for phosphate since this is the algal growth-limiting nutrient in the lake. Table 17.1 shows that people living in the surrounding area contributed 75% of the phosphate load during the plant growing season (June–September inclusive) but only 46% of the total annual input. The difference was caused mainly by a relatively lower input through the tributaries during the growing season and a higher retention in the lake during the same period.

Accumulation was due principally to sedimentation of algae. The phosphate input from the village was calculated as 4.8 g d^{-1} of P per person. The national average value for phosphate excretion is 2.0 g d^{-1} and for usage of detergents is 2.7 g d^{-1}. The agreement between these statistical values and the actual measured quantities is quite good (assuming that the village was a national average village). The data suggest that detergent phosphate accounts for 26% of the annual input and for 40% of the summer input. The phosphate contribution from direct land run-off was estimated at only 1 650 kg P per year or 53% of the tributary input. The authors calculated that 3% of the total cattle manure phosphate in the whole watershed would be sufficient to account for the missing 47% in the streams. If this was in fact the case, the relative contribution of the cattle was 24% of the total load.

The annual P input was 0.79 g m^{-2} yr^{-1} of P. It would be practicable to remove all but 0.26 g m^{-2} yr^{-1}. The authors assume that this reduction would improve the water quality in two to three years, but ignore a possible release from the sediments which could lengthen this period considerably.

The average loading of nitrogen was 136 000 kg yr^{-1}, 91% of which was from land run-off, 6% from waste water and 3% from rainfall.

17.2.3. Concluding remarks

It is extraordinary that although rivers transport large quantities of matter, including all plant nutrients, their chemical composition is not often closely monitored. A review by Golterman (1975) revealed that for the major river systems very few data on phosphate, nitrogen, iron and silicate concentrations are available and that the amount of particulate matter is rarely measured at all. Although in some cases the phosphate transported by clay seems to be unavailable for algal growth (e.g. Great Slave Lake and Lake Kinneret, see Chapter 5), this is not necessarily always so.

In the equations used for nutrient budgets a distinction must be made between that part of the phosphate input available for algae and that part which sinks immediately into the hypolimnion. The part taken up by the algae will eventually be mineralised again, while a (small) fraction will sink. The morphology and hydrodynamics of the epilimnion will determine to a large extent the amount of algal phosphate uptake.

Uhlmann and Albrecht (1968) found that in 1 m deep ponds the P retention was 10^2-10^5 times as great as it was in lakes between 10 and 100 m deep. Phosphate retention will also vary with season; proportionally less of the input will be retained in algae during winter than during summer. In shallow lakes it seems likely that the winter phosphate input will accumulate on the sediments and be released in summer time, when the demand by algal growth is larger than the supplies that are immediately available from input and internal recycling. For these and other reasons it is not yet possible to give a quantitative relationship between algal growth and the increasing amounts of nutrients entering lakes. At present it is probably more useful to compare lakes with different nutrient loading and to evaluate the degree of eutrophication.

17.3. EUTROPHICATION

17.3.1. Definitions and causes of eutrophication

Two different definitions of eutrophication exist. A clear distinction must be made between eutrophication meaning an increase in nutrient supply, and eutrophication meaning the automatic result of this increased supply, i.e. increased plant growth. Oxygen depletion either in hypolimnia or overnight in shallow lakes should never be called eutrophication.

The maximum possible amount of algal growth in a lake can be expressed as the trophic status of that lake (Greek trophos = food). A broad division of lakes can be made into *oligotrophic* (Greek oligos = little; thus little food giving few algae), *eutrophic* (well fed, giving easily noticeable algal growth) ar *hypertrophic* (excess food giving too many algae). These terms have no absolute meaning in terms of nutrient content of a lake and are used to mean different things in different situations. Those working in high alpine lakes will probably confer the eutrophic status on lakes with a lower nutrient content than that of, say, shallow lakes in delta areas. Eutrophication is a special case of water pollution; it often occurs simultaneously with organic pollution because human sewage and waste water of intensive husbandry ("bio-industry") are the most common causes of eutrophication.

The word eutrophication is used increasingly in the sense of artificial addition of plant nutrients to waters. Its original use was to describe the old German fish-pond cultures, which were often built on infertile acid soils. Addition of calcium carbonate (lime) increased the pH and thus the CO_2 supply; the ponds were said to have become eutrophic, and for a long time the calcium concentration was used as a measure of the trophic status.

The more modern usage is to apply the term eutrophic to waters with high concentrations of nutrients but especially of phosphate or nitrogen or both.

At present eutrophication is generally considered to be undesirable, although this is not always true. That which is undesirable in some waters may

TABLE 17.2

Relative quantities of essential elements in plant tissue (demand) and their supply in river-water (from Vallentyne, 1973)

Element	Demand plants (%)	Supply water (%)	Demand plants/ supply water approx.
Oxygen	80.5	89	1
Hydrogen	9.7	11	1
Carbon	6.5	0.0012	5 000
Silicon	1.3	0.00065	2 000
Nitrogen	0.7	0.000023	30 000
Calcium	0.4	0.0015	< 1 000
Potassium	0.3	0.00023	1 300
Phosphorus	0.08	0.000001	80 000
Magnesium	0.07	0.0004	< 1 000
Sulphur	0.06	0.0004	< 1 000
Chlorine	0.06	0.0008	< 1 000
Sodium	0.04	0.0006	< 1 000
Iron	0.02	0.00007	< 1 000
Manganese	0.0007	0.0000015	< 1 000
Boron	0.001	0.00001	< 1 000
Zinc	0.0003	0.000001	< 1 000
Copper	0.0001	0.000001	< 1 000
Molybdenum	0.00005	0.0000003	< 1 000
Cobalt	0.000002	0.000000005	< 1 000

be harmless or even desirable in others. Shallow ponds such as those used for recreational or even for commercial fishing may be profitable with rather ' dense algal crops. They would, however, be less suitable for other uses such as water sport or for aesthetic purposes. The clear deep lakes with aerobic hypolimnia are an opposite case. Increased algal growth will have the undesirable effect of reducing the O_2 content of the hypolimnion (see subsection 9.4.1).

About 15 to 20 elements are necessary for the growth of freshwater plants. Vallentyne (1973) showed the special significance of phosphorus and nitrogen by preparing a "demand/supply" table for all essential elements (see Table 17.2). The plant "demand" was approximated by doubling the quantities of elements of an average community of freshwater plants. The "supply" term was approximated from data on the chemical composition of mean world riverwater. It is clear that phosphate and nitrogen supplies are underestimated by this procedure, as is carbon (which diffuses from the air). Nevertheless the ratio is by far the highest for nitrogen and phosphate.

17.3.2. Nitrogen and phosphate supplies as causes of eutrophication

The present concern about eutrophication relates to the rapidly increasing amounts of these elements which are normally present at fairly low concentrations and which usually limit both algal growth rate and biomass. Phosphate is often considered to be the major cause of eutrophication. It was formerly the growth-limiting factor for algae in the vast majority of lakes, but its usage has increased tremendously during the last few decades both in agricultural fertilizers and, more significantly, in synthetic detergent.

Nitrogen is thought to be limiting for algal growth in a small number of lakes, especially in tropical regions. Talling (1965) considered algal growth in the lakes of the East African Rift Valley (Uganda) to be limited by nitrogen supplies as the result of nitrogen depletion of soils by intensive erosion in the past. It is possible of course that nitrogen may become secondarily limiting in lakes in which it was not originally so, as the result of the tremendous increases in phosphate concentrations. A further reason why nitrogen is less likely to be the primary cause of increasing amounts of algal blooms is the fact that, unlike phosphate, it does not accumulate in lakes and that considerable amounts of nitrogen are lost by denitrification.

Nitrogen cannot therefore lead to the cumulative phenomenon of ever-increasing algal blooms. The slow increase of algal populations due to phosphate enrichment may lead to an increasing efficiency with which the available nitrogen supplies are utilised and depleted.

Lund (1970) showed that chlorophyll a concentration of some English waters is strongly correlated with both the winter PO_4-P and NO_3-N concentrations (Figs. 17.1 and 17.2), which are themselves interrelated (Fig. 17.3). It is clear from Fig. 17.3 that in nutrient-poor lakes less phosphate than nitrogen is available for algae. The line in Fig. 17.3 indicates a N:P ratio of

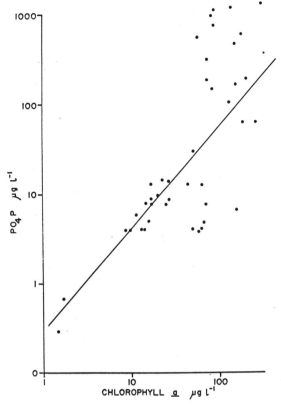

Fig. 17.1. Maximum winter PO_4-P content (vertical axis) compared with maximum summer algal biomass expressed as chlorophyll a (horizontal axis) for several lakes in England. The continuous line indicates an approximate average value for all but the uppermost eight points. (From Lund, 1970.)

10:1. This is the approximate mean algal composition. It can be seen that the points for the nutrient-poor lakes show a N:P ratio of about 100:1. About 90 units of N are therefore not used by the algae. On the other hand, nutrient-rich lakes receiving much sewage with a N:P ratio of 10:1 will have N:P values approaching this ratio, so that here all the nitrogen can be used. Natural and agricultural run-off from soil is the source of nitrogen in oligotrophic lakes. In the English lakes the chlorophyll a concentration never exceeds about 100 g m^{-3} (Figs. 17.1 and 17.2) probably due to "self-shading" effects (see Chapter 4).

Lund pointed out that other factors controlling the size of the algal populations are mean depth and water retention time. For example Esthwaite Water, which has a 2.5 times greater water retention time than Blelham Tarn, receives a greater PO_4-P concentration and has a lower midwinter concentration of phosphate but a greater average biomass. In Esthwaite Water the re-

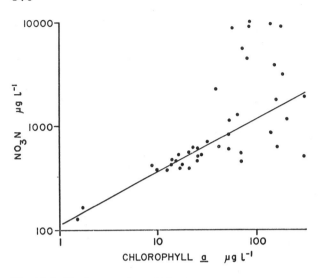

Fig. 17.2. Maximum winter NO_3-N content (vertical axis) compared with maximum summer algal biomass expressed as chlorophyll a (horizontal axis) for several lakes in England. The continuous line indicates an approximate average value for all but the uppermost eight points. (From Lund, 1970.)

newal of nutrients is thus slower and biological uptake greater than in Blelham Tarn. Therefore, the comparison between midwinter nutrient concentrations and chlorophyll a will be valid only in systems which do not differ too markedly in water retention time. Winter nutrient concentrations are not always as important as they are in the English Lake District; in other situations the input during the rest of the year may be relatively much more important and the hydrological regime may be very different. The observed correlation between chlorophyll a and nutrient concentrations causes the midwinter concentrations to be a good parameter on which to predict algal development in the next year. Depth is also an important factor. In two unstratified lakes of unequal depth but with the same light penetration a greater proportion of the algae will receive sufficient light for growth in the shallower lake than in the deeper lake. Thus shallower lakes are likely to have larger winter populations which may utilise a greater proportion of the incoming nutrients. Mortimer (1969) showed the vernal increase of diatoms to start and end later in the deepest of four neighbouring lakes. A better correlation between biomass and nutrients should be obtained if total phosphate and total nitrogen concentrations are correlated with chlorophyll because in this case the amount of nutrients in the organisms will be included. This correlation has no predictive value though, and serves only as a description of the situation (Fig. 17.4).

Typical examples where phosphate is the key factor for eutrophication are the Swiss alpine lakes, especially the Lake of Zürich. Thomas (1973) sum-

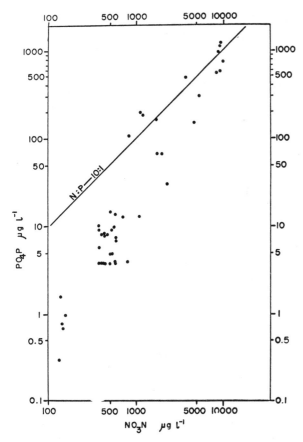

Fig. 17.3. Maximum winter PO_4-P (vertical axis) compared with maximum winter NO_3-N content for several lakes in England. (From Lund, 1970.)

marised the phosphate concentration in three of these lakes, Lake of Constance, Lake of Zürich and Greifensee (see Table 17.3).

In an earlier paper, Thomas (1968) had already demonstrated that since 1955 there had been a steep increase of the phosphate concentration in the Lake of Zürich (Fig. 17.5) and showed that adding phosphate without other nutrients, could cause increased algal growth. Thomas (1973) demonstrated the importance of the fact that the sewage input passed directly into the epilimnion (0—10 m layer) in the Lake of Zürich, whereas the natural input, mostly in autumn rain, mixes into the whole lake. The final proof that phosphate was the main factor in eutrophication came after 1967. In that year sewage works began removing phosphate from all sewage entering the main water body of the lake. Turbidity in the epilimnion has decreased and the oxygen concentrations in the hypolimnion have increased so much that by 1970 they were higher than in 1896 when measurements were begun. These

372

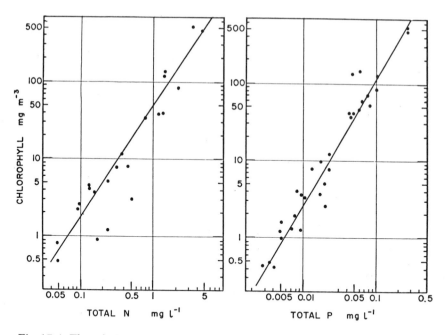

Fig. 17.4. The relationship between chlorophyll content and total nitrogen (left-hand graph) or total phosphorus (right-hand graph) in Japanese waters sampled in May, June and September. Based on figures from Sakamoto (1966) as cited in Lund (1970).

improvements were smaller in the area called "Untersee" where less phosphate was removed (Thomas, 1971) (see Table 17.4).

Lehn (1972a) showed that during winter circulation in the Lake of Constance the phytoplankton density increased in proportion to the phosphate concentration. When chlorophyll concentration was graphed against phosphate, two straight lines were obtained, the one for the shallower Gnadensee having a steeper gradient than the one for the deeper Obersee (Fig. 17.6). Only tentative inferences about the influence of depth on eutrophication may be

TABLE 17.3

Annual average PO_4-P concentration in three Swiss lakes (mg m^{-3})

Year	Lake of Constance	Lake of Zürich	Greifensee
1940	0		
1945		23	
1950	2	27	60
1955	4	42	180
1960	10	52	300
1965	18	77	430
1970	38	84	480

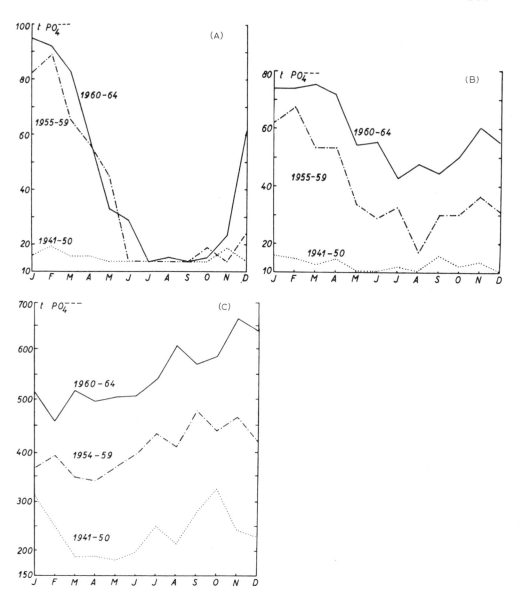

Fig. 17.5. Phosphate content in the Lake of Zürich, in: A, the 0—10-m layer; B, the 10—20-m layer; C, the 20-136-m layer. Mean values for different year classes. (From Thomas, 1968.)

made however, since no figures are given to enable a comparison of the phosphate loading of the two basins. Lehn (1972b) demonstrated that there was a higher density of Crustacea in Gnadensee than in Obersee and that oxygen concentrations in the hypolimnion decreased from 80% to 40%. At the same

374

TABLE 17.4

Oxygen concentration in the Lake of Zürich near Thalwil (ranges of October values in mg l^{-1})

Depth (m)	1960—1965	1966	1967—1968	1969—1970
0.3	8.4—9.8	8.6	11.8—12.8	9.9—10.7
5.0	8.4—9.6	8.4	11.7—11.8	9.0—9.9
10.0	7.7—9.1	2.8	7.0—8.6	4.4—7.4
15—20	2.2—4.3	3.8—4.4	3.4—4.2	4.9—7.3
25—50	3.5—7.4	4.9—6.4	5.0—7.7	6.5—9.0
50—100	1.2—7.7	4.4—7.3	7.2—8.5	6.9—9.6
110—120	0.3—4.1	0.5—3.5	5.9—6.9	3.4—8.2
Mud surface	0.2—1.3	0.5	1.3—2.7	0.0—0.4

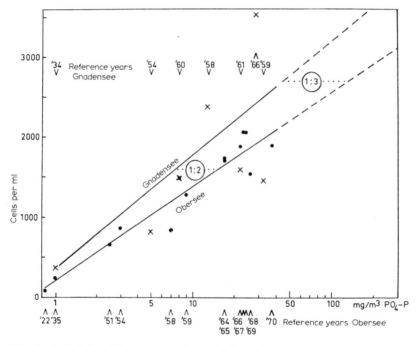

Fig. 17.6. Relationship between phytoplankton (cell numbers) and phosphate concentration in Gnadensee (= ×) and Obersee (= •). (Lehn, 1972a.)

time fish catches increased considerably, though this may have resulted partly from improved techniques. Wagner (1972) showed a nearly tenfold increase in the phosphate concentration of the sediments between 1935 and 1971 (Fig. 17.7).

The effect of phosphate can also be demonstrated by an approach such as

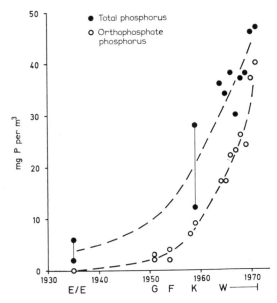

Fig. 17.7. Mean phosphate concentration in the Lake of Constance between 1935 and 1971. E/E: Elster and Einsele (1937); G: Grim (1955); F: Fast (1955); K: Klifmüller (1962); W: Wagner (1972).

that of Lund (1969, 1972b). He showed that in Blelham Tarn *Asterionella* took less time to reach its maximum density as phosphate concentrations increased. This indicates that eq. 14.3 is at least qualitatively correct. The cell density increased also, however. Apparently in Blelham Tarn silicate is not the yield-limiting factor as it is in Windermere. In Windermere South Basin, increasing phosphate concentrations did not shorten the time needed for *Asterionella* to reach its maximum density, but from the sharp decrease in this maximum density during the period in which phosphate increased it must be concluded that a mechanism began to operate which was unaccounted for by the relationships given in eq. 14.3. Lund noted that in the years when no large *Asterionella* populations occurred other diatoms became abundant, but not until much later in the year. This might be explained by assuming that these diatoms grew more slowly than *Asterionella* (see Chapter 14).

More experimental evidence for phosphate and nitrogen as key factors in eutrophication comes from the work of Schindler (1971) and Schindler et al. (1973). He fertilized a very oligotrophic lake (lake 227) in the experimental lake area of northwestern Ontario (Canada). There are a series of such lakes on the hard rocks of the Precambrian Shield. They all have rather low concentrations of dissolved solids. Schindler added 0.34 g m^{-2} yr^{-1} of P and 5.0 g m^{-2} yr^{-1} of N in 1969, followed by 0.48 g m^{-2} yr^{-1} of P and 6.29 g m^{-2} yr^{-1} of N in 1970. The applications were in 17—21 equal weekly doses.

The standing crop of phytoplankton (measured as chlorophyll a, sestonic P, N and C) increased in that period to concentrations typical for eutrophic lakes. These experiments are important because the concentrations of both gaseous CO_2 and total CO_2 in the euphotic waters were low and did not change much (Table 17.5). Secchi-disc visibility decreased from 3 to 1 m. The average C:N:P ratios for the seston in the epilimnion of the lake were 130: 13:1, while ratios of 4 unfertilized control lakes averaged to 118:11:1. The absence of carbon compounds in the added fertilizer did not appear to have restricted the increase in phytoplankton standing crop. Schindler demonstrated that diffusion of CO_2 from the air supplied sufficient CO_2 so that carbon is not the growth *yield*-limiting factor. (CO_2 influence on growth *rate* will be discussed below.) It must be mentioned though that by adding $NaNO_3$ (nearly 1 equivalent per m^2 during 2 years) Schindler must have considerably increased the alkalinity of the lake. Without suggesting that Schindler's results would necessarily have been radically different, it would have been much better if he had used NH_4NO_3 as nitrogen source and $NH_4H_2PO_4$ as phosphate source. In polythene tube (diameter 1 m) experiments Schindler showed that only phosphate enrichment (with or without nitrogen) was able to support phytoplankton standing crops higher than those of the control waters. In bioassays in bottles he showed that addition of bicarbonate did increase the $^{14}CO_2$ uptake in the 4-h experimental period and regarded this as a paradox. This paradox can however be easily explained (see p. 378). Schindler confirmed these results in further experiments (Schindler et al., 1973). The phytoplankton standing crop was increased by two orders of magnitude and had reached the theoretical maximum. Cryptophyceae and Chrysophyceae were replaced by Chlorophyceae and Cyanophyceae. The Cyanophyceae caused the nitrogen supply to increase twofold by N_2 fixation, so that the lake became nearly self-supporting for nitrogen. Although no extra CO_2 was added, diffusion from the air provided sufficient CO_2 to sustain the high yield. The PO_4-P concentration remained low in lake 227 and 80% of the added PO_4-P was taken up in the sediment. Sedimentation was measured as the difference between input and output plus change in concentration, ΔM:

$$S = I - \Delta M - E$$

The nearby lake 239 was used as a control for monitoring input via precipitation, etc. Silica in lake 227 became depleted earlier in the year because of greater abundance of diatoms. It would have been interesting to see whether addition of fertilizer silica could have maintained a large diatom crop and prevented the bloom of blue-green algae. Quantities of nutrients in the lake increased each summer averaging 2 kg yr^{-1} (about 7% of the amount added) but winter concentrations did not change significantly. These results do not support the statement that there was no release of phosphate from the sediments. It would be interesting to make a detailed phosphate balance sheet of the epilimnion by comparing ΔM with I in order to see whether diffusion of

TABLE 17.5

Chemical analyses and other observations from the epilimnion of lake 227 (Schindler, 1971)

	Before fertilization			After fertilization	
	June 1968	July 1968	June 1969	August 1969	August 1970
Chlorophyll a (μg l^{-1})	5.8	3.1	2–3	9–24	48–92
Phytoplankton cell volume (cm^3 m^3)	<1	<1	<1	2–7	10–16
Dominant phytoplankton genera	*Ochromonas* *Chromulina*	*Cryptomonas* *Mallomonas*		*Staurastrum* *Phacomyxa* *Spondylosium*	*Oscillatoria* *Lyngbya* *Pseudoanabaena*
Reactive phosphorus (μg l^{-1})	<1	<1	0–1	<1–3	<1–4
Total dissolved P (μg l^{-1})	no data	no data	3–5	2–8	3–16
Seston P (μg l^{-1})	no data	no data	1–5	15–29	26–50
NO$_3$-N (μg l^{-1})	7–8	7	0–10	70–117	<1–3
Total dissolved N (μg l^{-1})	no data	no data	130–160	210–310	270–565
Seston N (μg l^{-1})	no data	no data	50–100	180–320	320–600
CO$_2$ (μmol. l^{-1})	no data	no data	37–62	5–79	21–80
Seston C (μg l^{-1})	no data	no data	600–900	1 840–2 760	3 960–5 120
Secchi-disc visibility (m)	2.2	2.5	3.0	1.2–1.3	0.70–0.95
pH	no data	no data	6.3–6.8	9.2–9.5	9.3–10.2

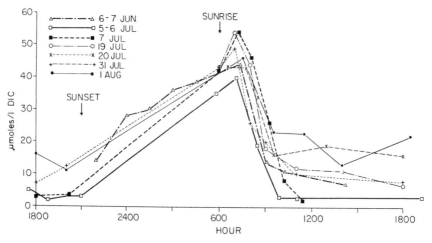

Fig. 17.8. Diurnal variations in concentration of dissolved inorganic carbon in lake 227 (0.5 m, summer of 1972). Note the regularity of the diurnal change, even though data span a 2-month period. (From Schindler et al., 1973.)

phosphate from the hypolimnion could be detected. It would also be extremely useful if a part of the lake could be left for "natural" recovery.

Schindler and Fee (1973) demonstrated the disadvantage of the ^{14}C bottle technique for eutrophic carbon-poor water and developed a new technique. They made continuous measurements (using gas chromatography) of dissolved inorganic carbon, community respiration and the inversion of CO_2 concentration. The diurnal variations were large (Fig. 17.8) demonstrating the depletion of CO_2 during the day time, which caused the photosynthetic rate to decline, and the replenishment of CO_2 during the night. This regularity of the diurnal change was maintained over at least a 2-month period during the summer months.

It is quite clear however that, although Schindler and his coworkers demonstrated that CO_2 diffusion may supply CO_2 in sufficient quantities to prevent it from becoming a *yield*-limiting factor, the growth *rate* may nevertheless be limited by the CO_2 supply, especially in a closed bottle. Because these experiments were performed after phosphate fertilization, carbon dioxide, rather than phosphate, was limiting the growth rate. It must be obvious, however, that no deductions may be made about the situation that existed in the *unfertilized* lake.

The bottle bioassay shows once more (see Chapter 10) how dangerous extrapolations from bioassays can be. Extrapolations of this kind had led to the invalid hypothesis — often called the "Lange-Kuentzel-Kerr thesis" — that carbon rather than nitrogen and phosphate is generally the factor limiting algal growth.

17.3.3. Other assumed causes of eutrophication

Lange (1970a) showed that when CO_2 is limiting growth then added organic carbon compounds have a stimulatory effect on the growth of a mixture of blue-green algae and bacteria. This was probably because bacteria produce CO_2 from the organic matter and since CO_2 was limiting, algal growth increased. Lange stated that "slow diffusion of atmospheric CO_2 and bicarbonate may make CO_2 a limiting factor during periods of vigorous photosynthesis". Some of his evidence for this statement was based on other people's work in cultures where it is obviously true. He also referred to other work, the authors of which made some wrong assumptions. Although Lange himself never claimed to have measured CO_2 or to have found it to be a limiting factor in lakes, he strongly suggested this and he has been frequently misquoted by both proponents and opponents of the "Lange-Kuentzel-Kerr thesis". Lange (1971b) found later that in filtered Lake Erie water growth of the blue-green algae *Anabaena circinalis*, *Microcystis aeruginosa* and *Nostoc muscorum* and of the green alga *Selenastrum capricornutum* was stimulated by nitrate in about two-thirds of the water samples. In the rest phosphate, cobalt and chelated iron stimulated growth. These experiments were made with *filtered* water though. In such bioassays one has removed just those algal cells whose growth one needs to study! Removal of the algae by filtration makes it impossible for example to take possible storage within the cells into account.

Schindler (1971) has criticised Lange's work on Lake Erie on several grounds. In those experiments for example where serial additions of PO_4-P were made to cultures, rather large quantities of PO_4-P were added. In Lake Erie it is clear that the bicarbonate concentration ($100—120$ mg l^{-1}) is sufficient for algal photosynthesis and the incidence of algal blooms has increased dramatically following increases in PO_4-P content (see below), while bicarbonate remained constant.

Kuentzel (1969) assumed that algae derive their supply of carbon only from the physically dissolved CO_2. Although it is an open question whether they can utilise bicarbonate as such (see Chapter 3), the dissolved CO_2 is in equilibrium both with the air and with the dissolved bicarbonate. The bicarbonate, at least, is present in many times greater amount than the physically dissolved CO_2. It is clear that Kuentzel made an erroneous assumption based perhaps on the earlier work with German fish ponds on acid soils. In these ponds addition of lime increased the alkalinity and therefore provided enough bicarbonate for rapid algal growth and thus for high fish production. All intensively cultured fish ponds receive extra phosphate however.

Kerr et al. (1972) found that the bacterial population increased within 12 h following the addition of nitrogen, phosphate and potassium chloride to Shriner's Pond (a small Georgia fish pond) but that the algal population of this pond increased only $36—48$ h after fertilization. From diurnal changes

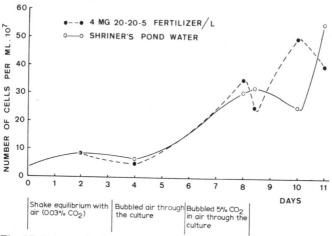

Fig. 17.9. Growth of *Anacystis nidulans* in Shriner's Pond water. When growth ceased at one level of CO_2, higher levels were supplied to each culture. In both cultures where air was the source of CO_2, growth ceased at pH 9.8—10.1. (From Kerr et al., 1970.

of CO_2, HCO_3^- phosphate and nitrogen removal rates and from the fact that algal growth was stimulated by bubbling CO_2 through water in two plastic enclosures (one with and one without fertilizer, see Fig. 17.9), Kerr et al. (1970) concluded that carbon dioxide regulated algal growth and that the N and P stimulation was owed to an increased bacterial production of CO_2. This paper was published in an inaccessible report (see below) and is much quoted by detergent manufacturers to counteract the limnologists' suggestion that phosphates be banned from future detergents. It is necessary therefore to criticise points of detail which would be otherwise out of place in a text-book such as this. The authors themselves stated that field data like theirs can not elucidate the mechanism responsible for algal blooms. They used their own laboratory data to support their present view (the data being published only in the report to the U.S. Department of the Interior, Federal Water Quality Administration), but these laboratory experiments do not in fact seem to show conclusively what the authors state. For example, enhance-ment of algal growth by increasing concentrations of CO_2 was not unequi-vocally shown because the experiments using increasing amounts of CO_2 were sequential and not simultaneous or random. It is possible, perhaps likely, that another factor was changing at the same time as the CO_2 concentration. Furthermore, the species of test alga used was *Anacystis nidulans* in cell densities, at the start, of $3 \cdot 10^{10}$ cells per litre, which increased to about $50 \cdot 10^{10}$ cells per litre during the experiments. These densities are extremely high from an ecological point of view. Growth during the experiment was very poor (a 16-fold increase or 4 cell divisions) and was not closely related to the amount of CO_2 available, which it should have been if CO_2 were operating as a limiting factor. Results obtained in the field experiments in

Shriner's Pond and the stimulation of growth in the plastic containers may well have been caused by the bubbling of gas which would induce more thorough mixing than would be natural. Other workers have criticised the same paper for basic faults in experimental design and interpretation (see discussion following Kerr et al.'s paper, 1972a).

Allen (1972) showed that in Star Lake (Vermont, U.S.A.) with low alkalinity and low calcium concentration primary production in 4-h bioassays could be stimulated by addition of molybdenum, zinc, cobalt, manganese and iron — either separately or together — with or without organic chelator. Phosphate did not stimulate photosynthesis, presumably because sufficient amounts were already available (11—25 μg l^{-1}) or because insufficient time was allowed for the induction and synthesis of the enzyme system needed to utilise the added phosphate. The lake is situated on Palaeozoic metamorphic rock and has no carbonate sediment. Both carbonate and autochthonous phosphate were probably in low concentrations. The phosphate probably originates from surface run-off from a forested area. Allen found that active photosynthesis occurred during periods of high pH, when mostly bicarbonate with some carbonate and no CO_2 was available.

Using the Wright and Hobbie technique (see section 16.6), Allen measured the instantaneous rate of uptake at naturally occurring inorganic carbon concentrations and found that photosynthesis was slowest when only bicarbonate was present. Allen did not, however, eliminate the possibility of light inhibition which may have had an influence both in the field and in his experiment. He suggested that bacterial production of carbon dioxide from extracellular products of macrophytes stimulated algal growth. In closed bottles, this may have been proportionally more important than in the open lake, where CO_2 diffusion from the air or sediments would provide an extra supply of carbon dioxide. It seems likely that the pH, which increased from 5.5 to 9.5 in the bottles during the experimental period, may have caused this effect as adaptation to such a pH shift may take a few hours. Allen did not study the influence of CO_2 supply from air or sediments but assumed it to be unimportant. In his comments following Allen's paper, Hobbie showed that diffusion could supply about 100 mg m^{-2} d^{-1} of inorganic carbon and criticised the uptake kinetic calculations. Furthermore, Hobbie made the valid point that CO_2 can be produced nearly instantaneously from HCO_3^- anions; an action which could be stimulated by the enzyme carbonic anhydrase in the cells. Our own results often show a significant decrease of $^{14}CO_2$ uptake in short experiments after doubling the inorganic phosphate concentration. As a stimulatory effect is found if a pre-incubation period with the phosphate is allowed, it seems likely that the decrease is an artefact. It could be caused for example by a limited uptake capacity, which may be used for phosphate instead of for CO_2 for a short period after the addition of the extra phosphate or by the influence of extra phosphate on the availability of iron.

A similar effect was found by Hamilton (1969) with phytoplankton from Cayuga Lake (New York, U.S.A.). In 8-h experiments he also found a phosphate inhibition but a silicate enhancement. He added more than 100 times the phosphate originally present, however, and the mechanism suggested above (PO_4 uptake instead of CO_2 uptake) may explain his results. Besides this, his cultures were at only 10% of I_0, which is far below the optimum for photosynthesis. Hamilton argued that the "light intensity" he used was the mean value occurring in the 23 m deep euphotic zone of Cayuga Lake, but it would seem better to have taken the weighted mean or even the light value where A_{max} occurred. The fact that silicate did cause greater CO_2 fixation rates could be explained by assuming that silicate limits the yield, while phosphate controls the growth rate (see section 14.2). The disadvantages of using short-duration bottle tests as a tool to detect possible nutrient deficiences or growth limitations have been mentioned already in Chapter 10.

It cannot be too strongly emphasized that short-term experiments on the influence of certain nutrients on photosynthesis can show only what may happen as the immediate result of the additional nutrient supply and may give no indication of effects which would result in the longer term under field conditions.

Glooschenko and Alvis (1973) used time intervals of 3 and 6 days and demonstrated that the addition of nitrate and phosphate either alone or in combination caused increased numbers of algae with differences in species composition in Lake Jackson (Florida, U.S.A.). This tendency to changes in species composition is often used as an argument against using long-lasting enrichment experiments, since changes induced in bottle cultures will almost certainly be different from any which would occur under natural conditions. However, they provide an indication of whether or not the total quantity of algae will increase as the result of fertilization and may give some hint of the direction in which any change in species composition is likely to be. Inexplicable phenomena such as the increase of *Anabaena cyanea* following the addition of silicate in Glooschenko and Alvis' experiments are probably due to secondary effects of pH change, competition, grazing, etc.

It has been suggested that the presence of humic substances in highly productive lakes may be an important factor leading to the nuisance growth of blue-green algae. Lange (1970b) found a stimulatory effect due to favourable chelation of iron by fulvic acid and suggested that the influx of sewage effluents with iron into lakes rich in humic substances may lead to excessive algal growth. He demonstrated the occurrence of a stimulatory effect of fulvic acid on the growth of several species of blue-green algae, but it is striking that as high a concentration as 100 mg l^{-1} was necessary to produce an effect; 10 mg l^{-1} had no effect, although this quantity would have been sufficient to act as a chelator in the required amounts. It can be argued that the effect found by Lange is due to a contaminant, e.g. phosphate, the presence of which has been neither discerned nor measured. The fulvic acid which Lange used did contain a percentage of unidentified mineral compounds. If humic substances are present in lakes in natural conditions, they will probably already contain iron (see Chapter 12), so that further addition will have little or no effect. There are no reports so far to suggest that iron is the algal growth-limiting factor in lakewater even in clear waters. In Chapter 11 it has been shown that iron does not need to be in solution in order to be available for algal uptake.

Prakash and Rashid (1968) claimed that the growth of marine dinoflagellates is enhanced by humic substances but these effects could easily be owed to metal chelation, and many of their experiments showed ambiguous results. Prakash et al. (1973) found that growth of some marine diatoms was stimulated by the addition of different humic-acid fractions. All the effects shown are probably due entirely to chelation as the basic culture medium used contained Fe-EDTA but also a high calcium concentration. The Ca^{2+} will have chelated the EDTA leaving no soluble iron. Interpretation of some reported experiments where growth enhancement of cultures resulted in as little as a 10-fold cell increase remains an open issue. If a substance is plausibly to be regarded as an algal growth enhancer, it seems desirable to consider a 100-fold (6—7 binary divisions) cell increase in the control as being a minimum criterion of success, unless the response to the addition of the enhancer is obvious, e.g. when there is no growth at all in the control.

In summary it seems that there is no strong evidence to suggest that carbon would ever limit algal yield in an unfertilized lake.

17.3.4. Nutrient losses from soils

Few data exist on the contribution of phosphate and nitrogen from soils (see also Chapter 6). A distinct difference exists though between the outwash of nitrogen and that of phosphate from agricultural land (Table 17.6).

Prochaczkova (1972) demonstrated for Slapy reservoir (Czechoslovakia) that the NO_3-N concentration increased with increasing amounts of fertilizer applied to land within the drainage area. Up to 25% of the nitrogen applied to the land appeared in the lake while much less of the phosphate was found, indicating a proportionally greater retention of phosphate in the soil. Retention of phosphate in the lake was also larger than nitrate. While the ratio of nitrogen to phosphate was nearly 2 in the fertilizer, it increased to 26 in the drainage water and to 30—50 in the outflow from the lake.

Phosphate movement from an agricultural watershed (Mahantango, Pennsylvania, U.S.A.) was measured by Kunishi et al. (1972), who found that total phosphate loss was 28—762 mg ha^{-1} d^{-1}, the amount depending on the rainfall. The greatest loss was during a rainstorm of 6.3 cm within about 12 h. The run-off contained roughly equal amounts of exchangeable sorbed phosphate and inorganic phosphate in true solution. During moderate rainfall (3.4 cm within 3 days) 28 mg ha^{-1} d^{-1} was removed, although the amount

TABLE 17.6

N and P loss from soils (kg km^{-2} yr^{-1})

	Forest regions		Agricultural regions	
	P	N	P	N
Lower alps of Switzerland	0—4	82	69	1 634
Swiss lowland	0—1	959	35	2 102

See also section 5.2.

of water passing by the run-off gauge for this period of measurement was five times less than in the storm. Phosphate removal results mainly from erosion of small particles of sediment, a process which depends on the speed of the water (see Chapter 18).

Gächter and Furrer (1972) and Furrer and Gächter (1972) concluded that the phosphate and nitrogen losses from soils depend on the method of cultivation, climatic conditions and topography. Their results are summarised in Table 17.6.

17.3.5. Conclusions and cures

The sequence of events leading to eutrophication has often been: before eutrophication oligotrophic waters often have a ratio of N:P greater than 10; the nitrogen concentration may be higher than "normal" owing to run-off from soils in the watershed and phosphate is then the growth-limiting factor; sewage discharges or perhaps some agricultural run-off, or both, cause nitrogen and phosphate concentrations to rise; the N:P ratio will approach 10 in the case of sewage effluent and then either N or P may be limiting; N_2 fixation by blue-green algae may make nitrogen less depleted than phosphate.

Moses Lake (described in Chapter 14) represents typically this situation. Another example is given by Gerloff (1969), who showed that in Lake Mendota greatest growth of *Microcystis aeruginosa* could be obtained by adding nitrate, phosphate and iron. There was sufficient nitrogen for only about 1% of maximum growth, phosphate for 7% and iron for 22%, which suggests that nitrogen was relatively least abundant. Gerloff summarised the deficiencies of bioassay as follows: it reflects the growth of a newly introduced alga during a brief interval only and after removal of the natural population. The natural daily input, which may have a larger influence than the amounts present at any one time is cut off in these bioassays, while filtration removes the cells grown already, thus underestimating the fertility.

If nitrogen alone is removed from sewage before it enters a lake, the N_2-fixing blue-green algae may be at an advantage. Dense growths of these algae are a particular nuisance. If fertilization is very heavy then CO_2 may eventually become growth rate-limiting in soft waters, while self-shading will limit algal growth in alkaline Ca-rich waters. This is the case now in many shallow (1—2 m) Dutch lakes, where chlorophyll concentrations of about 300 mg m^{-3} are reached and phosphate is quite often above 1 g m^{-3} of PO_4-P. The sediments in these lakes are already heavily loaded, so phosphate diversion will cause little improvement in the state of these lakes. Addition of organic substances to soft waters where CO_2 may have become growth limiting may increase the CO_2 supply and thus algal growth. There is no evidence yet, however, to suggest that this would ever happen under natural conditions in the lake itself. CO_2 supply from the air or from bottom sediments will be sufficient to maintain growth of relatively large algal crops since the supply will continue during darkness, thereby replenishing any deficit which may have developed during periods of intense photosynthesis. Another argument commonly used to dispute the role of phosphate as the key factor governing eutrophication is that the algae will also need several other elements for growth. However, a lack of only one essential nutrient is able to prevent algal growth. Thus no algae can grow without phosphate, although this does not mean that all excessive algal growth is due solely to input of phosphate.

The inescapable conclusion therefore is that there is only one remedy for the excessive growth of algae: diminished input of phosphates into lakes by removing them from sewage effluents and by replacing polyphosphates in detergents with a phosphate-free product. *It is not important whether phosphate is currently the limiting factor or not, or even that it has ever been so; it is the only essential element that can easily be made to limit algal growth.* Control can be achieved easily by precipitation of phosphates from sewage effluents and by replacing polyphosphates in detergents by phosphate-free chelators. When phosphate loadings are very high, both measures should be adopted. In those situations where only a part of the sewage is treated biologically, phosphate removal is difficult and the replacement of phosphates in detergents is the only way to stave off eutrophication in the short term. It has often been stated that this partial removal of phosphates would have no significant effect because in Europe only 40% and in the U.S.A. 70% might be prevented from reaching the lake. It should be emphasized, however, that eutrophication is a process which develops slowly. It is the result of a continuous process of nutrient enrichment. By replacing detergent phosphate the process can be slowed down so that there is time to build plants for the removal of phosphate from all sewage. A certain amount of phosphate does no harm. Removal of 40—70% of the excess loading will in many circumstances greatly diminish the nuisance effects. Eutrophication is not an "all or nothing" process, it happens by degrees and even a two- or threefold decrease in growth of algal crops would be a worthwile improvement.

Limnologists realise that very little of the process of eutrophication is understood at present. Several questions remain unanswered. Why does the process operate slowly? What level of algal growth will a certain lake produce at a given phosphate loading? Does the algal growth depend quantitatively on the phosphate concentration? What is the relationship between phosphate concentration, chlorophyll concentration and primary production? What determines the way in which different biological systems and organisms are produced from the more or less standard range of lake nutrients?

Some of the answers will perhaps be found by comparing lakes from all over the world or within the same lake district. Nevertheless the treatment of lakes cannot be postponed until answers have been found to all the fundamental questions. Considerable success can sometimes be achieved by a pragmatic approach, just as many human diseases were treated long before their specific cause or mode of operation were known.

The unnecessary delay in the replacement of phosphates by non-phosphate-containing chelators has probably done more harm to waterbodies than any other source of pollution.

17.4. CASE STUDIES

In the previous sections the problem of eutrophication has been considered mostly in general terms although with some reference to individual lakes. Many aspects of the problem are more clear if considered in the context of particular case histories, although sometimes certain features will be obscured by this approach. A few detailed examples from better documented case histories will therefore now be discussed.

17.4.1. The St. Lawrence Great Lakes

Beeton (1969) reviewed changes associated with accelerated eutrophication in these lakes. The changes are most conspicuous in Lake Erie and Lake Ontario. Changes in Lake Michigan are less severe but are still greater than those which have occurred in Lake Superior and Lake Huron.

The sequence of events occurring in Lake Erie has been the subject of both a special report (Burns and Ross, 1972a, b) and of a summary paper (Burns and Ross, 1972c). Massive algal blooms in the central basin were caused by inflow of increased amounts of many nutrients, with phosphate being often apparently the key factor. Fig. 17.10 shows a drastic increase between 1935 and 1965 in phosphate input due mainly to the almost exponential increase in the use of phosphate-containing detergents. The algal bloom caused a layer of algae, 2—3 cm thick, to be laid down on the floor of the basin, this being the major cause of anoxic conditions in 1971 during stratification of 4 200 km^2 of the 12 700 km^2 total hypolimnetic surface. By the end of September the hypolimnion was reduced to 6 600 km^2 due to the sinking of the thermocline; this whole area was now anoxic. Rate of phosphate release from sediments was estimated at $0.022 \cdot 10^{-3}$ mol. m^{-2} d^{-1} under oxygenated con-

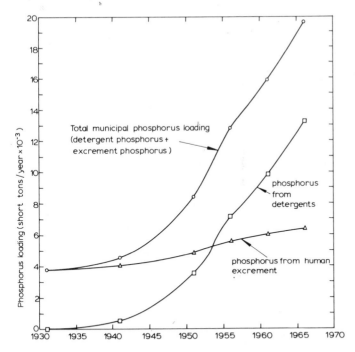

Fig. 17.10. Variation of two fractions of the phosphorus input into Lake Erie from principal sources. (Burns and Ross, 1972b.)

ditions in winter and $0.24 \cdot 10^{-3}$ mol. m^{-2} d^{-1} under anoxic summer conditions. These rates of release were estimated from concentrations measured in the hypolimnion. No attempts were made by Burns and Ross to explain the difference between oxygenated and anoxic concentrations, but a tentative explanation involving the occurrence of different systems of transport moving the phosphate away from the sediments is suggested here in Chapter 8. Burns and Ross did consider the fate of the regenerated quantities of phosphate and suggested that some phosphate from the hypolimnion rose into the epilimnion between September and October when mixing occurred and the hypolimnion was destroyed. Their main argument was that the proportion of soluble reactive phosphate to particulate phosphate increased rather than decreased with the overturn during this period (see Table 17.7).

It could be argued, however, that $157.9 - 126.9 = 31 \cdot 10^{6}$ mol. actually disappeared during this period — a quantity which is even greater than the soluble reactive phosphate in the hypolimnion — so no definite conclusion can be reached. The internal phosphate loading of the epilimnion due to anoxic regeneration in the hypolimnion may therefore be less than the authors estimated it to be.

Burns and Ross (1972c) mentioned the occurrence of a correlation be-

TABLE 17.7

Phosphate quantities in the Central Basin of Lake Erie, September and October 1970 (From Burns and Ross, 1972c) (mol. $\times 10^6$)

	Volume (km^3)	Soluble reactive phosphate	Particulate phosphate	Soluble organic phosphate	Total phosphate
September 1970					
Hypolimnion	9.4	18.8	7.6	3.7	30.1
Epilimnion	275.7	16.6	82.6	28.6	127.8
Total	285.1	35.4	90.2	32.3	157.9
October 1970					
Total	285.2	31.9	73.0	22.0	126.9

tween hypolimnion volume decrease and increase of chlorophyll concentration. But the increase in chlorophyll content takes place steadily after 3 July whilst the hypolimnion decreases in volume only slightly to begin with till 1 September after which the decrease is more pronounced.

A diminished external source of nutrients may result in a decrease in the algal blooms, which would help to maintain the hypolimnion oxygenated. This in its turn may reduce or remove altogether a possible extra anoxic internal loading. Thus Burns and Ross calculated that a reduction of the external loading by 1 378 tonnes may reduce the total loading by 3 122 tonnes. It is unwise to suppose that in all deep lakes internal anoxic loading will necessarily occur or to use this as a justification for not decreasing the external input of nutrients.

Braidech et al. (1972) found that the algae which were deposited and mineralised in the hypolimnion of Lake Erie were planktonic and of the genera *Tribonema* (Chrysophyta) and *Oedogonium* (Chlorophyta) which occur in the epilimnion. These organisms obtained some of their nutrients directly from the sediments by living and photosynthesising amongst the inorganic debris, thus contributing towards oxygen production in the hypolimnion. From mid-August onwards, this photosynthetic activity in the hypolimnion was reduced, probably as the result of shading effects from the plankton in the upper water layers and perhaps because of decreasing day length. Dobson and Gilbertson (1972) provided evidence that average oxygen depletion rate has more than doubled since 1929. It is now 3.6 mg l^{-1} of O_2 per month, which is well above the critical depletion rate of 3.0 mg l^{-1}, above which level anoxic conditions will develop.

Eutrophication effects have been less severe in Lake Michigan (U.S.A.). Schelske and Stoermer (1972) measured natural phytoplankton growth in plastic bags of 1 or 4 m^3 in situ following the addition of PO_4-P (20 mg m^{-3}), NO_3-N (200 mg m^{-3}) and SiO_2 (700 mg m^{-3}). A highly significant correla-

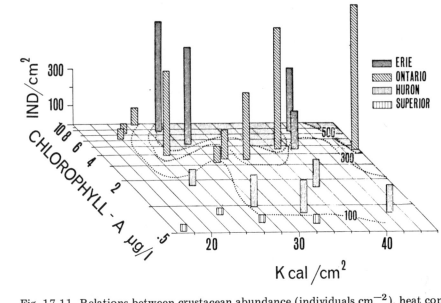

Fig. 17.11. Relations between crustacean abundance (individuals cm^{-2}), heat content (kcal. cm^{-2}) and chlorophyll a concentration in surface waters (μg l^{-1}) for four lakes. The bars represent mean crustacean abundance in different parts of the lakes: in June, July and August for lakes Ontario and Erie and during August for lakes Superior and Huron. Dotted lines are isolines for crustacean abundance. (From Patalas, 1972.)

tion was found between the rate of CO_2 fixation and particulate phosphate content. Increased phosphate loading (there are high phosphate to silicate ratios in the tributaries) causes silicate to be depleted earlier than without the extra phosphate input. Diatom growth will thus reach an earlier maximum (see Lund's work on Blelham Tarn) with a possibility that they may be replaced subsequently by non-siliceous species, probably blue-green algae. Although inshore samples contain more algal cells than offshore samples and primary production is much higher (10 mg m^{-3} hr^{-1} inshore compared with 3 mg m^{-3} hr^{-1} of C offshore) the overall effects of phosphate and nitrogen addition were found to be the same for both. The authors provide evidence that phosphate is probably the limiting factor for algal growth under most conditions in this lake.

Beeton and Edmondson (1972) pointed out that in very large lakes the inshore regions are the first to be affected by eutrophication. The inshore water of Lake Michigan contains greater concentrations of nitrogen and phosphate and a lower silicate concentration than does the open water. Diatoms are more abundant inshore and have shorter doubling times. The authors also report increasing concentrations of chloride, sulphate and calcium in some of the Great Lakes over a period of 35 years. Some of the **differences may be** attributed, at least in part, to changes in techniques. There have been changes in species composition too: the diatoms *Coscinodiscus radiatus, Diatoma tenue* var. *elongatum, Fragilaria capucina* and *Melosira binderana* were abundant species in 1960—65, whereas they were not reported or were not abundant either in 1938—40 or 1948—49. *Asterionella formosa, Fragilaria crotonensis, Melosira ambigua, M. granulata* and *Stephanodiscus tenuis* as a group were as abundant in 1965 as they were in 1938, although they showed slight increases or decreases for individual species.

Patalas (1972) calculated phosphate loadings for Lake Superior (0.03 g m^{-2} yr^{-1}), Lake Huron (0.15 g m^{-2} yr^{-1}), Lake Michigan (0.29 g m^{-2} yr^{-1}), Lake Ontario (0.86 g m^{-2} yr^{-1}), and Lake Erie (0.98 g m^{-2} yr^{-1}) and found that chlorophyll *a* concentrations and Secchi-disc visibility were closely related to these loadings. Crustacean abundance was related indirectly to the phosphate loading to the extent that a general trend was discernible from Lake Superior to Lake Erie: a diminishing proportion of calanoids and an increasing proportion of cyclopoids and cladocerans. Patalas' results are summarised in Fig. 17.11, which also illustrates the influence of temperature in the different lakes (see subsection 15.3.4). Temperature exerts a major controlling influence on the Crustacea since it controls the length of their life cycle. At higher temperatures therefore more generations per year can be produced. How many animals survive will, however, be determined by many other factors in the lake system.

17.4.2. Lake Washington

The best-documented case of a lake which has become eutrophic and has been subsequently restored is Lake Washington (Florida, U.S.A.). This lake began to deteriorate markedly, because of increasing sewage input, in 1955 when *Oscillatoria rubescens* became prominent in the plankton. Edmondson (1969a, 1970) described how the lake responded promptly to a major decrease in its nutrient income.

In 1957 sewage influx was contributing about 56% of the phosphate and 12% of the nitrogen income of the lake. Other sources of these nutrients were from water inputs of various origins, all with much lower concentrations. In 1963 the first reduction of about 25% of the sewage inflow was made by diverting it elsewhere. In 1965 the volume was reduced to 55% of the original 1957 value. The final diversion took place in 1968, although about 99% of it had already been sent elsewhere by March 1967. From Fig. 17.12 it can be seen how chlorophyll decrease corresponded with the phosphate decrease while nitrogen and CO_2 remained high. Edmondson pointed out how the lake started to improve even after the first diversion which removed only a quarter of the 56% input from the sewage component. The improvements caused by the 56% diversion might ultimately have been even greater than the decrease of chlorophyll and phosphate measured in 1965. The second diversion started in that year, possibly before the full benefit of the first had become apparent.

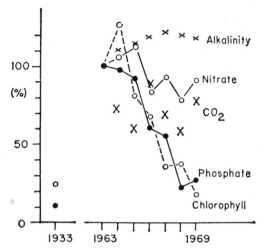

Fig. 17.12. Mean winter (January—April) values in surface water of phosphate phosphorus and nitrate nitrogen, and mean summer (July and August) values of chlorophyll in surface phytoplankton. The 1963 values, plotted as 100% were (in mg l^{-1}): P, 57; N, 428; and chlorophyll, 38. Unconnected points show winter means of bicarbonate alkalinity and free carbon dioxide in surface water (25.3 and 3.2 mg l^{-1} in 1963). (From Edmondson, 1970.)

Secchi-disc transparency increased from 1.0 m in 1963 to 2.8 m in 1969. Edmondson remarked that care should be taken in calculating budgets based on concentrations in highly concentrated but localised influx sources because diffuse but more dilute influx sources may contribute comparable total amounts. Success in the treatment of eutrophication in Lake Washington gives support to the statement made earlier (see subsection 17.3.5) that major benefits follow removal of *some* of the P load; it is not essential to try to deal with all the P influx sources in the system. It is also an example of how the enthusiasm of one man (Edmondson) may achieve results which would have been long delayed — if indeed they happened at all — when organised through the usual processes of water management policy.

17.4.3. Swiss and Austrian alpine lakes

Other examples of severe eutrophication are the Swiss and the Austrian alpine lakes. The Swiss lakes were mentioned earlier (subsection 17.3.2) when data from the Lake of Zürich were used to demonstrate the role of phosphate in a nutrient budget.

Pleisch (1970) constructed a detailed phosphate and nitrogen budget for the Pfäffikersee and Greifensee. During 1967—1968 Pfäffikersee received an inflow of 2.2 g m^{-2} yr^{-1} of PO_4-P (0.13 g m^{-3}) and 30.5 g m^{-2} yr^{-1} of N (1.77 g m^{-3}), and Greifensee received 6 g m^{-2} yr^{-1} of P (0.33 g m^{-3}) and 45.1 g m^{-2} yr^{-1} (2.5 g m^{-3}) of N.

392

Domestic and industrial water inflows were the main sources of phosphate at that time. During periods of flooding, large quantities of phosphate may be washed into the lakes, but most of these phosphates sediment within an hour or so and are probably not available to algae. Since 1951 the phosphate content of the inflows to the Pfäffikersee has increased 3.5 times and those to the Greifensee 5.6 times. These increases are owed (apparently) almost entirely to accelerating usage of phosphate-containing detergents. Populations around the lakes have increased by only 30% and 91%, respectively. The increasing use of polyphosphates in detergents in Switzerland may well be similar to that in The Netherlands, for which the figures are known (Fig. 17.13). The general shape of this curve is also similar to that of the P loading in Lake Erie, though the Lake Erie curve starts earlier. If one assumes that in 1951 no phosphate was used in detergents and that in 1968 the detergent component was equal to the human biochemical output (thus representing 50% of the total output, see Chapter 6), one can calculate that a doubling of the population in the watershed of the Greifensee will have caused the P load to increase 4 times. This figure is not very far below the actual increase of 5.6 times. The difference may be owed to increased agricultural use. The

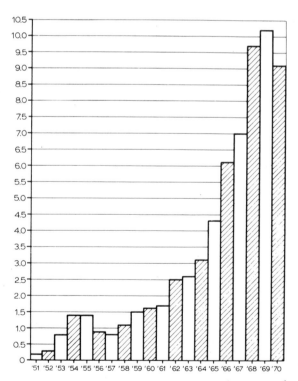

Fig. 17.13. Quantities of sodium triphosphate ($\times$ 10^{6}kg) used in The Netherlands in the detergent industry. Data from Central Office for Statistics (cited from Anonymous, 1973).

most important source of nitrate for these two Swiss lakes was the outwash from soil. Nitrate from forest soil accounted for 1.75 mg l^{-1} and that from marshy soil for 2.5 mg l^{-1} of NO_3-N. Domestic sewage was another source of nitrogen but Pleisch's study showed that the increased nitrate inflow was mainly attributable to increases in the riverwater and to the number of people in the catchment area.

In both lakes nitrogen was retained more than phosphate (85% and 65% for N; 52% and 4% for P). These differences in behaviour may be related to differences in the location of the nutrients. In the case of phosphates there is a relatively high concentration in the epilimnion. Considerable losses may therefore occur as the result of through-flow in these upper water layers. Nitrogen, on the other hand, tends to accumulate in the hypolimnion, which has no through-flow at all during the summer. Considerations such as these should serve as a caveat against measuring nutrient concentrations in localised inflows and using these figures as a basis for generalising about the whole lake system.

Findenegg (1971) studied the eutrophication of some Austrian alpine lakes similar to the Swiss ones. In the Wallersee a 12-fold increase of algal dry weight occurred between 1965 and 1969 due mainly to *Ceratium hirundinella*, a species which had previously been rare. In a similar lake (Obertrumersee) *Oscillatoria rubescens* became abundant and the algal dry weight increased 5-fold, while *Ceratium* was eliminated. In other lakes the increase was due to increased populations of existing algal species of such genera as *Oocystis* and *Ankistrodesmus*. During the same years the Wörthersee did not show a marked change, probably because *Oscillatoria rubescens* had already invaded in 1930 and persisted thereafter.

Phosphate and nitrogen concentrations in these lakes were all rather similar (5—30 μg l^{-1}) though somewhat higher values occurred in the Wallersee and Obertrumersee. Findenegg attributed the successful performance of *Oscillatoria* to its ability to migrate between metalimnion and epilimnion. In this way it can make best use of the nutrient input in the epilimnion. The cause of eutrophication in these Austrian lakes is increased use for recreational purposes (Findenegg, 1972).

Findenegg also demonstrated how increasing algal populations decrease light transmission. He recognised a group of lakes in which surface production increased two- to threefold and light transmission decreased 10—13%, and a group in which production increased more than threefold with a corresponding decrease in light transmission of more than 15%. By comparing changes in light transmission through the water in the blue-green with those in the red part of the spectrum Findenegg was able to determine whether green or other algae contributed most towards the increase in dry weight.

17.4.4. French alpine lakes

Balvay (1971), Lefebvre et al. (1972) and Laurent et al. (1972) described the eutrophication of Lac d'Annecy, a French alpine lake. *Tabellaria genestrata* and *Asterionella formosa* became the dominant species, while *Fragilaria crotonensis* decreased in abundance. In 1961 blue-green algae appeared in significant numbers and became dominant during summer as had been forseen by Dussart as early as 1952.

After a sewage diversion scheme had been carried out in stages between 1966 and 1971, the species composition of the net phytoplankton returned to what it had been before eutrophication. This happened even though 37% of the human sewage was still not.connected to the sewage system in 1971. Blue-green algae were no longer found and a new equilibrium between *F. crotonensis*, *Asterionella formosa* and *Dynobrion divergens* and *D. sociale* established itself.

Laurent (1972) and Laurent et al. (1972) described the eutrophication of Lac Léman (Lake of Geneva) where total phosphate concentrations increased to about 25 mg m^{-3} of PO$_4$-P (and were even higher in the hypolimnion). Later the concentrations declined markedly, probably due to an apatite precipitate (see subsection 9.4.4).

Oxygen concentrations in the bottom waters decreased gradually in the course of years. Laurent considered likely consequences of the effects of exploitation, eutrophication, and the introduction of salmonid fish, but the individual contribution of these and other effects has not yet been assessed. Apart from *Tribonema* sp. (Xanthophyceae), most of the other plankton species in the Lake of Geneva are also found in Lac d'Annecy. The sudden explosive increase in the Lake of Geneva of *Tribonema* sp. and of *Oscillatoria rubescens* is often attributed to eutrophication.

17.4.5. Lake Kinneret

Lake Kinneret — also known as the Sea of Galilee — is an example of a lake which receives considerable natural quantities of phosphate (Serruya and Pollingher, 1971). In rainy years the river Jordan supplies phosphate in quantities which average about 1.0 g m^{-2} yr^{-1} or more. Excessive eutrophication does not occur for two reasons. Firstly the inflowing riverwater plunges rather steeply into the hypolimnion, and secondly the phosphate carried is strongly bound within the clay structure. For this reason the phosphate is not immediately available to algae. *Scenedesmus* cells in cultures were found to be unable to use the particulate phosphate of the river Jordan for their growth.

Excessive growth of *Peridinium cinctum* f. *westii* develops in the lake in early spring but during summer the water is quite clear while later in the year some Cyanophyta and Chlorophyta occur (Serruya and Pollingher, 1971). Although *Peridinium* seems to have a remarkably high C:P ratio (C:N:P = 500:30:1), apparently due to storage of carbohydrates, this cannot be the only reason for its high production in conditions of low P concentrations. A controversy exists whether or not phosphate is a limiting factor for growth of Lake Kinneret algae. Berman (1969, 1970) indirectly argued that phosphate is not limiting for *Peridinium* growth, during the major period of the *Peridinium* growth, although it may often be a factor controlling the cessation of the bloom.

Considering the demand/supply ratio, it seems likely that phosphate is limiting certainly during the non-*Peridinium* period, when the Chlorophyceae are the predominant phytoplankton component. Berman suggested that phosphatase activity is an indication of the status of P nutrition. Halmann (1972) and Halmann and Stiller (1974) showed that phosphate concentration is the main factor limiting algal growth. Nitrogen concentration is high in March (about $1 \, \mathrm{mg} \, l^{-1}$ of NO_3-N), but sharply decreases during the bloom of *Peridinium*. Nutrients are supplied to the epilimnion in spring by the river Jordan and in fall from the hypolimnion as the thermocline sinks. Serruya and Pollingher demonstrated that the nature and timing of the *Peridinium* and the fall bloom of blue-green algae depend essentially on the date of the first flood of the river Jordan.

Serruya (1971) studied the occurrence of Fe, Mn, P, N, Org-P and $CaCO_3$ in the sediments. It seems likely that both clay and iron control the phosphate metabolism of the hypolimnion. The phosphate entering the lake adsorbed onto clay will sediment directly, while the ferric-hydroxide cycle (subsection 9.4.4) is also active. From this source phosphate may be released into the hypolimnetic waters and may enter the epilimnion by diffusion. As Lake Kinneret is not a very deep stratifying lake the sinking of the thermocline may enhance the upward transport of nutrients by reducing the diffusion pathway.

During the peak of the spring bloom (February—March), the *Peridinium* cells are dispersed homogeneously through the water column only at night. Toward the morning they may form a thin layer with very high densities. Using in vivo fluorimetry Berman (1972) found a sharp maximum of 300 $\mathrm{mg} \, m^{-3}$ of chlorophyll at 2—3 m depth and about 4 000 cells per ml. Later this peak sank to 3—5 m. This migration (studied also by Berman and Rodhe, 1971) of the *Peridinium* bloom may be one of the means by which this flagellate can compete favourably both by avoiding layers of a too high irradiance and by obtaining nutrients from deeper water layers. Lake Kinneret is an example of a lake with a high productivity based on natural processes. Obviously it will be very difficult to take measures against the algal growth. Sedimentation of silt and precipitation with ferric salt of the soluble phosphate in a pre-impoundment (just like the purification of Rhine water in Ketelmeer protects the IJsselmeer, see section 19.1) and restriction of upward diffusion from the deeper part of the hypolimnion by artificial aeration seem the only likely possibilities.

17.5. CONSEQUENCES AND CLASSIFICATIONS OF THE STATE OF EUTROPHICA-
TION

The increased input of algal growth-limiting nutrients results firstly in an
enhancement of plant growth. In the earlier stages of eutrophication this will
lead to slightly greater algal biomass and replacement of some species (e.g.
Desmidiaceae) by other algae. These changes will be hardly detectable at
first and present no problems in water management. The increase will continue
gradually and the greater algal productivity will lead to a noticeable production of O_2, with concentrations reaching well above 100% in day time. At the
same time mineralisation of increasing quantities of dead algae will result in
oxygen deficits in the water. In shallow lakes this will lead to decreased O_2
concentrations during the night, while in stratifying lakes the O_2 depletion
will become more serious in the hypolimnion. Many of the deep alpine lakes
have periods during which the hypolimnion is completely anoxic.

Mass development of blue-green algae (*Oscillatoria*, *Anabaena*, *Aphanizomenon*, etc.) will take place, while on the shore line development of
Cladophora and periphytic algae will increase. Floating masses of blue-green
algae and scum formation will prevent aeration even of surface waters, and
shallower bays in the larger lakes may become affected. The Lake of Zürich
even had bays where mass development of the anaerobic sulphur bacterium
(*Lamprocystis*) occurred. In The Netherlands it is generally true that shallow
lakes containing more than 200 mg m^{-3} of chlorophyll are avoided by the
general public. The oxygen depletion will affect normal fish life. As eutrophication is usually only one aspect of a more general pollution by sewage, the
O_2 disturbances will often be worsened by the input of organic matter as
well. The amount of organic matter produced per person is roughly equivalent to 100 g d^{-1} of O_2 and the amount of phosphate to 2—4 g d^{-1}. The
amount of organic algal matter formed from this phosphate will be orders
of magnitude higher than the organic matter per person, as the phosphate
will recycle several times a year (perhaps 10—20) and will thus form 10—20
times 200—400 g organic matter. In European waters a major effect of
eutrophication on fish populations is the replacement of salmonids and
coregones by cyprinids.

Vollenweider (1968) gave an extensive summary of work on eutrophication and reviewed different methods of classification. Most of these are of
more use to water managers than to scientists. Using Thomas' work on the
Swiss lakes as a guide, Vollenweider states that to describe the degree of
eutrophication "the only criterion which has been and could be used is the
difference between the alkalinity in summer and winter in the epilimnion".
Characteristics of some of these lakes are shown in Table 17.8.

Vollenweider pointed out that this tentative classification is not rigorous
enough to meet the demands of theoretical limnology (probably no classification ever will be), but it does provide rough guidelines for use in applied

TABLE 17.8

Trophic state and difference between summer and winter alkalinity in several alpine lakes (From Vollenweider, 1968)

Alkalinity (meq. l^{-1}):	Oligo-mesotrophic			Meso-eutrophic	Polytrophic	
	0–0.19	0.2–0.39	0.4–0.59	0.6–0.79	0.8–0.99	> 1.0
	Lago Maggiore	Lago di Como	Sempachersee	Lake Lugano	Lower Lake of Constance	Greifensee
	Lake of Lucerne		Lake of Zürich (Obersee)	Lake of Zürich	Pfäffikersee	Baldeggersee
	Windermere	Lac de Neuchâtel		Lac de St. Point		
P conc. (mg m^{-3})	< 10			10– 30	300– 100	> 100
N conc. (mg m^{-3})	<400			300–650	500–1 500	>1 500

398

limnology. Vollenweider regarded lakes with PO_4-P concentrations above 20 mg m^{-3} and total-N concentrations above 300 mg m^{-3} as likely to be in danger of eutrophication, with its ensuing problems, and with O_2 depletion of the hypolimnion. Vollenweider compiled a table of permissible loadings after a careful comparative study of the Swiss lakes using the O_2 depletion of the hypolimnion as main criterion for the degree of eutrophication (Table 17.9). He acknowledged that the table is based on inadequate evidence and should be used only as a very rough guide to be modified as soon as information on hydrology, morphometry, retention times and internal loading from the sediments becomes available. It should also be emphasised that this classification is specific for Swiss lakes with their own characteristics of water hardness, stratification, etc. It should not be applied to lakes in different regions (for example Windermere fits, but Esthwaite Water does not fit in the table) and certainly not to shallow non-stratifying lakes. For shallow lakes in near-equilibrium with the sediments the P concentrations of PO_4-P will rarely be below 10 mg m^{-3} but concentrations as high as 25—50 mg m^{-3} will not necessarily induce unwanted effects. It seems likely that the concept of "permissible loading" (a dangerous conception) must be related not only to depth, but also to a function of **depth and water retention time**.

Water colour should also be considered as an important factor because intense colour reduces light penetration significantly and does so selectively at different wavelengths. The depth of water in which light acts as a limiting factor is thus increased. In humus-rich shallow lakes, where phosphate will be released during peat decomposition, winter values may increase to 500 mg m^{-3} of PO_4-P, but hydrological conditions and water colour will operate to prevent excessive algal growth. Phosphate from such natural sources as those occurring in these lakes or from that bound closely in clay particles will not therefore always induce a natural (= endogenous) eutrophication. Lake George (Uganda) however is an example where severe natural endogenous eutrophication has occurred (see section 15.5).

TABLE 17.9

Permissible and dangerous loading levels for total nitrogen and biochemically active phosphate (g m^{-2} yr^{-1}); these levels apply only to Swiss lakes

Mean depth (m)	Permissible loading up to		Dangerous loading up to	
	N	P	N	P
5	1.0	0.07	2.0	0.13
10	1.5	0.10	3.0	0.20
50	4.0	0.25	8.0	0.50
100	6.0	0.40	12.0	0.80
150	7.5	0.50	15.0	1.00
200	9.0	0.60	18.0	1.20

Several works reviewing the problems and mechanisms of eutrophication have recently appeared, those edited by Rohlich (1969), Milway (1970), Allen and Kramer (1972) and Jenkins (1972) being the most important. Ample consideration is given in these works to the description of case histories, causes and effects of eutrophication, and possible precautions and corrective actions. All aspects of phosphate (mining, biology, inorganic chemistry, etc.) are reviewed in Griffith et al. (1973).

REFERENCES

Allen, H.E. and Kramer, J.R. (Editors), 1972. *Nutrients in Natural Waters*. John Wiley, New York, N.Y., 457 pp.

Allen, H.L., 1972. Phytoplankton photosynthesis, micronutrient interactions, and inorganic carbon availability in a soft-water Vermont lake. In: G.E. Likens (Editor), *Nutrients and Eutrophication: The Limiting-Nutrient Controversy. Proc. Symp. at the W.K. Kellogg Biological Station, Michigan State Univ., 1971. Am. Soc. Limnol. Oceanogr., Lawrence, Spec. Symp.*, 1: 63—83.

Anonymous, 1973. Memorandum over de Rijn en het eutrofiëringsvraagstuk in Nederland. H_2O, 6(19): 478—482.

Balvay, G., 1971. Eutrophisation et phytoplancton du lac d'Annecy. *Suppl. Terre Vive*, 20: 13—16.

Beeton, A.M., 1969. Changes in the environment and biota of the Great Lakes. In: *Eutrophication: Causes, Consequences, Correctives. Proc. Symp. Univ. of Wisconsin, Madison, Wisc., 1967.* Natl. Acad. Sci., Washington, D.C., pp. 150—187.

Beeton, A.M. and Edmondson, W.T., 1972. The eutrophication problem. *J. Fish. Res. Board Can.*, 29(6): 673—682.

Berman, T., 1969. Phosphatases release of inorganic phosphorus in Lake Kinneret. *Nature*, 224: 1231—1232.

Berman, T., 1970. Alkaline phosphatases and phosphorus availability in Lake Kinneret. *Limnol. Oceanogr.*, 15: 663—674.

Berman, T., 1972. Profiles of chlorophyll concentrations by *in vivo* fluorescence: Some limnological applications. *Limnol. Oceanogr.*, 17: 616—618.

Berman, T. and Rodhe, W., 1971. Distribution and migration of *Peridinium* in Lake Kinneret. *Mitt. Int. Ver. Theor. Angew. Limnol.*, 19: 266—276.

Biffi, F., 1963. Determinazione del fattore tempo come caratteristica del potere di autodepurazione del Lago d'Orta in relazione ad un inquinamento constante. *Atti Ist. Veneto Sci., Sci. Mat. Nat.*, 121: 131—136.

Braidech, T., Gehring, P. and Kleveno, C., 1972. Biological studies related to oxygen depletion and nutrient regeneration processes in the Lake Erie Central Basin. In: N.M. Burns and C. Ross (Editors), *Project Hypo.* Canada Centre for Inland Waters, Burlington, Ont., Paper no. 6: 51—70.

Burns, N.M. and Ross, C., 1972a. Project Hypo — discussion of findings. In: N.M. Burns and C. Ross (Editors), *Project Hypo.* Canada Centre for Inland Waters, Burlington, Ont. Paper no. 6: 120—126.

Burns, N.M. and Ross, C., 1972b. Oxygen-nutrient relationships within the Central Basin of Lake Erie. In: N.M. Burns and C. Ross (Editors), *Project Hypo.* Canada Centre for Inland Waters, Burlington, Ont., Paper no. 6: 85—119.

Burns, N.M. and Ross, C., 1972c. Oxygen-nutrient relationships within the central basin of Lake Erie. In: H.E. Allen and J.R. Kramer (Editors), *Nutrients in Natural Waters.* John Wiley, New York, N.Y., pp. 193—250.

Dobson, H.H. and Gilbertson, M., 1972. Oxygen depletion in the hypolimnion of the

Central Basin of Lake Erie, 1929—1970. In: N.M. Burns and C. Ross (Editors), *Project Hypo*. Canada Centre for Inland Waters, Burlington, Ont., Paper no. 6: 3—8.

Edmondson, W.T., 1969a. Eutrophication in North America. In: *Eutrophication: Causes, Consequences, Correctives. Proc. Symp. Univ. Wisconsin, Madison, Wisc., 1967*. Natl. Acad. Sci., Washington, D.C., pp. 124—149.

Edmondson, W.T., 1969b. Cultural eutrophication with special reference to Lake Washington. *Mitt. Int. Ver. Theor. Angew. Limnol.*, no. 17: 19—32.

Edmondson, W.T., 1970. Phosphorus, nitrogen, and algae in Lake Washington after diversion of sewage. *Science*, 169: 690—691.

Elster, H.J. and Einsele, W., 1937. Beiträge zur Hydrographie des Bodensees (Obersee). *Inst. Rev. Ges. Hydrobiol. Hydrogr.*, 35(4/6): 523—585.

Fast, H., 1955. Systematische Untersuchungen über den chemischen und bakteriologischen Zustand des Bodensees. *Jahrb. Wasser*, 22: 11—37.

Findenegg, I., 1971. Unterschiedliche Formen der Eutrophierung von Ostalpenseen. *Schweiz. Z. Hydrol.*, 33(1): 85—95.

Findenegg, I., 1972. Die Auswirkung der Eutrophierung einiger Ostalpenseen auf die Lichttransmission ihres Wassers. *Wetter Leben*, 24: 110—118.

Furrer, O.J. and Gächter, R., 1972. Der Beitrag der Landwirtschaft zur Eutrophierung der Gewässer in der Schweiz. II. Einfluss von Düngung und Nutzung des Bodens auf die Stickstoff- und Phosphormengen im Wasser. *Schweiz. Z. Hydrol.*, 34(1): 71—92.

Gächter, R., 1968. Phosphorhaushalt und planktische Primärproduktion im Vierwaldstättersee (Horwer Bucht). *Schweiz. Z. Hydrol.*, 30(1): 1—66.

Gächter, R. and Furrer, O.J., 1972 Der Beitrag der Landwirtschaft zur Eutrophierung der Gewässer in der Schweiz. I. Ergebnisse von direkten Messungen im Einzugsgebiet verschiedener Vorfluter. *Schweiz. Z. Hydrol.*, 34(1): 41—70.

Gerloff, G.C., 1969. Evaluating nutrient supplies for the growth of aquatic plants in natural waters. In: *Eutrophication: Causes, Consequences, Correctives. Proc. Symp. Univ. Wisconsin, Madison, Wisc., 1967*. Natl. Acad. Sci., Washington, D.C., pp. 537—555.

Glooschenko, W.A. and Alvis, C., 1973. Changes in species composition of phytoplankton due to enrichment by N, P, and Si of water from a North Florida lake. *Hydrobiologia*, 42(2/3): 285—294.

Golterman, H.L., 1973. Vertical-movement of phosphate in freshwater. In: E.J. Griffith, A. Beeton, J.M. Spencer and D.T. Mitchell (Editors), *Environmental Phosphorus Handbook*. John Wiley, New York, N.Y., pp. 509—538.

Golterman, H.L., 1975. Chemistry of running water. In: B. Whitton (Editor), *River Ecology*. Blackwell, Oxford (in press).

Griffith, E.J., Beeton, A.M., Spencer, J.M. and Mitchell, D.T. (Editors), 1973. *Environmental Phosphorus Handbook*. John Wiley, New York, N.Y., 718 pp.

Grim, J., 1955. Die chemischen und planktologischen Veränderungen des Bodensee—Obersees in den letzten 30 Jahren. *Arch. Hydrobiol. Suppl.*, 22: 310—322.

Halmann, M., 1972. Chemical ecology. Evidence for phosphate as the only factor limiting algal growth in Lake Kinneret. *Israel J. Chem.*, 10: 841—855.

Halmann, M. and Stiller, M., 1974. Turnover and uptake of the solved phosphate in freshwater. *Limnol. Oceanogr.*, 19: 774—784.

Hamilton, D.H., 1969. Nutrient limitation of summer phytoplankton growth in Cayuga Lake. *Limnol. Oceanogr.*, 14(4): 579—590.

Hetling, L.J. and Sykes, R.M., 1973. Sources of nutrients in Canadarago lake. *J. Water Pollut. Control Fed.*, 45(1): 145—156.

Jenkins, D. (Editor), 1972. Phosphorus in fresh water and the marine environment. Symp. University College, London, 1972. *Water Res.*, 7(1/2): 1—342.

Kerr, P.C., Paris, D.F. and Brockway, D.L., 1970. The interrelation of carbon and phosphorus in regulating heterotrophic and autotrophic populations in aquatic ecosystems. *U.S. Department of the Interior Federal Water Quality Administration. Natl. Pollut. Fate Res. Progr.*, 53 pp.

Kerr, P.C., Brockway, D.L., Paris, D.F. and Barnett, J.T., 1972a. The interrelation of carbon and phosphorus in regulating heterotrophic and autotrophic populations in an aquatic ecosystem, Shriner's Pond. In: G.E. Likens (Editor), *Nutrients and Eutrophication: The Limiting-Nutrient Controversy. Proc. Symp. at the W.K. Kellogg Biological Station, Michigan State Univ., 1971*. Am. Soc. Limnol. Oceanogr., Lawrence, Kansas, Spec. Symp., 1: 41—62.

Klifmüller, R., 1962. Der Anstieg des Phosphat-Phosphors als Ausdruck Fortschreitenden Eutrophierung im Bodensee. *Int. Rev. Ges. Hydrobiol. Hydrogr.*, 47: 118—122.

Kuentzel, L.E., 1969. Bacteria, carbon dioxide, and algal blooms. *J. Water Pollut. Control Fed.*, 41(10) 1737—1747.

Kunishi, H.M., et al., 1972. Phosphate movement from an agricultural watershed during two rainfall periods. *J. Agric. Food Chem.*, 20(4): 900—905.

Lange, W., 1970a. Cyanophyta—bacteria systems: Effects of added carbon compounds or phosphate on algal growth at low concentrations. *J. Phycol.*, 6: 230—234.

Lange, W., 1970b. Blue-green algae and humic substances. *Proc. Conf. Great Lakes Res., 13th*, pp. 58—70.

Lange, W., 1971a. Enhancement of algal growth in Cyanophyta—bacteria systems by carbonaceous compounds. *Can. J. Microbiol.*, 17(3): 303—314.

Lange, W., 1971b. Limiting nutrient elements in filtered Lake Erie water. *Water Res.*, 5: 1031—1048.

Laurent, P.J., 1972. Lac Léman: Effects of exploitation, eutrophication, and introductions on the salmonid community. *J. Fish. Res. Board Can.*, 29: 867—875.

Laurent, P.J., Garaucher, J. and Viviers, P., 1972. The conditions of lakes and ponds in relation to the carrying out of treatment measures. In: S.H. Jenkins (Editor), *Advanced Water Pollution Research*. Pergamon Press, Oxford, pp. III-23/1—III-23/10.

Lefebvre, J., Laurent, P. and Louis, J., 1972. Application de l'analyse des correspondances à l'étude de l'évolution du phytoplancton du lac d'Annecy. *Invest. Pesq.*, 36(1): 119—126.

Lehn, H., 1972a. Zur Beziehung Phytoplankton—Phosphat im Bodensee. *Arch. Hydrobiol.*, 70(4): 556—559.

Lehn, H., 1972b. Zur Trophie im Bodensee. *Verh. Int. Ver. Theor. Angew. Limnol.*, 18: 467—474.

Lund, J.W.G., 1969. Limnology and its application to potable water supplies. *J. Br. Water Works Assoc.*, 49: 14.

Lund, J.W.G., 1970. Primary production. *Water Treat. Exam.*, 19: 332—358.

Lund, J.W.G., 1972a. Preliminary observations on the use of large experimental tubes in lakes. *Verh. Int. Ver. Theor. Angew. Limnol.*, 18 (part 1): 71—77.

Lund, J.W.G., 1972b. Eutrophication. *Proc. R. Soc. Lond., Ser. B*, 180: 371—382.

Milway, C.P. (Editor), 1970. *Eutrophication in Large Lakes and Impoundments. Uppsala Symposium, 1968*. Organization for Economic Co-operation and Development (OECD), Paris, 560 pp.

Mortimer, C.H., 1969. Physical factors with bearing on eutrophication in lakes in general and in large lakes in particular. In: *Eutrophication: Causes, Consequences, Correctives. Proc. Symp. Univ. of Wisconsin, Madison, Wisc., 1967*. Natl. Acad. Sci., Washington, D.C., pp. 340—368.

O'Melia, Ch.R., 1972. Approach to the modeling of lakes. *Schweiz. Z. Hydrol.*, 34(1): 1—33.

Patalas, K., 1972. Crustacean plankton and the eutrophication of St. Lawrence Great Lakes. *J. Fish. Res. Board Can.*, 29: 1451—1462.

Piontelli, R., and Tonolli, V., 1964. Il tempo di residenza delle acque lacustri in relazioni ai fenomeni di arricchimento in sostanze immesse, con particolare riguardo al Lago Maggiore. *Mem. Ist. Ital. Idrobiol.*, 17: 247—266.

Pleisch, P., 1970. Die Herkunft eutrophierender Stoffe beim Pfäffiker- und Greifensee. *Vierteljahresschr. Naturforsch. Ges. Zürich*, 115(2): 127—129.

Prakash, A. and Rashid, M.A., 1968. Influence of humic substances on the growth of marine phytoplankton: Dinoflagellates. *Limnol. Oceanogr.*, 13: 598—606.

Prakash, A., Rashid, M.A., Jensen, A. and Subba Rao, D.V., 1973. Influence of humic sub-stances on the growth of marine phytoplankton: Diatoms. *Limnol. Oceanogr.*, 18(4): 516—524.

Prochaczkova, L., 1972. Some abiotic factors in relation to the nitrogen and phosphorus concentration of river reservoir water. *Fortschr. Wasserchem.*, 14: 119—124.

Rohlich, G.A., 1969. Engineering aspects of nutrient removal. In: *Eutrophication: Causes, Consequences, Correctives. Proc. Symp. Univ. of Wisconsin, Madison, Wisc., 1967.* Natl. Acad. Sci., Washington, D.C., pp. 340—368.

Sakamoto, M., 1966. Primary production by phytoplankton community in some Japanese lakes and its dependence on lake depth. *Arch. Hydrobiol.*, 62: 1—25.

Schelske, C.L. and Stoermer, E.F., 1972. Phosphorus, silica, and eutrophication of Lake Michigan. In: G.E. Likens (Editor). *Nutrients and Eutrophication. Proc. Symp. W.K. Kellogg Biological Station, Michigan State Univ., 1971.* Am. Soc. Limnol. Oceanogr., Lawrence, Kansas, Spec. Symp., 1(1972): 157—171.

Schindler, D.W., 1971. Carbon, nitrogen, and phosphorus and the eutrophication of fresh-water lakes. *J. Phycol.*, 7: 321—329.

Schindler, D.W. and Fee, E.J., 1973. Diurnal variation of dissolved inorganic carbon and its use in estimating primary production and CO_2 invasion in lake 227. *J. Fish. Res. Board Can.*, 30(10): 1501—1510.

Schindler, D.W., et al., 1973. Eutrophication of lake 227 by addition of phosphate and nitrate: the second, third, and fourth years of enrichment 1970, 1971, and 1972. *J. Fish. Res. Board Can.*, 30(10): 1415—1440.

Serruya, C., 1971. Lake Kinneret: The nutrient chemistry of the sediments. *Limnol. Oceanogr.*, 16: 510—521.

Serruya, C. and Pollingher, U., 1971. An attempt at forecasting the *Peridinium* bloom in Lake Kinneret (Lake Tiberias). *Mitt. Int. Ver. Theor. Angew. Limnol.*, No. 19: 277—291.

Talling, J.F., 1965. The photosynthetic activity of phytoplankton in East African lakes. *Int. Rev. Ges. Hydrobiol. Hydrogr.*, 50(1): 1—32.

Thomas, E.A., 1968. Die Phosphattrophierung des Zürichsees und anderer Schweizer Seen. *Mitt. Int. Ver. Theor. Agew. Limnol.*, 14: 231—242.

Thomas, E.A., 1971. Oligotrophierung des Zürichsees. *Vierteljahresschr. Naturforsch. Ges. Zürich*, 116(1): 165—179.

Thomas, E.A., 1973. Phosphorus and eutrophication. In: E.J. Griffith, A. Beeton, J.M. Spencer and D.T. Mitchell (Editors), *Environmental Phosphorus Handbook.* John Wiley, New York, N.Y., pp. 585—611.

Uhlmann, D. and Albrecht, E., 1968. Biogeochemische Faktoren der Eutrophierung von Trinkwasser-Talsperren. *Limnologica (Berlin)*, 6(2): 225—245.

Vallentyne, J.R., 1973. The algal bowl — a Faustian view of eutrophication. *Fed. Proc. Fed. Am. Soc. Exp. Biol.*, 32(7): 1754—1757.

Vollenweider, R.A., 1964. Über oligomiktrische Verhältnisse des Lago Maggiore und einiger anderer insubrischer Seen. *Mem. Ist. Ital. Idrobiol.*, 17: 191—206.

Vollenweider, R.A., 1968. *Water Management Research; Scientific Fundamentals of the Eutrophication of Lakes and Flowing Waters, With Particular Reference to Nitrogen and Phosphorus as Factors in Eutrophication.* Technical report of the Organization for Economic Co-operation and Development, Paris, DAS/CSI/68.27.

Vollenweider, R.A., 1969. Möglichkeiten und Grenzen elementarer Modelle der Stoffbilanz von Seen. *Arch. Hydrobiol.*, 66(1): 1—36.

Wagner, G., 1972. Stratifikation der Sedimente und Sedimentationsrate im Bodensee. *Verh. Int. Ver. Theor. Angew. Limnol.*, 18 (part 1): 475—481.

CHAPTER 18

SEDIMENTS

18.1. INTRODUCTION

All lake waters contain at least some suspended matter which settles slow-ly to the bottom of a lake and accumulates as sediment. This consists of in-organic and organic compounds coming from sources both outside (alloch-thonous) and within (autochthonous) the lake. The organic matter consists of phytoplankton cells and in some smaller lakes of macrophytes, together with the detritus derived from decaying material. Some larger-sized organisms (e.g. zooplankton and fish) also occur of course, and these will sink fairly rapidly after death because of their greater volume. The inorganic matter con-sists of erosion products from the rocks in the watershed (rock particles and clay), together with compounds such as $Fe(OH)_3$ and SiO_2, which may be brought into the lake or formed within it from soluble compounds. The latter processes have already been discussed in subsections 3.3.1, 9.4.2 and 9.4.4.

Suspended matter coming from inflowing rivers will be discussed in this chapter. It is the main source of all inorganic suspended compounds which arrive in most lakes.

18.2. EROSION AND CHEMICAL WEATHERING

18.2.1. Origin and chemical nature of erosion products

All suspended inorganic matter in rivers (usually referred to as gravel, sand or silt, depending on particle size, see subsection 18.2.3) originates from the weathering and erosion of rocks. Rocks are composed of a number of miner-als, while important species consist of only one (Table 18.1, p. 405). These minerals form the main material of which the earth's crust, i.e. the outermost 30—40 km of the earth, is made. Below this is a fluid stratum of molten "rocks", the magma. The main constituents of the crust forming the conti-nental masses are the Al silicates, hence the name Sial which is used for this part. The ocean floors consist of a material rich in silicon and magnesium (the Sima, or Mafic, the latter being used where Fe is also a significant component). The siallic continental masses occur to a depth of about 8 km and are then underlain by Sima. Siallic magma, containing over 50% silica, consolidates to form granitic rock, whereas mafic magma, containing less than 50% silica, consolidates to form basaltic rock. Granitic rocks are of deep-seated origin and their presence on the earth's surface is caused by tectonic movements pushing them up through the outer overlying layers. Basaltic rocks, however, form from lavas that spread from volcanoes on the earth's surface.

The rocks of the earth's crust are divided into three groups according to their origin. Table 18.1 summarises the chemical composition of some of these rocks arranged in order of decreasing solubility.

Igneous rocks

Igneous rocks are formed as the products of consolidation of a magma. If the magma cools slowly in the deeper regions of the earth's crust, the individual crystals grow to a large size, e.g. in granites. Upon more rapid cooling, especially after outflow at the earth's surface, finer-grained rocks or even volcanic glasses are formed. After an outburst of magmatic material previously under heavy pressure, layers of finely divided volcanic ashes and tuffs will result. The mode of formation of igneous rocks is important because glasses and fine crystals are more easily attacked by the processes of weathering and erosion than the more massive rocks consisting of large crystals.

During the process of cooling of a large magma reservoir, there may even occur a differentiation resulting in the successive formation of rocks of varying chemical composition. Originally heavy minerals, the olivines, are formed with a ratio of 1 SiO_2 to 2 (Mg, Fe)O. These sink to the bottom of the magmatic focus. As olivine contains less silica than the average magma, the remaining fluid magma will have been relatively enriched in silica and other elements. Progressively rocks with a higher content of silica solidify. They contain "dark" minerals such as olivine (in smaller quantities), pyroxenes (SiO_2 : (Ca, Mg, Fe)O = 1:1), amphiboles and glimmers ("mica"), together with increasing amounts of "light" minerals, especially the feldspars (plagioclase, orthoclase and microcline) and finally quartz, which fills the interstitial spaces between the earlier formed minerals. Consequently the major constituents of normal igneous rocks are silicates or quartz. Elementary composition of the commonest is:

olivine:	$(Fe,Mg)_2SiO_4$	
pyroxene:	$(Fe,Mg,Ca)SiO_3$	
feldspars:	$CaAl_2Si_2O_8$	anorthite } mixed crystals of plagioclases
	$NaAlSi_3O_8$	albite
	$KAlSi_3O_8$	orthoclase or microcline
quartz:	SiO_2	

These are the most important types of minerals. Others are the amphiboles and micas of much more complex structure.

Igneous rocks are the original as well as the main components of the earth's crust. In the "Sima layer" under the ocean floors, more homogeneous and undifferentiated rocks occur (mainly silica and magnesium with a certain amount of iron). The continents (the sial parts of the crust) consist partly of the products of a slow differentiation with accumulation of silica, aluminium and the other constituents mentioned above.

TABLE 18.1

Chemical composition of important rocks arranged in order of decreasing solubility

Non-silicates

Rock salt (Halite)	$NaCl$
Gypsum	$CaSO_4$
Calcite	$CaCO_3$ calc-spar
Dolomite	$MgCO_3$, $CaCO_3$ pearl-spar

Silicates

Feldspar	$KAlSi_3O_8$ orthoclase
Soda-lime feldspar	a series from
	$NaAlSi_3O_8$ plagioclase, albite to
	$CaAl_2Si_2O_8$ anorthite
Basalt	$NaAlSi_3O_8$ plagioclase
	$RSiO_3$ pyroxene R (= Ca^{2+}, Mg^{2+} or Fe)
	$NaAl(SiO_3)_2$ alkali-pyroxene
	$(Fe,Mg)_2SiO_4$ olivine
Granite	SiO_2 quartz
	$NaAlSi_3O_8$ } feldspar
	$KAlSi_3O_8$
	$Al_2Si_2O_5(OH)_4$ kaolin
	$K(Fe,Mg)_3Si_3AlO_{10}(OH)_2$ biotite
	$KAl_2AlSi_3O_{10}(OH,F)_2$ K mica or K musco-vite and more complex structures

Sedimentary rocks

By weathering of igneous rocks their components are set free: the alkaline elements (sodium and potassium) are easily dissolved and ultimately appear in the hydrosphere although K is also adsorbed on clay. Calcium in solution forms carbonate rocks with the carbon dioxide of the atmosphere. This happens especially through the intermediation of organisms. Carbonate rocks often contain a certain amount of magnesium, up to the composition of $CaMg(CO_3)_2$, dolomite. Under tropical conditions aluminium and iron may remain in the form of hydroxides during lateritic weathering, while in the temperate regions the main product of weathering is some form of hydrous aluminium silicate — the clay minerals. Quartz is mainly physically disintegrated, especially in temperate regions and may give rise to layers of sand.

All the formations under this caption belong to the types of sedimentary rocks. When these are buried under heavy layers of later sediments, a certain degree of recrystallisation will ensue giving rise to the formation of limestones (calcium carbonate and dolomite) shales and slates (clay matter) and sandstones. Occasionally salt deposits are formed by evaporation in closed sea basins. The main sedimentary rocks are: evaporites, carbonates, and hydrolysates.

Metamorphic rocks.

Metamorphic rocks are igneous or sedimentary rocks which have been altered by high temperature and/or by strong pressure. When cold rocks come in contact with magma at temperatures of several hundred degrees centigrade, they are altered, and changes occur in their texture, appearance and mineralogical composition. The metamorphic rocks contain in addition to feldspar and quartz such minerals as amphiboles, chlorites, micas. Some denser rocks are formed, containing such minerals as garnet and kyanite. They often have a layered structure and are called schists, which occur in a layered or "schistose" manner. If the layer of surface rocks becomes folded into undulating formations by lateral thrusting pressures, the deeper parts of the folds become permeated by the underlying magma to form gneisses. Gneiss differs from schist by having a coarser larger-scale banding of quartz or feldspar. Layers of limestone rocks undergo metamorphosis to become marbles, whilst quartz sands or sandstones are hardened into quartzites. The minerals produced by metamorphosis are generally resistant to weathering.

18.2.2. Erosion of silicates

In an uninhabited world, water and wind erosion would certainly be the main source of minerals for lakes, the rest coming from the rain. The quantities involved in the erosion process are inconceivably large. Goldschmidt (1937) estimated that during geological history (several hundred millions of years), 600 g of rock have been weathered per kg seawater or 160 kg of rock per cm^2 of the earth, which is equivalent to the denudation of a layer 400—600 m thick over the whole earth.

Modern geochemistry has shown that the amount of rock eroded must have been even greater but some was re-utilised during the formation of new rocks. Goldschmidt's estimate may be used therefore to indicate a minimal "net" quantity. All material within the upper layers of the lithosphere are thus participating in a complicated process of dissolution followed by transport.

The main materials produced by erosion are the silicates; the weathering process may be schematised as shown in Table 18.2. It should be realised that the arrows in this scheme do not symbolise distinct chemical reactions, but represent the general trend of a mixture of chemical reactions whose details are, as yet, little understood.

An atmospheric influence is necessary for dissolution to occur, since it provides H_2O and acids such as CO_2, H_2S, SO_2 or even HCl from volcanic action. The most important stage in the process is an incongruent dissolution of aluminium silicates, e.g.:

$$KAl\ silicate + H_2CO_3 + H_2O \rightarrow HCO_3^- + H_4SiO_4 + K^+ + Al\ silicate$$

i.e. a primary mineral is converted into a secondary mineral, which will eventually dissolve further. The breakdown process is accompanied by a release

TABLE 18.2

Schematic representation of the weathering process

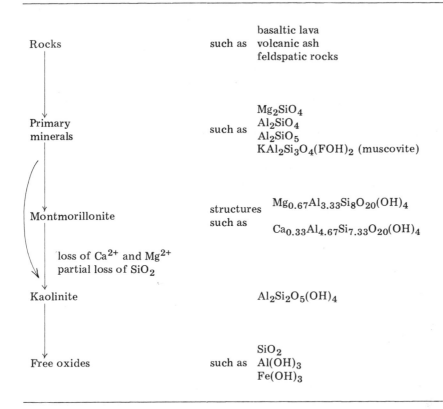

Rocks	such as	basaltic lava volcanic ash feldspatic rocks
Primary minerals	such as	Mg_2SiO_4 Al_2SiO_4 Al_2SiO_5 $KAl_2Si_3O_4(FOH)_2$ (muscovite)
Montmorillonite	structures such as	$Mg_{0.67}Al_{3.33}Si_8O_{20}(OH)_4$ $Ca_{0.33}Al_{4.67}Si_{7.33}O_{20}(OH)_4$
Kaolinite		$Al_2Si_2O_5(OH)_4$
Free oxides	such as	SiO_2 $Al(OH)_3$ $Fe(OH)_3$

loss of Ca^{2+} and Mg^{2+}
partial loss of SiO_2

of both cations and silica. The water becomes alkaline due to the reaction of H_2CO_3 as proton donor. The solid residue therefore becomes more acid than the original aluminium silicate was. In this way, feldspars produce kaolinites, montmorillonites and micas, all of which will be transported by the rivers as silt in addition to the soluble products.

In the next subsection on silt a few case studies will be mentioned where the different components of silt are analysed.

18.2.3. Chemical nature of silt

A satisfactory definition of silt does not exist and little is known about its chemical composition (Golterman, 1973a, b, c). Some consider all particulate matter in rivers to be silt, while others define silt as all particles with a grain size smaller than 16 μm (De Groot, 1970) or 50 μm (Terwindt, 1967). The latter author gives a summary of many, though mostly Dutch, definitions in use.

The United States Department of Agriculture uses the following arbitrary limiting diameters:

fine gravel	1000—2000 μm
coarse sand	100—1000 μm
medium sand	25— 500 μm
fine sand	100— 250 μm
very fine sand	50— 100 μm
silt	5— 50 μm
clay	<5 μm

Müller and Förstner (1968a) defined the different fractions as follows:

sand	>63 μm
silt	2—63 μm
clay	<2 μm

Dutch agriculturalists distinguish between clays with diameters of < 16 μm and < 2 μm. The influence of the diameter on the holding capacity of interstitial water is important; furthermore, clay particles have the capacity to bind water chemically. Sand — mainly SiO_2 — is often considered not to be a constituent of mud* and may be regarded as a separate entity. Since mud has a complicated and variable chemical composition, it is even more difficult to define the *chemical nature* of mud than to define its grain size.

Its main constituents are silicates, carbonates (mostly $CaCO_3$) and organic matter. Phosphates and (heavy) metals are important compounds adsorbed on the particles. In 40 mg of Rhine silt (filtered from 1 l of Rhine water), 10—17 mg of SiO_2-Si, 0.35 mg of PO_4-P, and 2 mg of Fe were found. The silica originates in most natural cases from the chemical weathering of silicate minerals as shown in Table 18.2. Not all the SiO_2 passes through all chemical weathering reactions of that scheme, however. Part of the silica is often considered to be derived from finely divided quartz mobilised after physical erosion of granite. No general scheme of weathering is available which would allow one to predict the type of clay minerals which will occur in silt transported by rivers.

Very little is known about variation in the types of clays which can occur in different rivers or in their tributaries. A few river clays have been analysed quantitatively and qualitatively (see review by Golterman, 1975). The quantity of silt carried and deposited depends on current velocity and on the presence of large lakes, which may serve as sediment traps. Thus, the amount of silt is small in the lower reaches of the rivers Rhine and Rhône, due to the presence of the Lake of Constance and the Lake of Geneva, which act as sediment traps.

---

* "Mud" is an even vaguer term than silt, though the two are sometimes used as synonyms.

Müller and Förstner (1968a, b), Müller and Gees (1970), Förstner et al. (1968) estimated that the upper alpine part of the river Rhine deposits $3 \cdot 10^6$ tonnes per year of sediments in the Lake of Constance. This represents an average load of 450 g m^{-3} in the riverwater, although it fluctuates with seasonal changes, being probably greatest during the spring melting of alpine snows.

Calcium carbonate comprises 8—89% of these sediments, while the residual non-calcareous material consists mainly of kaolinite and glimmerton — a mixture of illite (= hydroxy mica) plus glimmerton —, ledikit (= muskovite, an ideal K$^+$ mica), and biotite (= Mg mica), together with smaller amounts of chlorite. These various forms of mica together with chlorite are derived from the northern Alps and are the predominant clay minerals in the assemblage of sediments in the Rhine. Towards the west the waters bring in sediments containing smectite derived from surrounding volcanic tuff in addition to the micas and chlorite.

Kaolinite occurs in the sedimentary sandstone rocks of the Older Molasse but does not appear in Rhine sediments. The average approximate composition of the non-calcareous component of the Rhine sediments is: 10% clay (particle size < 2 μm: mainly illite and chlorite); 70% silt (2—63 μm); and 20% sand (> 63 μm). Calcite forms 26% and dolomite 11% of the mean carbonate content of this silt. Grain size, carbonate percentage, pH and conductivity are almost independent of volume of water discharge and flow velocity, and are related more to the supply areas.

The situation in the Rhône Valley is remarkably similar. Serruya (1969) gives a sediment discharge rate of $3—10 \cdot 10^6$ tonnes per year (1935—1965), with $2.7 \cdot 10^6$ tonnes in 1964—1965. This averages 85 kg sec^{-1}, from which value an erosion of 0.2 mm per year over the whole watershed can be calculated. The water flow rate fell within the range of 65—400 m^3 sec^{-1} (1935—1964), the average being 180 m^3 sec^{-1}.

Serruya found that quartz, epidote, amphibole and chlorite are the most frequently occurring minerals of the coarse non-calcareous fraction, whilst micro-quartz, illite and chlorite form most of the fine non-calcareous fraction. About 50% of the total sediments consists of CaCO$_3$, although local variations occur, for example sediments have been found near Mantua (Italy) which contain 90% of CaCO$_3$.

The Rhine and Rhône sediments seem very similar, both consisting of mica (illite), plus some chlorite. According to Gibbs this constitution implies the occurrence of a strong physical erosion factor which could well result from the similar relief of the catchment areas. The clay minerals that are deposited in the lower regions of the Rhine and Rhône probably originate from tributaries at lower elevation.

Serruya (1969) also studied the deeper lake sediments (i.e. the Postglacial sediments covering the Dryas, Boreal, and Atlantic periods) by use of both cores and seismic techniques. Her main conclusions are summarised in section

2.3 but she also gave detailed information on such factors as climatic influence, human activities in the watershed, erosion rates during the different geological periods, and data on organic carbon and nitrogen content in the whole series of sediments:

	% of C	% of N
Dryas	0.2—0.5	0.03—0.05
Preboreal	0.4—0.5	0.05—0.06
Boreal	0.5—0.7	0.06—0.075
Atlantic	1.0	0.8 —0.10
Subboreal	1.0—1.5	0.11—0.12
Subatlantic	1.0	—

Serruya also reports the concentrations of many trace-element metals (Chapter 11) and concludes that the intensity of drainage and thus the rate of sedimentation are dominant factors controlling trace-element composition.

Gonet (1971) studied the sediments of the Lake of Geneva sampling from the river Promenthouse near Nyon. The river transports 200—1500 mg of PO_4-P per second with the suspended matter. He used a geoelectric method based on the different conductivities of the sediment constituents to measure their displacement. The sediments contain 40% of $CaCO_3$, 38% of SiO_2, about 12% of Al_2O_3, and about 10% organic matter, with 0.15—0.25% of PO_4-P. Govet showed that the sediments from this river may settle initially near the river entry point but may be resuspended later again and transported into the deeper part of the lake. During this secondary transport the sediment loses part of the phosphate.

In compiling a mass balance sheet for the Lake of Geneva, Meybeck (1972) estimated that $1.8 \cdot 10^6$ tonnes of dissolved compounds and $7-9 \cdot 10^6$ tonnes of suspended matter reached the lake each year. The ratio between these values is high but not improbable. The values themselves imply an erosion rate throughout the watershed of 1 340 tonnes km^{-2} yr^{-1} or a reduction in rock depth of 0.7 mm per year. Amongst this eroded material about 7 000 tonnes of phosphorus enter the lake, which is equivalent to 10 g m^{-2} yr^{-1} of PO_4-P. Meybeck also provides information on the chemical composition and distribution of the sediments across the lake. Coarse detrital sediments (mainly sand) remain near the river inflow area, whereas the finer detrital sediments (particle size a few microns) spread homogeneously across the lake, although the sedimentation rate varies between 2 cm yr^{-1} near the river mouth and a few mm yr^{-1} in the centre of the lake basin. Sediments of intermediate-sized sand particles are transported by swiftly flowing turbid currents towards the central plain forming an underwater "fan delta". $CaCO_3$ is the main constituent of the sediments deposited away from the river mouths, e.g. in "Petit Lac".

Factors that control salinity and composition and concentration of the suspended matter in the Amazon have been studied by Gibbs (1967a, b). The salinity values are high in waters in the Andes region (120—140 ppm) and decrease to become very low (36 ppm) at the river mouth. This dilution is due to inflow of tributary rivers coming from the tropical savannah region which is underlain by Precambrian igneous and metamorphic rocks. The concentration of suspended solids in the Amazon waters at the river mouth was found by Gibbs to have a mean value of 90 ppm. He grouped suspended solids into three classes: (1) quartz, K feldspar and plagioclase; (2) clay minerals (montmorillonite, kaolinite, mica, chlorite); and (3) minor constituents (gibbsite, talc, amphibole and others).

Nine environmental parameters — including the type of rocks, climate, the percentage of the total tributary basin covered with broadleaf evergreen vegetation and relief in the source areas — were chosen as factors which may control both the salinity and the composition and concentration of the suspended solids. Relief (the weighted mean elevation of the area) proved to be the most significant parameter, which determined to a great extent the concentrations of dissolved salts, the nature of the suspended solids and the particle size. It also determined the concentration of many of the various minerals and exerted a strong influence on the relative importance of physical and chemical weathering. The low relief basins in the tropical belt showed a preponderance of chemical weathering, which resulted in the production of minerals such as kaolinite and gibbsite and also produced a relatively high ratio of dissolved to total substances in the load. Physical weathering was shown to be the principal erosion agent in the mountainous catchment areas within the Andes region, resulting in products such as mica, chlorite, quartz and feldspars and a low ratio of dissolved to total load.

In a discussion of Gibbs' paper Garner (1968) emphasised the importance of climatic influence and pointed out that Gibbs' definition of relief is essentially synonymous with mean elevation above sea level and not with local differences in elevation. Frost action and aridity — both factors which lead to physical weathering — are pronounced in high altitude regions but are not caused specifically by relief. Garner also placed greater emphasis than Gibbs on the importance of valley alluvium in the eastern Andes. Gibbs defended his original conclusions in a later paper (Gibbs, 1968), in which he replied to Garner's criticisms.

The highest percentage of montmorillonite in the load was derived from the tropical catchment area, but the mountainous environment yielded more total montmorillonite because of its higher erosion rate. Of the variability of the percentage of montmorillonite 96% was accounted for by the calcareous rocks being probably the parent material. It is not clear whether the presence of Ca^{2+} prevents the weathering of montmorillonite to kaolinite or whether this is due to a difference of source material. Kaolinite was negatively correlated with calcareous rock, but was derived from igneous and metamorphic rocks.

Little information is available on the adsorption of nutrients onto the sus-

TABLE 18.3

Hydrological data from the Rio de la Plata (from Bonetto, 1972)

	Upper Parana and Paraguay at confluence	Parana at mouth	Uruguay at mouth	Rio de la Plata at mouth
Discharge (m^3 sec^{-1})	17 000	18 000	5 000	23 000
Annual load of dissolved solids (× 10^6 tonnes)	50	62	8	70
Annual load of suspended matter (× 10^6 tonnes)	87	112	17	129
Total annual load (× 10^6 tonnes)	137	174	25	199
Dissolved load as percentage of total load	36	36	32	35

pended matter. Low concentrations of dissolved phosphate may give a wrong impression concerning the fertility of the water because, if much of the total phosphate present is adsorbed on clay, this form also can be available to algae (see Golterman, 1975).

In the Rio de la Plata Basin, the relative importance of the annual suspended load compared to the dissolved load is remarkable (Depetris and Griffin, 1968; Bonetto, 1972) (Table 18.3). Here the Upper Parana River and the Paraguay River come from the same plateau as does the tributary of the Amazon known as the Rio Tapajo. Deposits from these rivers dominate the mineralogy accounting for more than 67% of the total suspended solids, 72% of the total illite and 73% of the total montmorillonite — the two most abundant clay minerals coming from the catchment areas of these two tributaries (Depetris, 1968). The mineralogy and silt size of the suspended load indicate acid igneous and crystalline basement rocks as the main source. In contrast with the situation in the Amazon where Gibbs concluded that the two clay minerals kaolinite and montmorillonite come from two different sources, there are no indications that such a situation occurs in the case of the Rio de la Plata.

Bonetto (1972) reported that the Upper Parana—Paraguay River Basin furnishes clay abundant in illite and kaolinite with less montmorillonite and chlorite. He supposed illite to come from the moutainous Andean environment, whereas montmorillonite and kaolinite are probably supplied by the Paraguay and Upper Parana rivers. Bonetto's statement contradicts that of Depetris, who found montmorillonite to be the most abundant clay mineral.

It seems likely that the factors which determine the ratio of kaolinite to montmorillonite are not yet fully understood. No scheme has yet been devised to enable the type of clay likely to be found in any particular river to be predicted. Climate and relief (which were shown to be controlling factors in the Amazon watershed) are generally believed to have the most important influence. Thus Sherman (1952) found that it was not the total rainfall but its seasonal distribution that determined whether montmorillonite or kaolinite formed from basalts in the Hawaiian Islands. The relative amounts of these clays formed depended on whether the climate was continuously dry or alternately dry and wet. Recent thermodynamic studies have helped towards an understanding of how silicate minerals undergo change. Kramer (1968) found that at low pH, high partial pressure of CO_2, and low alkalinity, only soluble breakdown products and kaolinite are found, whereas in more alkaline waters with low SiO_2 concentrations, chlorite, kaolinite and montmorillonite are all found. The occurrence of high SiO_2 concentrations in alkaline water favours the formation of illite. Kramer thinks it possible that montmorillonites are metastable in a geological sense and may alter to feldspars, kaolinite or chlorite according to different circumstances. Stumm and Morgan (1970) gave equations to represent the equilibria between montmorillonite and kaolinite, but these equations have not yet been solved when applied to conditions as complicated as those found in the Rhine where, in addition to the silicate minerals, the calcium—carbonate—bicarbonate system is also important. $CaCO_3$ can form up to 18% of Rhine silt (De Groot, 1970). In natural situations insufficient time may have elapsed for the equilibrium to have become fully established. Knowledge of the chemical nature of silt is nevertheless essential because many chemical compounds are transported by silt by reason of the ion-exchange capacity of both kaolinite (12—13 meq. per 100 g) and montmorillonite (80—150 meq. per 100 g).

18.2.4. Adsorption of metals on silt

De Groot et al. (1971) found several metals in sediments from the rivers Rhine and Ems, the latter being regarded as an unpolluted river (see Table 18.4). Values of Ca and Mg are not given because the adsorbed component

TABLE 18.4

Heavy-metal content in Rhine and Ems sediments as mg kg^{-1} (ppm) of the < 16 μm fraction (after De Groot et al., 1971)

	Rhine	Ems		Rhine	Ems
Fe	54 000	112	As	310	60
Mn	2 600	3 300	La	80	80
Zn	3 900	700	Co	43	40
Cr	760	180	Hg	18	3
Pb	850	100	Sc	12	12
Cu	470	150	Sm	7	9

TABLE 18.5

Quantities of heavy metals transported annually by the Rhine (data from RIWA report, 1970)

Metal	Tonnes per year	Metal	Tonnes per year
Hg	85	Pb	1 500
As	1 000	Cu	2 900
Cd	200	Zn	9 000

cannot be distinguished from Ca and Mg present in carbonate particles.

De Groot concluded from the differences between Rhine and Ems that the presence of high concentrations of Zn, Cr, Pb, Cu, As and Hg in Rhine silt should be regarded as indications of external influences, loosely termed pollutants. Some of the Ems values are certainly not minimal because Serruya (1969) measured much lower values for Zn, Pb, and Cu in the (Sub)atlantic and (Sub)boreal sediments from the Lake of Geneva (see Table 11.2). The Ems, therefore, may be more polluted than De Groot believed it to be. Table 18.5 shows the amounts of heavy metals transported each year by the Rhine, most probably by adsorption on the silt.

In view of the different ion-exchange capacities of the clay minerals it is important, for Dutch scientists in particular, to know whether Rhine silt is predominantly montmorillonite or kaolinite. Heavy metals which are at present adsorbed only on the silt might eventually increase in quantity sufficiently to exceed the total ion-exchange capacity of the silt system and thereafter could enter The Netherlands in a more soluble form. A proportion of adsorbed metals is mobilised after sedimentation whilst the rest remains in the silts.

18.2.5. Quantities and transport of silt

Although the amount of suspended eroded material transported by rivers is probably greater than the amount transported as dissolved compounds, few quantitative estimates of the amount of suspended matter are available.

In the Amazon, the Rio de la Plata, and rivers in the U.S.A. the proportion of suspended matter seems to be 1.5—2.0 times that of the dissolved material, this being a fairly conservative estimate. Only in hard-water rivers like the Rhine the dissolved components do appear to be relatively more important. Some idea of the relative importance of suspended matter may be obtained by comparing the $0.6 \cdot 10^9$ tonnes transported by the Amazon with the $4 \cdot 10^9$ tonnes of solutes transported by all rivers. If the factor is indeed 1.5—2.0, this could mean a total quantity of silt of $6—8 \cdot 10^9$ tonnes per year. It is regrettable that the silt load of relatively few rivers is monitored but fortunately information is available for the Rhine and the Amazon.

Müller and Förstner (1968a, b) found that the amount of suspended matter transported by a river was dependent upon flow and could be described by the relation:

$$C_s = a \cdot Q^b$$

where C_s = suspended sediment load (mg l^{-1}), Q = flow (in m^3 sec^{-1}), and a and b are constants. This equation can be used for predicting rate and quantity of long-term sediment transport. The constant a can vary between 0.004 and 80 000 (the latter value being valid in Rio Puerco at Bermudo, northern Mexico) and depends on climate, vegetation, rock erodibility, and the morphological stability of the valley system. The constant b varies between 0.0 and 2.5 and is a measure of erosional forces (Müller and Förstner, 1968a, b). The relationship given by the equation does not hold for the lower reaches of rivers, where the decrease in flow rate and the occurrence of flocculation cause the suspended matter to settle out.

Müller found that the constants a and b are 0.004 and 2.2 respectively in the Rhine waters near the alpine town of Lustenau, where the river enters the Lake of Constance. In the outflowing Rhine waters from the Lake of Constance, Müller found a = 20 and b = 0.0, indicating a rather constant load of 20 mg l^{-1}.

Terwindt (1967), studying the transport of mud along the Dutch coast, estimated the quantity of silts transported by the rivers Rhine and Meuse. The silt content of the two Rhine branches, the Lower Rhine and Waal, varied between 50 and 500 mg l^{-1} (dry weight), with average contents of 57 and 67 mg l^{-1} respectively, while that of the Meuse was 64 mg l^{-1}. From these figures he calculated that the Rhine transported $3.2 \cdot 10^6$ tonnes and the Meuse $0.7 \cdot 10^6$ tonnes of dry matter per year.

Santema (1953) has given mean values of $4 \cdot 10^6$ tonnes per year for the period 1870—1885, $2.8 \cdot 10^6$ tonnes for 1928—1931, and $2.3 \cdot 10^6$ tonnes for 1940—1950. If these quantities were spread over The Netherlands, they would form a 1-cm layer every century.

Some of the eroded material carried by a river will sediment en route whereas the rest will reach the river mouth resulting in the formation of a delta if hydrological conditions allow. Sioli (1966a) attributed the absence of a real delta at the mouth of the Amazon to the existence of an ocean current which runs along the Brazilian coast. The sedimentation of large quantities of mud (estimated to be of the order of about 10^9 tonnes per year) takes place therefore mainly up towards the coast of French Guiana (Sioli, 1966b; Golterman, 1973c). The low sedimentation rate at the Amazon mouth may also be due partly because no flocculation occurs in the unpolluted freshwater of the Amazon (Golterman, 1973c). In contrast to the Amazon situation, large amounts of sediments are deposited at the mouth of the Rhine, and only small quantities are transported along the Dutch coast. They move in a narrow seawater current travelling parallel to the coast and are deposited

416

partly in the Dutch Wadden Sea (De Groot, 1964, 1970; Terwindt, 1967).

Given suitable geological conditions, the natural sedimentation in rivers is stimulated by two processes, namely decrease in water velocity and increase in salinity. Particles which arrive at a river mouth do not usually settle easily. Waves, tidal water movement and freshwater outflow will transport the particles backwards and forwards. Eventually most settle and the rest are carried out along the estuary towards the sea. The process has been reviewed by Postma (1967), who considered the two factors from a quantitative viewpoint.

From Fig. 18.1 it can be seen that unconsolidated sediment particles of a size between 10 and 50 μm are deposited at flow velocities below 10 cm per sec, while for consolidated clay and silt with a water content of 50% or less the deposition velocity is near 100 cm per sec. Erosion velocity for unconsolidated sediment is about 1.5—2 times the deposition velocity. According to Postma this difference is of great importance in tidal areas. The consolidation process, i.e. squeezing out of water due to the superposed sediment load is irreversible in the case of clay. The second factor in the sedimentation process is flocculation, which may occur simultaneously with a change in water velocity in the tidal areas. Flocculation is of less importance where a

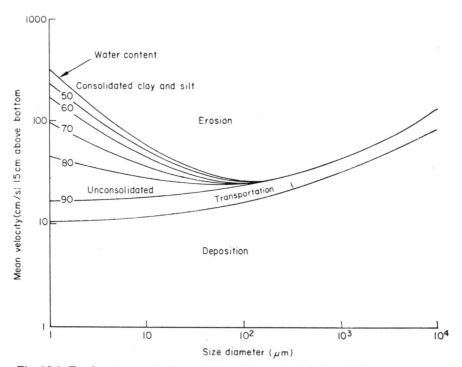

Fig. 18.1. Erosion, transportation and deposition velocities for different grain sizes. Possible values for various stages of consolidation are indicated. (From Postma, 1967.)

river ends in a freshwater lake, e.g. the Rhône in the Lake of Geneva (Postma, 1967). In some cases, e.g. where the Rhine enters the IJsselmeer, flocculation does not occur, probably because the process is likely to have taken place in the river further upstream because of the high calcium pollution of the Rhine. It has not yet been shown whether this was also the case in earlier periods; for weathering of $CaCO_3$ rocks has probably been significant over a long period. Different layers of Rhine silt are to be found in the IJsselmeer representing situations when the lake was either salt, brackish or fresh. The degree of flocculation has not yet been estimated, however, and neither has the influence of chemical changes on the composition of the mud. The flocculation may be caused by disturbance of the electric charge of the colloidal, or semi-colloidal particles. For the clays this is mostly a negative charge. The hydroxyl ions and the ion-binding capacity of the clay are also important for this process.

The negative charge on the clay particle is balanced by a locally higher concentration of hydrated cations. The thickness of this layer depends on the total ion concentrations in the water, pH, temperature, etc., and on the chemical nature of the cations adsorbed and comprising the layer. Flocculation occurs when the thickness of this layer falls below a certain value. As the result of flocculation, the particle size and thus the sedimentation velocity increases. Particular clays such as kaolinite, illite and montmorillonite, each behave differently with regard to the flocculation factors such as chloride concentrations (Whitehouse et al., 1960). With increasing Cl^- concentration (from 0.5 to 18.0‰), the settling velocity of illite in their experiments increased from 89 to 100 cm sec^{-1}, for montmorillonite from 0.2 to 18.8 cm sec^{-1}, but for kaolinite it did not change. Light transmission showed similar differences for these clays in Whitehouse et al.'s experiments. A second factor in the flocculation process may be the change in Ca^{2+} and Mg^{2+} concentration, where the river silt passes into seawater. It is expected that montmorillonite and kaolinite (from which more of these cations are leached out) will behave differently. Since Whitehouse carried out his experiments with seawater dilutions, possible cation and chlorinity effects worked together as seawater contains more Ca and Mg than most freshwaters. The flocculation process has mainly been studied in seawater areas. It seems likely that similar processes also occur if a river enters a lake with a markedly different chemical composition.

In addition to the inorganic constituents of the silt, organic compounds also occur. They originate from leaching of solids or from the organisms living in the river. It is not known what influence any of these compounds have on sedimentation. Their presence is important, however, because after sedimentation has occurred they are the substrate for bacterial growth and thus they lower the redox potential. According to De Groot et al. (1971) chelating compounds may be produced during this breakdown process and these may form stable soluble metal complexes with cations such as Fe^{3+}, Cu^{2+}, Zn^{2+},

Fe^{2+}, and Pb^{2+}. Manganese forms much less stable compounds than iron does but is mobilised more easily than Fe^{3+} under reducing conditions. De Groot (1964) showed that Mn in river muds is relatively immobile — a fact which enabled him to use it as a tracer element for the different muds in the North Sea region. This immobility shows that in the Rhine Delta region the chelate effect more than counters any effects of reducing conditions.

In a very interesting study of the mobility of heavy metals, De Groot et al. (1971) showed that Hg in particular becomes mobile after sedimentation. This they explained by the strong tendency of Hg to form chelates with N- or S-containing ligands, and to the formation of complexes between Hg and the Cl ion. It is likely that the formation of methyl-mercury should also be taken into account. Of all the metals studied and measured by De Groot, Hg mobilises most readily. Other metals that are easily mobilised are As, Cr, Pb, Zn and Cu. Those not mobilised are: Sm, Mn, Sc and La, while Fe and Co are mobilised to an intermediate extent. In addition to acknowledging De Groot's findings, it should be realised that the settlement of silt and the breakdown of organic matter are not the sole factors leading to production of important metal-chelating agents. Lewis and Broadbent (1961) showed that copper ions in particular are bound mainly by chelation with the organic compounds of the silt. It was demonstrated experimentally that these cations may be held very firmly and in large quantities by organic matter, the binding strength and adsorption capacity varying greatly according to the kind of organic matter present.

18.3. LAKE SEDIMENT PROFILES

A study of lake sediment profiles, considering such details as variation in depth of the various layers, can yield much information about the history of a particular lake, and its inflows. The composition of the sediment layers does not appear to be related directly to events occurring in the lake itself. Exceptions are the organic-matter content and those elements, such as iron and manganese, which may leach out after reducing conditions have become established.

Mackereth (1966) studied the lake sediments in a number of different lakes of contemporary origin but with a present variation in primary production in the English Lake District. The lakes were formed during the glaciation of a dome-shaped area of Ordovician and Silurian rocks (see Chapter 2). A clearly recognisable distinction can be seen in the lake sediments between a lower layer of unknown depth containing mainly inorganic material deposited during periods of active glaciation and more recent layers which are rich in organic matter because of biological activities which have occurred during the Postglacial period. The time boundary between the glacial clays and Postglacial organic deposits has been dated at about 10 000 years B.P. (before the present radio carbon dating gave a figure of 11 878 ± 120 B.P.). The depth of the organic mud deposits varies from lake to lake and probably in different regions of the same lake but is generally between 4.5 and 6 m. Sediment depth bears no apparent relation to the internal productivity of the lake, since Ennerdale, which is much less productive than Esthwaite, has a thicker Post-glacial deposit. It is more likely to be related to the degree of erosion occurring in the watershed because 70—80% of the deposits consists of clays and silt. Mackereth assumed that all photosynthetically grown organic matter from the lake itself is oxidised in the hypolimnion, an assumption which is now known to be untrue, and argued that the organic matter found in the sediments must therefore have been derived from surrounding soils during the erosion process. Mackereth tried to distinguish between types of organic matter by comparing loss on ignition values with the true carbon, hydrogen and NH_4-N content of the sediments. In muds containing only small quantities of organic matter this approach gives inaccurate measurements because some water is bound amongst the clay particles and is not removed at the temperature at which the mud is dried ($110°C$). The large variation in C:H ratio found from measurements made in this way masked any variation due to differences in the organic-matter content itself. The C:N ratios found were similarly disturbed by some NH_4^+ adsorbed on the clay. This adsorbed amount could represent a quantity of nitrogen equivalent to about 0.2% of dry weight and must have derived from a secondary addition to the lake, because igneous rocks contain only 0.005% and the Borrowdale volcanic rock only 0.007% of nitrogen. Mackereth found that the C:N ratio ranged between 10 and 15, but because he did not measure the inorganic nitrogen directly, no correction could be made to account for inaccuracies in the method. He found that the ash content was roughly inversely proportional to the organic-carbon content. In fact the mineral percentage plus twice the carbon percentage was found in most cases to approach 90—100%, the remainder apparently being water.

Mackereth found a striking similarity between the shapes of the carbon content profiles of Ennerdale and Esthwaite sediments and suggested that the organic matter in them must have been derived from surrounding soils because the lake productivities are so different. As the carbon content of the sediments lies between 5—10% for Ennerdale (unproductive) and around 15% for Esthwaite (productive), one could argue that the algal production does contribute and that the changes in the sediment profile at different depths are due more to differences in quantities of eroded minerals at different periods of the lake history, since these changes are the same in both lakes. Evidence to support this argument can be derived from a study of the sodium and potassium contents of the mineral sediment. A linear relationship exists between the sodium and potassium values and the percentage of mineral matter in the sediment (Fig. 18.2). Moreover, a direct relationship also exists between mineral content of the sediment (representing soil erosion) and the sodium and potassium content of this mineral matter (Fig. 18.3). Such a rela-

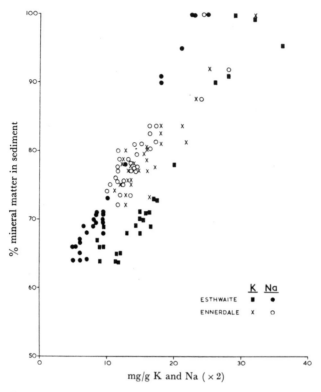

Fig. 18.2. Relationship between total mineral content and the concentration of sodium and potassium in the dry sediments of Ennerdale and Esthwaite. (From Mackereth, 1966.)

tionship can be explained by assuming that during periods of rapid erosion the length of time when chemical leaching could occur was too short, so that potassium and sodium were not removed during the erosion of the soil. The concentration of alkali metals in the mineral matter diminishes progressively with diminishing total mineral content. Mackereth concluded that the mineral content of the sediment is related directly to the intensity of erosion, since a variable rate of oranic deposition could not be expected to produce the observed relationship. It is also possible though that a constant supply of minerals with a varying production of organic matter could also produce results showing the same relationship as that shown in Fig. 18.3.

In Fig. 18.4 it is perhaps significant that the points for Esthwaite lie both above and below the points for Ennerdale. This may indicate that the present organic production has now diminished proportionally the influence exerted by materials originating from erosion products. The whole situation is quite complex because the erosion products provided most of the nutrients for organic production. A high erosion rate may thus induce a high rate of production of organic matter unless so much suspended matter is washed into the

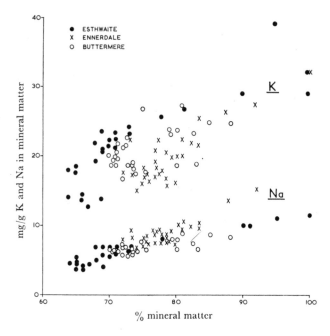

Fig. 18.3. Relationship between the sodium and potassium content of the mineral matter, and the total mineral content in the sediments, of Esthwaite, Ennerdale and Buttermere. (From Mackereth, 1966.)

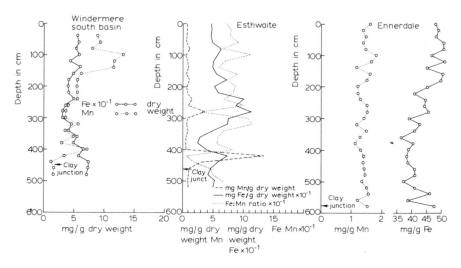

Fig. 18.4. Distribution in depth of the concentration of iron and manganese in the dry sediments of Windermere South Basin, Esthwaite and Ennerdale. (From Mackereth, 1966.)

lake that turbidity results in light becoming a growth-limiting factor. In such conditions much phosphate will also be washed into the lake, but this will probably be bound within the lattice of the clay particles and thus will not be available for algal growth. It should be noted that clays adsorb more potassium than sodium in contrast to the water that normally contains more soluble sodium than potassium. This effect is a result of the strong adsorption of potassium on the clay.

The distribution of halogens in the profiles resembles that of potassium, although halogen content is not directly related to the deposition of inorganic material. It is strongly associated with the organic carbon, although the distribution does not resemble that of organic carbon. Mackereth suggested that the deposition is related to a climatic factor, e.g. the "oceanicity". Increasing oceanicity implies higher rainfall and would then be associated with increasing erosion. Magnesium content is clearly associated with that of mineral products like potassium, while calcium is evidently more easily leached from the soil, except during periods of very intense erosion.

Iron and manganese concentrations are influenced by the intensity of input of eroding particles and by the redox potential in the sediments. It seems unlikely that Fe or Mn would normally arrive in a lake as soluble ions and then be precipitated to form sedimentary particles. Their transport during erosion would not separate the two elements; iron and manganese occur in the lithosphere in concentrations of 50 and 1 $mg\ g^{-1}$, respectively, and should remain in this proportion. Production processes in the soils in the watershed will decrease this ratio, however, as manganese can be more easily reduced. The distribution patterns of Fe and Mn in the lake sediments are shown in Fig. 18.4.

In the glacial sediments of Ennerdale the concentration of Mn and Fe approaches the values found in the lithosphere but in the Postglacial sediments dilution by organic matter occurred. The distribution follows that of potassium which points to the conclusion that the input is controlled mainly by erosion. In Windermere the glacial clay has a higher iron content than the lithosphere value of 50 $mg\ g^{-1}$, this being due to mechanical separation of fine particulate material to form an iron-rich clay. Mackereth interpreted the occurrence of larger amounts of manganese in the upper layers in terms of its preferential removal from soils in the catchment area. This could be brought about by the onset of conditions which were sufficient to reduce manganese but not iron in the soils. On arriving in the lake, however, the manganese would be deposited again because aerobic conditions always prevail there. A similar increase in manganese concentration as found for Windermere is also found in the deeper strata of the Esthwaite sediments (at 430 cm) but above these strata much lower concentrations occur. Iron content at 330 cm depth also reaches a value of twice that found in the lithosphere, this probably being also due to intensified leaching from soil. In strata overlying the 400-cm zone manganese content is lower than the

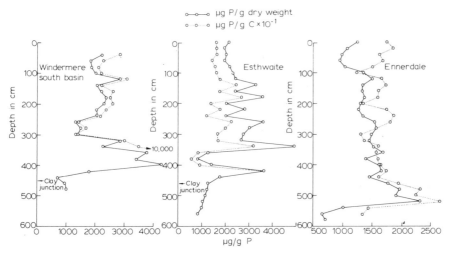

Fig. 18.5. Distribution in depth of the concentration of phosphorus in the dry sediments of Windermere, South Basin, Esthwaite and Ennerdale (solid line), and the relationship between carbon and phosphorus (dotted line). (From Mackereth, 1966.)

normal lithosphere value. Iron content is at a maximum at 300 cm.

Anaerobic conditions in the hypolimnion occur currently in Esthwaite but not in the other two lakes, and the diminishing Fe and Mn content in the upper strata of the Esthwaite sediments can be explained by supposing such conditions to have operated during the later deposition periods.

Mackereth pointed out that the situation in Esthwaite resembles that in Linsley Pond. He also discussed in detail the influence of reducing conditions on the behaviour of Fe and Mn, both in the soils of the drainage basin and in the hypolimnion and sediments of lakes.

The distribution of phosphate is shown in Fig. 18.5. The lower ends of the curves indicate where the glacial clays join the overlying deposits. The phosphate concentration there resembles that found in igneous rocks (0.1%) and this is strong evidence to suggest that fine rock particles with their own inorganic phosphate content were eroded and then deposited to form the glacial clays. The phosphate concentration increases to 3—5 times this value in the lower strata of the Postglacial deposits, while in the upper zone it varies considerably falling in an irregular way. The phosphate enrichment in the Postglacial sediments probably results from more intensive leaching of soluble phosphate from surrounding soils followed by precipitation in the lake (in addition to the continuing input of mineral clay particles). In the Ennerdale Postglacial sediments the phosphate distribution pattern is different from that of iron, and Mackereth concluded that the precipitation of phosphate was biological and occurred at a constant rate. Assuming that the rate of phosphate supply has been constant throughout the Postglacial period, Mackereth argued that the concentration of phosphate in a sediment would

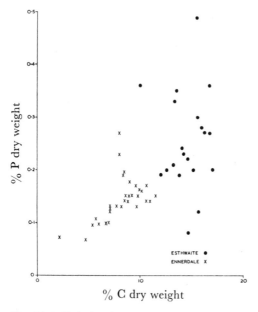

Fig. 18.6. Relationship between carbon and phosphorus in sediment samples from Esthwaite and Ennerdale. (From Mackereth, 1966.)

be expected to be inversely proportional to the rate of deposition of the sediment, a situation corresponding with that of carbon (see above), with the difference however that the sediment itself does contain phosphate. He did in fact detect an approximately constant relationship between carbon and phosphate (Fig. 18.6) and assumed that both elements were reflecting the changing deposition rate of the sediments. It should be noted that Mackereth did not argue that the similar behaviour of carbon and phosphate necessarily indicated that the precipitation mechanism was primarily biological. After subtracting the amount of lattice-bound phosphate from the total found in the sediments, Mackereth calculated a residual phosphate concentration of 8 mg P per g of C, which resembles that of algae. In contrast to the Ennerdale sediments, neither in Windermere nor in Esthwaite does phosphate show any relation with carbon (but it does show a relation with iron and manganese).

In Windermere (and in Ennerdale) no losses of phosphate have occurred from the sediment. In Esthwaite phosphate shows a sharp rise from the low value at the glacial clay junction at 410 cm together with the iron maximum, while at 380 cm both phosphate and iron show minimal values, due to the loss of iron from the lake. Mackereth showed that in addition to being precipitated and sedimented by biological processes phosphate is precipitated to a significant extent by both iron and manganese, although only the iron serves to retain phosphate within the sediments after it has been precipitated there. It is probable that it is mainly organically bound phosphate which sediments but that after mineralisation has occurred within the sediments the inorganic phosphate formed becomes bound secondarily to the ferric hydroxide. Mackereth has produced some evidence which shows that the release of phosphate after reducing conditions have become established is not due solely to the dissolution of ferrous ions; he pointed out that vivianite, $Fe_3^{2+}(PO_4)_2 \cdot 8 H_2O$, is often found in these sediments as large particles (see also Chapter 5). He did not try to separate the different phosphate components. Other elements studied by him were sulphur, zinc, copper, cobalt and nickel. Mackereth pointed out that his profiles of chemical content from lake district sediment cores showed similarities to those from Linsley Pond (Hutchinson and Wolleck, 1940; Livingstone and Boykin, 1962).

Hutchinson and Wolleck (1940) discerned similar changes in the concentrations of several elements in the sediments of Linsley Pond. An increase in the inorganic content of the uppermost recent unconsolidated sediment layer was attributed to recent erosion of (agri)cultural origin, while the higher organic-matter content of underlying layers was believed to have been caused by a change from oligotrophy to eutrophy. Changes in iron, calcium and manganese content were thought to have been due to biochemical processes. Crude protein was determined in the sediments (N $\times$ 6.25) but no correction for NH_3 was made.

Livingstone and Boykin (1962), studying the phosphate distribution in the sediments of Linsley Pond, suggested, probably correctly, that most of the phosphate is bound by sorption reactions with mineral material. They believe that the increasing productivity of the lake is not part of a general trend which all lakes are tending to show (as suggested by Hutchinson and Wolleck) but is caused by a changing rate of mineral sedimentation. Reduction in the amount of precipitation of phosphate clay may easily have made the lake more eutrophic because more phosphate would then remain in a form available to algae. Although not all phosphate bound onto clay is unavailable to algae, it may happen that phosphate together with the clay particles has been transported away from the epilimnion before the algae could make use of it.

Shapiro et al. (1971) studied the phosphate content in sediment profiles from Lake Washington in an attempt to discern changes in primary productivity following sewage diversion (see Chapter 17). They encountered difficulties due to the problems of diagnosis of the different phosphate components. The lake receives phosphate-containing minerals in the inflow and these release soluble phosphates into the waters of the hypolimnion. Cores taken in 1958 showed the uppermost values in the 1958 strata contained in cores collected in nitrogen and phosphate. Concentrations of all three components had decreased to lower values in the 1958 strata contained in cores collected in 1970. By this time of course the 1958 strata were no longer uppermost. The surface layers in the 1970 cores did not yet reflect the decrease in algal growth which has occurred recently in the lake. Shapiro et al. suggested that the phosphate is converted from iron phosphate to aluminium or calcium phosphate during the process of ageing, and this together with other time-related changes which occur in the sediments is likely to stultify any attempts to reconstruct lake conditions. This process of post-sedimentation conversion of phosphates into calcium phosphate has been described for Lake Erie in Chapter 5.

An early method for deducing an approximate dating of core samples involved a consideration of the relative amounts and kinds of pollen present at different depths. The pattern of qualitative and quantitative change in pollen content enables a depth/time curve correlation to be ascertained, and thus it could be demonstrated that the cessation of active glaciation occurred approximately 10 500 years ago in the English Lake District. Core strata representing the phase of minimum erosion intensity (380—260 cm) covered the period from 9 000 to 5 000 B.P., after which erosion occurred at a steadily increasing rate (Franks and Pennington, 1961, cited in Mackereth). This increasing rate of erosion is associated with a reduction in the tree/non-tree pollen ratio (Franks, 1956).

Pearsall and Pennington (1947) attributed changes in pollen composition to

TABLE 18.6

Division of Postglacial (Flandrian) sections of the profiles (Pennington et al., 1972)

Regional pollen zone	^{14}C dates (Lochs Sionascaig and Clair)	Major human interference episodes	Pollen zone definitions
VI	950 ± 100 B.C.	ca. 950 B.C.	Calluna zone (local: Calluna—Myrica—
		ca. 1 550 B.C.	Cyperaceae or Calluna—birch—pine)
	2 070 ± 100 B.C.	ca. 2 070 B.C.	
V ii	2 750 ± 100 B.C.		pine—birch—alder
			end of elm in regional
	3 410 ± 110 B.C.	ca. 3 410 B.C.	pollen
V i			pine—birch—alder, + elm
	4 300 ± 140 B.C.		in regional pollen
	4 570 ± 145 B.C.		
IV			pine—birch
	5 930 ± 160 B.C.		
III			birch—hazel
	6 960 ± 130 B.C.		
II			juniper
I			Empetrum
transition			Rumex—Lycopodium selago

——————— ca. 8 300 B.C. (beginning of Flandrian) ———————

forest clearance probably initially by members of a Megalithic culture and later by Norse settlers. Mackereth pointed out that it is possible that the changing levels of human activity as inferred from the pollen composition were associated with climatic changes. Indications that major climatic changes may have occurred are indeed found in the distribution patterns of halogens and boron. A detailed study of pollen distribution and zonation in sediments from some Scottish lochs was made by Pennington et al. (1972). These authors showed that the pollen spectrum was very consistently related to changes in sediment composition. Using the ^{14}C-dating method, an accurate time-scale for the sediments was established. The paper gives interesting data on soil history and provides evidence to suggest that peat has developed on the watershed since about 5 000 B.P. It also demonstrates how an interaction has occurred between soil history and lake sediment genesis. Although it is somewhat beyond the scope of this book, their table 2 is given here as an indication of how an interdisciplinary approach can help in the study of lake sediments (Table 18.6).

REFERENCES

Bonetto, A.A., 1972. *Report on IBP/PF Projects*. Instituto Nacional de Limnologia, Santo Tomé, Argentina, 55 pp.
De Groot, A.J., 1964. Origin and transport of mud in coastal waters from the western Scheldt to the Danish frontier. In: L.M.J.U. van Straaten (Editor), *Deltaic and Shallow Marine Deposits, Developments in Sedimentology, 1*. Elsevier, Amsterdam, pp. 93—100.
De Groot, A.J., 1970. Geochemisch onderzoek in deltagebieden. *Natuurk. Voordr., N.S.*, 48: 61—75.
De Groot, A.J., De Goeij, J.J.M. and Zegers, C., 1971. Contents and behaviour of mercury as compared with other heavy metals in sediments from the rivers Rhine and Ems. *Geol. Mijnb.*, 50(3): 393—398.
Depetris, P.J., 1968. Mineralogia de algunos sedimentos fluviales de la cuenca del Rio de la Plata. *Rev. Asoc. Geol. Argent.*, 23(4): 317—325.
Depetris, P.J. and Griffin, J.J., 1968. Suspended load in the Rio de la Plata drainage basin. *Sedimentology*, 11: 53—60.
Förstner, U., Müller, G. and Reineck, H.E., 1968. Sedimente und Sedimentgefüge des Rheindeltas im Bodensee. *Neues Jahrb. Miner. Abh.*, 109: 33—62.
Franks, J.W., 1956. *Pollen Analytical Studies of the Esthwaite Basin*. Thesis, London University, London.
Franks, J.W. and Pennington, W., 1961. The late-glacial and post-glacial deposits of the Esthwaite basin, North Lancashire. *New Phytol.*, 60: 27—42.
Garner, H.F., 1968. Geochemistry of the Amazon River system: Discussion. *Bull. Geol. Soc. Am.*, 79: 1081—1086.
Gibbs, R.J., 1967a. Amazon river: Environmental factors that control its dissolved and suspended load. *Science*, 156: 1734—1737.
Gibbs, R.J., 1967b. The geochemistry of the Amazon river system: Part I. The factors that control the salinity and the composition and concentration of the suspended solids. *Bull. Geol. Soc. Am.*, 78: 1203—1232.
Gibbs, R.J., 1968. Geochemistry of the Amazon river system: Reply. *Bull. Geol. Soc. Am.*, 79: 1087—1092.

Goldschmidt, V.M., 1937. The principles of distribution of chemical elements in minerals and rocks. *J. Chem. Soc. Lond.*, 1937: 655—673.

Golterman, H.L., 1973a. Vertical movement of phosphate in freshwater. In: E.J. Griffith, A. Beeton, J.M. Spencer and D.T. Mitchell, *Environmental Phosphorus Handbook.* John Wiley, New York, N.Y., pp. 509—538.

Golterman, H.L., 1973b. Natural phosphate sources in relation to phosphate budgets: A contribution to the understanding of eutrophication. *Water Res.*, 7: 3—17.

Golterman, H.L., 1973c. Deposition of river silts in the Rhine and Meuse Delta. *Freshwater Biol.*, 3: 267—281.

Golterman, H.L., 1975. Chemistry of running water. In: B. Whitton (Editor), *River Ecology.* Blackwell, Oxford (in press).

Gonet, O., 1971. Introduction à l'étude des relations chimiques entre les sédiments du fond et l'eau du Léman. *Bull. Soc. Vaudoise Sci. Nat.*, 71(337): 131—156.

Hutchinson, G.E. and Wolleck, A., 1940. Studies on Connecticut lake sediments. II. Chemical analyses of a core from Linsley Pond, North Branford. *Am. J. Sci.*, 238(7): 493—517.

Kramer, J.R., 1968. Mineral-water equilibria in silicate weathering. *Int. Geol. Congr., 23rd, Prague, 1968,* 6(Geochemistry): 149—160.

Lewis, T.E. and Broadbent, F.E., 1961. Soil organic matter—metal complexes: 4. Nature and properties of exchange sites. *Soil Sci.*, 91: 393—399.

Livingstone, D.A. and Boykin, J.C., 1962. Vertical distribution of phosphorus in Linsley Pond mud. *Limnol. Oceanogr.*, 7: 57—62.

Mackereth, F.J.H., 1966. Some chemical observations on post-glacial lake sediments. *Philos. Trans. R. Soc. Lond., Ser. B*, 250: 165—213.

Meybeck, M., 1972. Bilan hydrochimique et géochimique du lac Léman. *Verh. Int. Ver. Theor. Angew. Limnol.*, 18(part 1): 442—453.

Müller, G. and Förstner, U., 1968a. Sedimenttransport im Mündungsgebiet des Alpenrheins. *Geol. Rundsch.*, 58(1): 229—259.

Müller, G. and Förstner, U., 1968b. General relationship between suspended sediment concentration and water discharge in the Alpenrhein and some other rivers. *Nature,* 217 (5125): 244—245.

Müller, G. and Gees, R.A., 1970. Distribution and thickness of Quaternary sediments in the Lake Constance Basin. *Sediment. Geol.*, 4: 81—87.

Pearsall, W.H. and Pennington, W., 1947. Ecological history of the English Lake District. *J. Ecol.*, 34: 134—148.

Pennington, W., Haworth, E.Y., Bonny, A.P. and Lishman, J.P., 1972. Lake sediments in northern Scotland. *Philos. Trans. R. Soc. Lond., Ser. B*, 264(861): 191—294.

Postma, H., 1967. Sediment transport and sedimentation in the estuarine environment. In: G.H. Lauff (Editor), *Estuaries.* Am. Assoc. Adv. Sci., Washington, D.C., pp. 158—179.

RIWA report, 1970. Rijncommissie Waterleidingsbedrijven. Jaarverslag '70. RIWA, Amsterdam.

Santema, P., 1953. Enkele beschouwingen over het slibtransport van de Rijn. *Ingenieur,* 7: B37—B40.

Serruya, C., 1969. Problems of sedimentation in the Lake of Geneva. *Verh. Int. Ver. Theor. Angew. Limnol.*, 17: 209—218.

Shapiro, J., Edmondson, W.T. and Allison, D.E., 1971. Changes in the chemical composition of sediments of Lake Washington, 1958—1970. *Limnol. Oceanogr.*, 16: 437—452.

Sherman, G.D., 1952. The genesis and morphology of the alumina-rich laterite clays. Clay and laterite genesis. *Am. Inst. Mineral. Metall.*, 154—161.

Sioli, H., 1966a. General features of the delta of the Amazon. In: *Humid Tropics Research; Proc. Dacca Symp. UNESCO, 1966,* pp. 381—390.

Sioli, H., 1966b. Soils in the estuary of the Amazon. In: *Humid Tropics Research; Proc. Dacca Symp. UNESCO, 1966,* pp. 89—96.

Stumm, W. and Morgan, J.J., 1970. *Aquatic Chemistry; An Introduction Emphasizing Chemical Equilibria in Natural Waters*. John Wiley, New York, N.Y., 583 pp.

Terwindt, J.H.J., 1967. Mud transport in the Dutch Delta area and along the adjacent coastline. *Neth. J. Sea Res.*, 3(4): 505—531.

Whitehouse, U.G., Jeffrey, L.M. and Debrecht, J.D., 1960. Differential settling tendencies of clay minerals in saline waters. *Proc. Natl. Conf. Clays Clay Min.*, *7th*, pp. 1—79.

CHAPTER 19

WATER MANAGEMENT PROBLEMS — WATER SUPPLY, RECREATION, SEWAGE DISPOSAL, AND MAN-MADE LAKES

19.1. WATER BALANCE AND QUALITY PROBLEMS

The amounts of water on earth are large: 70% of the earth's surface area is covered with $1.4 \cdot 10^9$ km³ of water (1 km³ = 10^9 m³), which if spread over the whole globe would give a 2 400 m deep layer or, allowing for the continents, a mean oceanic depth of 3 600 m. Rain will supply the land with about 100 000 km³ per year of fresh water, 37 500 km³ of which returns to the oceans by river flow, the remaining two-thirds returning to the atmosphere by evaporation from open water surfaces or as a result of plant transpiration. As rivers contain only 1 000 km³ of water, their volume must be renewed on average every ten days. Water incorporated in glaciers may have a hydrological cycle of several hundreds of years.

The quantity of freshwater on the earth is estimated to be between 2% and 2.5% of that of salt water. This gives a total quantity of about $29 \cdot 10^6$ km³ or a layer about 50 m thick if uniformly spread out. Of this 50 m about 49.5 m is present as glaciers and polar ice. Of the remaining 0.5 m 50% is in rivers and ground water, while 50% is in lakes (125 000—150 000 km³). Of this latter quantity 31 500 km³ is in Lake Baikal, 34 000 km³ in Lake Tanganyika and about 25 000 km³ in the St. Lawrence lakes. Lake Victoria, the lake with largest surface, contains only 2 700 km³. For comparison a few "great" European lakes are: Lago Maggiore 35 km³, Lago di Como 20 km³, Lake of Geneva 90 km³ and Windermere 0.7 km³.

Another 100 000 km³ of water is in saline lakes such as the Caspian Sea, the Dead Sea, Great Salt Lake, etc. For their water supply human populations depend on lakes, rivers and ground waters. The yearly river flow of 37 500 km³ would supply more than sufficient water for the $3.7 \cdot 10^9$ humans on earth, namely 10 000 m³ per person per year. Supply problems arise because people have not distributed themselves uniformly or in proportion to river flow, and periods of low flow occur in hot dry weather. In addition human populations have tended to settle along rivers and have polluted these rivers — beyond any hope of recovery in some cases.

Man demands water for domestic use, for industrial consumption, for agriculture, recreation and for transport. The quantities used vary with technological development and agricultural practice. Agricultural and industrial use is much larger than domestic use. Domestic use is estimated to take between 40 and 250 litre per day. Water use is expected to increase in future. In Table 19.1 the quantities which will be used in the year 2050 in France are com-

TABLE 19.1

Water demand in m^3 per person per day (France)

	1955	2050 (predicted)	Increase (%)
Public water supply	0.12	0.34	180
Industrial use	0.41	1.64	300
Irrigation	0.63	1.84	190
Navigation	0.08	0.09	10
Total	1.24	3.91	

TABLE 19.2

Water use in the United States in 1965 and predicted use in the year 2000 (in rounded figures), m^3 per person per day (From U.S. Water Resources Council, 1968, Water Assessment)

Use	1965	2000 (predicted)	Increase (%)
Rural domestic	0.045	0.04	—10
Municipal domestic plus some industrial	0.460	0.60	3
Industrial (self-supplied)	0.900	1.50	70
Electricity generation	1.60	5.60	250
agriculture	2.20	1.80	—20
Total use	5 20	9.60	85
Consumptive	1.50	1.60	—
Non-consumptive	3.70	8.00	110

pared with those used in 1955. In the U.S.A. water is used at a greater rate and the predicted use is even greater (Table 19.2). When comparing Tables 19.1 and 19.2, it should be noted that use of water for cooling in electricity generation is not included in Table 19.1, because the water can re-enter the hydrological cycle after being cooled to normal temperature without much cost. If this use is ignored the daily U.S.A. use is similar to that in France — about 4 m^3. It is obvious that 4 m^3 (or even 9.6 m^3) ought to cause no problems for water supply. Nevertheless problems *do* arise and two examples will be mentioned.

The city of Los Angeles is an example of how a settlement in the wrong place can cause serious problems. Nowadays the sources of water supply for Los Angeles are in other states, several hundreds of kilometres away; even the transport of water from Canada has been considered. At present drinking water is no longer on tap but is supplied in bottles. The second and much

TABLE 19.3

Water balance of The Netherlands

	Needs (km³ yr⁻¹)		Sources (km³ yr⁻¹)
Domestic public supply	1.9	Rhine	70
Industry	5.5	Meuse	8
Maintainance of water table	3.3	Other	1
"Flushing"	12.2	Precipitation minus evaporation	10
Loss by River mouth	9.3		
Total	31.5		89

worse example, however, is the situation in the lower reaches of the Rhine. The Netherlands are situated in the delta sediments and are one-third below sea level, thus they are very much dependent on water from the Rhine. This has accumulated the pollution of the Ruhr Valley and of the French potassium mines. In the water-demand table for The Netherlands (Table 19.3) a special item for agricultural use appears, i.e. flushing of the canals and ditches in order to remove the upwelling salt water. For example the polders around Leiden (Hoogheemraadschap Leiden, 980 km²), receive about 100 000 tonnes of chloride annually. This used to be flushed towards the sea with 0.5 km³ per year of Rhine water. Nowadays the chloride concentration is prohibitive: it ranges between 200 and 400 mg l⁻¹, mainly coming from effluents from the French potassium mines. Agriculture is therefore quite often seriously affected. In other polders the same problem arises. In spite of international efforts the chloride concentration is still increasing. For example the chloride loads (kg sec⁻¹) near Kembs, where the Rhine leaves Switzerland (see Fig. 19.1) near Seltz, downstream of the French potassium mines, and near Lobith, where the Rhine enters The Netherlands are:

	1955	1965	1970
Kembs	8	9	14.5
Seltz	90	160	165
Lobith	244	337	365

From the Dutch water balance (Table 19.3) it can be seen that human usage is near that in France and U.S.A.; 31 km³ per year for a population of $13 \cdot 10^6$ humans is quite close to 4 m³ per day per caput.

Another item on the balance is the outflow from the Rhine mouth. A minimum outflow of 9.3 km³ must be maintained, to prevent salt water penetrating. It is obvious that in dry summers there are problems especially in providing water for drinking and agriculture.

Due to the continuing over-exploitation of underground freshwater more and more Rhine water must be used for drinking water though the quality is poor due to the pollution from the Ruhr Valley. The mean yearly oxygen content is less than 50%, while during summer values of 25% are quite common. The combination of organic domestic pollution, industrial products such as heavy metals, phenols, oil compounds, and pesticides renders the water almost unsuitable as a supply of drinking water, even after intensive treatment. To cope with periods of rather poor quality large reservoirs are now being built, but since in a delta area no valleys are available they can only be dug below the soil surface. One large reservoir of freshwater has been created — the IJsselmeer which is a freshwater lake that was originally seawater, from which water can be taken for agriculture when the chloride concentration of the Rhine is too high (Fig. 19.1). The chloride concentration of the IJsselmeer, even though it is fed by the Rhine, is lower owing to dilution by rain. It eventually reaches high values much later than the Rhine, which then already shows the seasonally decreasing chloride concentration.

In winter, Rhine water is transported from the polders towards the IJsselmeer or ultimately the North Sea because there is excess rainfall. In summer Rhine water is taken into the polders through the IJsselmeer to compensate for evaporation. The flow through the polder reservoirs such as Tjeukemeer therefore reverses seasonally (Fig. 19.1). The chemical changes (for example in concentrations of phosphate, nitrogen, silicate, and iron compounds) resulting from this hydrology are partly described in the chapters related to these elements and more in detail in Golterman (1973).

The IJsselmeer is protected against some of the pollutants of the Rhine, because heavy metals and phosphate accumulate in the Ketelmeer (Fig. 19. 1A), an artificial pre-impoundment.

19.2. DRINKING OR TAPWATER

The water used for domestic purposes is commonly called potable water, although tapwater would be a better name. The normal household uses the water roughly as shown in Table 19.4. This usage varies in different countries, and it changes with time. Car washing is one of the increasing items.

Criteria for drinking water vary from country to country. In some countries certain minimum criteria must be met. These include the absence of colour, taste and smell, absence of *Escherichia coli* (< 1 per litre), and the turbidity and reducing power must be virtually absent ($KMnO_4 < 0.1$ mg l^{-1}). Temperature must be within a defined range, normally between 5 and 15 up to 20°C. The chloride concentration must often be less than 150 mg l^{-1}, originally for public health reasons, though 300—500 mg l^{-1} would now be allowed. Nowadays industry asks for a low chloride concentration because this makes the preparation of boiler water (demineralised water) cheaper. Toxic substances should be virtually absent. The Federal Water Pollution Control Administra-

TABLE 19.4

Domestic use of tapwater in The Netherlands and England (litre per day per person, 1967)

	The Netherlands	England
Drinking and cooking	15—20	5
Dish washing and cleaning	15—20	14
Wash and bathwater	60	64
Flushing of toilets	40	50
Gardening and car washing		9
Total	130—140	142

tion (U.S.A.) uses permissible criteria and desirable criteria. Some examples are given in Table 19.5. These criteria deal with the raw water quality. The committee assumed that the common treatment processes in use in the U.S.A. even ⸱⸱ ⸱⸱ ⸱ʳmplest form would provide a potable water. The processes used are ⸱ᵗʰ aluminium or ferric hydroxide, sedimentation (6 h or less), on and disinfection with chlorine. It is obvious that in cases ᵃter must be obtained from sources such as the Rhine, more ᵗtment must be used. This includes centrifugal filtration, sand on over activated charcoal, and treatment with ozone.

ᵗia for public water supply (from Report of the Committee on Water 168)

ᵗacteristic	Permissible	Desirable
Colour (Pt-Co standard)	75	<10
Odour	discussable	virtually absent
Turbidity	discussable	virtually absent
Coliform organisms (per 100 ml)	10 000	<100
Fecal coliforms (per 100 ml)	2 000	< 20
Boron (mg l^{-1})	1.0	1.0
Cadmium (mg l^{-1})	0.01	0.01.
Chloride (mg l^{-1})	250	<25
Lead (mg l^{-1})	0.05	absent
Nitrates (mg l^{-1} of N)	10	virtually absent
Sulphate (mg l^{-1})	250	<50
Total dissolved solids (mg l^{-1})	500	<200
Several pesticides (mg l^{-1})	0.001—0.1	0.001—0.1
Radioactivity (pc l^{-1})		
Gross β	1 000	<100
Radium-226	3	<1
Strontium-90	10	<2

19.3. WASTE WATER

Ultimately all the domestic and industrial water becomes waste water containing organic matter, inorganic solutes, bacteria, and other waste products. Organic matter comes from households and industry. Its quantity is expressed as BOD_5 units and is equal to the amount of oxygen used in five days if the water is left in closed bottles and at constant temperature.

Values of BOD_5 range between 50 and 75 g of O_2 per person per day, but after 25 days 1.5 times as much oxygen is taken up, so a good approximation to the amount of organic waste is an amount of organic carbon equal to 100 g of O_2 per day per person. This quantity of oxygen is the total in 10 m³ of water if fully saturated in equilibrium with air.

A second source of organic matter is industry. Milk factories have a waste water with BOD_5 concentrations between 500 and 2 000 mg l⁻¹. Liquor from silage processes can have BOD_5 values of above 10 000 mg l⁻¹. In paper mills 2 tonnes of wood produce 1 tonne of paper, the remaining tonne being washed away in the different production stages. A factory producing 100 tonnes of paper has the same organic output as a city of 50 000—100 000 inhabitants. Industries making starch and cardboard have during their season a production of organic waste equal to that of several millions of people. In the province of Groningen, where little or no treatment is given to the effluent, these industries produce such quantities of organic waste that children can light the methane bubbles which form in canals and ditches.

Large quantities of organic matter are produced during intensive animal farming (bio-industry). Large numbers of cattle are "grown" on such a small surface that the manure cannot be used on the farm itself. Owens (1973) compared the BOD values of the faeces of several animals with those of man. (Table 19.6).

All organic matter from the sources mentioned above ought to be oxidised in a treatment plant before these waste waters are allowed to flow back into nature.

TABLE 19.6

A comparison of the BOD_5 of excreta produced by man and other animals (the quantities of water required to meet that demand are also included; Owens, 1973)

Source	BOD of excreta (day⁻¹)	Daily flow of fully oxygenated water required (m³ d⁻¹)
Cow	500	50
Pig	140	14
Sheep	100	10
Hen	10	1
Man	60	6

A second purpose of treatment is to remove bacteria mainly originating in household sewage. The commonest bacterium is *E. coli*, but the typhoid bacterium *Salmonella typhosa* may be present in a ratio of above 1:100 000.

Treatment of normal household waste water and sewage occurs in several steps. The water or sewage is first kept in a funnel-like tank in which suspended matter settles and can easily be removed, while floating material can be skimmed off. The effluent from this tank is oxidised, either in an activated sludge tank or in a sprinkler.

The activated sludge is a vast mass of bacteria that grow aerobically on organic waste. The oxidation takes place in a large basin with intensive aeration. Roughly 5 m^3 of air are required per 1 m^3 waste water but the quantities blown through are several times greater. After this treatment the water is left behind to settle again. Part of the sludge sedimenting on the bottom is mixed with the incoming waste water in order to maintain high bacterial density, and part is digested in an anaerobic digestor where it is converted into methane thus reducing the amount of solid matter. The cost of removing the excess solid sludge is one of the largest items in the total cost in developed countries so the principle of sewage treatment is to have a system with low efficiency, i.e. as great a loss of organic matter as possible (high maintenance energy of the bacteria).

In the sprinkler method waste water is sprinkled over large porous stones which act as a solid base for the bacteria. The organic matter is oxidised while trickling over the stones. Because the bacteria will die and fall off or will be washed off the stones into the effluent, a final settlement tank finishes the operations.

Because sewage treatment is an oxidative mineralisation, organic nitrogen and phosphorus are converted into inorganic compounds, of which only a small proportion is retained in the treatment plant. *When these compounds are released into lakes or rivers, nearly the same amount of organic matter will be formed as that which has been broken down at such great cost.* In cases where this causes concern, biological treatment should be combined with a chemical treatment in order to remove phosphate at least. This may be effected by coagulation and precipitation with lime, or iron and aluminium hydroxide. Chemical treatment is also indicated in cases where high costs or shortage of capital delay the building of a biological plant. During the oxidative treatment up to 99% of the faecal *E. coli* bacteria are destroyed. However, if the effluent which still contains several millions of coliform bacteria per litre is to be released in open water where swimming takes place then it is often chlorinated to kill the bacteria. During this chlorination chloramine may be formed (ammonia is of course present in excess) and as chloramine is toxic, it is better to use ozone.

It is remarkable how effective biological treatment can be, remembering that the supply of influent to the plant is rather variable during the day, with sudden peaks and periods of dryness. The biological treatment has however

TABLE 19.7

Average composition of sewage water and composition after primary and secondary treatment of Amsterdam sewage, 1964

Feature	Ranges reported in literature	Before treatment	After primary settlement	After secondary treatment
BOD_5 (mg l^{-1})	300—600	400	400	7
COD (mg l^{-1})	500—1 500	600—2 000		
N (mg l^{-1})	50—100		55	50
Suspended matter (mg l^{-1})	250—400			virtually absent
Chloride (mg l^{-1})	100—150		200	200
Fe(mg l^{-1})	1—5		2	0.6
PO_4-P (mg l^{-1})	10—20		16	12
E. coli (10^6 l^{-1})	500—1 000	500		5
Detergents			21	7[*]

[*]1965. In 1966: 4.7; 1967: 4.0; 1968: 2.1; 1969: 1.3.

a balanced output which a chemical technological process could emulate only with difficulty.

Table 19.7 gives some mean values for untreated and treated sewage water indicating the success of the operation. The process is not infallible, however, as was demonstrated by the case of the hard detergents. From the Table it can also be seen how the replacement of hard (= non-biodegradable) detergents by soft (biodegradable) detergents has decreased the concentration of detergents from 7 to 1.3 mg l^{-1}. The principle that prevention is better than cure could be applied more often though.

Another development is the so-called oxidation ditch. This is an arena-like ditch in which the water slowly circulates. Originally they were built for smaller villages and camping sites, oxidation of organic matter proceeded naturally without artificial aeration. BOD reduction is excellent, while due to sorption on sediments much greater quantities of phosphorus and nitrogen are

TABLE 19.8

Purification efficiency in oxidation ditch of Santa Maria (1 500 human equivalents)

	Influent	Effluent
COD (mg l^{-1})	873	51
BOD (mg l^{-1})	485	4
Cl^- (mg l^{-1})	148	147
Tot-N	64.5	3.5
Tot-P	13.2	5.4

removed than in the active sludge process. Nitrogen is also lost by denitrification (Table 19.8). The plant is operated intermittently, waste being pumped into the ditch for 4 h at a rate of 15 m³ h⁻¹ followed by 4 h passive aeration. By remixing the effluent containing nitrate with the incoming anaerobic influent, nitrogen removal can be high. Anaerobic denitrification (see subsection 6.2.3) can remove as much as 95% of the nitrogen. Nowadays much larger plants for up to 400 000 people are being built with complete artificial aeration. They are also used for industrial organic waste. Dijkstra (1971) has described an oxidation ditch treating effluent from a factory producing caprolactam and acrylonitrile. Owing to its long residence time the system was not sensitive to peak concentrations. Sudden peaks of phenol, sulphuric acid and cyanide in the influent were not reflected in the effluent. Aeration was provided by 100 rotating brushes. Although an oxidation ditch normally produces very little excess sludge, this plant was so heavily loaded that a final settlement tank was necessary. The advantages of the oxidation ditch compared with other forms of sewage treatment are the high efficiency, short lag phase, no excess sludge formation, and low operating cost. Disadvantages are the large area of land required and the need for energy input. Large active sludge plants can be self-supporting for energy.

Industry produces large amounts of inorganic waste in addition to true organic matter. While most of the organic matter, if not toxic, can be oxidised in sewage works, most of the inorganic matter cannot, and pretreatment at the factory itself is essential. The types of industrial waste are shown in Table 19.9. Discharge of industrial inorganic toxic materials directly to public sewers should never be allowed for the toxins may prevent biological oxidation or may accumulate in the excess sludge rendering it unsuitable for agricultural purposes.

In technological circles water pollution is often defined by relating it to the human use of the water. The uses normally considered are: potable supply, recreation and other amenity use, irrigation water, and navigation. Some of these uses demand less purification of inflowing sewage than does the protection of the ecosystem.

TABLE 19.9

Classification of typical industrial wastes

Organic	(a) non-toxic but degradable and requiring oxygen for oxidation; if present in excess can cause anaerobic conditions and formation of H_2S (b) toxic material e.g. phenolic compounds, pesticides such as D.D.T., dieldrin and others, herbicides, pathogenic viruses and bacteria (c) non-biodegradable, e.g. hard detergents
Inorganic	(a) reducing compounds such as ferrous and sulfite solutions (b) inorganic acids, bases and Ca^{2+} and Mg^{2+} solutions, which influence the pH of the receiving water (c) inorganic toxic materials, such as Cu, Ni, Cr, Zn, CN^-, CrO_4^{2-} (d) inorganic nutrients

From a limnological point of view water discharged to rivers and lakes should have a quality such that it does not change quantitatively or qualitatively the naturally occurring populations.

19.4. RECREATIONAL USE OF WATER BODIES

In the developed countries outdoor recreation on or near water is increasing rapidly. Swimming is still the most popular outdoor activity while boating and fishing appear among the ten top activities and are increasing rapidly. In The Netherlands the number of small sailing boats increases by about 10% per year, while the number passing through locks or sluices doubles every one to three years. In several lakes near the larger cities boat density has reached its maximum potential of 12 boats per ha, a density at which mutual interference occurs. Although interference becomes serious at 4 boats per ha, it is assumed that 50% of the boats do not actually sail but lie ashore, while even on days of maximum use one-third do not leave the harbour or their moorings.

Swimming and playing on beaches remain however the most important open air recreation. If there is a choice between freshwater and seawater beaches, the former are preferred because they are generally warmer and are considered to be safer. During one summer one of the dikes of the newly reclaimed polders in The Netherlands, on which there are 3—4 beaches each 5—10 km long was visited by $1.2 \cdot 10^6$ humans, while 70 000 people were found on a beach of 7 800 m length on one peak day; about 10 persons per metre length, or 1 person per 7 m^2. During holidays or days of peak demand, 20% of a city population may go to a beach if it is nearby. The numbers of visitors per beach is related to the distance from the larger cities and quite often reaches maximal densities. About 5 000 people per day visit the sandpit "Zandenplas" (12 000 m^2; 1.5 m deep), i.e. about 3 m^3 of water per person if all were in the water at the same time. The public health authorities consider 15 m^3 of water to be the minimum water quantity per bather.

Recreation and water quality influence each other strongly. High densities such as those mentioned above have a devastating effect. Bathers leave behind a large amount of waste. If one compares the number of parking places (2 000 for one of the beaches mentioned above) plus the number of people not coming by private car with the number of available toilets (2 × 15), it becomes quite clear that much organic matter, nutrients and E. coli will remain in such a lake after a sunny day. After a few warm days the "Zandenplas" has often to be closed, sometimes for several days, owing to the high number of bacteria. Eutrophication is also a typical result of such mass recreation. Some evidence is available that E. coli in eutrophic waters have a longer survival rate, owing to the shallower illuminated depth and because flagellates, which eat the bacteria, disappear as a result of the predominance of blue-green algae.

Another source of (water) pollution comes from the many camping sites which too often release their sewage into lakes. Because these camping sites are only used for short periods in a year provision of full-scale treatment works is not practicable, but simple oxidation ditches with chemical treatment to remove phosphate are probably the solution.

The influence of water quality on the recreation seekers is probably less, because even when bacterial numbers are high, there is little evidence that people get infected easily. Swimmers itch has been attributed to eutrophic waters, but this does not seem to happen often. The green colour of the water is commonly considered by the recreation seeker as being unpleasant (above 200 mg l^{-1} of chlorophyll); the colour is evident at 20—50 mg m^{-3}. Colour seems to be the only real aesthetic objection, but turbidity makes bathing and boating unsafe. Surface scums of blue-green algae may make recreation impossible for all but the most determined pleasure seeker. Fishing for sport may also affect the aquatic ecosystem.

Worthington (1949) has described how the construction of the railway to Windermere brought more fishermen there and the increased fishing pressure caused trout and char to decrease in numbers. In 1940 the perch population was artificially reduced in the hope that the competitive populations of char and trout would increase. However, the number of salmonids decreased even further because pike, deprived of one food, changed its diet and started eating salmonids (see Chapter 15).

In The Netherlands commercial freshwater fishing for all fish except eels is forbidden, all scale fishes being reserved for sport. Restocking with large numbers of young fish, mostly predators, has resulted in an unbalanced population. Pike (*Esox lucius*) after being introduced into Ireland became a pest and resulted in the disappearance of some indigenous fish species.

"Over" fishing and "over" stocking have done much harm in causing population imbalances.

High boat densities have a devastating effect on floating macrophytes and reed beds. Finally destruction of littoral vegetation by constructing angling places such as jetties or even wooden stages or by making open places in the reed vegetation either by boat or from the shoreline has a greater effect on the aquatic environment than might be thought likely.

19.5. MAN-MADE LAKES

19.5.1. Introduction

When man gets too much or too little water he builds dams to control water flow. One of the oldest dams of which remains are still visible is that in the Lunzer Mittler See, where monks in the middle ages formed a reservoir for their brewery. It is a remarkable fact that in many cities (e.g. Amsterdam) the public water supply was started by the beer breweries. Nowadays dams are built for water control in all parts of the world and not only for the water itself but often for the production of electricity.

Optimal water control has been achieved in some situations. In Egypt the Aswan dam prevents floods and stores the water for dry periods, but it collects the silt on which the people downstream once relied. Gains are then twofold, though this is not always the case. The multipurpose reservoir gives the greatest economic benefit. Attendant problems may, however, arise such as increase in bilharzia (a disease related to swimmers itch in temperate waters) and loss of silt.

Fishing is an extra benefit from man-made lakes, being important as an extra source of proteins, especially in those tropical countries where cattle farming is difficult. Man-made lakes are reviewed by Lowe-McConnell (1966), Obeng (1969) and in the proceedings of a general symposium held in 1971 (Ackermann et al., 1973). Although such lakes are found nearly everywhere, a few cases will be discussed in this chapter, each case representing a special situation.

In Russia (including Siberia) enormous reservoirs have been constructed, mainly for the production of electricity. These are water bodies of temperate character. Tropical reservoirs are found in Africa. They are constructed for production of electricity, water control (and storage) and in addition create important fisheries for the local populations. Spain has several potable supply reservoirs. In the U.S.A. reservoirs serve for the production of electric power, flood control and navigation.

19.5.2. Russian and East-European reservoirs

In Siberia one of the biggest dams (5 km long) is found near the Padum rapids creating the Bratsk Reservoir. The dam was finished in 1961, and during the Limnological Conference in Russia when the participants of the field excursion visited the dam, the hydro-electric power station was then the biggest in the world producing about 5 000 MW of electricity. This reservoir is probably one of the biggest artificial water bodies in the world.

The production of bacterial biomass in the water in 1963—1964 was comparable to that of the primary production, due to the high density of saprophytic bacteria which were mineralising allochthonous organic matter. High bacterial production is essential during the first years after filling a reservoir; it would be interesting to compare present photosynthetic and bacterial production in this reservoir. The reservoir has been described (in English) by Kalashnikova and Sorokin in Shtegman (1969).

Other studies deal with the Rybinsk Reservoir, the Svir Reservoir, and with the Volga Reservoir.

The Rybinsk Reservoir was built on the Upper Volga in 1941 and covers an area of 4 550 km^2, has a volume of 25 km^3 and an average depth of 5.6 m The lake shores, formed of sands and loams, are level and are occupied by forests and arable land (Butorin et al., 1973). Qualitative changes in organic matter in the Rybinsk Reservoir (since 1941) have been described by

Skopintsev and Bakulina (1969) and appear to be due to the entrance of snow melt water into the reservoir. This study includes an evaluation of the different methods for measuring COD. Permanganate oxidisability was on the average about 40% of that determined using dichromate, which is exceptional. It may be due to the particular nature of the lake. Dynamics of bacterial populations in the Rybinsk Reservoir in 1961 and 1962 were studied by Kuznetsov and Karpova (1969) and in 1963 and 1964 by Kuznetsov et al. (1969). In 1961 and 1962 the waters of the Rybinsk Reservoir, which is older than the Bratsk Reservoir, were characterised by a low population of saprophytic bacteria; i.e. 100—1 000 per ml. These bacteria depended on photosynthesis by algae and on incoming easily decomposable organic matter. In 1964 photosynthetic assimilation of CO_2 was still 2—3 times greater than that by bacteria because growth of bacteria continued using incoming organic matter as substrate. Unfortunately the influence of ageing of these reservoirs on their biology is confounded with effects of changes in hydrology such as changing water level. The enormous bacterial production in the first years after the filling of these reservoirs is typical. The inundation of considerable areas of terrestrial vegetation and soils rich in organic matter causes a temporary eutrophication due to bacterial mineralisation and extra input of nutrients. This temporary increase of productivity of microorganisms causes an increase in the growth rate of fish which may also feed on the submerged, dying terrestrial fauna. Later an equilibrium will be reached.

Many other studies deal with Russian reservoirs too, but are unfortunately though understandably, written in Russian. Productivity studies are summarised by Winberg (1972), whereas a brief survey of a small part of the Russian literature has been published by Rzoska (1966). Many data on fauna and flora of rivers, lakes and reservoirs of the U.S.S.R. are summarised by Zhadin and Gerd (1970).

The practice of building reservoirs in Eastern Europe goes back to perhaps Roman times. There are now many artificial lakes. In Romania for example well over 1 000 reservoirs have been built, mainly for fish production and water storage (Gâştescu and Breier, 1973).

The Slapy Reservoir in Czechoslovakia is another reservoir which has been well studied. (A series of studies has been published in two volumes edited by Hrbáček (1966, 1973).) In East-Germany some lakes have been created simply to assist in sewage purification (Uhlmann, 1968; Uhlmann et al., 1971). One wonders how long it will be before the lakes will be used for additional purposes and purification of their influent will be found necessary. Although water bodies can be used for water supply, flood control, recreation and navigation, none of these are compatible with sewage purification. If the lakes are going to be used for tertiary treatment, however, they could be used for recreation and navigation.

19.5.3. African reservoirs

Large reservoirs are being built in Africa, and some of them are being fairly intensively studied. The major problems which occur during the construction of these reservoirs are the resettlement of the populations (Scudder, 1966) and the medical problems related to the creation of new habitats for insects. The main diseases which must be considered are malaria (especially during the construction of the reservoirs), Schistosomiasis and Onchocerciasis. Schistosomiasis (= Bilharzia) is caused by *Schistosoma haematobium* and *S. mansoni*, the hosts of which are small water snails of the genera *Bulinus* and *Biophaiaria*. The transmission of schistosomiasis depends on concurrent high densities both of humans and snails in slow-flowing water. Onchocerciasis is caused by the worm *Onchocerca volvulus* following its transmission from man to man by the bites of the fly *Simulium*. The disease is endemic in Africa, Central America and Venezuela and may be promoted by man-made lakes. *Simulium* breeds in running, highly oxygenated waters. Larval forms are highly susceptible to DDT, with which a long reach of the White Nile was cleared (Lewis 1966; Waddy, 1966). Some morphometric data of the main African lakes and reservoirs are given in Table 19.10.

Lake Kariba, formed by a dam across the Zambesi, was completed in 1958. The main purpose was to produce hydro-electric power (800 MW) but it was realised that a fishery industry could be developed too. Physical and chemical aspects of Lake Kariba are described by Coche (1969). Fish and fisheries, together with physical and chemical background information, are described by Harding (1966). Salinity decreased in the first seven years, mainly due to a decrease in concentration of bicarbonate, resulting from decreasing mineralisation. Fish production in Zambia on the northern bank of Kariba increased from virtually nothing in 1959 to about 4 000 tonnes in 1963.

One of the problems common in newly created tropical lakes is the prolific growth of macrophytes. Thus Lake Kariba showed a most distinct explosion in the quantity of the water fern *Salvinia auriculata*, which became one of the more spectacular but menacing features of the lake (Mitchell, 1973).

Recently Balon and Coche (1974) made a comprehensive study of the limnology and ichthyobiology of Lake Kariba, a large man-made lake in the River Zambesi. They give many comparisons with other African man-made lakes and provide useful information about these lakes.

The establishment of fisheries in man-made lakes in the tropics has been reviewed by Jackson (1966) and the prolific growth of macrophytes by Little (1966). Balon (1973) estimated the net production of fish in Lake Kainji (Nigeria) at 686 kg ha^{-1} yr^{-1} and a maximum sustainable yield at 219 kg ha^{-1} yr^{-1}. This is a very high efficiency of yield (219/686 = ca. 30%). The highest mean fish biomasses (standing crops) were found for *Tilapia mossambica*, *Cyphomyrus discorhynchus* and *Mormyrops deliciosus*, each being more than 90 kg ha^{-1}. Total fish biomass was 533 kg ha^{-1} equivalent to 17 800 tonnes of fish in the lake.

TABLE 19.10

Morphometric data of some natural and man-made African lakes (from Worthington, 1966)

Lake	Area at maximum depth (km²)	Altitude (m)	Maximum depth (m)	Maximum drawdown (m)	Mean annual ratio of out-flow to volume	Nominal retention period (years)
Man-made						
Lake Kariba	4 300	530	125	3	1:9	9
Lake Volta	8 500	92	70	3	1:4	4
Kainji Lake	1 280	155	55	10	4:1	0.25
Lake Nasser	5 250	185	97		1:2	2
Jebel Auliya River	600	377	12	ca. 6	8:1	0.125
Sennar River	140	422	16	ca. 17(?)	70:1	0.014
Roseires River	200	480		ca. 42	16:1	0.06
Natural lakes						
Lake Albert	5 283	672	48		?	
Lake Tanganyika	39 000	844	1 500		1:1 500	1 500
Lake Victoria	68 000	1 234	90		1: 120	120

Denyoh (1969; personal communication, 1971) attempted to quantify the fisheries of Lake Volta. He estimated that there were at least 1 000 villages containing 20 000 fishermen scattered along the shore; these used 11 000—14 000 fishing canoes and caught 50 000—70 000 tonnes of fish per year as compared with an annual catch of about 10 000 tonnes from the previous riverine fishery. *Tilapia* species now form more than 50% of processed fish, other commercially important fishes being *Lates niloticus* and species of *Labeo distichodus*, *Citharinus* and cat fishes. Completely new fish markets have come into being with the formation of Lake Volta. If the sustainable yield of Lake Volta could be increased to that of Lake Kainji, the total fish production would be increased by a factor of 3. (A series of papers describing Lake Volta have been published in Ackermann et al., 1973.)

Early descriptions of the biology of Lake Volta were published by Ewer (1966) and in a series of papers by Petr (1968a, b, 1969, 1970a, b, 1971, 1972), Entz (1969, 1972), and Viner (1969, 1970a, b). In this lake *Pistia* developed as a pest with deoxygenated water below the weed mats.

The great man-made lake formed by the high Aswan dam on the Nile comprises two lakes: Lake Nasser (Egypt) and Lake Nubia (Sudan), and is one of the largest in Africa. The two surfaces are 5 250 and 1 000 km^2, with a water volume of 132 km^3. The lake is surrounded by desert and the Nile is the only source of water. Air humidity is always low (summer under 35%), so the climatic situation is entirely different from the other large African lakes such as Lake Volta (Entz and Ramsey, personal communication, 1973). The large fish production in the man-made lakes may be detrimental to coastal fisheries. The creation of Lake Nasser destroyed the Mediterranean sardine fisheries, which however were largely compensated by the fish production in Lake Nasser. The influence of Lake Nasser on the biology of the Nile system is discussed by Hammerton (1972), while early results about the lake are given by Entz and Ramsey. The chemistry of the Nile system has been reviewed by Golterman (1975).

19.5.4. Spanish reservoirs

About 1 700 reservoirs have been built in Spain for drinking water supply. Vidal-Celma (1969, 1971, 1973) investigated the first eight years of the Sau Reservoir. This reservoir (7.5 km^3) is situated near Barcelona at 420 m above sea level and has a maximal depth of 75 m. Stratification occurs, and due to a primary production of 0.5—6 g m^{-2} d^{-1} of carbon, with a mean value of 2 g m^{-2} d^{-1}, an anaerobic hypolimnion developed with high concentrations of phosphate and manganese. Due to erosion in the watershed 0.37 g sec^{-1} of Mn enters the lake, but the phosphate concentration is sufficiently high to sustain abundant primary production. The reservoir is an excellent site to study manganese metabolism. Iron concentration in the hypolimnion is low due to the presence of H_2S (3.2 mg l^{-1}), in spite of the concentration of

1.38 mg l^{-1} of iron in the incoming water. Phytoplankton growth seems to be more regulated by nitrogen than by phosphate shortage, resembling the nitrogen deficiency in tropical lakes.

Margalef et al. (1973) have started a survey of the water characteristics and their effects on production and composition of plankton of nearly one hundred reservoirs throughout Spain. The work is related to Margalef's concept of species diversity and productivity (see section 13.3). Margalef suggested this limnological synthesis may lead to a typology valid for practical purposes. In Spain, which broadly is a tertiary basin limited by sierras of mesozoic limestones, about 1 000 mm of rain falls on the largest part of the siliceous country. In the rest of Spain, where more soluble rocks are present, less rain falls and there are periods of drought that extend from one to six months. Fluctuations in the level of reservoirs are important and are fairly regular depending on human decisions and climate; the fluctuations are always detrimental to benthic life, but can indirectly enhance plankton development. The reservoirs investigated are in two chemically deifferent groups The first group, found mainly in western Spain, has calcium-poor water, with silica concentrations mainly between 1.5 and 3.0 mg l^{-1} of SiO_2-Si. The second group has calcium-rich waters, of which most have the same range of concentration of silica as the first group. It is not known how these concentrations vary with season.

Other chemical variations may be caused by precipitation of calcium and iron phosphates. Some differences in the plankton have already been found, but the influence of the age of the reservoirs on the immigration of organisms cannot be excluded from the diversity caused by the environment.

Small green and blue-green algae are accepted as indicators of eutrophy and Margalef has classified reservoirs on the basis of their phytoplankton populations into two categories of eutrophy. He described a lower and higher degree of eutrophy; the lower having green and the upper having blue-green algae. As expected the more eutrophic reservoirs have a higher chlorophyll concentration and a higher rate of primary production per unit of chlorophyll than the less eutrophic reservoirs. Margalef considers that net production per unit of chlorophyll is higher in eutrophic than oligotrophic reservoirs. This may be due to a more rapid recycling of nutrients in eutrophic waters because the density of dead algae may be a limiting factor for mineralising bacteria (see Chapter 8). It appears that the older reservoirs are on the average more eutrophic, and that there is no clear correlation between the residence time of the water and the degree of eutrophy.

19.5.5. U.S.A. reservoirs

An impressive programme of river regulation and recreational development has been carried out in the Tennessee River Valley. Originally the river Tennessee flooded in winter and reached rather low levels during summer. Nowadays more than thirty dams keep control and have created a series of beautiful lakes. Besides the control of floods, production of electricity and recreation have developed rapidly. The boating and fishing capacity is tremendous, owing to a very long shore line. Several of the dams are 20—30 m high, while the Chickamauga dam rises 40 m above its foundation. Dam and reservoir represented an investment of about $ 42 million. A lock well over 100 m long can raise or lower commercial tugs and recreational craft about 15 m, using over $300 \cdot 10^3$ m^3 of water each time. A whole series of such locks in these multi-purpose dams have created a series of lakes with a 3-m navigable depth for 1 000 km, on which river traffic (km tonnes) was in 1967 about 75 times as great as it was in 1933. The Tennessee River waterway is part of the interconnected inland waterway system of the U.S.A. extending from the Great Lakes to the Gulf of Mexico.

Power generation is a second major achievement. The powerhouse at Chickamauga dam contains four units of 2.7 MW capacity and generated nearly 10^9 kWh in 1967. The recreation significance is indicated in the value of recreational development and its associated equipment on the lake at Chickamauga which is estimated to be over $ 28 million.

Along Chickamauga's shoreline (ca. 1 300 km) are 19 boat docks and resorts, six state and local parks, 74 public access areas, 30 clubhouses and over 1 000 private residences. The lake (140 km^2) attracts nearly 4 million visitors a year and is only one in a whole series of which the true economic value can only be guessed at. One of the environmental impacts of the TVA scheme could of course have been malaria, which was prevalent in parts of the valley. A combination of shoreline preparation, larviciding and water-level management together with improved sanitation has practically eliminated malaria in this region.

19.5.6. Mismanagement and concluding remarks

The creation of man-made lakes has a great beneficial effect and is probably one of the most positive impacts of man on his environment. The freshwater bodies that have been created can compete with natural lakes in their good effects on water supply, recreational value and prevention of floods. Hydroelectricity is probably the cleanest and the cheapest form of energy. Early recognition of possible disadvantages can nearly always prevent damage. The most important ones are concerned with public health and resettlement. Resettlement needs an intensive study of the psychological, physiological and medical condition of the people and the new area to be resettled. White (1969, 1973) gave useful information on the cost of preliminary studies for Lakes Kariba and Kainji, which amounted to £ 420,000 and £ 750,000 respectively, and he suggested that cooperation with university-based research teams could increase scientific output without increasing demands for money. Besides medico-sociological problems, smaller management problems must be solved.

Clearance of potential shorelines before submersion is essential if the lakes have a recreational value and clearance of the bottom may be necessary if net fishing is to be possible. Man-made lakes , just like natural water bodies, must be protected against eutrophication.

A typical example of mismanagement is Lake Apopka, the third largest freshwater lake in Florida (Scheider and Little, 1973). The construction of an outlet converted Lake Apopka from a natural water body to a managed reservoir which passed through extremes from recreational popularity and from having water of high quality to having highly eutrophic, poor quality conditions, which still prevail. Dense algal blooms and coarse fish dominance caused by agricultural, municipal and industrial waste reduced fishing pressure by orders of magnitude. An unconsolidated bottom sediment (muck) now covers 90% of the lake bottom and is more than 10 m thick in some areas, with an average depth of about 1.5 m. Anaerobic conditions and the flocculent consistency of these sediments made these areas unsuitable for propagation of fish-food organisms and game fish. The lake restoration plan is based on the removal of nutrients already accumulated by dredging the richest deposits, oxidising the remainder by lake drawdown, and by diverting agricultural waste. If sufficient care had been taken to make this diversion before the lake was converted into a reservoir, the cost would have been considerably less.

Special cases of man-made lakes are found in The Netherlands, where the former Zuiderzee was closed from the sea by a 30 km long dike. It was predicted that the lake, now called IJsselmeer, would become fresh in 5—6 years

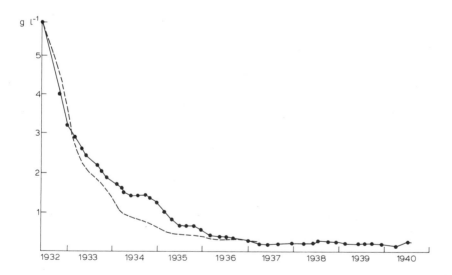

Fig. 19.2. Course of the fall of the salinity in the IJsselmeer, since 1932, when the dam was closed. Dotted line: predicted values; drawn line: measured values.

as a result of inflowing Rhine water (Fig. 19.2). The predicted values (made without aid of electronic computers) agreed very well with the measured ones. The differences that were found were due to deviations of rainfall from the expected normal values and to higher than expected concentrations of chloride in the Rhine, owing to pollution. The lake now demonstrates the serious consequences of eutrophication with chlorophyll values around 100 mg m^{-3}. Phosphate is accumulating in the lake and has reached values up to 0.2 mg l^{-1}.

After the closing of the dike enormous populations of chironomids occurred for several years, but later these returned to normal values. Eel fisheries developed quite well in the lake, as the young eels passed through the locks with ships. The drastic changes that took place in the populations of most organisms were not well documented, however. Therefore a special institute has been established to study the changes that will occur in the new freshwater lakes being formed in the south-west of The Netherlands as a result of closing the other river mouths of the Rhine (Delta plan). Biological consequences are summarised in the Progress Reports of the Delta Institute for Hydrobiological Research[*].

In the new lake IJsselmeer new polders are being created which will reduce the size of the (new) lake to about 1 200 km^2. As the new polders (5 m below sea level) have a lower water table than the surrounding old land, a series of "border lakes" have been formed to protect the old land against drainage (Fig. 19.1). The lakes thus formed for hydrological reasons are now used mainly for recreation, but unfortunately the water quality is very poor as cities on the old land still discharge their partially treated sewage into them. Eutrophication has caused algal growth to create the theoretical maximum concentrations of chlorophyll of over 300 mg m^{-3}. Restoration plans exist including phosphate removal from all sewage effluents and flushing of the lake with phosphate-poor groundwater. These lakes may serve as another example of the principle that protection (by prevention before harm has been done) is always cheaper than cure by restoration afterwards. The costs of such protective prevention are low compared to the costs of making the lake and should be included automatically in any feasibility study. Fundamental knowledge of the aquatic ecosystem is an essential tool if water management is ever to progress.

REFERENCES

Ackermann, W.C., White, G.F. and Worthington, E.B. (Editors), 1973. *Man-Made Lakes: Their Problems and Environmental Effects.* Symposium Knoxville, Tenn., 1971. American Geophysical Union, Washington, D.C., 847 pp.

---

[*]These have been published since 1971 in the *Verh. K. Ned. Akad. Wet., Afd. Natuurk.*

Balon, E.K., 1973. Results of fish population size assessments in Lake Kariba coves (Zambia), a decade after their creation. In: W.C. Ackermann, G.F. White and E.B. Worthington (Editors): *Man-Made Lakes: Their Problems and Environmental Effects*. American Geophysical Union, Washington, D.C., pp. 149—158.

Balon, E.K and Coche, A.G. (Editors), 1974. *Lake Kariba: A Man-Made Tropical Ecosystem in Central Africa*. Junk, The Hague, 767 pp. (Monographiae Biologicae, vol. 24).

Butorin, N.V., et al., 1973. Effects of the Rybinsk Reservoir on the surrounding area. In: W.C. Ackermann, G.F. White and E.B. Worthington (Editors): *Man-Made Lakes: Their Problems and Environmental Effects*. American Geophysical Union, Washington, D.C., pp. 246—250.

Coche, A.G., 1969. Aspects of physical and chemical limnology of Lake Kariba, Africa. A general outline. In: L.E. Obeng (Editor), *Man-Made Lakes: The Accra Symposium*. Ghana Universities Press, Accra, pp. 116—122.

Denyoh, F.M.K., 1969. Changes in fish population and gear selectivity in the Volta Lake. In: L.E. Obeng (Editor), *Man-Made Lakes: The Accra Symposium*. Ghana Universities Press, Accra, pp. 206—219.

Dijkstra, F., 1971. Industrieel afvalwater: Sanerings- and zuiveringsmaatregelen bij de chemische bedrijven van DSM. H_2O, 4: 331—335.

Entz, B., 1969. Observations on the limnochemical conditions of the Volta Lake. In: L.E. Obeng (Editor), *Man-Made Lakes: The Accra Symposium*. Ghana Universities Press, Accra, pp. 105—115.

Entz, B., 1972. Comparison of the physical and chemical environments of the Volta Lake and Lake Nasser. In: Z. Kajak and A. Hillbricht-Ilkowska (Editors), *Productivity Problems of Freshwaters*. Warszawa—Krakow, pp. 883—891.

Ewer, D.W., 1966. Biological investigations on the Volta Lake, May 1964 to May 1965. In: R.H. Low-McConnell, *Man-Made Lakes, Proc. Symp. held at the R. Geogr. Soc, London, 1965*. Academic Press, London, pp. 21—31.

Gâştescu, P. and Breier, A., 1973. Artificial lakes in Romania. In: W.C. Ackermann, G.F. White and E.G. Worthington (Editors), *Man-Made Lakes: Their Problems and Environmental Effects*. American Geophysical Union, Washington, D.C., pp. 50—55.

Golterman, H.L., 1973. Deposition of river silts in the Rhine and Meuse Delta. *Freshwater Biol.*, 3: 267—281.

Golterman, H.L., 1975. Chemistry of running water. In: B. Whitton (Editor), *River Ecology*. Blackwell, Oxford (in press).

Hammerton, D., 1972. The Nile river — a case history. In: R.T. Oglesby, C.A. Carlson and J.A. McCann (Editors), *Proc. Int. Symp. on River Ecology and the Impact of Man, University of Massachusetts, Amherst, 1971*. Academic Press, New York, N.Y., pp. 171—214.

Harding, D., 1966. Lake Kariba; The hydrology and development of fisheries. In: R.H. Lowe-McConnell (Editor), *Man-Made Lakes, Proc. Symp. held at the R. Geogr. Soc., London, 1965*. Academic Press, London, pp. 7—20.

Hrbácek, J. (Editor), 1966, 1973. *Hydrobiological Studies*. Academia, Prague (1966: part 1; 1973: parts 2 and 3).

Jackson, P.B.N., 1966. The establishment of fisheries in man-made lakes in the tropics. In: R.H. Lowe-McConnell (Editor), *Man-Made Lakes, Proc. Symp. held at the R. Geogr. Soc., London, 1965*. Academic Press, London, pp. 53—73.

Kalashnikova, E.P. and Sorokin, Yu. I., 1969. Microflora in the Bratsk reservoir. In: B.K. Shtegman (Editor), *Production and Circulation of Organic Matter in Inland Waters*. Israel Program for Scientific Translations, Jerusalem, pp. 177—184.

Kuznetsov, S.I. and Karpova, N.S., 1969. Dynamics of bacterial population in the Rybinsk reservoir in 1961 and 1962. In: B.K. Shtegman (Editor), *Production and Circulation of Organic Matter in Inland Waters*. Israel Program for Scientific Translations, Jerusalem, pp. 121—126.

Kuznetsov, S.I., Romanenko, V.I. and Karpova, N.S., 1969. Bacterial population and production of organic matter in the Rybinsk reservoir in 1963 and 1964. In: B.K. Shtegman (Editor), *Production and Circulation of Organic Matter in Inland Waters.* Israel Program for Scientific Translations, Jerusalem, pp. 127—136.

Lewis, D.J., 1966. Nile control and its effects on insects of medical importance. In: R.H. Lowe-McConnell (Editor), *Man-Made Lakes, Proc. Symp. held at the R. Geogr. Soc., London, 1965.* Academic Press, London, pp. 43—45.

Little, E.C.S., 1966. The invasion of man-made lakes by plants. In: R.H. Lowe-McConnell (Editor), *Man-Made Lakes, Proc. Symp. held at the R. Geogr. Soc., London, 1965.* Academic Press, London, pp. 75—86.

Lowe-McConnell, R.H. (Editor), 1966. *Man-Made Lakes, Proc. Symp. held at the R. Geogr. Soc., London, 1965.* Academic Press, London, 281 pp.

Margalef, R., et al., 1973. Plankton production and water quality in Spanish reservoirs. First report on a research project. In: *Congr. Int. Commission on Large Dams, 11th, Madrid, 1973,* 21 pp.

Mitchell, D.S., 1973. Supply of plant nutrient chemicals in Lake Kariba. In: W.C. Ackermann, G.F. White and E.B. Worthington (Editors), *Man-Made Lakes: Their Problems and Environmental Effects.* American Geophysical Union, Washington, D.C., pp. 165—169.

Obeng, L.E. (Editor), 1969a. *Man-Made Lakes: The Accra Symposium.* Ghana Universities Press, Accra, 398 pp.

Obeng, L.E., 1969b. The invertebrate fauna of aquatic plants of the Volta Lake in relation to the spread of helminth parasites. In: L.E. Obeng (Editor), *Man-Made Lakes: The Accra Symposium.* Ghana Universities Press, Accra, pp. 320—325.

Owens, M., 1973. Resource under pressure — Water. In: *Intensive Agriculture and the Environment.* An Foras Taluntais, Dublin, pp. 33—39.

Petr, T., 1968a. The establishment of lacustrine fish population in the Volta Lake in Ghana during 1964—1966. *Bull. I.F.A.N., Ser. A,* 30(1): 257—269.

Petr, T., 1968b. Distribution, abundance and food of commercial fish in the Black Volta and the Volta man-made lake in Ghana during the first period of filling (1964—1966). I. Mormyridae. *Hydrobiologia,* 32(3—4): 417—448.

Petr, T., 1969. Fish population changes in the Volta Lake over the period January 1965 — September 1966. In: L.E. Obeng (Editor), *Man-Made Lakes: The Accra Symposium.* Ghana Universities Press, Accra, pp. 220—234.

Petr, T., 1970a. Chironomidae (Diptera) from light catches on the man-made lake in Ghana. *Hydrobiologia,* 35(3—4): 449—468.

Petr, T., 1970b. Macroinvertebrates of flooded trees in the man-made Volta Lake (Ghana) with special reference to the burrowing mayfly *Povilla adusta* Navas. *Hydrobiologia,* 36(3—4): 373—398.

Petr, T., 1971. Lake Volta — a progress report. *New Sci. and Sci. J.,* January 1971: 178—180.

Petr, T., 1972. Benthic fauna of a tropical man-made lake (Volta Lake, Ghana 1965—1968). *Arch. Hydrobiol.,* 70(4): 484—533.

Rzoska, J., 1966. The biology of reservoirs in the U.S.S.R. In: R.H. Lowe-McConnell (Editor), *Man-Made Lakes, Proc. Symp. held at the R. Geogr. Soc., London, 1965.* Academic Press, London, pp. 149—157.

Schneider, R. and Little, J.A., 1973. Rise and fall of Lake Apopka: A case study on reservoir mismanagement. In: W.C. Ackermann, G.F. White and E.B. Worthington (Editors), *Man-Made Lakes: Their Problems and Environmental Effects.* American Geophysical Union, Washington, D.C., pp. 690—694.

Scudder, T., 1966. Man-made lakes and population resettlement in Africa. In: R.H. Lowe-McConnell (Editor), *Man-Made Lakes, Proc. Symp. held at the R. Geogr. Soc., London, 1965.* Academic Press, London, pp. 99—108.

Shtegman, B.K. (Editor), 1969. *Production and Circulation of Organic Matter in Inland Waters.* Israel Program for Scientific Translations, Jerusalem, 287 pp.

Skopintsev, V.A. and Bakulina, A.G., 1969. Organic matter in the Rybinsk reservoir in 1964. In: B.K. Shtegman (Editor), *Production and Circulation of Organic Matter in Inland Waters.* Israel Program for Scientific Translations, Jerusalem, pp. 1—31.

Uhlmann, D., 1968. Modellversuche über die Abhängigkeit der planktischen Bioaktivität von der Verweilzeit des Wassers. *Int. Rev. Ges. Hydrobiol.,* 53(1): 101—139.

Uhlmann, D., Benndorf, J. and Albert, W., 1971. Prognose des Stoffhaltes von Steugewässern mit Hilfe kontinuierlicher oder semikontinuierlicher biologischer Modelle, I: Grundlagen. *Int. Rev. Ges. Hydrobiol.,* 56(4): 513—539.

Vidal-Celma, A., 1969. Evolution d'un lac de barrage dans le NE de l'Espagne pendant les quatre premières années de service. *Verh. Int. Ver. Theor. Angew. Limnol.,* 17: 191—200.

Vidal-Celma, A., 1971. Evolucion d'un poblament. El llac artificial de Sau. *Trab. Soc. Catal. Biol. Barcelona,* 31: 15—27.

Vidal-Celma, A., 1973. Développement et évaluation du phytoplancton dans le reservoir de Sau. In: *Congr. Int. Commission on Large Dams, 11th, Madrid, 1973,* 23 pp.

Viner, A.B., 1969. Observation of the hydrobiology of the Volta Lake, April 1965—April 1966. In: L.E. Obeng (Editor), *Man-Made Lakes: the Accra Symposium.* Ghana Universities Press, Accra, pp. 133—143.

Viner, A.B., 1970a. Hydrobiology of Lake Volta, Ghana. I. Stratification and circulation of water. *Hydrobiologia,* 35: 209—229.

Viner, A.B., 1970b. Hydrobiology of Lake Volta, Ghana. II. Some observations on biological features associated with the morphology and water stratification. *Hydrobiologia,* 35: 230—248.

Waddy, B.B., 1966. Medical problems arising from the making of lakes in the tropics. In: R.H. Lowe-McConnell (Editors), *Man-Made Lakes, Proc. Symp. held at the R. Geogr. Soc., London, 1965.* Academic Press, London, pp. 87—94.

White, E., 1969. The place of biological research in the development of the resources of man-made lakes. In: L.E. Obeng (Editor), *Man-Made Lakes: the Accra Symposium.* Ghana Universities Press, Accra, pp. 37—49.

White, E., 1973. Zambia's Kafue hydroelectric scheme and its biological problems. In: W.C. Ackermann, G.F. White and E.B. Worthington (Editors), *Man-Made Lakes: Their Problems and Environmental Effects.* American Geophysical Union, Washington, D.C., pp. 620—628.

Winberg, G.G., 1972. Etudes sur le bilan biologique énergetique et la productivité des lacs en Union Soviètique. Edgardo Baldi Memorial Lecture. *Verh. Int. Ver. Theor. Angew. Limnol.,* 18(1): 39—64.

Worthington, E.B., 1949. An experiment with populations of fish in Windermere, 1939—1948. *Proc. Zool. Soc. Lond.,* 120: 113—149.

Worthington, E.B., 1966. Introductory survey. In: R.H. Lowe-McConnell (Editor), *Man-Made Lakes, Proc. Symp. held at the R. Geogr. Soc., London, 1965.* Academic Press, London, pp. 3—6.

Zhadin, V.I. and Gerd, S.V., 1970. *Fauna and Flora of the Rivers, Lakes and Reservoirs of the U.S.S.R.* Israel Program for Scientific Translations, Jerusalem, 625 pp.

APPENDIX I

List of algae often found in lakes or referred to in this book*

PROCARYOTA
    I. Cyanophyta
        Chroococcales (solitary cells, no heterocysts): *Aphanocapsa, Chroococcus, Coelo-sphaerium, Gloeocapsa, Holopedia, Microcystis* (= *Anacystis*), *Synechococ-cus*
        Hormogonales (= Oscillatoriales; filamentous, heterocysts sometimes present)
            Nostocinales (filaments simple): e.g., *Anabaena, Aphanizomenon, Lyngbya, Nostoc, Oscillatoria, Phormidium, Plectonema*
            Stigonematinales (filaments heterotrichous, with connecting pits): e.g. *Stigonema*

EUCARYOTA

    I. Chrysophyta
        e.g. Chrysomonadales (unicellular or colonial flagellates): *Mallomonas, Ochro-monas, Synura, Uroglena*
    II. Xanthophyta
        e.g. Heterococcales (unicellular or colonial with rigid cell_wall): *Monodus*
    III. Haptophyta
        e.g. Prymnesiales (unicellular flagellates with a haptomena and flagella): *Prym-nesium*
    IV. Bacillariophyta (unicellular or colonial algae; cell wall of two distinct halves com-posed of silicon)
        Centrobacillariophyceae (Centrales): *Cyclotella, Melosira, Sceletonema, Stephano-discus, Thalassiosira*
        Pennatibacillariophyceae (Pennales): *Asterionella, Diatoma, Fragilaria, Synedra, Tabellaria, Gomphonema, Navicula, Nitzschia, Phaeodactylum*
    V. Chlorophyta
        Chlorophyceae
            Volvocales (flagellates green): *Chlamydomonas, Eudorina, Gonium, Phacotus, Volvox*
            Chlorococcales: *Ankistrodesmus, Chlorella, Chlorococcum, Coelastrum, Dictyosphaerium, Oocystis, Pediastrum, Scenedesmus, Tetraedon*
            Ulotrichales: *Stichococcus, Ulothrix*
        Zygnemaphyceae
            Desmidiales: *Closterium, Cosmarium, Desmidium, Staurastrum, Xanthidium*
    VI. Euglenophyta
        Euglenales: *Euglena, Phacus*
    VII. Dinophyta
        Dynophyceae: *Amphidinium, Ceratium, Gymnodinium, Peridinium*
    VIII. Cryptophyta
        Cryptomonadales: *Chroomonas, Cryptomonas*
    IX. Rhodophyta
        Bangiales: *Bangia, Porphyridium*

---

*From: Round, F.E., 1973. *The Biology of the Algae*. Edward Arnold, London, 2nd ed., 278 pp.

APPENDIX II

Geological time-scale

Era	Period	Epoch	Millions of years ago (approx.)
Cenozoic	Quaternary	Holocene (Recent)*	0.01
		Pleistocene	2
	Tertiary	Pliocene	10
		Miocene	27
		Oligocene	38
		Eocene	55
		Paleocene	65—70
Mesozoic	Cretaceous		130
	Jurassic		180
	Triassic		225
Paleozoic	Permian		260
	Carboniferous { Pennsylvanian / Mississippian		340
	Devonian		400
	Silurian		430
	Ordovician		480
	Cambrian		570
Proterozoic	Precambrian	Upper	1 900
		Middle	2 700
		Lower	3 500

*Subdivided in the geologic climate units: Preboreal, Boreal, Atlantic, Subboreal and Subatlantic.

APPENDIX III

Photosynthetic and respiratory pathways or carbon cycles (not all cycles occur in all organisms; for further explanation see Chapters 4, 6 and 16)

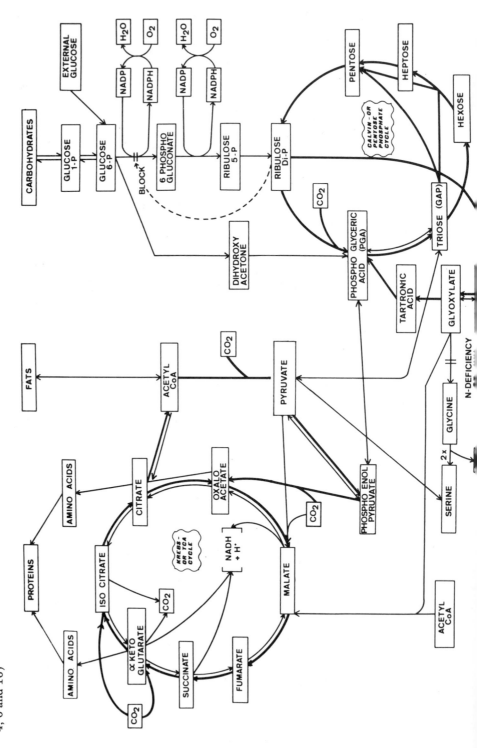

REFERENCES INDEX

460

466

SUBJECT INDEX

abiotic features, 1
accessory growth factors, 217, 218
— pigments, 239, 246
accumulation in sediments, 357
acetate, turnover rate of, 219, 353
acetyl CoA, 326
acetylene reduction, 105, 106, 107
Achnanthes microcephala, 8
Achromobacter, 330, 333, 339
acidity, 32, 164
acids, fulvic, 102, 204, 382
—, humic, 204
Acinetobacter, 339
activated sludge, 437
activity coefficient, 265, 266
—, glacier, 17, 73
—, river, 20
adaptation, 7, 61, 69, 191, 246, 347, 348
— to copper, 211
adsorption of ammonia, 168
— of phosphate, 168, 169
— — on clay, 168
Aegerisee, 362
Aerobacter, 338
age, ice, 16
— specific growth function, 307
— specific survivorship, 307
Agmenellum, 111
— *quadruplicatum*, 109
agriculture, phosphate in, 89
α-ketoglutaric acid, 71
Albert, Lake, 1, 14, 29, 88, 132
albite, 404, 405
Alcaligenes, 330, 333
algae, dry weight of, 264
—, sedimentation of, 359, 365
—, sinking rate of, 139
algal blooms, 380, 386
— cells (see also cells), 174
— cultures, 174
— development, prediction of, 370
— growth, 69
— —, enhancement of, 380
— —, excessive, 398
— life cycle 191, 192
— successions in Loch Leven, 314

alkalinity, 29, 39, 41, 73
—, difference, 396
— in hypolimnion, 163
—, total, 42
Allen curve, 307, 316
allochthonous material, 72, 141
— N-sources, 99
— organic carbon, 362
— organic matter, 329, 442
— phosphate, 94
— — input, 96
— sediments, 403
alluvial deposits, 22
alpine lakes, French, 393, 394
— —, Swiss and Austrian, 391—393
aluminium silicate, 124, 403, 405
— —, incongruent dissolution of, 406
A_{max}, 73, 74, 77
Amazon, 410, 411, 414, 415
Amazonia, 29
amino-acid imbalance, 331
amino acids of *Calanus*, 284
ammonia, adsorption of, 168
— in hypolimnion, 167, 168
— in lakewater, 99, 100, 101
— oxydation, 100, 101, 333, 348
— uptake, 101, 105, 118
amphiboles, 404, 406, 409, 411
Amphidinium carteri, 182
amylase producers, 339
Anabaena, 3, 41
— *affinis*, 304
— *circinalis*, 268, 269
— *cylindrica*, 61, 67, 68, 107, 119, 182
— *flos aquae*, 8, 82, 93, 106, 107, 113,
    196, 237, 266, 272, 273, 292, 304, 396
— *planctonica*, 266
— *variabilis*, 109, 110
Anabaenopsis, 273
Anacystis, 1, 106
— *nidulans*, 111, 182, 292, 380
anaerobic mineralisation, 140
animal husbandry, intensive, 367
Ankistrodesmus, 195, 200, 292
— *acicularis*, 266
— *arcuatus*, 8
— *falcatus*, 249

480